T0275605

Logical Foundations of Proof Complexity

This book treats bounded arithmetic and propositional proof complexity from the point of view of computational complexity. The first seven chapters include the necessary logical background for the material and are suitable for a graduate course.

Associated with each of many complexity classes are both a two-sorted predicate calculus theory, with induction restricted to concepts in the class, and a propositional proof system. The complexity classes range from AC^0 for the weakest theory up to the polynomial hierarchy. Each bounded theorem in a theory translates into a family of (quantified) propositional tautologies with polynomial size proofs in the corresponding proof system. The theory proves the soundness of the associated proof system.

The result is a uniform treatment of many systems in the literature, including Buss's theories for the polynomial hierarchy and many disparate systems for complexity classes such as AC^0, $AC^0(m)$, TC^0, NC^1, L, NL, NC, and P.

Stephen Cook is a professor at the University of Toronto. He is author of many research papers, including his famous 1971 paper "The Complexity of Theorem Proving Procedures," and the 1982 recipient of the Turing Award. He was awarded a Steacie Fellowship in 1977 and a Killam Research Fellowship in 1982 and received the CRM/Fields Institute Prize in 1999. He is a Fellow of the Royal Society of London and the Royal Society of Canada and was elected to membership in the National Academy of Sciences (United States) and the American Academy of Arts and Sciences.

Phuong Nguyen is a postdoctoral researcher at McGill University. He received his MSc and PhD degrees from University of Toronto in 2004 and 2008, respectively. He has been awarded postdoctoral fellowships by the Eduard Čech Center for Algebra and Geometry (the Czech Republic) and by the Natural Sciences and Engineering Research Council of Canada (NSERC).

PERSPECTIVES IN LOGIC

The *Perspectives in Logic* series publishes substantial, high-quality books whose central theme lies in any area or aspect of logic. Books that present new material not now available in book form are particularly welcome. The series ranges from introductory texts suitable for beginning graduate courses to specialized monographs at the frontiers of research. Each book offers an illuminating perspective for its intended audience.

The series has its origins in the old *Perspectives in Mathematical Logic* series edited by the Ω-Group for "Mathematische Logik" of the Heidelberger Akademie der Wissenchaften, whose beginnings date back to the 1960s. The Association for Symbolic Logic has assumed editorial responsibility for the series and changed its name to reflect its interest in books that span the full range of disciplines in which logic plays an important role.

For more information, see http://www.aslonline.org/books_perspectives.html

PERSPECTIVES IN LOGIC

Logical Foundations of Proof Complexity

STEPHEN COOK

University of Toronto

PHUONG NGUYEN

McGill University

ASSOCIATION FOR SYMBOLIC LOGIC

CAMBRIDGE
UNIVERSITY PRESS

CAMBRIDGE
UNIVERSITY PRESS

32 Avenue of the Americas, New York NY 10013-2473, USA

Cambridge University Press is part of the University of Cambridge.

It furthers the University's mission by disseminating knowledge in the pursuit of education, learning and research at the highest international levels of excellence.

www.cambridge.org
Information on this title: www.cambridge.org/9781107694118

© Association for Symbolic Logic 2010

First published 2010, 2011
Second Edition 2012
Reprinted 2013
First paperback edition 2014

A catalogue record for this publication is available from the British Library

ISBN 978-0-521-51729-4 Hardback
ISBN 978-1-107-69411-8 Paperback

CONTENTS

PREFACE

"Proof complexity" as used here has two related aspects: (i) the complexity of proofs of propositional formulas, and (ii) the study of weak (i.e., "bounded") theories of arithmetic. Aspect (i) goes back at least to Tseitin [109], who proved an exponential lower bound on the lengths of proofs in the weak system known as regular resolution. Later Cook and Reckhow [46] introduced a general definition of propositional proof system and related it to mainstream complexity theory by pointing out that such a system exists in which all tautologies have polynomial length proofs iff the two complexity classes *NP* and *co-NP* coincide.

Aspect (ii) goes back to Parikh [88], who introduced the theory known as $I\Delta_0$, which is Peano Arithmetic with induction restricted to bounded formulas. Paris and Wilkie advanced the study of $I\Delta_0$ and extensions in a series of papers (including [90, 89]) which relate them to complexity theory. Buss's seminal book [20] introduced the much-studied interleaved hierarchies S_2^i and T_2^i of theories related to the complexity classes Σ_i^p making up the polynomial hierarchy. Clote and Takeuti [38] and others introduced a host of theories related to other complexity classes.

The notion of propositional translation, which relates aspects (i) and (ii), goes back to [39], which introduced the equational theory *PV* for polynomial time functions and showed how theorems of *PV* can be translated into families of tautologies which have polynomial length proofs in the extended Frege proof system. Later (and independently) Paris and Wilkie [90] gave an elegant translation of bounded theorems in the relativized theory $I\Delta_0(R)$ to polynomial length families of proofs in the weak propositional system bounded-depth Frege. Krajíček and Pudlák [73] introduced a hierarchy of proof systems $\langle G_i \rangle$ for the quantified propositional calculus and showed how bounded theorems in Buss's theory T_2^i translate into polynomial length proofs in G_i.

The aim of the present book is, first of all, to provide a sufficient background in logic for students in computer science and mathematics to understand our treatment of bounded arithmetic, and then to give an original treatment of the subject which emphasizes the three-way relationship among complexity classes, weak theories, and propositional proof systems.

Our treatment is unusual in that after Chapters 2 and 3 (which present Gentzen's sequent calculus LK and the bounded theory $I\Delta_0$) we present our theories using the two-sorted vocabulary of Zambella [112]: one sort for natural numbers and the other for binary strings (i.e., finite sets of natural numbers). Our point of view is that the objects of interest are the binary strings: they are the natural inputs to the computing devices (Turing machines and Boolean circuits) studied by complexity theorists. The numbers are there as auxiliary variables, for example, to index the bits in the strings and measure their length. One reason for using this vocabulary is that the weakest complexity classes (such as AC^0) that we study do not contain integer multiplication as a function, and since standard theories of arithmetic include multiplication as a primitive function, it is awkward to turn them into theories for these weak classes. In fact, our theories are simpler than many of the usual single-sorted theories in bounded arithmetic, because there is only one primitive function $|X|$ (the length of X) for strings X, while the axioms for the number sort are just those for $I\Delta_0$.

Another advantage of using the two-sorted systems is that our propositional translations are especially simple: they are based on the Paris-Wilkie method [90]. The propositional atoms in the translation of a bounded formula $\varphi(X)$ with a free string variable X simply represent the bits of X.

Chapter 5 introduces our base theory V^0, which corresponds to the smallest complexity class AC^0 which we consider. All two-sorted theories we consider are extensions of V^0. Chapter 6 studies V^1, which is a two-sorted version of Buss's theory S_2^1 and is related to the complexity class P (polynomial time). Chapter 7 introduces propositional translations for some theories. These translate bounded predicate formulas to families of quantified Boolean formulas. Chapter 8 introduces "minimal" theories for polynomial time by a method which is used extensively in Chapter 9. Chapter 8 also presents standard results concerning Buss's theories S_2^i and T_2^i, but in the form of the two-sorted versions V^i and TV^i of these theories. Chapter 9 is based on the second author's PhD thesis, and uses an original uniform method to introduce minimal theories for many complexity classes between AC^0 and P. Some of these are related to single-sorted theories in the literature. Chapter 10 gives more examples of propositional translations and gives evidence for the thesis that each theory has a corresponding propositional proof system which serves as a kind of nonuniform version of the theory.

One purpose of this book is to serve as a basis for a program we call "Bounded Reverse Mathematics". This is inspired by the Friedman/Simpson program Reverse Mathematics [101], where now "Bounded" refers to bounded arithmetic. The goal is to find the weakest theory capable of proving a given theorem. The theorems in question are those of interest in computer science, and in general these can be proved in weak theories. From the complexity theory point of view, the idea is to find the smallest complexity class such that the theorem can be proved using concepts in that class. This activity not only sheds light on the role of complexity classes in proofs, it can also lead to simplified proofs. A good example is Razborov's [96] greatly simplified proof of Hastad's Switching Lemma, which grew out of his attempt to formalize the lemma using only polynomial time concepts. His new proof led to important new results in propositional proof complexity. Throughout the book we give examples of theorems provable in the theories we describe.

The first seven chapters of this book grew out of notes for a graduate course taught several times beginning in 1998 at the University of Toronto by the first author. The prerequisites for the course and the book are some knowledge of both mathematical logic and complexity theory. However, Chapters 2 and 3 give a complete treatment of the necessary logic, and the Appendix together with material scattered throughout should provide sufficient background in complexity theory. There are exercises sprinkled throughout the text, which are intended both to supplement the material presented and to help the reader master the material. The more difficult exercises are marked with an asterisk.

Two sources have been invaluable to the authors in writing this book. The first is Krajíček's monograph [72], which is an essential possession for anyone working in this field. The second source is Buss's chapters [27, 28] in the *Handbook of Proof Theory*. His chapter I provides an excellent introduction to the proof theory of *LK*, and his chapter II provides a thorough introduction to the first-order theories of bounded arithmetic. And of course Buss's monograph [20] *Bounded Arithmetic* was the origin of much of the material in our book.

We are grateful to Sam Buss and Jan Krajíček not only for their books but also for their considerable encouragement and help during the lengthy process of writing our book.

This book includes valuable input from several students of the first author as well as material from their PhD theses. The students include (besides the second author) Antonina Kolokolova, Tsuyoshi Morioka, Steven Perron, and Michael Soltys.

We are indebted to many others who have provided us with feedback on earlier versions of the book. These include Noriko Arai, Toshi Arai, Anton Belov, Mark Braverman, Timothy Chow, Lila Fontes, Kaveh Ghasemloo, Remo Goetschi, Daniel Ivan, Emil Jeřábek, Akitoshi Kawamura, Markus Latte, Dai Tri Man Le, Leonid Libkin, Dieter van Melkebeek, Toni Pitassi, Francois Pitt, Pavel Pudlák, Alan Skelley, Robert Solovay, Neil Thapen, Alasdair Urquhart, and Daniel Weller.

Stephen Cook
Phuong Nguyen

Chapter I

INTRODUCTION

This book studies logical systems which use restricted reasoning based on concepts from computational complexity. The complexity classes of interest lie mainly between the basic class AC^0 (whose members are computed by polynomial-size families of bounded-depth circuits), and the polynomial hierarchy PH, and include the sequence

$$AC^0 \subset AC^0(m) \subseteq TC^0 \subseteq NC^1 \subseteq L \subseteq NL \subseteq P \subseteq PH \qquad (1)$$

where P is polynomial time. (See the Appendix for definitions.)

We associate with each of these classes a logical theory and a proof system for the (quantified) propositional calculus. The proof system can be considered a nonuniform version of the universal (or sometimes the bounded) fragment of the theory. The functions definable in the logical theory are those associated with the complexity class, and (in some cases) the lines in a polynomial size proof in the propositional system express concepts in the complexity class. Universal (or bounded) theorems of the logical theory translate into families of valid formulas with polynomial size proofs in the corresponding proof system. The logical theory proves the soundness of the proof system.

Conceptually the theory VC associated with a complexity class $\mathbf{C}$ can prove a given mathematical theorem if the induction hypotheses needed in the proof can be formulated using concepts from $\mathbf{C}$. We are interested in trying to find the weakest class $\mathbf{C}$ needed to prove various theorems of interest in computer science.

Here are some examples of the three-way association among complexity classes, theories, and proof systems:

class	AC^0	TC^0	NC^1	P	PH	
theory	V^0	VTC^0	VNC^1	VP	V^∞	(2)
system	AC^0-Frege	TC^0-Frege	Frege	eFrege	$\langle G_i \rangle$.	

Consider for example the class NC^1. The uniform version is $ALogTime$, the class of problems solvable by an alternating Turing machine in time $\mathcal{O}(\log n)$. The definable functions in the associated theory VNC^1 are the NC^1 functions, i.e., those functions whose bit graphs are NC^1 relations.

A problem in nonuniform NC^1 is defined by a polynomial-size family of log-depth Boolean circuits, or equivalently a polynomial-size family of propositional formulas. The corresponding propositional proof systems are called *Frege* systems, and are described in standard logic textbooks: a *Frege* proof of a tautology A consists of a sequence of propositional formulas ending in A, where each formula is either an axiom or follows from earlier formulas by a rule of inference. Universal theorems of VNC^1 translate into polynomial-size families of *Frege* proofs. Finally VNC^1 proves the soundness of *Frege* systems, and any proof system whose soundness is provable in VNC^1 can be p-simulated by a *Frege* system (Theorem X.3.11).

The famous open question in complexity theory is whether the conjecture that P is a proper subset of NP is in fact true (we know $P \subseteq NP \subseteq PH$). If $P = NP$ then the polynomial hierarchy PH collapses to P, but it is possible that PH collapses only to NP and still $P \neq NP$. What may be less well known is that not only is it possible that $PH = P$, but it is consistent with our present knowledge that $PH = AC^0(6)$, so that all classes in (1) might be equal except for AC^0 and $AC^0(p)$ for p prime. This is one motivation for studying the theories associated with these complexity classes, since it ought to be easier to separate the theories corresponding to the complexity classes than to separate the classes themselves (but so far the theories in (2) have not been separated, except for V^0).

A common example used to illustrate the complexity of the concepts needed to prove a theorem is the Pigeonhole Principle (PHP). Our version states that if $n + 1$ pigeons are placed in n holes, then some hole has two or more pigeons. We can present an instance of the PHP using a Boolean array $\langle P(i, j)\rangle$ $(0 \leq i \leq n, 0 \leq j < n)$, where $P(i, j)$ asserts that pigeon i is placed in hole j. Then the PHP can be formulated in the theory V^0 by the formula

$$\forall i \leq n \,\exists j < n \, P(i, j) \supset \exists i_1, i_2 \leq n \,\exists j < n \, (i_1 \neq i_2 \wedge P(i_1, j) \wedge P(i_2, j)). \tag{3}$$

Ajtai [5] proved (in effect) that this formula is not a theorem of V^0, and also that the propositional version (which uses atoms p_{ij} to represent $P(i, j)$ and finite conjunctions and disjunctions to express the bounded universal and existential number quantifiers) does not have polynomial size AC^0-*Frege* proofs. The intuitive reason for this is that a counting argument seems to be required to prove the PHP, but the complexity class AC^0 cannot count the number of ones in a string of bits. On the other hand, the class NC^1 can count, and indeed Buss proved that the propositional PHP does have polynomial size *Frege* proofs, and his method shows that (3) is a theorem of the theory VNC^1. (In fact it is a theorem of the apparently weaker theory VTC^0.)

A second example comes from linear algebra. If A and B are $n \times n$ matrices over some field, then

$$AB = I \supset BA = I. \tag{4}$$

A standard proof of this uses Gaussian elimination, which is a polynomial-time process. Indeed Soltys showed that (4) is a theorem of the theory **VP** corresponding to polynomial-time reasoning, and it follows that its propositional translation (say over the field of two elements) has polynomial-size proofs in the corresponding proof system *eFrege*. It is an open question whether (4) over **GF(2)** (or any field) can be proved in **VNC^1**, or whether the propositional version has polynomial-size **Frege** proofs.

The preceding example (4) is a universal theorem, in the sense that its statement has no existential quantifier. Another class of examples comes from existential theorems. From linear algebra, a natural example about $n \times n$ matrices is

$$\forall A \exists B \neq 0 (AB = I \lor AB = 0). \tag{5}$$

The complexity of finding B for a given A, even over **GF(2)**, is thought not to be in **NC^1** (it is hard for log space). Assuming that this is the case, it follows that (5) is not a theorem of **VNC^1**, since only **NC^1** functions are definable in that theory. This conclusion is the result of a general witnessing theorem, which states that if the formula $\forall x \exists y \varphi(x, y)$ (for suitable formulas φ) is provable in the theory associated with complexity class **C**, then there is a Skolem function $f(x)$ whose complexity is in **C** and which satisfies $\forall x \varphi(x, f(x))$.

The theory **VNC^1** proves that (4) follows from (5). Both (4) and (5) are theorems of the theory **VP** associated with polynomial time.

Another example of an existential theorem is "Fermat's Little Theorem", which states that if n is a prime number and $1 \leq a < n$, then $a^{n-1} \equiv 1 \pmod{n}$. Its existential content is captured by its contrapositive form

$$(1 \leq a < n) \land (a^{n-1} \not\equiv 1 \pmod{n}) \supset \exists d (1 < d < n \land d | n). \tag{6}$$

It is not hard to see that the function $a^{n-1} \bmod n$ can be computed in time polynomial in the lengths of a and n, using repeated squaring. If (6) is provable in **VP**, then by the witnessing theorem mentioned above it would follow that there is a polynomial time function $f(a, n)$ whose value $d = f(a, n)$ provides a proper divisor of n whenever a, n satisfy the hypothesis in (6). With the exception of the so-called Carmichael numbers, which can be factored in polynomial time, every composite n satisfies the hypothesis of (6) for at least half of the values of a, $1 \leq a < n$. Hence $f(a, n)$ would provide a probabilistic polynomial time algorithm for integer factoring. Such an algorithm is not known to exist, and would provide a method for breaking the RSA public-key encryption scheme.

Thus Fermat's Little Theorem is not provable in *VP*, assuming that there is no probabilistic polynomial time factoring algorithm.

Propositional tautologies can be used to express universal theorems such as (3) (in which the Predicate *P* is implicitly universally quantified and the bounded number quantifiers can be expanded in translation) and (4), but are not well suited to express existential theorems such as (5) and (6). However the latter can be expressed using formulas in the quantified propositional calculus (QPC), which extends the propositional calculus by allowing quantifiers $\forall p$ and $\exists p$ over propositional variables p. Each of the complexity classes in (2) has an associated QPC system, and in fact the systems $\langle G_i \rangle$ mentioned for *PH* form a hierarchy of QPC systems.

Most of the theories presented in this book, including those in (2), have the same "second-order" underlying vocabulary $\mathcal{L}_A^2$, introduced by Zambella. The vocabulary $\mathcal{L}_A^2$ is actually a vocabulary for the two-sorted first-order predicate calculus, where one sort is for numbers in $\mathbb{N}$ and the second sort is for finite sets of numbers. Here we regard an object of the second sort as a finite string over the alphabet $\{0, 1\}$ (the i-th bit in the string is 1 iff i is in the set). The strings are the objects of interest for the complexity classes, and serve as the main inputs for the machines or circuits that determine the class. The numbers serve a useful purpose as indices for the strings when describing properties of the strings. When they are used as machine or circuit inputs, they are presented in unary notation.

In the more common single-sorted theories such as Buss's hierarchies S_2^i and T_2^i the underlying objects are numbers which are presented in binary notation as inputs to Turing machines. Our two-sorted treatment has the advantage that the underlying vocabulary has no primitive operations on strings except the length function $|X|$ and the bit predicate $X(i)$ (meaning $i \in X$). This is especially important for studying weak complexity classes such as AC^0. The standard vocabulary for single-sorted theories includes number multiplication, which is not an AC^0 function on binary strings.

Chapter II provides a sufficient background in first-order logic for the rest of the book, including Gentzen's proof system *LK*. An unusual feature is our treatment of anchored (or "free-cut-free") *LK*-proofs. The completeness of these restricted systems is proved directly by a simple term-model construction as opposed to the usual syntactic cut-elimination method. The second form of the Herbrand Theorem proved here has many applications in later chapters for witnessing theorems.

Chapter III presents the necessary background on Peano Arithmetic (the first-order theory of $\mathbb{N}$ under $+$ and $\times$) and its subsystems, including the bounded theory $I\Delta_0$. The functions definable in $I\Delta_0$ are precisely those in the complexity class known as *LTH* (the Linear Time Hierarchy). An important theorem needed for this result is that the predicate $y = 2^x$ is definable in the vocabulary of arithmetic using a bounded formula

(Section III.3.3). The universal theory $\overline{I\Delta_0}$ has function symbols for each function in the Linear Time Hierarchy, and forms a conservative extension of $I\Delta_0$. This theory serves as a prototype for universal theories defined in later chapters for other complexity classes.

Chapter IV introduces the syntax and intended semantics for the two-sorted theories, which will be used throughout the remaining chapters. Here Σ_0^B is defined to be the class of formulas with no string quantifiers, and with all number quantifiers bounded. The Σ_1^B-formulas begin with zero or more bounded existential string quantifiers followed by a Σ_0^B-formula, and more generally Σ_i^B-formulas begin with at most i alternating blocks of bounded string quantifiers $\exists \forall \exists \ldots$. Representation theorems are proved which state that formulas in the syntactic class Σ_0^B represent precisely the (two-sorted) AC^0 relations, and for $i \geq 1$, formulas in Σ_i^B represent the relations in the i-th level of the polynomial hierarchy.

Chapter V introduces the hierarchy of two-sorted theories $V^0 \subset V^1 \subseteq V^2 \subseteq \cdots$. For $i \geq 1$, V^i is the two-sorted version of Buss's single-sorted theory S_2^i, which is associated with the ith level of the polynomial hierarchy. In this chapter we concentrate on V^0, which is associated with the complexity class AC^0. All two-sorted theories considered in later chapters are extensions of V^0. A Buss-style witnessing theorem is proved for V^0, showing that the existential string quantifiers in a Σ_1^B-theorem of V^0 can be witnessed by AC^0-functions. Since Σ_1^B-formulas have all string quantifiers in front, both the statement and the proof of the theorem are simpler than for the usual Buss-style witnessing theorems. (The same applies to the witnessing theorems proved in later chapters.) The final section proves that V^0 is finitely axiomatizable.

Chapter VI concentrates on the theory V^1, which is associated with the complexity class P. All (and only) polynomial time functions are Σ_1^B-definable in V^1. The positive direction is shown in two ways: by analyzing Turing machine computations and by using Cobham's characterization of these functions. The witnessing theorem for V^1 is shown using (two-sorted versions of) the anchored proofs described in Chapter II, and implies that only polynomial time functions are Σ_1^B-definable in V^1.

Chapter VII gives a general definition of propositional proof system. The goal is to associate a proof system with each theory so that each Σ_0^B-theorem of the theory translates into a polynomial size family of proofs in the proof system. Further, the theory should prove the soundness of the proof system, but this is not shown until Chapter X. In Chapter VII, translations are defined from V^0 to bounded-depth PK-proofs (i.e. bounded-depth Frege proofs), and also from V^1 to extended Frege proofs. Systems G_i and $G_i^\star$ for the quantified propositional calculus are defined, and for $i \geq 1$ we show how to translate bounded theorems of V^i

to polynomial size families of proofs in the system $G_i^\star$. The two-sorted treatment makes these translations simple and natural.

Chapter VIII begins by introducing other two-sorted theories associated with polynomial time. The finitely axiomatized theory VP and its universal conservative extension VPV both appear to be weaker than V^1, although they have the same Σ_1^B theorems as V^1. $VP = TV^0$ is the base of the hierarchy of theories $TV^0 \subseteq TV^1 \subseteq \cdots$, where for $i \geq 1$, TV^i is isomorphic to Buss's single-sorted theory T_2^i. The definable problems in TV^1 have the complexity of Polynomial Local Search. A form of the Herbrand Theorem known as KPT Witnessing is proved and applied to show independence of the Replacement axiom scheme from some theories, and to relating the collapse of the V^∞ hierarchy with the provable collapse of the polynomial hierarchy. The Σ_j^B-definable search problems in V^i and TV^i are characterized for many i and j. The RSUV isomorphism theorem between S_2^i and V^i is proved.

See Table 3 on page 250 for a summary of which search problems are definable in V^i and TV^i.

Chapter IX gives a uniform way of introducing minimal canonical theories for many complexity classes between AC^0 and P, including those mentioned earlier in (1). Each finitely axiomatized theory is defined as an extension of V^0 obtained by adding a single axiom stating the existence of a computation solving a complete problem for the associated complexity class. Evidence for the "minimality" of each theory is presented by defining a universal theory whose axioms are simply a set of basic axioms for V^0 together with the defining axioms for all the functions in the associated complexity class. These functions are defined as the function AC^0-closure of the complexity class, or (as is the case for P) using a recursion-theoretic characterization of the function class. The main theorem in each case is that the universal theory is a conservative extension of the finitely axiomatized theory.

Table 1 on page 7 gives a summary of the two-sorted theories presented in Chapter IX and elsewhere, and Table 2 on page 8 gives a list of some theorems provable (or possibly not provable) in the various theories.

Chapter X extends Chapter VII by presenting quantified propositional proof systems associated with various complexity classes, and defining translations from the bounded theorems of the theories introduced in Chapter IX to the appropriate proof system. Witnessing theorems for subsystems of G (quantified propositional calculus) are proved. The notion of *reflection principle* (soundness of a proof system) is defined, and many results showing which kinds of reflection principle for various systems can (or probably cannot) be proved in various theories. It is shown how reflection principles can be used to axiomatize some of the theories.

CLASS	THEORY	SEE
AC^0	V^0	Section V.1
	$\overline{V}^0$	Section V.6
$AC^0(2)$	$V^0(2), \widehat{V^0(2)}, \overline{V^0(2)}$	Section IX.4.2
	$VAC^0(2)V$	Section IX.4.4
$AC^0(m)$	$V^0(m), \widehat{V^0(m)}, \overline{V^0(m)}$	Section IX.4.6
$AC^0(6)$	$VAC^0(6)V$	Section IX.4.8
ACC	$VACC$	Section IX.4.6
TC^0	$VTC^0, \widehat{VTC^0}, \overline{VTC}^0$	Section IX.3.2
	VTC^0V	Section IX.3.4
NC^1	$VNC^1, \widehat{VNC^1}, \overline{VNC}^1$	Section IX.5.3
	VNC^1V	Section IX.5.5
L	$VL, \widehat{VL}, \overline{VL}$	Section IX.6.3
	VLV	Section IX.6.4
NL	$VNL, \widehat{VNL}, \overline{VNL}$	Section IX.6.1
	$V^1\text{-}KROM$	Section IX.6.2
AC^k $(k \geq 1)$	VAC^k	Section IX.5.6
NC^{k+1} $(k \geq 1)$	VNC^{k+1}	Section IX.5.6
NC	VNC	Section IX.5.6
	U^1	Section IX.5.6
P	VP	Section VIII.1
	VPV	Section VIII.2
	TV^0	Section VIII.3
	$V^1\text{-}HORN$	Section VIII.4
	V^1	Chapter VI
C (for $C \subseteq P$)	$VC, \widehat{VC}, \overline{VC}$	Section IX.2.1
$CC(PLS)$	TV^1	Section VIII.5
	V^2	Section VIII.7.2

TABLE 1. Theories and their Σ_1^B-definable classes.

THEORY	(NON)THEOREM(?)	SEE
V^0	(seq.) Jordan Curve Theorem	[84]
	$\not\vdash$ ***PHP***	Corollary VII.2.4
	$\not\vdash$ onto ***PHP***, $\not\vdash$ ***Count***$_m$	Section IX.4.3
$V^0(2)$	onto ***PHP***, ***Count***$_2$	Section IX.4.3
	(set) Jordan Curve Theorem	Section IX.4.5
	PHP?, ***Count***$_3$?	Section IX.7.4
$V^0(m)$	***Count***$_{m'}$ (if $\gcd(m, m') > 1$)	Section IX.4.7
	Count$_{m'}$? (if $\gcd(m, m') = 1$)	Section IX.7.4
	PHP?	Section IX.7.4
VTC^0	sorting	Exercise IX.3.9
	Reflection Principles for d-***PTK***	Section X.4.2
	PHP	Section IX.3.5
	Finite Szpilrajn's Theorem	Section IX.3.7
	Bondy's Theorem	Section IX.3.8
	define $\lfloor X/Y \rfloor$?	Section IX.7.3
VNC^1	Reflection Principle for ***PK***	Theorem X.3.9
	Barrington's Theorem	Sec. IX.5.5 & [82]
	NUMONES	Section IX.5.4
VL	Lind's characterization of ***L***	Section IX.6.4
	Reingold's Theorem?	Section IX.7.2
VNL	Grädel's Theorem (for ***NL***)	Theorem IX.6.24
VNC^2	Cayley–Hamilton Theorem?	Section IX.7.1
$VP = TV^0$	Reflection Principle for ***ePK***	Exercise X.2.22
	Grädel's Theorem (for ***P***)	Theorem VIII.4.8
	$\not\vdash$ Fermat's Little Theorem (cond.)	page 3
V^1	Prime Factorization Theorem	Exercise VI.4.4
V^i ($i \geq 1$)	Π_i^q-$RFN_{G_{i-1}}$, Π_{i+2}^q-$RFN_{G_i^*}$	Theorem X.2.17
TV^i ($i \geq 0$)	Π_{i+2}^q-$RFN_{G_{i+1}^*}$, Π_{i+1}^q-RFN_{G_i}	Theorem X.2.20

TABLE 2. Some theories and their (non)theorems/solvable problems (and open questions). ("cond." stands for conditional.) Many theorems of ***VP***, such as Kuratowski's Theorem, Hall's Theorem, Menger's Theorem are not discussed here.

Chapter II

THE PREDICATE CALCULUS AND THE SYSTEM *LK*

In this chapter we present the logical foundations for theories of bounded arithmetic. We introduce Gentzen's proof system *LK* for the predicate calculus, and prove that it is sound, and complete even when proofs have a restricted form called "anchored". We augment the system *LK* by adding equality axioms. We prove the Compactness Theorem for predicate calculus, and the Herbrand Theorem.

In general we distinguish between syntactic notions and semantic notions. Examples of syntactic notions are variables, connectives, formulas, and formal proofs. The semantic notions relate to meaning; for example truth assignments, structures, validity, and logical consequence.

The first section treats the simple case of propositional calculus.

II.1. Propositional Calculus

Propositional formulas (called simply *formulas* in this section) are built from the logical constants $\bot$, $\top$ (for False, True), propositional variables (or atoms) $P_1, P_2, \ldots$, connectives $\neg, \vee, \wedge$, and parentheses $($, $)$. We use $P, Q, R, \ldots$ to stand for propositional variables, $A, B, C, \ldots$ to stand for formulas, and $\Phi, \Psi, \ldots$ to stand for sets of formulas. When writing formulas such as $(P \vee (Q \wedge R))$, our convention is that $P, Q, R, \ldots$ stand for distinct variables.

Formulas are built according to the following rules:

- $\bot$, $\top$, P, are formulas (also called *atomic formulas*) for any variable P.
- If A and B are formulas, then so are $(A \vee B)$, $(A \wedge B)$, and $\neg A$.

The implication connective $\supset$ is not allowed in our formulas, but we will take $(A \supset B)$ to stand for $(\neg A \vee B)$. Also $(A \leftrightarrow B)$ stands for $((A \supset B) \wedge (B \supset A))$.

We sometimes abbreviate formulas by omitting parentheses, but the intended formula has all parentheses present as defined above.

A *truth assignment* is an assignment of truth values F, T to atoms. Given a truth assignment τ, the truth value A^τ of a formula A is defined inductively as follows: $\bot^\tau = F$, $\top^\tau = T$, $P^\tau = \tau(P)$ for atom P, $(A \wedge B)^\tau = T$ iff both $A^\tau = T$ and $B^\tau = T$, $(A \vee B)^\tau = T$ iff either $A^\tau = T$ or $B^\tau = T$, $(\neg A)^\tau = T$ iff $A^\tau = F$.

Definition II.1.1. A truth assignment τ *satisfies* A iff $A^\tau = T$; τ *satisfies* a set Φ of formulas iff τ satisfies A for all $A \in \Phi$. Φ is *satisfiable* iff some τ satisfies Φ; otherwise Φ is *unsatisfiable*. Similarly for A. $\Phi \models A$ (i.e., A is a *logical consequence* of Φ) iff τ satisfies A for every τ such that τ satisfies Φ. A formula A is *valid* iff $\models A$ (i.e., $A^\tau = T$ for all τ). A valid propositional formula is called a *tautology*. We say that A and B are *equivalent* (written $A \Longleftrightarrow B$) iff $A \models B$ and $B \models A$.

Note that $\Longleftrightarrow$ refers to semantic equivalence, as opposed to $=_{\text{syn}}$, which indicates syntactic equivalence. For example, $(P \vee Q) \Longleftrightarrow (Q \vee P)$, but $(P \vee Q) \neq_{syn} (Q \vee P)$.

II.1.1. Gentzen's Propositional Proof System *PK*. We present the propositional part *PK* of Gentzen's sequent-based proof system *LK*. Each line in a proof in the system *PK* is a *sequent* of the form

$$A_1, \ldots, A_k \longrightarrow B_1, \ldots, B_\ell \tag{7}$$

where $\longrightarrow$ is a new symbol and $A_1, \ldots, A_k$ and $B_1, \ldots, B_\ell$ are sequences of formulas ($k, \ell \geq 0$) called *cedents*. We call the cedent $A_1, \ldots, A_k$ the *antecedent* and $B_1, \ldots, B_\ell$ the *succedent* (or *consequent*).

The semantics of sequents is given as follows. We say that a truth assignment τ *satisfies* the sequent (7) iff either τ falsifies some A_i or τ satisfies some B_i. Thus the sequent is equivalent to the formula

$$\neg A_1 \vee \neg A_2 \vee \cdots \vee \neg A_k \vee B_1 \vee B_2 \vee \cdots \vee B_\ell. \tag{8}$$

(Here and elsewhere, a disjunction $C_1 \vee \cdots \vee C_n$ indicates parentheses have been inserted with association to the right. For example, $C_1 \vee C_2 \vee C_3 \vee C_4$ stands for $(C_1 \vee (C_2 \vee (C_3 \vee C_4)))$. Similarly for a disjunction $C_1 \wedge \cdots \wedge C_n$.) In other words, the conjunction of the A's implies the disjunction of the B's. In the cases in which the antecedent or succedent is empty, we see that the sequent $\longrightarrow A$ is equivalent to the formula A, and $A \longrightarrow$ is equivalent to $\neg A$, and just $\longrightarrow$ (with both antecedent and succedent empty) is false (unsatisfiable). We say that a sequent is *valid* if it is true under all truth assignments (which is the same as saying that its corresponding formula is a tautology).

Definition II.1.2. A *PK* *proof* of a sequent S is a finite tree whose nodes are (labeled with) sequents, whose root (called the *endsequent*) is S and is written at the bottom, whose leaves (or *initial sequents*) are logical axioms (see below), such that each non-leaf sequent follows from the sequent(s) immediately above by one of the rules of inference given below.

The *logical axioms* are of the form

$$A \longrightarrow A, \qquad \bot \longrightarrow, \qquad \longrightarrow \top$$

where A is any formula. (Note that we differ here from most other treatments, which require that A be an atomic formula.) The rules of inference are as follows (here Γ and Δ denote finite sequences of formulas).

weakening rules

$$\text{left: } \frac{\Gamma \longrightarrow \Delta}{A, \Gamma \longrightarrow \Delta} \qquad\qquad \text{right: } \frac{\Gamma \longrightarrow \Delta}{\Gamma \longrightarrow \Delta, A}$$

exchange rules

$$\text{left: } \frac{\Gamma_1, A, B, \Gamma_2 \longrightarrow \Delta}{\Gamma_1, B, A, \Gamma_2 \longrightarrow \Delta} \qquad\qquad \text{right: } \frac{\Gamma \longrightarrow \Delta_1, A, B, \Delta_2}{\Gamma \longrightarrow \Delta_1, B, A, \Delta_2}$$

contraction rules

$$\text{left: } \frac{\Gamma, A, A \longrightarrow \Delta}{\Gamma, A \longrightarrow \Delta} \qquad\qquad \text{right: } \frac{\Gamma \longrightarrow \Delta, A, A}{\Gamma \longrightarrow \Delta, A}$$

$\neg$ introduction rules

$$\text{left: } \frac{\Gamma \longrightarrow \Delta, A}{\neg A, \Gamma \longrightarrow \Delta} \qquad\qquad \text{right: } \frac{A, \Gamma \longrightarrow \Delta}{\Gamma \longrightarrow \Delta, \neg A}$$

$\wedge$ introduction rules

$$\text{left: } \frac{A, B, \Gamma \longrightarrow \Delta}{(A \wedge B), \Gamma \longrightarrow \Delta} \qquad\qquad \text{right: } \frac{\Gamma \longrightarrow \Delta, A \qquad \Gamma \longrightarrow \Delta, B}{\Gamma \longrightarrow \Delta, (A \wedge B)}$$

$\vee$ introduction rules

$$\text{left: } \frac{A, \Gamma \longrightarrow \Delta \qquad B, \Gamma \longrightarrow \Delta}{(A \vee B), \Gamma \longrightarrow \Delta} \qquad\qquad \text{right: } \frac{\Gamma \longrightarrow \Delta, A, B}{\Gamma \longrightarrow \Delta, (A \vee B)}$$

cut rule

$$\frac{\Gamma \longrightarrow \Delta, A \qquad A, \Gamma \longrightarrow \Delta}{\Gamma \longrightarrow \Delta}$$

The formula A in the cut rule is called the *cut* formula. A proof that does not use the cut rule is called *cut-free*. The new formulas in the bottom sequents of the introduction rules are called *principal formulas* and the formula(s) in the top sequent(s) that are used to form the principal formulas are called *auxiliary formulas*.

Note that there is one left introduction rule and one right introduction rule for each of the three logical connectives $\wedge, \vee, \neg$. Further, these rules seem to be the simplest possible, given the fact that in each case the bottom sequent is valid iff all top sequents are valid.

Note that repeated use of the exchange rules allows us to execute an arbitrary reordering of the formulas in the antecedent or succedent of a sequent. In presenting a proof in the system *PK*, we will usually omit mention of the steps requiring the exchange rules, but of course they are there implicitly.

DEFINITION II.1.3. A *PK* proof of a formula A is a *PK* proof of the sequent $\longrightarrow A$.

As an example, we give a *PK* proof of one of De Morgan's laws:

$$\neg(P \wedge Q) \longrightarrow \neg P \vee \neg Q.$$

To find this (or any) proof, it is a good idea to start with the conclusion at the bottom, and work up by removing the connectives one at a time, outermost first, by using the introduction rules in reverse. This can be continued until some formula A occurs on both the left and right side of a sequent, or $\top$ occurs on the right, or $\bot$ occurs on the left. Then this sequent can be derived from one of the axioms $A \longrightarrow A$ or $\longrightarrow \top$ or $\bot \longrightarrow$ using weakenings and exchanges. The cut and contraction rules are not necessary, and weakenings are only needed immediately below axioms. (The cut rule can be used to shorten proofs, and contraction will be needed later for the predicate calculus.)

$$
\cfrac{
  \cfrac{
    \cfrac{
      \cfrac{P \longrightarrow P}{P \longrightarrow P, \neg Q}\text{(weakening)}
    }{\longrightarrow P, \neg P, \neg Q}(\neg\text{ right})
    \qquad
    \cfrac{
      \cfrac{Q \longrightarrow Q}{Q \longrightarrow Q, \neg P}\text{(weakening)}
    }{\longrightarrow Q, \neg P, \neg Q}(\neg\text{ right})
  }{\longrightarrow P \wedge Q, \neg P, \neg Q}(\wedge\text{ right})
}{
  \cfrac{
    \cfrac{\longrightarrow P \wedge Q, \neg P \vee \neg Q}{\neg(P \wedge Q) \longrightarrow \neg P \vee \neg Q}(\neg\text{ left})
  }{}(\vee\text{ right})
}
$$

EXERCISE II.1.4. Give *PK* proofs for each of the following valid sequents:

(a) $\neg P \vee \neg Q \longrightarrow \neg(P \wedge Q)$.
(b) $\neg(P \vee Q) \longrightarrow \neg P \wedge \neg Q$.
(c) $\neg P \wedge \neg Q \longrightarrow \neg(P \vee Q)$.

EXERCISE II.1.5. Show that the contraction rules can be derived from the cut rule (with weakenings and exchanges).

EXERCISE II.1.6. Suppose that we allowed $\supset$ as a primitive connective, rather than one introduced by definition. Give the appropriate left and right introduction rules for $\supset$.

II.1.2. Soundness and Completeness of *PK*. Now we prove that *PK* is both sound and complete. That is, a propositional sequent is provable in *PK* iff it is valid.

THEOREM II.1.7 (Soundness). *Every sequent provable in **PK** is valid.*

PROOF. We show that the endsequent in every **PK** proof is valid, by induction on the number of sequents in the proof. For the base case, the proof is a single line: a logical axiom. Each logical axiom is obviously valid. For the induction step, one needs only verify for each rule that the bottom sequent is a logical consequence of the top sequent(s).   □

THEOREM II.1.8 (Completeness). *Every valid propositional sequent is provable in **PK** without using cut or contraction.*

PROOF. The idea is discussed in the example proof above of De Morgan's laws. We need to use the inversion principle.

LEMMA II.1.9 (Inversion Principle). *For each **PK** rule except for weakenings, if the bottom sequent is valid, then all top sequents are valid.*

This principle is easily verified by inspecting each of the eleven rules in question.

Now for the completeness theorem: We show that every valid sequent $\Gamma \longrightarrow \Delta$ has a **PK** proof, by induction on the total number of logical connectives $\wedge, \vee, \neg$ occurring in $\Gamma \longrightarrow \Delta$. For the base case, every formula in Γ and Δ is an atom or one of the constants $\bot, \top$, and since the sequent is valid, some atom P must occur in both Γ and Δ, or $\bot$ occurs in Γ or $\top$ occurs in Δ. Hence $\Gamma \longrightarrow \Delta$ can be derived from one of the logical axioms by weakenings and exchanges.

For the induction step, let A be any formula which is not an atom and not a constant in Γ or Δ. Then by the definition of propositional formula A must have one of the forms $(B \wedge C)$, $(B \vee C)$, or $\neg B$. Thus $\Gamma \longrightarrow \Delta$ can be derived from $\wedge$ introduction, $\vee$ introduction, or $\neg$ introduction, respectively, using either the left case or the right case, depending on whether A is in Γ or Δ, and also using exchanges, but no weakenings. In each case, each top sequent of the rule will have at least one fewer connective than $\Gamma \longrightarrow \Delta$, and the sequent is valid by the inversion principle. Hence each top sequent has a **PK** proof, by the induction hypothesis.   □

The soundness and completeness theorems relate the semantic notion of validity to the syntactic notion of proof.

II.1.3. *PK* Proofs from Assumptions. We generalize the (semantic) definition of logical consequence from formulas to sequents in the obvious way: A sequent S is a *logical consequence* of a set Φ of sequents iff every truth assignment τ that satisfies Φ also satisfies S. We generalize the (syntactic) definition of a **PK** proof of a sequent S to a **PK** proof of S *from a set Φ of sequents* (also called a **PK**-Φ proof) by allowing sequents in Φ to be leaves (called *nonlogical axioms*) in the proof tree, in addition to the logical axioms. It turns out that soundness and completeness generalize to this setting.

THEOREM II.1.10 (Derivational Soundness and Completeness). *A propositional sequent S is a logical consequence of a set Φ of sequents iff S has a **PK**-Φ proof.*

Derivational soundness is proved in the same way as simple soundness: by induction on the number of sequents in the **PK**-Φ proof, using the fact that the bottom sequent of each rule is a logical consequence of the top sequent(s).

A remarkable aspect of derivational completeness is that a finite proof exists even in case Φ is an infinite set. This is because of the compactness theorem (below) which implies that if S is a logical consequence of Φ, then S is a logical consequence of some finite subset of Φ.

In general, to prove S from Φ the cut rule is required. For example, there is no **PK** proof of $\longrightarrow P$ from $\longrightarrow P \wedge Q$ without using the cut rule. This follows from the *subformula property*, which states that in a cut-free proof π of a sequent S, every formula in every sequent of π is a subformula of some formula in S. This is stated more generally in the Proposition II.1.15.

EXERCISE II.1.11. Let A_S be the formula giving the meaning of a sequent S, as in (8). Show that there is a cut-free **PK** derivation of $\longrightarrow A_S$ from S.

PROOF OF THEOREM II.1.10 (Completeness). From the above easy exercise and from the earlier Completeness Theorem and from Theorem II.1.16, Form 2 (compactness), we obtain an easy proof of derivational completeness. Suppose that the sequent $\Gamma \longrightarrow \Delta$ is a logical consequence of sequents $S_1, \ldots, S_k$. Then by the above exercise we can derive each of the sequents $\longrightarrow A_{S_1}, \ldots, \longrightarrow A_{S_k}$ from the sequents $S_1, \ldots, S_k$. Also the sequent

$$A_{S_1}, \ldots, A_{S_k}, \Gamma \longrightarrow \Delta \tag{9}$$

is valid, and hence has a **PK** proof by Theorem II.1.8. Finally from (9) using successive cuts with cut formulas $A_{S_1}, \ldots, A_{S_k}$ we obtain the desired **PK** derivation of $\Gamma \longrightarrow \Delta$ from the the sequents $S_1, \ldots, S_k$.   □

We now wish to show that the cut formulas in the derivation can be restricted to formulas occurring in the hypothesis sequents.

DEFINITION II.1.12 (Anchored Proof). An instance of the cut rule in a **PK**-Φ proof π is *anchored* if the cut formula A (also) occurs as a formula (rather than a subformula) in some nonlogical axiom of π. A **PK**-Φ proof π is *anchored* if every instance of cut in π is anchored.

Our *anchored* proofs are similar to *free-cut-free* proofs in [72] and elsewhere. Our use of the term *anchored* is inspired by [27].

The derivational completeness theorem can be strengthened as follows.

THEOREM II.1.13 (Anchored Completeness). *If a propositional sequent S is a logical consequence of a set Φ of sequents, then there is an anchored PK-Φ proof of S.*

We illustrate the proof of the anchored completeness theorem by proving the special case in which Φ consists of the single sequent $A \longrightarrow B$. Assume that the sequent $\Gamma \longrightarrow \Delta$ is a logical consequence of $A \longrightarrow B$. Then both of the sequents $\Gamma \longrightarrow \Delta, A$ and $B, A, \Gamma \longrightarrow \Delta$ are valid (why?). Hence by Theorem II.1.8 they have **PK** proofs π_1 and π_2, respectively. We can use these proofs to get a proof of $\Gamma \longrightarrow \Delta$ from $A \longrightarrow B$ as shown below, where the double line indicates the rules weakening and exchange have been applied.

$$
\cfrac{
\cfrac{\vdots\ \pi_1}{\Gamma \longrightarrow \Delta, A}
\qquad
\cfrac{\cfrac{A \longrightarrow B}{A, \Gamma \longrightarrow \Delta, B} \qquad B, A, \Gamma \overset{\vdots\ \pi_2}{\longrightarrow} \Delta}{A, \Gamma \longrightarrow \Delta}\text{(cut)}
}{\Gamma \longrightarrow \Delta}\text{(cut)}
$$

Next consider the case in which Φ has the form

$$\{\longrightarrow A_1, \longrightarrow A_2, \ldots, \longrightarrow A_k\}$$

for some set $\{A_1, \ldots, A_k\}$ of formulas. Assume that $\Gamma \longrightarrow \Delta$ is a logical consequence of Φ in this case. Then the sequent

$$A_1, A_2, \ldots, A_k, \Gamma \longrightarrow \Delta$$

is valid, and hence has a **PK** proof π. Now we can use the assumptions Φ and the cut rule to successively remove $A_1, A_2, \ldots, A_k$ from the above sequent to conclude $\Gamma \longrightarrow \Delta$. For example, A_1 is removed as follows (the double line represents applications of the rule weakening and exchange):

$$
\cfrac{
\cfrac{\longrightarrow A_1}{A_2, \ldots, A_k, \Gamma \longrightarrow \Delta, A_1}
\qquad
A_1, A_2, \ldots, A_k, \Gamma \overset{\vdots\ \pi}{\longrightarrow} \Delta
}{A_2, \ldots, A_k, \Gamma \longrightarrow \Delta}\text{(cut)}
$$

EXERCISE II.1.14. Prove the anchored completeness theorem for the more general case in which Φ is any finite set of sequents. Use induction on the number of sequents in Φ.

A nice property of anchored proofs is the following.

PROPOSITION II.1.15 (Subformula Property). *If π is an anchored PK-Φ proof of S, then every formula in every sequent of π is a subformula of a formula either in S or in some nonlogical axiom of π.*

PROOF. This follows by induction on the number of sequents in π, using the fact that for every rule other than cut, every formula on the top is a subformula of some formula on the bottom. For the case of cut we use the fact that every cut formula is a formula in some nonlogical axiom of π. $\qquad\square$

The Subformula Property can be generalized in a way that applies to cut-free *LK* proofs in the predicate calculus, and this will play an important role later in proving witnessing theorems.

II.1.4. Propositional Compactness. We conclude our treatment of the propositional calculus with a fundamental result which also plays an important role in the predicate calculus.

THEOREM II.1.16 (Propositional Compactness). *We state three different forms of this result. All three are equivalent.*

FORM 1: *If* Φ *is an unsatisfiable set of propositional formulas, then some finite subset of* Φ *is unsatisfiable.*

FORM 2: *If a formula A is a logical consequence of a set* Φ *of formulas, then A is a logical consequence of some finite subset of* Φ.

FORM 3: *If every finite subset of a set* Φ *of formulas is satisfiable, then* Φ *is satisfiable.*

EXERCISE II.1.17. Prove the equivalence of the three forms. (Note that Form 3 is the contrapositive of Form 1.)

PROOF OF FORM 1. Let Φ be an unsatisfiable set of formulas. By our definition of propositional formula, all propositional variables in Φ come from a countable list $P_1, P_2, \ldots$. (See Exercise II.1.19 for the uncountable case.) Organize the set of truth assignments into an infinite rooted binary tree B. Each node except the root is labeled with a literal P_i or $\neg P_i$. The two children of the root are labeled P_1 and $\neg P_1$, indicating that P_1 is assigned T or F, respectively. The two children of each of these nodes are labeled P_2 and $\neg P_2$, respectively, indicating the truth value of P_2. Thus each infinite branch in the tree represents a complete truth assignment, and each path from the root to a node represents a truth assignment to the atoms $P_1, \ldots, P_i$, for some i.

Now for every node v in the tree B, prune the tree at v (i.e., remove the subtree rooted at v, keeping v itself) if the partial truth assignment τ_v represented by the path to v falsifies some formula A_v in Φ, where all atoms in A_v get values from τ_v. Let B' be the resulting pruned tree. Since Φ is unsatisfiable, every path from the root in B' must end after finitely many steps in some leaf v labeled with a formula A_v in Φ. It follows from König's Lemma below that B' is finite. Let Φ' be the finite subset of Φ consisting of all formulas A_v labeling the leaves of B'. Since every truth assignment τ determines a path in B' which ends in a leaf A_v falsified by τ, it follows that Φ' is unsatisfiable.   □

LEMMA II.1.18 (König's Lemma). *Suppose T is a rooted tree in which every node has only finitely many children. If every branch in T is finite, then T is finite.*

PROOF. We prove the contrapositive: If T is infinite (but every node has only finitely many children) then T has an infinite branch. We can define an infinite path in T as follows: Start at the root. Since T is infinite but the root has only finitely many children, the subtree rooted at one of these children must be infinite. Choose such a child as the second node in the branch, and continue.   □

EXERCISE II.1.19. (*For those with some knowledge of set theory or point set topology*) The above proof of the propositional compactness theorem only works when the set of atoms is countable, but the result still holds even when Φ is an uncountable set with an uncountable set $\mathcal{A}$ of atoms. Complete each of the two proof outlines below.

(a) Prove Form 3 using Zorn's Lemma as follows: Call a set Ψ of formulas *finitely satisfiable* if every finite subset of Ψ is satisfiable. Assume that Φ is finitely satisfiable. Let $\mathcal{C}$ be the class of all finitely satisfiable sets $\Psi \supseteq \Phi$ of propositional formulas using atoms in Φ. Order these sets Ψ by inclusion. Show that the union of any chain of sets in $\mathcal{C}$ is again in the class $\mathcal{C}$. Hence by Zorn's Lemma, $\mathcal{C}$ has a maximal element Ψ_0. Show that Ψ_0 has a unique satisfying assignment, and hence Φ is satisfiable.

(b) Show that Form 1 follows from Tychonoff's Theorem: The product of compact topological spaces is compact. The set of all truth assignments to the atom set $\mathcal{A}$ can be given the product topology, when viewed as the product for all atoms P in $\mathcal{A}$ of the two-point space $\{T, F\}$ of assignments to P, with the discrete topology. By Tychonoff's Theorem, this space of assignments is compact. Show that for each formula A, the set of assignments falsifying A is open. Thus Form 1 follows from the definition of compact: every open cover has a finite subcover.

II.2. Predicate Calculus

In this section we present the syntax and semantics of the predicate calculus (also called first-order logic). We show how to generalize Gentzen's proof system **PK** for the propositional calculus to the system **LK** for the predicate calculus, by adding quantifier introduction rules. We show that **LK** is sound and complete. We prove an anchored completeness theorem which limits the need for the cut rule in the presence of nonlogical axioms.

II.2.1. Syntax of the Predicate Calculus. A *first-order vocabulary* (or just *vocabulary*, or *language*) $\mathcal{L}$ is specified by the following:

1) For each $n \geq 0$ a set of n-ary function symbols (possibly empty). We use $f, g, h, \ldots$ as meta-symbols for function symbols. A zero-ary function symbol is called a constant symbol.
2) For each $n \geq 0$, a set of n-ary predicate symbols (which must be nonempty for some n). We use $P, Q, R, \ldots$ as meta-symbols for predicate symbols. A zero-ary predicate symbol is the same as a propositional atom.

In addition, the following symbols are available to build first-order terms and formulas:

1) An infinite set of variables. We use $x, y, z, \ldots$ and sometimes $a, b, c, \ldots$ as meta-symbols for variables.
2) Connectives $\neg, \wedge, \vee$ (not, and, or); logical constants $\bot, \top$ (for False, True).
3) Quantifiers $\forall, \exists$ (for all, there exists).
4) $(,)$ (parentheses).

Given a vocabulary $\mathcal{L}$, *$\mathcal{L}$-terms* are certain strings built from variables and function symbols of $\mathcal{L}$, and are intended to represent objects in the universe of discourse. We will drop mention of $\mathcal{L}$ when it is not important, or clear from context.

DEFINITION II.2.1 (*$\mathcal{L}$-Terms*). Let $\mathcal{L}$ be a first-order vocabulary.

1) Every variable is an $\mathcal{L}$-term.
2) If f is an n-ary function symbol of $\mathcal{L}$ and $t_1, \ldots, t_n$ are $\mathcal{L}$-terms, then $f t_1 \ldots t_n$ is an $\mathcal{L}$-term.

Recall that a 0-ary function symbol is called a constant symbol (or sometimes just a *constant*). Note that all constants in $\mathcal{L}$ are $\mathcal{L}$-terms.

DEFINITION II.2.2 (*$\mathcal{L}$-Formulas*).   Let $\mathcal{L}$ be a first-order vocabulary. First-order formulas in $\mathcal{L}$ (or *$\mathcal{L}$-formulas*, or just *formulas*) are defined inductively as follows:

1) $P t_1 \cdots t_n$ is an *atomic* $\mathcal{L}$-formula, where P is an n-ary predicate symbol in $\mathcal{L}$ and $t_1, \ldots, t_n$ are $\mathcal{L}$-terms. Also each of the logical constants $\bot, \top$ is an atomic formula.
2) If A and B are $\mathcal{L}$-formulas, so are $\neg A$, $(A \wedge B)$, and $(A \vee B)$.
3) If A is an $\mathcal{L}$-formula and x is a variable, then $\forall x A$ and $\exists x A$ are $\mathcal{L}$-formulas.

Examples of formulas: $(\neg \forall x P x \vee \exists x \neg P x)$, $(\forall x \neg P x y \wedge \neg \forall z P f y z)$.

As in the case of propositional formulas, we use the notation $(A \supset B)$ for $(\neg A \vee B)$ and $(A \leftrightarrow B)$ for $((A \supset B) \wedge (B \supset A))$.

It can be shown that no proper initial segment of a term is a term, and hence every term can be parsed uniquely according to Definition II.2.1. A similar remark applies to formulas, and Definition II.2.2.

NOTATION. $r = s$ stands for $= rs$, and $r \neq s$ stands for $\neg(r = s)$.

DEFINITION II.2.3 (The Vocabulary of Arithmetic).

$$\mathcal{L}_A = [0, 1, +, \cdot \; ; \; =, \leq].$$

Here $0, 1$ are constants; $+, \cdot$ are binary function symbols; $=, \leq$ are binary predicate symbols. In practice we use infix notation for $+, \cdot, =, \leq$. Thus, for example, $(t_1 \cdot t_2) =_{syn} \cdot t_1 t_2$ and $(t_1 + t_2) =_{syn} + t_1 t_2$.

DEFINITION II.2.4 (Free and Bound Variables). An occurrence of x in A is *bound* iff it is in a subformula of A of the form $\forall x B$ or $\exists x B$. Otherwise the occurrence is *free*.

Notice that a variable can have both free and bound occurrences in one formula. For example, in $Px \wedge \forall x Qx$, the first occurrence of x is free, and the second occurrence is bound.

DEFINITION II.2.5. A formula is *closed* if it contains no free occurrence of a variable. A term is *closed* if it contains no variable. A closed formula is called a *sentence*.

II.2.2. Semantics of Predicate Calculus.

DEFINITION II.2.6 ($\mathcal{L}$-Structure). If $\mathcal{L}$ is a first-order vocabulary, then an $\mathcal{L}$-structure $\mathcal{M}$ consists of the following:

1) A nonempty set M called the *universe*. (Variables in an $\mathcal{L}$-formula are intended to range over M.)
2) For each n-ary function symbol f in $\mathcal{L}$, an associated function $f^{\mathcal{M}} : M^n \to M$.
3) For each n-ary predicate symbol P in $\mathcal{L}$, an associated relation $P^{\mathcal{M}} \subseteq M^n$. If $\mathcal{L}$ contains $=$, then $=^{\mathcal{M}}$ must be the true equality relation on M.

Notice that the predicate symbol $=$ gets special treatment in the above definition, in that $=^{\mathcal{M}}$ must always be the true equality relation. Any other predicate symbol may be interpreted by an arbitrary relation of the appropriate arity.

Every $\mathcal{L}$-sentence becomes either true or false when interpreted by an $\mathcal{L}$-structure $\mathcal{M}$, as explained below. If a sentence A becomes true under $\mathcal{M}$, then we say $\mathcal{M}$ *satisfies* A, or $\mathcal{M}$ is a *model* for A, and write $\mathcal{M} \models A$.

If A has free variables, then these variables must be interpreted as specific elements in the universe M before A gets a truth value under the structure $\mathcal{M}$. For this we need the following:

DEFINITION II.2.7 (Object Assignment). An *object assignment* σ for a structure $\mathcal{M}$ is a mapping from variables to the universe M.

Below we give the formal definition of notion $\mathcal{M} \models A[\sigma]$, which is intended to mean that the structure $\mathcal{M}$ satisfies the formula A when the free variables of A are interpreted according to the object assignment σ. First it is necessary to define the notation $t^{\mathcal{M}}[\sigma]$, which is the element of universe M assigned to the term t by the structure $\mathcal{M}$ when the variables of t are interpreted according to σ.

NOTATION. If x is a variable and $m \in M$, then the object assignment $\sigma(m/x)$ is the same as σ except it maps x to m.

DEFINITION II.2.8 (Basic Semantic Definition). Let $\mathcal{L}$ be a first-order vocabulary, let $\mathcal{M}$ be an $\mathcal{L}$-structure, and let σ be an object assignment for $\mathcal{M}$. Each $\mathcal{L}$-term t is assigned an element $t^{\mathcal{M}}[\sigma]$ in M, defined by structural induction on terms t, as follows (refer to the definition of $\mathcal{L}$-term):

(a) $x^{\mathcal{M}}[\sigma]$ is $\sigma(x)$, for each variable x.
(b) $(ft_1 \cdots t_n)^{\mathcal{M}}[\sigma] = f^{\mathcal{M}}(t_1^{\mathcal{M}}[\sigma], \ldots, t_n^{\mathcal{M}}[\sigma])$.

For A an $\mathcal{L}$-formula, the notion $\mathcal{M} \models A[\sigma]$ ($\mathcal{M}$ *satisfies A under* σ) is defined by structural induction on formulas A as follows (refer to the definition of formula):

(a) $\mathcal{M} \models \top$ and $\mathcal{M} \not\models \bot$.
(b) $\mathcal{M} \models (Pt_1 \cdots t_n)[\sigma]$ iff $\langle t_1^{\mathcal{M}}[\sigma], \ldots, t_n^{\mathcal{M}}[\sigma] \rangle \in P^{\mathcal{M}}$.
(c) If $\mathcal{L}$ contains $=$, then $\mathcal{M} \models (s = t)[\sigma]$ iff $s^{\mathcal{M}}[\sigma] = t^{\mathcal{M}}[\sigma]$.
(d) $\mathcal{M} \models \neg A[\sigma]$ iff $\mathcal{M} \not\models A[\sigma]$.
(e) $\mathcal{M} \models (A \lor B)[\sigma]$ iff $\mathcal{M} \models A[\sigma]$ or $\mathcal{M} \models B[\sigma]$.
(f) $\mathcal{M} \models (A \land B)[\sigma]$ iff $\mathcal{M} \models A[\sigma]$ and $\mathcal{M} \models B[\sigma]$.
(g) $\mathcal{M} \models (\forall x A)[\sigma]$ iff $\mathcal{M} \models A[\sigma(m/x)]$ for all $m \in M$.
(h) $\mathcal{M} \models (\exists x A)[\sigma]$ iff $\mathcal{M} \models A[\sigma(m/x)]$ for some $m \in M$.

Note that item (c) in the definition of $\mathcal{M} \models A[\sigma]$ follows from (b) and the fact that $=^{\mathcal{M}}$ is always the equality relation.

If t is a closed term (i.e., contains no variables), then $t^{\mathcal{M}}[\sigma]$ is independent of σ, and so we sometimes just write $t^{\mathcal{M}}$. Similarly, if A is a sentence, then we sometimes write $\mathcal{M} \models A$ instead of $\mathcal{M} \models A[\sigma]$, since σ does not matter.

DEFINITION II.2.9 (Standard Model). The *standard model* $\underline{\mathbb{N}}$ for the vocabulary $\mathcal{L}_A$ is a structure with universe $M = \mathbb{N} = \{0, 1, 2, \ldots\}$, where $0, 1, +, \cdot, =, \leq$ get their usual meanings on the natural numbers.

As an example, $\underline{\mathbb{N}} \models \forall x \forall y \exists z (x + z = y \lor y + z = x)$ (since either $y - x$ or $x - y$ exists) but $\underline{\mathbb{N}} \not\models \forall x \exists y (y + y = x)$ since not all natural numbers are even.

In the future we sometimes assume that there is some first-order vocabulary $\mathcal{L}$ in the background, and do not necessarily mention it explicitly.

NOTATION. In general, Φ denotes a set of formulas, $A, B, C, \ldots$ denote formulas, $\mathcal{M}$ denotes a structure, and σ denotes an object assignment.

DEFINITION II.2.10.   (a) $\mathcal{M} \models \Phi[\sigma]$ iff $\mathcal{M} \models A[\sigma]$ for all $A \in \Phi$.
(b) $\mathcal{M} \models \Phi$ iff $\mathcal{M} \models \Phi[\sigma]$ for all σ.
(c) $\Phi \models A$ iff for all $\mathcal{M}$ and all σ, if $\mathcal{M} \models \Phi[\sigma]$ then $\mathcal{M} \models A[\sigma]$.
(d) $\models A$ (A is *valid*) iff $\mathcal{M} \models A[\sigma]$ for all $\mathcal{M}$ and σ.
(e) $A \Longleftrightarrow B$ (A and B are *logically equivalent*, or just *equivalent*) iff for all $\mathcal{M}$ and all σ, $\mathcal{M} \models A[\sigma]$ iff $\mathcal{M} \models B[\sigma]$.

$\Phi \models A$ is read "A is a logical consequence of Φ". Do not confuse this with our other use of the symbol $\models$, as in $\mathcal{M} \models A$ ($\mathcal{M}$ satisfies A). In the latter, $\mathcal{M}$ is a structure, rather than a set of formulas.

If Φ consists of a single formula B, then we write $B \models A$ instead of $\{B\} \models A$.

DEFINITION II.2.11 (Substitution). Let s, t be terms, and A a formula. Then $t(s/x)$ is the result of replacing all occurrences of x in t by s, and $A(s/x)$ is the result of replacing all *free* occurrences of x in A by s.

LEMMA II.2.12. *For each structure $\mathcal{M}$ and each object assignment σ,*

$$(s(t/x))^{\mathcal{M}}[\sigma] = s^{\mathcal{M}}[\sigma(m/x)]$$

where $m = t^{\mathcal{M}}[\sigma]$.

PROOF. Structural induction on s. □

DEFINITION II.2.13. A term t is *freely substitutable for x in A* iff no free occurrence of x in A is in a subformula of A of the form $\forall y B$ or $\exists y B$, where y occurs in t.

THEOREM II.2.14 (Substitution). *If t is freely substitutable for x in A then for all structures $\mathcal{M}$ and all object assignments σ, $\mathcal{M} \models A(t/x)[\sigma]$ iff $\mathcal{M} \models A[\sigma(m/x)]$, where $m = t^{\mathcal{M}}[\sigma]$.*

PROOF. Structural induction on A. □

REMARK (Change of Bound Variable). If t is not freely substitutable for x in A, it is because some variable y in t gets "caught" by a quantifier, say $\exists y B$. Then replace $\exists y B$ in A by $\exists z B$, where z is a new variable. Then the meaning of A does not change (by the Formula Replacement Theorem below), but by repeatedly changing bound variables in this way t becomes freely substitutable for x in A.

THEOREM II.2.15 (Formula Replacement). *If B and B' are equivalent and A' results from A by replacing some occurrence of B in A by B', then A and A' are equivalent.*

PROOF. Structural induction on A relative to B. □

II.2.3. The First-Order Proof System *LK*. We now extend the propositional proof system *PK* to the first-order sequent proof system *LK*. For this it is convenient to introduce two kinds of variables: *free variables* denoted by $a, b, c, \ldots$ and *bound variables* denoted by $x, y, z, \ldots$. A first-order sequent has the form

$$A_1, \ldots, A_k \longrightarrow B_1, \ldots, B_\ell$$

where now the A_i and B_j are first-order formulas satisfying the restriction that they have no free occurrences of the "bound" variables $x, y, z, \ldots$ and no bound occurrences of the "free" variables $a, b, c, \ldots$.

The sequent system *LK* is an extension of the propositional system *PK*, where now all formulas are first-order formulas satisfying the restriction explained above.

In addition to the rules given for *PK*, the system *LK* has four rules for introducing the quantifiers.

IMPORTANT REMARK. In the rules below, t is any term not involving any bound variables $x, y, z, \ldots$ and $A(t)$ is the result of substituting t for all free occurrences of x in $A(x)$. Similarly $A(b)$ is the result of substituting b for all free occurrences of x in $A(x)$. Note that t and b can always be freely substituted for x in $A(x)$ when $\forall x A(x)$ or $\exists x A(x)$ satisfy the free/bound variable restrictions described above.

$\forall$ introduction rules

$$\text{left: } \frac{A(t), \Gamma \longrightarrow \Delta}{\forall x A(x), \Gamma \longrightarrow \Delta} \qquad \text{right: } \frac{\Gamma \longrightarrow \Delta, A(b)}{\Gamma \longrightarrow \Delta, \forall x A(x)}$$

$\exists$ introduction rules

$$\text{left: } \frac{A(b), \Gamma \longrightarrow \Delta}{\exists x A(x), \Gamma \longrightarrow \Delta} \qquad \text{right: } \frac{\Gamma \longrightarrow \Delta, A(t)}{\Gamma \longrightarrow \Delta, \exists x A(x)}$$

Restriction. The free variable b is called an *eigenvariable* and must not occur in the conclusion in $\forall$-right or $\exists$-left. Also, as remarked above, the term t must not involve any bound variables $x, y, z, \ldots$.

The new formulas in the bottom sequents $(\exists x A(x)$ or $\forall x A(x))$ are called *principal formulas*, and the corresponding formulas in the top sequents $(A(b)$ or $A(t))$ are called *auxiliary formulas*.

DEFINITION II.2.16 (Semantics of first-order sequents). The semantics of first-order sequents is a natural generalization of the semantics of propositional sequents. Again the sequent $A_1, \ldots, A_k \longrightarrow B_1, \ldots, B_\ell$ has the same meaning as its associated formula

$$\neg A_1 \vee \neg A_2 \vee \cdots \vee \neg A_k \vee B_1 \vee B_2 \vee \cdots \vee B_\ell.$$

In particular, we say that the sequent is *valid* iff its associated formula is valid.

THEOREM II.2.17 (Soundness for *LK*). *Every sequent provable in *LK* is valid.*

PROOF. This is proved by induction on the number of sequents in the *LK* proof, as in the case of *PK*. However, unlike the case of *PK*, not all of the four new quantifier rules satisfy the condition that the bottom sequent is a logical consequence of the top sequent. In particular this may be false for $\forall$-right and for $\exists$-left. However it is easy to check that each rule satisfies the weaker condition that if the top sequent is valid, then the bottom sequent is valid, and this suffices for the proof. □

Exercise II.2.18. Give examples to show that the restriction given on the quantifier rules, that b must not occur in the conclusion in $\forall$-right and $\exists$-left, is necessary to ensure that these rules preserve validity.

Example of an LK proof. An **LK** proof of a valid first-order sequent can be obtained using the same method as in the propositional case: Write the goal sequent at the bottom, and move up by using the introduction rules in reverse. A good heuristic is: if there is a choice about which quantifier to remove next, choose $\forall$-right and $\exists$-left first (working backward), since these rules carry a restriction.

Here is an **LK** proof of the sequent $\forall x Px \lor \forall x Qx \longrightarrow \forall x (Px \lor Qx)$.

$$
\dfrac{\dfrac{\dfrac{Pb \longrightarrow Pb}{Pb \longrightarrow Pb, Qb}\text{ (weakening)}}{\forall x Px \longrightarrow Pb, Qb}\text{ ($\forall$ left)} \qquad \dfrac{\dfrac{\dfrac{Qb \longrightarrow Qb}{Qb \longrightarrow Pb, Qb}\text{ (weakening)}}{\forall x Qx \longrightarrow Pb, Qb}\text{ ($\forall$ left)}}{}}{\dfrac{\dfrac{\forall x Px \lor \forall x Qx \longrightarrow Pb, Qb}{\forall x Px \lor \forall x Qx \longrightarrow Pb \lor Qb}\text{ ($\lor$ right)}}{\forall x Px \lor \forall x Qx \longrightarrow \forall x (Px \lor Qx)}\text{ ($\forall$ right)}}\text{ ($\lor$ left)}
$$

Exercise II.2.19. Give **LK** proofs for the following valid sequents:

(a) $\forall x Px \land \forall x Qx \longrightarrow \forall x (Px \land Qx)$.

(b) $\forall x (Px \land Qx) \longrightarrow \forall x Px \land \forall x Qx$.

(c) $\exists x (Px \lor Qx) \longrightarrow \exists x Px \lor \exists x Qx$.

(d) $\exists x Px \lor \exists x Qx \longrightarrow \exists x (Px \lor Qx)$.

(e) $\exists x (Px \land Qx) \longrightarrow \exists x Px \land \exists x Qx$.

(f) $\exists y \forall x Pxy \longrightarrow \forall x \exists y Pxy$.

(g) $\forall x Px \longrightarrow \exists x Px$.

Check that the rule restrictions seem to prevent generating **LK** proofs for the following invalid sequents:

(h) $\exists x Px \land \exists x Qx \longrightarrow \exists x (Px \land Qx)$.

(i) $\forall x \exists y Pxy \longrightarrow \exists y \forall x Pxy$.

II.2.4. Free Variable Normal Form. In future chapters it will be useful to assume that **LK** proofs satisfy certain restrictions on free variables.

Definition II.2.20 (Free Variable Normal Form). Let π be an **LK** proof with endsequent S. A free variable in S is called a *parameter variable* of π. We say π is in *free variable normal form* if (1) no free variable is completely eliminated from any sequent in π by any rule except possibly $\forall$-right and $\exists$-left, and in these cases the eigenvariable which is eliminated is not a parameter variable, and (2) every nonparameter free variable appearing in π is used exactly once as an eigenvariable.

Thus if a proof is in free variable normal form, then any occurrence of a parameter variable persists until the endsequent, and any occurrence of a nonparameter free variable persists until it is eliminated as an eigenvariable in $\forall$-right or $\exists$-left.

We now describe a simple procedure for transforming an *LK* proof π to a similar proof of the same endsequent in free variable normal form, assuming that the underlying vocabulary $\mathcal{L}$ has at least one constant symbol e. Note that the only rules other than $\forall$-right and $\exists$-left which can eliminate a free variable from a sequent are cut, $\exists$-right, and $\forall$-left. It is important that π have a tree structure in order for the procedure to work.

Transform π by repeatedly performing the following operation until the resulting proof is in free variable normal form. Select some upper-most rule in π which eliminates a free variable from a sequent which violates free variable normal form. If the rule is $\forall$-right or $\exists$-left, and the eigenvariable b which is eliminated occurs somewhere in the proof other than above this rule, then replace b by a new variable b' (which does not occur elsewhere in the proof) in every sequent above this rule. If the rule is cut, $\exists$-right, or $\forall$-left, then replace every variable eliminated by the rule by the same constant symbol e in every sequent above the rule (so now the rule does not eliminate any free variable).

II.2.5. Completeness of *LK* without Equality.

NOTATION. Let Φ be a set of formulas. Then $\longrightarrow \Phi$ is the set of all sequents of the form $\longrightarrow A$, where A is in Φ.

DEFINITION II.2.21. Assume that the underlying vocabulary does not contain $=$. If Φ is a set of formulas, then an *LK*-Φ proof is an *LK* proof in which sequents at the leaves may be either logical axioms or nonlogical axioms of the form $\longrightarrow A$, where A is in Φ.

Notice that a structure $\mathcal{M}$ satisfies $\longrightarrow \Phi$ iff $\mathcal{M}$ satisfies Φ. Also a sequent $\Gamma \longrightarrow \Delta$ is a logical consequence of $\longrightarrow \Phi$ iff $\Gamma \longrightarrow \Delta$ is a logical consequence of Φ.

We would like to be able to say that a sequent $\Gamma \longrightarrow \Delta$ is a logical consequence of a set Φ of formulas iff there is an *LK*-Φ proof of $\Gamma \longrightarrow \Delta$. Unfortunately the soundness direction of the assertion is false. For example, using the $\forall$-right rule we can derive $\longrightarrow \forall x P x$ from $\longrightarrow P b$, but $\longrightarrow \forall x P x$ is not a logical consequence of $P b$.

We could correct the soundness statement by asserting it true for sentences, but we want to generalize this a little by introducing the notion of the universal closure of a formula or sequent.

DEFINITION II.2.22. Suppose that A is a formula whose free variables comprise the list $a_1, \ldots, a_n$. Then the *universal closure* of A, written $\forall A$, is the sentence $\forall x_1 \ldots \forall x_n A(x_1/a_1, \ldots, x_n/a_n)$, where $x_1, \ldots, x_n$ is a list of new (bound) variables. If Φ is a set of formulas, then $\forall \Phi$ is the set of all sentences $\forall A$, for A in Φ.

Notice that if A is a sentence (i.e., it has no free variables), then $\forall A$ is the same as A.

Initially we study the case in which the underlying vocabulary does not contain $=$. To handle the case in which $=$ occurs we must introduce equality axioms. This will be done later.

THEOREM II.2.23 (Derivational Soundness and Completeness of $\boldsymbol{LK}$). *Assume that the underlying vocabulary does not contain $=$. Let Φ be a set of formulas and let $\Gamma \longrightarrow \Delta$ be a sequent. Then there is an $\boldsymbol{LK}$-Φ proof of $\Gamma \longrightarrow \Delta$ iff $\Gamma \longrightarrow \Delta$ is a logical consequence of $\forall\Phi$. The soundness (only if) direction holds also when the underlying vocabulary contains $=$.*

PROOF OF SOUNDNESS. Let π be a $\boldsymbol{LK}$-Φ proof of $\Gamma \longrightarrow \Delta$. We must show that $\Gamma \longrightarrow \Delta$ is a logical consequence of $\forall\Phi$. We want to prove this by induction on the number of sequents in the proof π, but in fact we need a stronger induction hypothesis, to the effect that the "closure" of $\Gamma \longrightarrow \Delta$ is a logical consequence of $\forall\Phi$. So we first have to define the closure of a sequent.

Thus we define the closure $\forall S$ of a sequent S to be the closure of its associated formula A_S (Definition II.2.16). Note that if $S =_{syn} \Gamma \longrightarrow \Delta$, then $\forall S$ is *not* equivalent to $\forall\Gamma \longrightarrow \forall\Delta$ in general.

We now prove by induction on the number of sequents in π, that if π is an $\boldsymbol{LK}$-Φ proof of a sequent S, then $\forall S$ is a logical consequence of $\forall\Phi$. Since $\forall S \models S$, it follows that S itself is a logical consequence of $\forall\Phi$, and so Soundness follows.

For the base case, the sequent S is either a logical axiom, which is valid and hence a consequence of $\forall\Phi$, or it is a nonlogical axiom $\longrightarrow A$, where A is a formula in Φ. In the latter case, $\forall S$ is equivalent to $\forall A$, which of course is a logical consequence of $\forall\Phi$.

For the induction step, it is sufficient to check that for each rule of $\boldsymbol{LK}$, the closure of the bottom sequent is a logical consequence of the closure(s) of the sequent(s) on top. With two exceptions, this statement is true when the word "closure" is omitted, and adding back the word "closure" does not change the argument much. The two exceptions are the rules $\forall$-right and $\exists$-left. For these, the bottom is *not* a logical consequence of the top in general, but an easy argument shows that the closures of the top and bottom are equivalent.   □

The proof of completeness is more difficult and more interesting than the proof of soundness. The following lemma lies at the heart of this proof.

LEMMA II.2.24 (Completeness). *Assume that the underlying vocabulary does not contain $=$. If $\Gamma \longrightarrow \Delta$ is a sequent and Φ is a (possibly infinite) set of formulas such that $\Gamma \longrightarrow \Delta$ is a logical consequence of Φ, then there is a finite subset $\{C_1, \ldots, C_n\}$ of Φ such that the sequent*

$$C_1, \ldots, C_n, \Gamma \longrightarrow \Delta$$

has an $\boldsymbol{LK}$ proof π which does not use the cut rule.

Note that a form of the Compactness Theorem for predicate calculus sentences without equality follows from the above lemma. See Theorem II.4.2 for a more general form of compactness.

PROOF OF DERIVATIONAL COMPLETENESS. Let Φ be a set of formulas such that $\Gamma \longrightarrow \Delta$ is a logical consequence of $\forall\Phi$. By the completeness lemma, there is a finite subset $\{C_1, \ldots, C_n\}$ of Φ such that

$$\forall C_1, \ldots, \forall C_n, \Gamma \longrightarrow \Delta$$

has a cut-free *LK* proof π. Note that for each $i, 1 \le i \le n$, the sequent $\longrightarrow \forall C_i$ has an *LK*-Φ proof from the nonlogical axiom $\longrightarrow C_i$ by repeated use of the rule $\forall$-right. Now the proof π can be extended, using these proofs of the sequents

$$\longrightarrow \forall C_1, \ \ldots, \ \longrightarrow \forall C_n$$

and repeated use of the cut rule, to form an *LK*-Φ proof $\Gamma \longrightarrow \Delta$. $\square$

PROOF OF THE COMPLETENESS LEMMA. We loosely follow the proof of the Cut-free Completeness Theorem, pp. 33–36 of Buss [27]. (Warning: our definition of logical consequence differs from Buss's when the formulas in the hypotheses have free variables.) We will only prove it for the case in which the underlying first-order vocabulary $\mathcal{L}$ has a countable set (including the case of a finite set) of function and predicate symbols; i.e., the function symbols form a list $f_1, f_2, \ldots$ and the predicate symbols form a list $P_1, P_2, \ldots$. This may not seem like much of a restriction, but for example in developing the model theory of the real numbers, it is sometimes useful to introduce a distinct constant symbol e_c for every real number c; and there are uncountably many real numbers. The completeness theorem and lemma hold for the uncountable case, but we shall not prove them for this case.

For the countable case, we may assign a distinct binary string to each function symbol, predicate symbol, variable, etc., and hence assign a unique binary string to each formula and term. This allows us to enumerate all the $\mathcal{L}$-formulas in a list $A_1, A_2, \ldots$ and enumerate all the $\mathcal{L}$-terms (which contain only free variables $a, b, c, \ldots$) in a list $t_1, t_2, \ldots$. The free variables available to build the formulas and terms in these lists must include all the free variables which appear in Φ, together with a countably infinite set $\{c_0, c_1, \ldots\}$ of new free variables which do not occur in any of the formulas in Φ. (These new free variables are needed for the cases $\exists$-left and $\forall$-right in the argument below.) Further we may assume that every formula occurs infinitely often in the list of formulas, and every term occurs infinitely often in the list of terms. Finally we may enumerate all pairs $\langle A_i, t_j \rangle$, using any method of enumerating all pairs of natural numbers.

We are trying to find an *LK* proof of some sequent of the form

$$C_1, \ldots, C_n, \Gamma \longrightarrow \Delta$$

for some n. Starting with $\Gamma \longrightarrow \Delta$ at the bottom, we work upward by applying the rules in reverse, much as in the proof of the propositional completeness theorem for **PK**. However now we will add formulas C_i to the antecedent from time to time. Also unlike the **PK** case we have no inversion principle to work with (specifically for the rules $\forall$-left and $\exists$-right). Thus it may happen that our proof-building procedure may not terminate. In this case we will show how to define a structure which shows that $\Gamma \longrightarrow \Delta$ is not a logical consequence of Φ.

We construct our cut-free proof tree π in stages. Initially π consists of just the sequent $\Gamma \longrightarrow \Delta$. At each stage we modify π by possibly adding a formula from Φ to the antecedent of every sequent in π, and by adding subtrees to some of the leaves.

NOTATION. A sequent in π is said to be *active* provided it is at a leaf and cannot be immediately derived from a logical axiom (i.e., no formula occurs in both its antecedent and succedent, the logical constant $\top$ does not occur in its succedent, and $\bot$ does not occur in its antecedent).

Each stage uses one pair in our enumeration of all pairs $\langle A_i, t_j \rangle$. Here is the procedure for the next stage, in general.

Let $\langle A_i, t_j \rangle$ be the next pair in the enumeration. We call A_i the *active* formula for this stage.

Step 1. If A_i is in Φ, then replace every sequent $\Gamma' \longrightarrow \Delta'$ in π with the sequent $\Gamma', A_i \longrightarrow \Delta'$.

Step 2. If A_i is atomic, do nothing and proceed to the next stage. Otherwise, modify π at the active sequents which contain A_i by applying the appropriate introduction rule in reverse, much as in the proof of propositional completeness (Theorem II.1.8). (It suffices to pick any one occurrence of A_i in each active sequent.) For example, if A_i is of the form $B \vee C$, then every active sequent in π of the form $\Gamma', B \vee C, \Gamma'' \longrightarrow \Delta'$ is replaced by the derivation

$$\frac{\Gamma', B, \Gamma'' \longrightarrow \Delta' \qquad \Gamma', C, \Gamma'' \longrightarrow \Delta'}{\Gamma', B \vee C, \Gamma'' \longrightarrow \Delta'}$$

Here the double line represents a derivation involving the rule $\vee$-left, together with exchanges to move the formulas B, C to the left end of the antecedent and move $B \vee C$ back to the right. The treatment is similar when $B \vee C$ occurs in the succedent, only the rule $\vee$-right is used.

If A_i is of the form $\exists x B(x)$, then every active sequent of π of the form $\Gamma', \exists x B(x), \Gamma'' \longrightarrow \Delta'$ is replaced by the derivation

$$\frac{\Gamma', B(c), \Gamma'' \longrightarrow \Delta'}{\Gamma', \exists x B(x), \Gamma'' \longrightarrow \Delta'}$$

where c is a new free variable, not used in π yet. (Also c may not occur in any formula in Φ, because otherwise at a later stage, *Step* 1 of the procedure might cause the variable restriction in the $\exists$-left rule to be violated.) In addition, any active sequent of the form $\Gamma' \longrightarrow \Delta', \exists x B(x), \Delta''$ is replaced by the derivation

$$\frac{\Gamma' \longrightarrow \Delta', \exists x B(x), B(t_j), \Delta''}{\Gamma' \longrightarrow \Delta', \exists x B(x), \Delta''}$$

Here the term t_j is the second component in the current pair $\langle A_i, t_j \rangle$. The derivation uses the rule $\exists$-right to introduce a new copy of $\exists x B(x)$, and then the rule contraction-right to combine the two copies of $\exists x B(x)$. This and the dual $\forall$-left case are the only two cases that use the term t_j, and the only cases that use the contraction rule.

The case where A_i begins with a universal quantifier is dual to the above existential case.

Step 3. If there are no active sequents remaining in π, then exit from the algorithm. Otherwise continue to the next stage.

Exercise II.2.25. Carry out the case above in which A_i begins with a universal quantifier.

If the algorithm constructing π ever halts, then π gives a cut-free proof of $\Gamma, C_1, \ldots, C_n \longrightarrow \Delta$ for some formulas $C_1, \ldots, C_n$ in Φ. This is because the nonactive leaf sequents all can be derived from the logical axioms using weakenings and exchanges. Thus π can be extended, using exchanges, to a cut-free proof of $C_1, \ldots, C_n, \Gamma \longrightarrow \Delta$, as desired.

It remains to show that if the above algorithm constructing π never halts, then the sequent $\Gamma \longrightarrow \Delta$ is not a logical consequence of Φ. So suppose the algorithm never halts, and let π be the result of running the algorithm forever. In general, π will be an infinite tree, although in special cases π is a finite tree. In general the objects at the nodes of the tree will not be finite sequents, but because of *Step* 1 of the algorithm above, they will be of the form $\Gamma', C_1, C_2, \ldots \longrightarrow \Delta'$, where $C_1, C_2, \ldots$ is an infinite sequence of formulas containing all formulas in Φ, each repeated infinitely often (unless Φ is empty). We shall refer to these infinite pseudo-sequents as just "sequents".

If π has only finitely many nodes, then at least one leaf node must be active (and contain only atomic formulas), since otherwise the algorithm would terminate. In this case, let β be a path in π from the root extending up to this active node. If on the other hand π has infinitely many nodes, then by Lemma II.1.18 (König), there must be an infinite branch β in π starting at the root and extending up through the tree. Thus in either case, β is a branch in π starting at the root, extending up through the tree, and such that all sequents on β were once active, and hence have no formula occurring on both the left and right, no $\top$ on the right and no $\bot$ on the left.

We use this branch β to construct a structure $\mathcal{M}$ and an object assignment σ which satisfy every formula in Φ, but falsify the sequent $\Gamma \longrightarrow \Delta$ (so $\Gamma \longrightarrow \Delta$ is not a logical consequence of Φ).

DEFINITION II.2.26 (Construction of the "Term Model" $\mathcal{M}$). The universe M of $\mathcal{M}$ is the set of all $\mathcal{L}$-terms t (which contain only "free" variables $a, b, c, \dots$). The object assignment σ just maps every variable a to itself.

The interpretation $f^{\mathcal{M}}$ of each k-ary function symbol f is defined so that $f^{\mathcal{M}}(r_1, \dots, r_k)$ is the term $f r_1 \dots r_k$, where $r_1, \dots, r_k$ are any terms (i.e., any members of the universe). The interpretation $P^{\mathcal{M}}$ of each k-ary predicate symbol P is defined by letting $P^{\mathcal{M}}(r_1, \dots, r_k)$ hold iff the atomic formula $P r_1 \dots r_k$ occurs in the antecedent (left side) of some sequent in the branch β.

EXERCISE II.2.27. Prove by structural induction that for every term t, $t^{\mathcal{M}}[\sigma] = t$.

CLAIM. For every formula A, if A occurs in some antecedent in the branch β, then $\mathcal{M}$ and σ satisfy A, and if A occurs in some succedent in β, then $\mathcal{M}$ and σ falsify A.

Since the root of π is the sequent $\Gamma, C_1, C_2, \dots \longrightarrow \Delta$, where $C_1, C_2, \dots$ contains all formulas in Φ, it follows that $\mathcal{M}$ and σ satisfy Φ and falsify $\Gamma \longrightarrow \Delta$.

We prove the Claim by structural induction on formulas A. For the base case, if A is an atomic formula, then by the definition of $P^{\mathcal{M}}$ above, A is satisfied iff A occurs in some antecedent of β or $A = \top$. But no atomic formula can occur both in an antecedent of some node in β and in a succedent (of possibly some other node) in β, since then these formulas would persist upward in β so that some particular sequent in β would have A occurring both on the left and on the right. Thus if A occurs in some succedent of β, it is not satisfied by $\mathcal{M}$ and σ (recall that $\top$ does not occur in any succedent of β).

For the induction step, there is a different case for each of the ways of constructing a formula from simpler formulas (see Definition II.2.2). In general, if A occurs in some sequent in β, then A persists upward in every higher sequent of β until it becomes the active formula ($A =_{\text{syn}} A_i$). Each case is handled by the corresponding introduction rule used in the algorithm. For example, if A is of the form $B \vee C$ and A occurs on the left of a sequent in β, then the rule $\vee$-left is applied in reverse, so that when β is extended upward either it will have some antecedent containing B or one containing C. In the case of B, we know that $\mathcal{M}$ and σ satisfy B by the induction hypothesis, and hence they satisfy $B \vee C$. (Similarly for C.)

Now consider the interesting case in which A is $\exists x B(x)$ and A occurs in some succedent of β. (See Step 2 above to find out what happens when A becomes active in this case.) The path β will hit a succedent with $B(t_j)$ in the succedent, and by the induction hypothesis, $\mathcal{M}$ and σ falsify $B(t_j)$. But this succedent still has a copy of $\exists x B(x)$, and in fact this copy will be in *every* succedent of β above this point. Hence *every* $\mathcal{L}$-term t will eventually be of the form t_j and so the formula $B(t)$ will occur as a succedent on β. (This is why we assumed that every term appears infinitely often in the sequence $t_1, t_2, \ldots$.) Therefore $\mathcal{M}$ and σ falsify $B(t)$ for every term t (i.e., for every element in the universe of $\mathcal{M}$). Therefore they falsify $\exists x B(x)$, as required.

This and the dual case in which A is $\forall x B(x)$ and occurs in some antecedent of β are the only subtle cases. All other cases are straightforward.   □

We now wish to strengthen the derivational completeness of *LK* and show that cuts can be restricted so that cut formulas are in Φ. The definition of *anchored **PK*** proof (Definition II.1.12) can be generalized to *anchored **LK*** proof. We will continue to restrict our attention to the case in which all nonlogical axioms have the simple form $\longrightarrow A$, although an analog of the following theorem does hold for an arbitrary set of nonlogical axioms, provided they are closed under substitution of terms for variables.

THEOREM II.2.28 (Anchored *LK* Completeness). *Assume that the underlying vocabulary does not contain* $=$. *Suppose that* Φ *is a set of formulas closed under substitution of terms for variables. (I.e., if* $A(b)$ *is in* Φ, *and* t *is any term not containing "bound" variables* $x, y, z, \ldots$, *then* $A(t)$ *is also in* Φ.) *Suppose that* $\Gamma \longrightarrow \Delta$ *is a sequent that is a logical consequence of* $\forall \Phi$. *Then there is an* ***LK**-Φ proof of* $\Gamma \longrightarrow \Delta$ *in which the cut rule is restricted so that the only cut formulas are formulas in* Φ.

Note that if all formulas in Φ are sentences, then the above theorem follows easily from the Completeness Lemma, since in this case $\forall \Phi$ is the same as Φ. However if formulas in Φ have free variables, then apparently the cut rule must be applied to the closures $\forall C$ of formulas C in Φ (as opposed to C itself) in order to get an *LK*-Φ proof of $\Gamma \longrightarrow \Delta$. It will be important later, in our proof of witnessing theorems, that cuts can be restricted to the formulas C.

EXERCISE II.2.29. Show how to modify the proof of the Completeness Lemma to obtain a proof of the Anchored *LK* Completeness Theorem. Explain the following modifications to that proof.

(a) The definition of *active sequent* on page 27 must be modified, since now we are allowing nonlogical axioms in π. Give the precise new definition.

(b) *Step* 1 of the procedure on page 27 must be modified, because now we are looking for a derivation of $\Gamma \longrightarrow \Delta$ from nonlogical axioms, rather than a proof of $C_1, \ldots, C_n, \Gamma \longrightarrow \Delta$. Describe the modification. (We still need to bring formulas A_i of Φ somehow into the proof, and your modification will involve adding a short derivation to π.)

(c) The restriction given in Step 2 for the case in which $\exists x B(x)$ is in the antecedent, that the variable c must not occur in any formula in Φ, must be dropped. Explain why.

(d) Explain why the term model $\mathcal{M}$ and object assignment σ, described on page 29 (Definition II.2.26), satisfy $\forall \Phi$. This should follow from the Claim on page 29, and your modification of *Step* 1, which should ensure that each formula in Φ occurs in the antecedent of some sequent in every branch in π. Conclude that $\Gamma \longrightarrow \Delta$ is not a logical consequence of $\forall \Phi$ (when the procedure does not terminate).

II.3. Equality Axioms

DEFINITION II.3.1. A *weak* $\mathcal{L}$-structure $\mathcal{M}$ is an $\mathcal{L}$-structure in which we drop the requirement that $=^{\mathcal{M}}$ is the equality relation (i.e., $=^{\mathcal{M}}$ can be any binary relation on M.)

Are there sentences $\mathcal{E}$ (axioms for equality) such that a weak structure $\mathcal{M}$ satisfies $\mathcal{E}$ iff $\mathcal{M}$ is a (proper) structure? It is easy to see that no such set $\mathcal{E}$ of axioms exists, because we can always inflate a point in a weak model to a set of equivalent points.

Nevertheless every vocabulary $\mathcal{L}$ has a standard set $\mathcal{E}_{\mathcal{L}}$ of equality axioms which satisfies the Equality Theorem below.

DEFINITION II.3.2 (Equality Axioms of $\mathcal{L}$ ($\mathcal{E}_{\mathcal{L}}$)).

EA1. $\forall x(x = x)$ (reflexivity);
EA2. $\forall x \forall y(x = y \supset y = x)$ (symmetry);
EA3. $\forall x \forall y \forall z((x = y \wedge y = z) \supset x = z)$ (transitivity);
EA4. $\forall x_1 \ldots \forall x_n \forall y_1 \ldots \forall y_n(x_1 = y_1 \wedge \cdots \wedge x_n = y_n) \supset f x_1 \ldots x_n = f y_1 \ldots y_n$ for each $n \geq 1$ and each n-ary function symbol f in $\mathcal{L}$.
EA5. $\forall x_1 \ldots \forall x_n \forall y_1 \ldots \forall y_n(x_1 = y_1 \wedge \cdots \wedge x_n = y_n) \supset (P x_1 \ldots x_n \supset P y_1 \ldots y_n)$ for each $n \geq 1$ and each n-ary predicate symbol P in $\mathcal{L}$ other than $=$.

Axioms **EA1**, **EA2**, **EA3** assert that $=$ is an equivalence relation. Axiom **EA4** asserts that functions respect the equivalence classes, and Axiom **EA5** asserts that predicates respect equivalence classes. Together the axioms assert that $=$ is a congruence relation with respect to the function and predicate symbols.

Note that the equality axioms are all valid, because of our requirement that $=$ be interpreted as equality in any (proper) structure.

Theorem II.3.3 (Equality). *Let Φ be any set of $\mathcal{L}$-formulas. Then Φ is satisfiable iff $\Phi \cup \mathcal{E}_{\mathcal{L}}$ is satisfied by some weak $\mathcal{L}$-structure.*

Corollary II.3.4. $\Phi \models A$ *iff for every weak $\mathcal{L}$-structure $\mathcal{M}$ and every object assignment σ, if $\mathcal{M}$ satisfies $\Phi \cup \mathcal{E}_{\mathcal{L}}$ under σ then $\mathcal{M}$ satisfies A under σ.*

Corollary II.3.5. $\forall \Phi \models A$ *iff A has an LK-Ψ proof, where $\Psi = \Phi \cup \mathcal{E}_{\mathcal{L}}$.*

Corollary II.3.4 follows immediately from the Equality Theorem and the fact that $\Phi \models A$ iff $\Phi \cup \{\neg A\}$ is unsatisfiable. Corollary II.3.5 follows from Corollary II.3.4 and the derivational soundness and completeness of *LK* (page 25), where in applying that theorem we treat $=$ as just another binary relation (so we can assume $\mathcal{L}$ does not have the official equality symbol).

Proof of Equality. The ONLY IF ($\Longrightarrow$) direction is obvious, because every structure $\mathcal{M}$ must interpret $=$ as true equality, and hence $\mathcal{M}$ satisfies the equality axioms $\mathcal{E}_{\mathcal{L}}$.

For the IF ($\Longleftarrow$) direction, suppose that $\mathcal{M}$ is a weak $\mathcal{L}$-structure with universe M, such that $\mathcal{M}$ satisfies $\Phi \cup \mathcal{E}_{\mathcal{L}}$. Our job is to construct a proper structure $\hat{\mathcal{M}}$ such that $\hat{\mathcal{M}}$ satisfies Φ. The idea is to let the elements of $\hat{\mathcal{M}}$ be the equivalence classes under the equivalence relation $=^{\mathcal{M}}$. Axioms **EA**4 and **EA**5 insure that the interpretation of each function and predicate symbol under $\mathcal{M}$ induces a corresponding function or predicate in $\hat{\mathcal{M}}$. Further each object assignment σ for $\mathcal{M}$ induces an object assignment $\hat{\sigma}$ on $\hat{\mathcal{M}}$. Then for every formula A and object assignment σ, we show by structural induction on A that $\mathcal{M} \models A[\sigma]$ iff $\hat{\mathcal{M}} \models A[\hat{\sigma}]$.   □

II.3.1. Equality Axioms for *LK*. For the purpose of using an *LK* proof to establish $\Phi \models A$, we can replace the standard equality axioms **EA**1, ..., **EA**5 by the following quantifier-free sequent schemes, where we must include an instance of the sequent for all terms t, u, v, t_i, u_i (not involving "bound" variables $x, y, z, \ldots$).

Definition II.3.6 (Equality Axioms for *LK*).

E1. $\longrightarrow t = t$;
E2. $t = u \longrightarrow u = t$;
E3. $t = u, u = v \longrightarrow t = v$;
E4. $t_1 = u_1, \ldots, t_n = u_n \longrightarrow f t_1 \ldots t_n = f u_1 \ldots u_n$, for each f in $\mathcal{L}$;
E5. $t_1 = u_1, \ldots, t_n = u_n, P t_1 \ldots t_n \longrightarrow P u_1 \ldots u_n$, for each P in $\mathcal{L}$ (here P is not $=$).

Note that the universal closures of **E1**, ..., **E5** are semantically equivalent to **EA**1, ..., **EA**5, and in fact using the *LK* rule $\forall$-right repeatedly, $\longrightarrow$ **EA**i is easily derived in *LK* from **E**i (with terms t, u, etc., taken to be distinct variables), $i = 1, \ldots, 5$. Thus Corollary II.3.5 above still holds when $\Psi = \Phi \cup \{\textbf{E1}, \ldots, \textbf{E5}\}$.

DEFINITION II.3.7 (Revised Definition of **LK** with $=$). If Φ is a set of $\mathcal{L}$-formulas, where $\mathcal{L}$ includes $=$, then by an **LK**-Φ proof we now mean an **LK**-Ψ proof in the sense of the earlier definition, page 24, where Ψ is Φ together with all instances of the equality axioms E1, ..., E5. If Φ is empty, we simply refer to an **LK**-proof (but allow E1, ..., E5 as axioms).

II.3.2. Revised Soundness and Completeness of **LK**.

THEOREM II.3.8 (Revised Soundness and Completeness of **LK**). *For any set Φ of formulas and sequent S,*

$$\forall \Phi \models S \textit{ iff } S \textit{ has an } \textbf{LK-}\Phi \textit{ proof.}$$

NOTATION. $\Phi \vdash A$ means that there is an **LK**-Φ proof of $\longrightarrow A$.

Recall that if Φ is a set of sentences, then $\forall \Phi$ is the same as Φ. Therefore

$$\Phi \models A \textit{ iff } \Phi \vdash A, \quad \textit{if } \Phi \textit{ is a set of sentences.}$$

Restricted use of cut. Note that E1, ..., E5 have no universal quantifiers, but instead have instances for all terms $t, u, \ldots$. Recall that in an anchored **LK** proof, cuts are restricted so that cut formulas must occur in the nonlogical axioms. In the presence of equality, the nonlogical axioms must include E1, ..., E5, but the only formulas occurring here are equations of the form $t = u$. Since the Anchored **LK** Completeness Theorem (page 30) still holds when Φ is a set of sequents rather than a set of formulas, and since E1, ..., E5 are closed under substitution of terms for variables, we can extend this theorem so that it works in the presence of equality.

DEFINITION II.3.9 (Anchored **LK** Proof). An **LK**-Φ proof π is *anchored*[1] provided every cut formula in π is a formula in some nonlogical axiom of π (including possibly E1, ..., E5).

THEOREM II.3.10 (Anchored **LK** Completeness with Equality). *Suppose that Φ is a set of formulas closed under substitution of terms for variables and that the sequent S is a logical consequence of $\forall \Phi$. Then there is an anchored **LK**-Φ proof of S.*

The proof is immediate from the Anchored **LK** Completeness Theorem (page 30) and the above discussion about axioms E1, ..., E5.

We are interested in anchored proofs because of their subformula property. The following result generalizes Proposition II.1.15.

THEOREM II.3.11 (Subformula Property of Anchored **LK** Proofs). *If π is an anchored **LK**-Φ proof of a sequent S, then every formula in every sequent of π is a term substitution instance of a subformula of a formula either in S or in a nonlogical axiom of π (including E1, ..., E5).*

---

[1] The definition of *anchored* in [27] is slightly stronger and more complicated.

PROOF SKETCH. The proof is by induction on the number of sequents in π. The induction step is proved by inspecting each *LK* rule. The case of the cut rule uses the fact that every cut formula in an anchored proof is a formula in some nonlogical axiom. The reason that we must consider term substitutions is because of the four quantifier rules. For example, in $\exists$-right, the formula $A(t)$ occurs on top, and this is a substitution instance of a subformula of $\exists x A(x)$, which occurs on the bottom.   □

II.4. Major Corollaries of Completeness

THEOREM II.4.1 (Löwenheim/Skolem). *If a set Φ of formulas from a countable vocabulary is satisfiable, then Φ is satisfiable in a countable (possibly finite) universe.*

PROOF. Suppose that Φ is a satisfiable set of sentences. We apply the proof of the Completeness Lemma (Lemma II.2.24), treating $=$ as any binary relation, replacing Φ by $\Phi' = \Phi \cup \mathcal{E}_{\mathcal{L}}$, and taking $\Gamma \longrightarrow \Delta$ to be the empty sequent (always false). In this case $\Gamma \longrightarrow \Delta$ is not a logical consequence of Φ', so the proof constructs a term model $\mathcal{M}$ satisfying Φ' (see page 29). This structure has a countable universe M consisting of all the $\mathcal{L}$-terms. By the proof of the Equality Theorem, we can pass to equivalence classes and construct a countable structure $\hat{\mathcal{M}}$ which satisfies Φ (and interprets $=$ as true equality).   □

As an application of the above theorem, we conclude that no countable set of first-order sentences can characterize the real numbers. This is because if the field of real numbers forms a model for the sentences, then there will also be a countable model for the sentences. But the countable model cannot be isomorphic to the field of reals, because there are uncountably many real numbers.

THEOREM II.4.2 (Compactness). *If Φ is an unsatisfiable set of predicate calculus formulas then some finite subset of Φ is unsatisfiable.*

(See also the three alternative forms in Theorem II.1.16.)

PROOF. First note that we may assume that Φ is a set of sentences, by replacing the free variables in Φ by distinct new constant symbols. The resulting set of sentences is satisfiable iff the original set of formulas is satisfiable. Since Φ is unsatisfiable iff the empty sequent $\longrightarrow$ is a logical consequence of Φ, and since *LK*-Ψ proofs are finite, the theorem now follows from Corollary II.3.5.   □

THEOREM II.4.3. *Suppose $\mathcal{L}$ has only finitely many function and predicate symbols (or recursively enumerable sets of function and predicate symbols.) Then the set of valid $\mathcal{L}$-sentences is recursively enumerable. Similarly for the set of unsatisfiable $\mathcal{L}$-sentences.*

Concerning this theorem, a set is *recursively enumerable* if there is an algorithm for enumerating its members. To enumerate the valid formulas, enumerate finite **LK** proofs. To enumerate the unsatisfiable formulas, note that A is unsatisfiable iff $\neg A$ is valid.

EXERCISE II.4.4 (Application of Compactness). Show that if a set Φ of sentences has arbitrarily large finite models, then Φ has an infinite model. (Hint: For each n construct a sentence A_n which is satisfiable in any universe with n or more elements but not satisfiable in any universe with fewer than n elements.)

II.5. The Herbrand Theorem

The Herbrand Theorem provides a complete method for proving the unsatisfiability of a set of universal sentences. It can be extended to a complete method for proving the unsatisfiability of an arbitrary set of first-order sentences by first converting the sentences to universal sentences by introducing "Skolem" functions for the existentially quantified variables. This forms the basis of the resolution proof method, which is used extensively by automated theorem provers.

DEFINITION II.5.1. A formula A is *quantifier-free* if A has no occurrence of either of the quantifiers $\forall$ or $\exists$. A $\forall$-sentence is a sentence of the form $\forall x_1 \ldots \forall x_k B$ where $k \geq 0$ and B is a quantifier-free formula. A *ground instance* of this sentence is a sentence of the form $B(t_1/x_1)(t_2/x_2)\ldots(t_k/x_k)$, where $t_1, \ldots, t_k$ are ground terms (i.e., terms with no variables) from the underlying vocabulary.

Notice that a ground instance of a $\forall$-sentence A is a logical consequence of A. Therefore if a set Φ_0 of ground instances of A is unsatisfiable, then A is unsatisfiable. The Herbrand Theorem implies a form of the converse.

DEFINITION II.5.2 ($\mathcal{L}$-Truth Assignment). An $\mathcal{L}$-*truth assignment* (or just *truth assignment*) is a map

$$\tau : \{\mathcal{L}\text{-atomic formulas}\} \to \{T, F\}.$$

We extend τ to the set of all quantifier-free $\mathcal{L}$-formulas by applying the usual rules for propositional connectives.

The above definition of truth assignment is the same as in the propositional calculus, except now we take the set of atoms to be the set of $\mathcal{L}$-atomic formulas. Thus we say that a set Φ_0 of quantifier-free formulas is *propositionally unsatisfiable* if no truth assignment satisfies every member of Φ_0.

LEMMA II.5.3. *If a set Φ_0 of quantifier-free sentences is propositionally unsatisfiable, then Φ_0 is unsatisfiable (in the first-order sense).*

Proof. We prove the contrapositive: Suppose that Φ_0 is satisfiable, and let $\mathcal{M}$ be a first-order structure which satisfies Φ_0. Then $\mathcal{M}$ induces a truth assignment τ by the definition $B^\tau = T$ iff $\mathcal{M} \models B$ for each atomic sentence B. Then $B^\tau = T$ iff $\mathcal{M} \models B$ for each quantifier-free sentence B, so τ satisfies Φ_0. □

We can now state our simplified proof method, which applies to sets of ∀-sentences: Simply take ground instances of sentences in Φ together with the equality axioms $\mathcal{E}_{\mathcal{L}}$ until a propositionally unsatisfiable set Φ_0 is found. The method does not specify how to check for propositional unsatisfiability: any method (such as truth tables) for that will do. Notice that by propositional compactness, it is sufficient to consider finite sets Φ_0 of ground instances. The Herbrand Theorem states that this method is sound and complete.

Theorem II.5.4 (Herbrand Theorem, Form 1). *Suppose that the underlying vocabulary $\mathcal{L}$ has at least one constant symbol, and let Φ be a set of ∀-sentences. Then Φ is unsatisfiable iff some finite set Φ_0 of ground instances of sentences in $\Phi \cup \mathcal{E}_{\mathcal{L}}$ is propositionally unsatisfiable.*

Corollary II.5.5 (Herbrand Theorem, Form 2). *Let Φ be a set of ∀-sentences and let $A(\vec{x}, y)$ be a quantifier-free formula with all free variables indicated such that*

$$\Phi \models \forall \vec{x} \exists y A(\vec{x}, y).$$

Then there exist finitely many terms $t_1(\vec{x}), \ldots, t_k(\vec{x})$ in the vocabulary of Φ and $A(\vec{x}, y)$ such that

$$\Phi \models \forall \vec{x} \big(A(\vec{x}, t_1(\vec{x})) \vee \cdots \vee A(\vec{x}, t_k(\vec{x})) \big).$$

We will use Form 2 in later chapters to prove "witnessing theorems" for various theories. The idea is that one of the terms $t_1(\vec{x}), \ldots, t_k(\vec{x})$ "witnesses" the existential quantifier $\exists y$ in the formula $\forall \vec{x} \exists y A(\vec{x}, y)$.

Exercise II.5.6. Prove Form 2 from Form 1. Start by showing that under the hypotheses of Form 2, $\Phi \cup \{\forall y \neg A(\vec{c}, y)\}$ is unsatisfiable, where $\vec{c}$ is a list of new constants.

Example II.5.7. Let c be a constant symbol, and let

$$\Phi = \{\forall x (Px \supset Pfx), Pc, \neg Pffc\}.$$

Then the set $\mathcal{H}$ of ground terms is $\{c, fc, ffc, \ldots\}$. We can take the set Φ_0 of ground instances to be

$$\Phi_0 = \{(Pc \supset Pfc), (Pfc \supset Pffc), Pc, \neg Pffc\}.$$

Then Φ_0 is propositionally unsatisfiable, so Φ is unsatisfiable.

Proof of the Soundness direction of Theorem II.5.4. If Φ_0 is propositionally unsatisfiable, then Φ is unsatisfiable. This follows easily from Lemma II.5.3, since Φ_0 is a logical consequence of Φ. □

PROOF OF THE COMPLETENESS DIRECTION OF THEOREM II.5.4. This follows from the Anchored *LK* Completeness Theorem (see Exercise II.5.9 below). Here we give a direct proof.

We prove the contrapositive: If every finite set of ground instances of $\Phi \cup \mathcal{E}_\mathcal{L}$ is propositionally satisfiable, then Φ is satisfiable. By Corollary II.3.4, we may ignore the special status of $=$.

Let Φ_0 be the set of *all* ground instances of $\Phi \cup \mathcal{E}_\mathcal{L}$ (using ground terms from $\mathcal{L}$). Assuming that every finite subset of Φ_0 is propositionally satisfiable, it follows from propositional compactness (Theorem II.1.16, Form 3) that the entire set Φ_0 is propositionally satisfiable. Let τ be a truth assignment which satisfies Φ_0. We use τ to construct an $\mathcal{L}$-structure $\mathcal{M}$ which satisfies Φ. We use a term model, similar to that used in the proof of the Completeness Lemma (Definition II.2.26).

Let the universe M of $\mathcal{M}$ be the set $\mathcal{H}$ of all ground $\mathcal{L}$-terms.

For each n-ary function symbol f define

$$f^{\mathcal{M}}(t_1, \ldots, t_n) = f t_1 \ldots t_n.$$

(In particular, $c^{\mathcal{M}} = c$ for each constant c, and it follows by induction that $t^{\mathcal{M}} = t$ for each ground term t.)

For each n-ary predicate symbol P of $\mathcal{L}$, define

$$P^{\mathcal{M}} = \{\langle t_1, \ldots, t_n \rangle : (P t_1 \ldots t_n)^\tau = T\}.$$

This completes the specification of $\mathcal{M}$. It follows easily by structural induction that $\mathcal{M} \models B$ iff $B^\tau = T$, for each quantifier-free $\mathcal{L}$-sentence B with no variables. Thus $\mathcal{M} \models B$ for every ground instance B of any sentence in Φ. Since every member of Φ is a $\forall$-sentence, and since the elements of the universe are precisely the ground terms, it follows that $\mathcal{M}$ satisfies every member of Φ. (A formal proof would use the Basic Semantic Definition (Definition II.2.8) and the Substitution Theorem (Theorem II.2.14).) □

EXERCISE II.5.8. Show (from the proof of the Herbrand Theorem) that a satisfiable set of $\forall$ sentences without $=$ and without function symbols except the constants $c_1, \ldots, c_n$ for $n \geq 1$ has a model with exactly n elements in the universe. Give an example showing that $n - 1$ elements would not suffice in general.

EXERCISE II.5.9. Show that the completeness direction of the Herbrand Theorem (Form 1) follows from the Anchored *LK* Completeness Theorem (with equality, Definition II.3.9 and Theorem II.3.10) and the following syntactic lemma.

LEMMA II.5.10. *Let Φ be a set of formulas closed under substitution of terms for variables. Let π be an LK-Φ proof in which all formulas are quantifier-free, let t be a term and let b be a variable, and let $\pi(t/b)$ be the* *result of replacing every occurrence of b in π by t. Then $\pi(t/b)$ is an **LK**-Φ proof.*

DEFINITION II.5.11 (Prenex Form). We say that a formula A is in *prenex form* if A has the form $Q_1 x_1 \ldots Q_n x_n B$, where each Q_i is either $\forall$ or $\exists$, and B is a quantifier-free formula.

THEOREM II.5.12 (Prenex Form). *There is a simple procedure which, given a formula A, produces an equivalent formula A' in prenex form.*

PROOF. First rename all quantified variables in A so that they are all distinct (see the remark on page 21). Now move all quantifiers out past the connectives $\wedge, \vee, \neg$ by repeated use of the equivalences below. (Recall that by the Formula Replacement Theorem (Theorem II.2.15), we can replace a subformula in A by an equivalent formula and the result is equivalent to A.)

Note. In each of the following equivalences, we must assume that x does not occur free in C.

$$(\forall x B \wedge C) \Longleftrightarrow \forall x (B \wedge C) \qquad (\forall x B \vee C) \Longleftrightarrow \forall x (B \vee C)$$
$$(C \wedge \forall x B) \Longleftrightarrow \forall x (C \wedge B) \qquad (C \vee \forall x B) \Longleftrightarrow \forall x (C \vee B)$$
$$(\exists x B \wedge C) \Longleftrightarrow \exists x (B \wedge C) \qquad (\exists x B \vee C) \Longleftrightarrow \exists x (B \vee C)$$
$$(C \wedge \exists x B) \Longleftrightarrow \exists x (C \wedge B) \qquad (C \vee \exists x B) \Longleftrightarrow \exists x (C \vee B)$$
$$\neg \forall x B \Longleftrightarrow \exists x \neg B \qquad\qquad \neg \exists x B \Longleftrightarrow \forall x \neg B. \qquad \square$$

II.6. Notes

Our treatment of **PK** in sections II.1.1 and II.1.2 is adapted from Section 1.2 of [27].

Sections II.2.1 to II.2.3 roughly follow Sections 2.1 and 2.2 of [27]. However an important difference is that the definition of $\Phi \models A$ in [27] treats free variables as though they are universally quantified, but our definition does not.

The proof of the Anchored **LK** Completeness Theorem outlined in Exercise II.2.29 grew out of discussions with S. Buss.

Chapter III

PEANO ARITHMETIC AND ITS SUBSYSTEMS

Peano Arithmetic is the first order theory of $\mathbb{N}$ with simple axioms for $+, \cdot, \leq$, and the induction axiom scheme. Here we focus on the subsystem $I\Delta_0$ of Peano Arithmetic, in which induction is restricted to bounded formulas. This subsystem plays an essential role in the development of the theories in later chapters: All (two-sorted) theories introduced in this book extend V^0, which is a conservative extension of $I\Delta_0$. At the end of the chapter we briefly discuss Buss's hierarchy $S_2^1 \subseteq T_2^1 \subseteq S_2^2 \ldots$. These single-sorted theories establish a link between bounded arithmetic and the polynomial time hierarchy, and have played a central role in the study of bounded arithmetic. In later chapters we introduce their two-sorted versions, including V^1, a theory that characterizes P. The theories considered in this chapter are singled-sorted, and the intended domain is $\mathbb{N} = \{0, 1, 2, \ldots\}$.

Subsection III.3.3 shows that the relation $y = 2^x$ is definable by a bounded formula in the vocabulary of $I\Delta_0$, and in Section III.4 this is used to show that bounded formulas represent precisely the relations in the Linear Time Hierarchy (*LTH*).

III.1. Peano Arithmetic

See Section II.2 for notions such as *vocabulary*, *formula*, and *logical consequence*.

DEFINITION III.1.1. A *theory* over a vocabulary $\mathcal{L}$ is a set $\mathcal{T}$ of formulas over $\mathcal{L}$ which is closed under logical consequence and universal closure.

We often specify a theory by a set Γ of *axioms* for $\mathcal{T}$, where Γ is a set of $\mathcal{L}$-formulas. In that case

$$\mathcal{T} = \{A : A \text{ is an } \mathcal{L}\text{-formula and } \forall\Gamma \models A\}.$$

Here $\forall\Gamma$ is the set of universal closures of formulas in Γ (Definition II.2.22).

Note that it is more usual to require that a theory be a set of sentences, rather than formulas. Our version of a usual theory $\mathcal{T}$ is $\mathcal{T}$ together with all formulas (with free variables) which are logical consequences of $\mathcal{T}$. Recall $\forall A \models A$, for any formula A.

NOTATION. We sometimes write $\mathcal{T} \vdash A$ to mean $A \in \mathcal{T}$. If $\mathcal{T} \vdash A$ we say that A is a *theorem* of $\mathcal{T}$.

The theories that we consider in this section have the vocabulary of arithmetic

$$\mathcal{L}_A = [0, 1, +, \cdot \; ; =, \leq]$$

as the underlying vocabulary (Definition II.2.3).

Recall that the *standard model* $\underline{\mathbb{N}}$ for $\mathcal{L}_A$ has universe $M = \mathbb{N}$ and $0, 1, +, \cdot, =, \leq$ get their standard meanings in $\mathbb{N}$.

NOTATION. $t < u$ stands for $(t \leq u \wedge t \neq u)$. For each $n \in \mathbb{N}$ we define a term $\underline{n}$ called the *numeral for n* inductively as follows:

$$\underline{0} = 0, \; \underline{1} = 1, \qquad \text{for } n \geq 1, \; \underline{n+1} = (\underline{n} + 1).$$

For example, $\underline{3}$ is the term $((1+1)+1)$. In general, the term $\underline{n}$ is interpreted as n in the standard model.

DEFINITION III.1.2. **TA** (True Arithmetic) is the theory over $\mathcal{L}_A$ consisting of all formulas whose universal closures are true in the standard model:

$$\mathbf{TA} = \{A : \underline{\mathbb{N}} \models \forall A\}.$$

It follows from Gödel's Incompleteness Theorem that **TA** has no computable set of axioms. The theories we define below are all proper subtheories of **TA** with nice, computable sets of axioms.

Note that by Definition II.2.6, $=$ is interpreted as true equality in all $\mathcal{L}_A$-structures, and hence we do not need to include the Equality Axioms in our list of axioms. (Of course **LK** proofs still need equality axioms: see Definition II.3.7 and Corollaries II.3.4, II.3.5).

We start by listing nine "basic" quantifier-free formulas **B1**, ..., **B8** and **C**, which comprise the axioms for our basic theory. See Figure 1 below.

B1. $x + 1 \neq 0$	**B5.** $x \cdot 0 = 0$
B2. $x + 1 = y + 1 \supset x = y$	**B6.** $x \cdot (y + 1) = (x \cdot y) + x$
B3. $x + 0 = x$	**B7.** $(x \leq y \wedge y \leq x) \supset x = y$
B4. $x + (y + 1) = (x + y) + 1$	**B8.** $x \leq x + y$
C. $0 + 1 = 1$	

FIGURE 1. 1-**BASIC**.

These axioms provide recursive definitions for $+$ and $\cdot$, and some basic properties of $\leq$. Axiom **C** is not necessary in the presence of induction, since it then follows from the theorem $0 + x = x$ (see Example III.1.8, **O2**). However we put it in so that $\forall \mathbf{B1}, \ldots, \forall \mathbf{B8}, \forall \mathbf{C}$ alone imply all true quantifier-free sentences over $\mathcal{L}_A$.

Lemma III.1.3. *If φ is a quantifier-free sentence of $\mathcal{L}_A$, then*

$$TA \vdash \varphi \qquad iff \qquad 1\text{-}BASIC \vdash \varphi.$$

Proof. The direction $\Longleftarrow$ holds because the axioms of 1-*BASIC* are valid in $\mathbb{N}$.

For the converse, we start by proving by induction on m that if $m < n$, then 1-*BASIC* $\vdash \underline{m} \neq \underline{n}$. The base case follows from **B1** and **C**, and the induction step follows from **B2** and **C**.

Next we use **B3**, **B4** and **C** to prove by induction on n that if $m + n = k$, then 1-*BASIC* $\vdash \underline{m} + \underline{n} = \underline{k}$. Similarly we use **B5**, **B6** and **C** to prove that if $m \cdot n = k$ then 1-*BASIC* $\vdash \underline{m} \cdot \underline{n} = \underline{k}$.

Now we use the above results to prove by structural induction on t, that if t is any term without variables, and t is interpreted as n in the standard model $\mathbb{N}$, then 1-*BASIC* $\vdash t = \underline{n}$.

It follows from the above results that if t and u are any terms without variables, then $TA \vdash t = u$ implies 1-*BASIC* $\vdash t = u$, and $TA \vdash t \neq u$ implies 1-*BASIC* $\vdash t \neq u$.

Consequently, if $m \leq n$, then for some k, 1-*BASIC* $\vdash \underline{n} = \underline{m} + \underline{k}$, and hence by **B8**, 1-*BASIC* $\vdash \underline{m} \leq \underline{n}$. Also if not $m \leq n$, then $n < m$, so by the above 1-*BASIC* $\vdash \underline{m} \neq \underline{n}$ and 1-*BASIC* $\vdash \underline{n} \leq \underline{m}$, so by **B7**, 1-*BASIC* $\vdash \neg \underline{m} \leq \underline{n}$.

Finally let φ be any quantifier-free sentence. We prove by structural induction on φ that if $TA \vdash \varphi$ then 1-*BASIC* $\vdash \varphi$ and if $TA \vdash \neg\varphi$ then 1-*BASIC* $\vdash \neg\varphi$. For the base case φ is atomic and has one of the forms $t = u$ or $t \leq u$, so the base case follows from the above. The induction step involves the three cases $\land$, $\lor$, and $\neg$, which are immediate. $\square$

Definition III.1.4 (Induction Scheme). If Φ is a set of formulas, then Φ-*IND* axioms are the formulas

$$[\varphi(0) \land \forall x, \; \varphi(x) \supset \varphi(x+1)] \supset \forall z \varphi(z) \tag{10}$$

where φ is a formula in Φ. Note that $\varphi(x)$ is permitted to have free variables other than x.

Definition III.1.5 (Peano Arithmetic). The theory *PA* has as axioms **B1**, ..., **B8**, together with the Φ-*IND* axioms, where Φ is the set of all $\mathcal{L}_A$ formulas.

(As we noted earlier, **C** is provable from the other axioms in the presence of induction.)

PA is a powerful theory capable of formalizing the major theorems of number theory. We define subsystems of *PA* by restricting the induction axiom to certain sets of formulas. We use the following notation.

Definition III.1.6 (Bounded Quantifiers). If the variable x does not occur in the term t, then $\exists x \leq tA$ stands for $\exists x(x \leq t \land A)$, and $\forall x \leq tA$

stands for $\forall x(x \leq t \supset A)$. Quantifiers that occur in this form are said to be *bounded*, and a *bounded formula* is one in which every quantifier is bounded.

NOTATION. Let $\exists \vec{x}$ stand for $\exists x_1 \exists x_2 \ldots \exists x_k, k \geq 0$.

DEFINITION III.1.7 (*IOPEN*, *IΔ_0*, *IΣ_1*). *OPEN* is the set of open (i.e., quantifier-free) formulas; Δ_0 is the set of bounded formulas; and Σ_1 is the set of formulas of the form $\exists \vec{x} \varphi$, where φ is bounded and $\vec{x}$ is a possibly empty vector of variables. The theories *IOPEN*, *IΔ_0*, and *IΣ_1* are the subsystems of *PA* obtained by restricting the induction scheme so that Φ is *OPEN*, Δ_0, and Σ_1, respectively.

Note that the underlying vocabulary of the theories defined above is $\mathcal{L}_A$.

EXAMPLE III.1.8. The following formulas (and their universal closures) are theorems of *IOPEN*:
O1. $(x + y) + z = x + (y + z)$ (Associativity of $+$);
O2. $x + y = y + x$ (Commutativity of $+$);
O3. $x \cdot (y + z) = (x \cdot y) + (x \cdot z)$ (Distributive law);
O4. $(x \cdot y) \cdot z = x \cdot (y \cdot z)$ (Associativity of $\cdot$);
O5. $x \cdot y = y \cdot x$ (Commutativity of $\cdot$);
O6. $x + z = y + z \supset x = y$ (Cancellation law for $+$);
O7. $0 \leq x$;
O8. $x \leq 0 \supset x = 0$;
O9. $x \leq x$;
O10. $x \neq x + 1$.

PROOF. **O1.** Induction on z.
O2. Induction on y, first establishing the special cases $y = 0$ and $y = 1$.
O3. Induction on z.
O4. Induction on z, using **O3**.
O5. Induction on y, after establishing $(y + 1) \cdot x = y \cdot x + x$ by induction on x.
O6. Induction on z.
O7. **B8**, **O2**, **B3**.
O8. **O7**, **B7**.
O9. **B8**, **B3**.
O10. Induction on x and **B2**.   □

Recall that $x < y$ stands for $(x \leq y \wedge x \neq y)$.

EXAMPLE III.1.9. The following formulas (and their universal closures) are theorems of *IΔ_0*:
D1. $x \neq 0 \supset \exists y \leq x(x = y + 1)$ (Predecessor);
D2. $\exists z(x + z = y \vee y + z = x)$;
D3. $x \leq y \leftrightarrow \exists z(x + z = y)$;
D4. $(x \leq y \wedge y \leq z) \supset x \leq z$ (Transitivity);

D5. $x \leq y \vee y \leq x$ (Total order);
D6. $x \leq y \leftrightarrow x + z \leq y + z$;
D7. $x \leq y \supset x \cdot z \leq y \cdot z$;
D8. $x \leq y + 1 \leftrightarrow (x \leq y \vee x = y + 1)$ (Discreteness 1);
D9. $x < y \leftrightarrow x + 1 \leq y$ (Discreteness 2);
D10. $x \cdot z = y \cdot z \wedge z \neq 0 \supset x = y$ (Cancellation law for $\cdot$).

PROOF. **D1.** Induction on x.
D2. Induction on x. Base case: **B2, O2**. Induction step: **B3, B4, D1**.
D3. $\Longrightarrow$: **D2, B3** and **B7**; $\Longleftarrow$: **B8**.
D4. **D3, O1**.
D5. **D2, B8**.
D6. $\Longrightarrow$: **D3, O1, O2**; $\Longleftarrow$: **D3, O6**.
D7. **D3** and algebra (**O1, ... , O8**).
D8. $\Longrightarrow$: **D3, D1**, and algebra; $\Longleftarrow$: **O9, B8, D4**.
D9. $\Longrightarrow$: **D3, D1**, and algebra; $\Longleftarrow$: **D3** and algebra.
D10. Exercise.   □

Taken together, these results show that all models of $I\Delta_0$ are commutative discretely-ordered semi-rings.

EXERCISE III.1.10. Show that $I\Delta_0$ proves the division theorem:

$$I\Delta_0 \vdash \forall x \forall y (0 < x \supset \exists q \, \exists r < x, \, y = x \cdot q + r).$$

It follows from Gödel's Incompleteness Theorem that there is a bounded formula $\varphi(x)$ such that $\forall x \varphi(x)$ is true but $I\Delta_0 \nvdash \forall x \varphi(x)$. However if φ is a true sentence in which all quantifiers are bounded, then intuitively φ expresses information about only finitely many tuples of numbers, and in this case we can show $I\Delta_0 \vdash \varphi$. The same applies more generally to true Σ_1 sentences φ.

LEMMA III.1.11. *If φ is a Σ_1 sentence, then $TA \vdash \varphi$ iff $I\Delta_0 \vdash \varphi$.*

PROOF. The direction $\Longleftarrow$ follows because all axioms of $I\Delta_0$ are true in the standard model.

For the converse, we prove by structural induction on bounded sentences φ that if $TA \vdash \varphi$ then $I\Delta_0 \vdash \varphi$, and if $TA \vdash \neg \varphi$ then $I\Delta_0 \vdash \neg \varphi$. The base case is φ is atomic, and this follows from Lemma III.1.3. For the induction step, the cases $\vee$, $\wedge$, and $\neg$ are immediate. The remaining cases are φ is $\forall x \leq t \psi(x)$ and φ is $\exists x \leq t \psi(x)$, where t is a term without variables, and $\psi(x)$ is a bounded formula with no free variable except possibly x. These cases follow from Lemma III.1.3 and Lemma III.1.12 below.

Now suppose that φ is a true Σ_1 sentence of the form $\exists \vec{x} \psi(\vec{x})$, where $\psi(\vec{x})$ is a bounded formula. Then $\psi(\vec{n})$ is a true bounded sentence for some numerals $\underline{n}_1, \ldots, \underline{n}_k$, so $I\Delta_0 \vdash \psi(\vec{n})$. Hence $I\Delta_0 \vdash \varphi$.   □

LEMMA III.1.12. *For each $n \in \mathbb{N}$,*

$$I\Delta_0 \vdash x \leq \underline{n} \leftrightarrow (x = \underline{0} \vee x = \underline{1} \vee \cdots \vee x = \underline{n}).$$

PROOF. Induction on n. The base case $n = 0$ follows from **O7** and **O8**, and the induction step follows from **D8**. □

III.1.1. Minimization.

DEFINITION III.1.13 (Minimization). The minimization axioms (or *least number principle* axioms) for a set Φ of formulas are denoted Φ-*MIN* and consist of the formulas

$$\exists z \varphi(z) \supset \exists y \big(\varphi(y) \wedge \neg \exists x (x < y \wedge \varphi(x)) \big)$$

where φ is a formula in Φ.

THEOREM III.1.14. *$I\Delta_0$ proves Δ_0-MIN.*

PROOF. The contrapositive of the minimization axiom for $\varphi(z)$ follows from the induction axiom for the bounded formula $\psi(z) \equiv \forall y \leq z(\neg\varphi(y))$. □

EXERCISE III.1.15. Show that $I\Delta_0$ can be alternatively axiomatized by **B1**, . . . , **B8**, **O10** (Example III.1.8), **D1** (Example III.1.9), and the axiom scheme Δ_0-*MIN*.

III.1.2. Bounded Induction Scheme. The Δ_0-*IND* scheme of $I\Delta_0$ can be replaced by the following *bounded induction scheme* for Δ_0 formulas, i.e.,

$$\big(\varphi(0) \wedge \forall x < z(\varphi(x) \supset \varphi(x+1)) \big) \supset \varphi(z) \tag{11}$$

where $\varphi(x)$ is any Δ_0 formula. (Note that the *IND* formula (10) for $\varphi(x)$ is a logical consequence of the universal closure of this.)

EXERCISE III.1.16. Prove that $I\Delta_0$ remains the same if the Δ_0-*IND* scheme is replaced by the above bounded induction scheme for Δ_0 formulas. (It suffices to show that the new scheme is provable in $I\Delta_0$.)

III.1.3. Strong Induction Scheme. The *strong induction axiom* for a formula $\varphi(x)$ is the following formula:

$$\forall x \big((\forall y < x \varphi(y)) \supset \varphi(x) \big) \supset \forall z \varphi(z). \tag{12}$$

EXERCISE III.1.17. Show that $I\Delta_0$ proves the strong induction axiom scheme for Δ_0 formulas.

III.2. Parikh's Theorem

By the results in the previous section, $I\Delta_0$ can be axiomatized by a set of bounded formulas. We say that it is a *polynomial-bounded theory*, a concept we will now define.

In general, a theory T may have symbols other than those in $\mathcal{L}_A$. We say that a term $t(\vec{x})$ is a *bounding term* for a function symbol $f(\vec{x})$ in T if

$$T \vdash \forall \vec{x} \; f(\vec{x}) \le t(\vec{x}). \tag{13}$$

We say that f is *polynomially bounded* (or just *p-bounded*) in T if it has a bounding term in the vocabulary $\mathcal{L}_A$.

EXERCISE III.2.1. Let T be an extension of $I\Delta_0$ and let $\mathcal{L}$ be the vocabulary of T. Suppose that the functions of $\mathcal{L}$ are polynomially bounded in T. Show that for each $\mathcal{L}$-term $s(\vec{x})$, there is an $\mathcal{L}_A$-term $t(\vec{x})$ such that

$$T \vdash \forall \vec{x} \; s(\vec{x}) \le t(\vec{x}).$$

Suppose that a theory T is an extension of $I\Delta_0$. We can still talk about bounded formulas φ in T using the same definition (Definition III.1.6) as before, but now φ may have function and predicate symbols not in the vocabulary $[0, 1, +, \cdot; =, \le]$ of $I\Delta_0$, and in particular the terms t bounding the quantifiers $\exists x \le t$ and $\forall x \le t$ may have extra function symbols. Note that by the exercise above, in the context of polynomial-bounded theories (defined below) we may assume without loss of generality that the bounding terms are $\mathcal{L}_A$-terms.

DEFINITION III.2.2 (Polynomial-Bounded Theory). Let T be a theory with vocabulary $\mathcal{L}$. Then T is a *polynomial-bounded theory* (or just *p-bounded theory*) if (i) it extends $I\Delta_0$; (ii) it can be axiomatized by a set of bounded formulas; and (iii) each function $f \in \mathcal{L}$ is polynomially bounded in T.

Note that $I\Delta_0$ is a polynomial-bounded theory.

Theories which satisfy (ii) are often called *bounded theories*.

THEOREM III.2.3 (Parikh's Theorem). *If T is a polynomial-bounded theory and $\varphi(\vec{x}, y)$ is a bounded formula with all free variables displayed such that $T \vdash \forall \vec{x} \exists y \varphi(\vec{x}, y)$, then there is a term t involving only variables in $\vec{x}$ such that T proves $\forall \vec{x} \exists y \le t \varphi(\vec{x}, y)$.*

It follows from Exercise III.2.1 that the bounding term t can be taken to be an $\mathcal{L}_A$-term. In fact, Parikh's Theorem can be generalized to say that if φ is a bounded formula and $T \vdash \exists \vec{y} \varphi$, then there are $\mathcal{L}_A$-terms $t_1, \ldots, t_k$ not involving any variable in $\vec{y}$ or any variable not occurring free in φ such that T proves $\exists y_1 \le t_1 \ldots \exists y_k \le t_k \varphi$. This follows from the above remark, and the following lemma.

LEMMA III.2.4. *Let T be an extension of $I\Delta_0$. Let z be a variable distinct from $y_1, \ldots, y_k$ and not occurring in φ. Then*

$$T \vdash \exists \vec{y} \varphi \leftrightarrow \exists z \exists y_1 \le z \ldots \exists y_k \le z \; \varphi.$$

EXERCISE III.2.5. Give a careful proof of the above lemma, using the theorems of $I\Delta_0$ described in Example III.1.9.

In section III.3.3 we will show how to represent the relation $y = 2^x$ by a bounded formula $\varphi_{exp}(x, y)$. It follows immediately from Parikh's Theorem that

$$I\Delta_0 \nvdash \forall x \exists y \varphi_{exp}(x, y).$$

On the other hand PA easily proves the $\exists y \varphi_{exp}(x, y)$ by induction on x. Therefore $I\Delta_0$ is a proper sub-theory of PA.

Our proof of Parikh's Theorem will be based on the Anchored LK Completeness Theorem with Equality (II.3.10). Let T be a polynomial-bounded theory and $\forall \vec{x} \exists y \varphi(\vec{x}, y)$ a theorem of T. We will look into an anchored proof of $\forall \vec{x} \exists y \varphi(\vec{x}, y)$ and show that a term t (not involving y) can be constructed so that $\forall \vec{x} \exists y \leq t \varphi(\vec{x}, y)$ is also a theorem of T. In order to apply the Anchored LK Completeness Theorem (with Equality), we need to find an axiomatization of T which is closed under substitution of terms for variables. Note that T is already axiomatized by a set of bounded formulas (Definition III.2.2). The desired axiomatization of T is obtained by substituting terms for all the free variables. We will consider the example where T is $I\Delta_0$. The general case is similar.

Recall that the axioms for $I\Delta_0$ consist of **B1**–**B8** (page 40) and the Δ_0-**IND** scheme, which can be replaced by the Bounded Induction Scheme (11).

DEFINITION III.2.6 (ID_0). ID_0 is the set of all term substitution instances of **B1**–**B8** and the Bounded Induction Scheme, where now the terms contain only "free" variables $a, b, c, \ldots$.

Note that all formulas in ID_0 are bounded.
For example $(c \cdot b) + 1 \neq 0$ is an instance of **B1**, and hence is in ID_0. Also

$$a + 0 = 0 + a \wedge \forall x < b(a + x = x + a \supset a + (x + 1) = (x + 1) + a)$$
$$\supset a + b = b + a$$

is an instance of (11) useful in proving the commutative law $a + b = b + a$ by induction on b, and is in ID_0.

The following is an immediate consequence of the Anchored LK Completeness Theorem II.3.10 and Derivational Soundness of LK (Theorem II.2.23).

THEOREM III.2.7 (LK-ID_0 Adequacy). *Let A be an $\mathcal{L}_A$ formula satisfying the LK constraint that only variables $a, b, c, \ldots$ occur free and only $x, y, z, \ldots$ occur bound. Then $I\Delta_0 \vdash A$ iff A has an anchored LK-ID_0 proof.*

PROOF OF PARIKH'S THEOREM. Suppose that T is a polynomial-bounded theory which is axiomatized by a set of bounded axioms such that $T \vdash \forall \vec{x} \exists y \varphi(\vec{x}, y)$, where $\varphi(\vec{x}, y)$ is a bounded formula. Let T be the set of all term substitution instances of the axioms of T. By arguing as above in the case $T = I\Delta_0$, we can assume that $\longrightarrow \exists y \varphi(\vec{a}, y)$ has an anchored LK-T proof π. Further we may assume that π is in free variable normal form (Section II.2.4). By the sub-formula property of anchored proofs (II.3.11), every formula in every sequent of π is either bounded, or a substitution instance of the endsequent $\exists y \varphi(\vec{a}, y)$. But in fact the proof of the sub-formula property actually shows more: Every formula in π is either bounded or it must be syntactically identical to $\exists y \varphi(\vec{a}, y)$, and in the latter case it must occur in the consequent (right side) of a sequent. The reason is that once an unbounded quantifier is introduced in π, the resulting formula can never be altered by any rule, since cut formulas are restricted to the bounded formulas occurring in T, and since no altered version of $\exists y \varphi(\vec{a}, y)$ occurs in the endsequent. (We may assume that $\exists y \varphi(\vec{a}, y)$ is an unbounded formula, since otherwise there is nothing to prove.)

We will convert π to an LK-T proof π' of $\exists y \le t \varphi(y)$ for some term t not containing y, by replacing each sequent S in π by a suitable sequent S', sometimes with a short derivation $D(S)$ of S' inserted.

Here and in general we treat the cedents Γ and Δ of a sequent $\Gamma \longrightarrow \Delta$ as multi-sets in which the order of formulas is irrelevant. In particular we ignore instances of the exchange rule.

The conversion of a sequent S in π to S', and the associated derivation $D(S)$, are defined by induction on the depth of S in π such that the following is satisfied:

Induction Hypothesis. If S has no occurrence of $\exists y \varphi$, then $S' = S$. If S has one or more occurrences of $\exists y \varphi$, then S' is a sequent which is the same as S except all occurrences of $\exists y \varphi$ are replaced by a single occurrence of $\exists y \le t \varphi$, where the term t depends on S and the placement of S in π. Further t satisfies the condition

$$\text{Every variable in } t \text{ occurs free in the original sequent } S. \tag{14}$$

Thus the endsequent of π' has the form $\longrightarrow \exists y \le t \varphi$, where every variable in t occurs free in $\exists y \varphi$.

In order to maintain the condition (14) we use our assumption that π is in free variable normal form. Thus if the variable b occurs in t in the formula $\exists y \le t \varphi$, so b occurs in S, then b cannot be eliminated from the descendants of S except by the rule $\forall$-right or $\exists$-left. These rules require special attention in the argument below.

We consider several cases, depending on the inference rule in π forming S, and whether $\exists y \varphi$ is the principle formula of that rule.

Case I. S is the result of $\exists$-right applied to $\varphi(s)$ for some term s, so the inference has the form

$$\frac{\Gamma \longrightarrow \Delta, \varphi(s)}{\Gamma \longrightarrow \Delta, \exists y \varphi(y)} \tag{15}$$

where S is the bottom sequent. Suppose first that Δ has no occurrence of $\exists y\varphi$. Since $\pmb{ID}_0$ proves $s \leq s$ there is a short $\pmb{LK\text{-}T}$ derivation of

$$\Gamma \longrightarrow \Delta, \exists y \leq s\varphi(y) \tag{16}$$

from the top sequent. Let $\pmb{D}(S)$ be that derivation and let S' be the sequent (16).

If Δ has one or more occurrence of $\exists y\varphi$, then by the induction hypothesis the top sequent S_1 of (15) was converted to a sequent S_1' in which all of these occurrences have been replaced by a single occurrence of the form $\exists y \leq t\varphi$. We proceed as before, producing a sequent of the form

$$\Gamma \longrightarrow \Delta', \exists y \leq t\varphi, \exists y \leq s\varphi. \tag{17}$$

Since $\pmb{ID}_0$ proves the two sequents $\longrightarrow s \leq s + t$ and $\longrightarrow t \leq s + t$, it follows that T proves

$$\exists y \leq s\varphi \longrightarrow \exists y \leq (s + t)\varphi$$

and

$$\exists y \leq t\varphi \longrightarrow \exists y \leq (s + t)\varphi.$$

We can use these and (17) with two cuts and a contraction to obtain a derivation of

$$\Gamma \longrightarrow \Delta', \exists y \leq (s + t)\varphi(y). \tag{18}$$

Let $\pmb{D}(S)$ be this derivation and let S' be the resulting sequent (18).

Case II. S is the result of weakening right, which introduces $\exists y\varphi$. Thus the inference has the form

$$\frac{\Gamma \longrightarrow \Delta}{\Gamma \longrightarrow \Delta, \exists y\varphi} \tag{19}$$

where S is the bottom sequent. If Δ does not contain $\exists y\varphi$, then define S' to be

$$\Gamma \longrightarrow \Delta, \exists y \leq 0\, \varphi$$

(introduced by weakening). If Δ contains one or more occurrences of $\exists y\varphi$, then take $S' = S_1'$, where S_1 is the top sequent of (19).

Case III. S is the result of $\forall$-right or $\exists$-left. We consider the case $\exists$-left. The other case is similar and we leave it as an exercise. The new quantifier introduced must be bounded, since all formulas in π except $\exists y\varphi$ are bounded, and the latter must occur on the right. Thus the inference has the form

$$\frac{b \leq r \wedge \psi(b), \Gamma \longrightarrow \Delta}{\exists x \leq r\psi(x), \Gamma \longrightarrow \Delta} \tag{20}$$

where S is the bottom sequent. If Λ has no occurrence of $\exists y\varphi$, then define $S' = S$ and let $D(S)$ be the derivation (20). Otherwise, by the induction hypothesis, the top sequent was converted to a sequent of the form

$$b \leq r \wedge \psi(b), \Gamma \longrightarrow \Delta', \exists y \leq s(b)\varphi(y). \tag{21}$$

Note that b may appear on the succedent and thus violate the Restriction of the $\exists$-left rule (page 22).

In order to apply the $\exists$-left rule (and continue to satisfy the condition (14)), we replace the bounding term $s(b)$ by an $\mathcal{L}_A$-term t that does not contain b. This is possible since the functions of $\mathcal{T}$ are polynomially bounded in $\mathcal{T}$. In particular, by Exercise III.2.1, we know that there are $\mathcal{L}_A$-terms r', $s'(b)$ such that $\mathcal{T}$ proves both

$$r \leq r' \qquad \text{and} \qquad s(b) \leq s'(b).$$

Let $t = s'(r')$. Then by the monotonicity of $\mathcal{L}_A$-terms, $\mathcal{T}$ proves $b \leq r \longrightarrow s(b) \leq t$. Thus $\mathcal{T}$ proves

$$b \leq r, \exists y \leq s(b)\varphi(y) \longrightarrow \exists y \leq t\varphi(y)$$

(i.e., the above sequent has an **LK-T** derivation). From this and (21) applying cut with cut formula $\exists y \leq s(b)\varphi$ we obtain

$$b \leq r \wedge \psi(b), \Gamma \longrightarrow \Delta', \exists y \leq t\varphi(y)$$

where t does not contain b. We can now apply the $\exists$-left rule to obtain

$$\exists x \leq r\psi(x), \Gamma \longrightarrow \Delta', \exists y \leq t\varphi(y). \tag{22}$$

Let $D(S)$ be this derivation and let S' be the resulting sequent (22).

Case IV. S results from a rule with two parents. Note that if this rule is cut, then the cut formula cannot be $\exists y\varphi$, because π is anchored. The only difficulty in converting S is that the two consequents Δ' and Δ'' of the parent sequents may have been converted to consequents with different bounded formulas $\exists y \leq t_1\varphi$ and $\exists y \leq t_2\varphi$. In this case proceed as in the second part of *Case I* to combine these two formulas to the single formula $\exists y \leq (t_1 + t_2)\varphi$.

Case V. All remaining cases. The inference is of the form derive S from the single sequent S_1. Then take S' to be the result of applying the same rule in the same way to S'_1, except in the case of contraction right when the principle formula is $\exists y\varphi$. In this case take $S' = S'_1$. □

EXERCISE III.2.8. Work out the sub-case $\forall$-right in *Case III*.

III.3. Conservative Extensions of $I\Delta_0$

In this section we occasionally present simple model-theoretic arguments, and the following standard definition from model theory is useful.

DEFINITION III.3.1 (Expansion of a Model). Let $\mathcal{L}_1 \subseteq \mathcal{L}_2$ be vocabularies and let $\mathcal{M}_i$ be an $\mathcal{L}_i$ structure for $i = 1, 2$. We say $\mathcal{M}_2$ is an *expansion* of $\mathcal{M}_1$ if $\mathcal{M}_1$ and $\mathcal{M}_2$ have the same universe and the same interpretation for symbols in $\mathcal{L}_1$.

III.3.1. Introducing New Function and Predicate Symbols. In the following discussion we assume that all predicate and function symbols have a standard interpretation in the set $\mathbb{N}$ of natural numbers. A theory $\mathcal{T}$ which extends $I\Delta_0$ has defining axioms for each predicate and function symbol in its vocabulary which ensure that they receive their standard interpretations in a model of $\mathcal{T}$ which is an expansion of the standard model $\underline{\mathbb{N}}$. We often use the same notation for both the function symbol and the function that it is intended to represent. For example, the predicate symbol P might be *Prime*, where *Prime*(x) is intended to mean that x is a prime number. Or f might be *LPD*, where *LPD*(x) is intended to mean the least prime number dividing x (or x if $x \le 1$).

NOTATION (unique existence). $\exists!x\varphi(x)$ stands for $\exists x, \varphi(x) \wedge \forall y(\varphi(y) \supset x = y)$, where y is a new variable not appearing in $\varphi(x)$. $\exists!x \le t\varphi(x)$, where t does not involve x, stands for

$$\exists x \le t, \varphi(x) \wedge \forall y \le t(\varphi(y) \supset x = y)$$

where y is a new variable not appearing in $\varphi(x)$ or t.

DEFINITION III.3.2 (Definable Predicates and Functions). Let $\mathcal{T}$ be a theory with vocabulary $\mathcal{L}$, and let Φ be a set of $\mathcal{L}$-formulas.
(*a*) We say that a predicate symbol $P(\vec{x})$ not in $\mathcal{L}$ is Φ-*definable* in $\mathcal{T}$ if there is an $\mathcal{L}$-formula $\varphi(\vec{x})$ in Φ such that

$$P(\vec{x}) \leftrightarrow \varphi(\vec{x}). \tag{23}$$

(*b*) We say that a function symbol $f(\vec{x})$ not in $\mathcal{L}$ is Φ-definable in $\mathcal{T}$ if there is a formula $\varphi(\vec{x}, y)$ in Φ such that

$$\mathcal{T} \vdash \forall \vec{x} \exists! y \varphi(\vec{x}, y), \tag{24}$$

and that

$$y = f(\vec{x}) \leftrightarrow \varphi(\vec{x}, y). \tag{25}$$

We say that (23) is a *defining axiom* for $P(\vec{x})$ and (25) is a *defining axiom* for $f(\vec{x})$. We say that a symbol is *definable* in $\mathcal{T}$ if it is Φ-definable in $\mathcal{T}$ for some Φ.

Although the choice of φ in the above definition is not uniquely determined by the predicate or function symbol, we will assume that a specific φ has been chosen, so we will speak of *the* defining axiom for the symbol.

For example, the defining axiom for the predicate *Prime*(x) (in any theory whose vocabulary contains $\mathcal{L}_A$) might be

$$Prime(x) \leftrightarrow 1 < x \wedge \forall y < x \forall z < x(y \cdot z \neq x).$$

NOTATION. Note that Δ_0 and Σ_1 (Definition III.1.7) are sets of $\mathcal{L}_A$-formulas. In general, given a vocabulary $\mathcal{L}$ the sets $\Delta_0(\mathcal{L})$ and $\Sigma_1(\mathcal{L})$ are defined as in Definition III.1.7 but the formulas are from $\mathcal{L}$. In this case we require that the terms bounding the quantifiers are $\mathcal{L}_A$-terms.

In Definition III.3.2, if $\Phi = \Delta_0(\mathcal{L})$ (resp. $\Phi = \Sigma_1(\mathcal{L})$) then we sometimes omit mention of $\mathcal{L}$ and simply say that the symbols P, f are Δ_0-definable (resp. Σ_1-definable) in $\mathcal{T}$.

In the case of functions, the choice $\Phi = \Sigma_1(\mathcal{L})$ plays a special role. A Σ_1-definable function in $\mathcal{T}$ is also called a *provably total function* in $\mathcal{T}$. For example one can show that the provably total functions of **TA** are precisely all total computable functions. The provably total functions of $I\Sigma_1$ are precisely the primitive recursive functions, and of S_2^1 (see Section III.5) the polytime functions. In Section III.4 we will show that the provably total functions of $I\Delta_0$ are precisely the functions of the Linear Time Hierarchy.

EXERCISE III.3.3. Suppose that the functions
$$f(x_1, \ldots, x_m) \text{ and } h_i(x_1, \ldots, x_n) \text{ (for } 1 \leq i \leq m)$$
are Σ_1-definable in a theory $\mathcal{T}$. Show that the function $f(h_1(\vec{x}), \ldots, h_m(\vec{x}))$ (where $\vec{x}$ stands for $x_1, \ldots, x_n$) is also Σ_1-definable in $\mathcal{T}$. (In other words, show that Σ_1-definable functions are closed under composition.)

DEFINITION III.3.4 (Conservative Extension). Suppose that $\mathcal{T}_1$ and $\mathcal{T}_2$ are two theories, where $\mathcal{T}_1 \subseteq \mathcal{T}_2$, and the vocabulary of $\mathcal{T}_2$ may contain function or predicate symbols not in $\mathcal{T}_1$. We say $\mathcal{T}_2$ *is a conservative extension of* $\mathcal{T}_1$ if for every formula A in the vocabulary of $\mathcal{T}_1$, if $\mathcal{T}_2 \vdash A$ then $\mathcal{T}_1 \vdash A$.

THEOREM III.3.5 (Extension by Definition). *If $\mathcal{T}_2$ results from $\mathcal{T}_1$ by expanding the vocabulary of $\mathcal{T}_1$ to include definable symbols, and by adding the defining axioms for these symbols, then $\mathcal{T}_2$ is a conservative extension of $\mathcal{T}_1$.*

PROOF. We give a simple model-theoretic argument. Suppose that A is a formula in the vocabulary of $\mathcal{T}_1$ and suppose that $\mathcal{T}_2 \vdash A$. Let $\mathcal{M}_1$ be a model of $\mathcal{T}_1$. We expand $\mathcal{M}_1$ to a model $\mathcal{M}_2$ of $\mathcal{T}_2$ by interpreting each new predicate and function symbol so that its defining axiom (23) or (25) is satisfied. Notice that this interpretation is uniquely determined by the defining axiom, and in the case of a function symbol the provability condition (24) is needed (both existence and uniqueness of y) in order to ensure that both directions of the equivalence (25) hold.

Since $\mathcal{M}_2$ is a model of $\mathcal{T}_2$, it follows that $\mathcal{M}_2 \models A$, and hence $\mathcal{M}_1 \models A$. Since $\mathcal{M}_1$ is an arbitrary model of $\mathcal{T}_1$, it follows that $\mathcal{T}_1 \vdash A$.   □

COROLLARY III.3.6. *Let $\mathcal{T}$ be a theory and $\mathcal{T}_0 = \mathcal{T} \subset \mathcal{T}_1 \subset \cdots$ be a sequence of extensions of $\mathcal{T}$ where each $\mathcal{T}_{n+1}$ is obtained by adding to $\mathcal{T}_n$ a definable symbol (in the vocabulary of $\mathcal{T}_n$) and its defining axiom. Let $\mathcal{T}_\infty = \bigcup_{n \geq 0} \mathcal{T}_n$. Then $\mathcal{T}_\infty$ is a conservative extension of $\mathcal{T}$.*

EXERCISE III.3.7. Prove the corollary using the Extension by Definition Theorem and the Compactness Theorem.

As an application of the Extension by Definition Theorem, we can conservatively extend **PA** to include symbols for all the *arithmetical* predicates (i.e., predicates definable by $\mathcal{L}_A$-formulas). In fact, the extension of **PA** remains conservative even if we allow induction on *formulas over the expanded vocabulary.*

Similarly we can also obtain a conservative extension of $I\Delta_0$ by adding to it predicate symbols and their defining axioms for all arithmetical predicates. However such a conservative extension of $I\Delta_0$ no longer proves the induction axiom scheme on bounded formulas over the expanded vocabulary. It does so if we only add Δ_0-definable symbols, and in fact we may add both Δ_0-definable predicate and function symbols. To show this, we start with the following important application of Parikh's Theorem.

THEOREM III.3.8 (Bounded Definability). *Let T be a polynomial-bounded theory. A function $f(\vec{x})$ (not in T) is Σ_1-definable in T iff it has a defining axiom*

$$y = f(\vec{x}) \leftrightarrow \varphi(\vec{x}, y)$$

where φ is a bounded formula with all free variables indicated, and there is an $\mathcal{L}_A$-term $t = t(\vec{x})$ such that T proves $\forall \vec{x} \exists! y \le t\varphi(\vec{x}, y)$.

PROOF. The IF direction is immediate from Definition III.3.2. The ONLY IF direction follows from the discussion after Parikh's Theorem III.2.3.   □

COROLLARY III.3.9. *If T is a polynomial-bounded theory, then a function f is Σ_1-definable in T iff f is Δ_0-definable in T.*

From the above theorem we see that the function 2^x is not Σ_1-definable in any polynomial-bounded theory, even though we shall show in Section III.3.3 that the *relation* $(y = 2^x)$ is Δ_0-definable in $I\Delta_0$. Since the function 2^x is Σ_1-definable in **PA**, it follows that $I\Delta_0 \subsetneq$ **PA**.

LEMMA III.3.10 (Conservative Extension). *Suppose that T is a polynomial-bounded theory and T^+ is the conservative extension of T obtained by adding to T a Δ_0-definable predicate or a Σ_1-definable function symbol and its defining axiom. Then T^+ is a polynomial-bounded theory and every bounded formula φ^+ in the vocabulary of T^+ can be translated into a bounded formula φ in the vocabulary of T such that*

$$T^+ \vdash \varphi^+ \leftrightarrow \varphi.$$

The following corollary follows immediately from the lemma.

COROLLARY III.3.11. *Let T and T^+ be as in the Conservative Extension Lemma. Let $\mathcal{L}$ and $\mathcal{L}^+$ denote the vocabulary of T and T^+, respectively. Assume further that T proves the $\Delta_0(\mathcal{L})$-**IND** axiom scheme. Then T^+ proves the $\Delta_0(\mathcal{L}^+)$-**IND** axiom scheme.*

PROOF OF THE CONSERVATIVE EXTENSION LEMMA.   First, suppose that $\mathcal{T}^+$ is obtained from $\mathcal{T}$ by adding to it a Δ_0-definable predicate symbol P and its defining axiom (23). That $\mathcal{T}^+$ is polynomial-bounded is immediate from Definition III.2.2. Now each bounded formula in the vocabulary of $\mathcal{T}^+$ can be translated to a bounded formula in the vocabulary of $\mathcal{T}$ simply by replacing each occurrence of a formula of the form $P(\vec{t})$ by $\varphi(\vec{t})$ (see the Formula Replacement Theorem, II.2.15). Note that the defining axiom (23) becomes the valid formula $\varphi(\vec{x}) \leftrightarrow \varphi(\vec{x})$.

Next suppose that $\mathcal{T}^+$ is obtained from $\mathcal{T}$ by adding to it a Σ_1-definable function symbol f and its defining axiom (25). That $\mathcal{T}^+$ is polynomial-bounded follows from Theorem III.3.8.

Start translating φ^+ by replacing every bounded quantifier $\forall x \leq u\psi$ by $\forall x \leq u'(x \leq u \supset \psi)$, where u' is obtained from u by replacing every occurrence of every function symbol other than $+, \cdot$ by its bounding term in $\mathcal{L}_A$. Similarly replace $\exists x \leq u\psi$ by $\exists x \leq u'(x \leq u \wedge \psi)$.

Now we may suppose by Theorem III.3.8 that f has a bounded defining axiom

$$y = f(\vec{x}) \leftrightarrow \varphi_1(\vec{x}, y)$$

and $f(\vec{x})$ has an $\mathcal{L}_A$ bounding term $t(\vec{x})$. Repeatedly remove occurrences of f in an atomic formula $\theta(s(f(\vec{u})))$ by replacing this with

$$\exists y \leq t(\vec{u}), \ \varphi_1(\vec{u}, y) \wedge \theta(s(y)). \qquad \square$$

Now we summarize the previous results.

THEOREM III.3.12 (Conservative Extension). *Let $\mathcal{T}_0$ be a polynomial-bounded theory over a vocabulary $\mathcal{L}_0$ which proves the $\Delta_0(\mathcal{L}_0)$-IND axioms. Let $\mathcal{T}_0 \subset \mathcal{T}_1 \subset \mathcal{T}_2 \subset \ldots$ be a sequence of extensions of $\mathcal{T}_0$ where each $\mathcal{T}_{i+1}$ is obtained from $\mathcal{T}_i$ by adding a Σ_1-definable function symbol f_{i+1} (or a Δ_0-definable predicate symbol P_{i+1}) and its defining axiom. Let*

$$\mathcal{T} = \bigcup_{i \geq 0} \mathcal{T}_i.$$

Then $\mathcal{T}$ is a polynomial-bounded theory and is a conservative extension of $\mathcal{T}_0$. Furthermore, if $\mathcal{L}$ is the vocabulary of $\mathcal{T}$, then $\mathcal{T}$ proves the equivalence of each $\Delta_0(\mathcal{L})$ formula with some $\Delta_0(\mathcal{L}_0)$ formula, and $\mathcal{T} \vdash \Delta_0(\mathcal{L})$-IND.

PROOF. First, we prove by induction on i that

1) $\mathcal{T}_i$ is a polynomial-bounded theory;
2) $\mathcal{T}_i$ is a conservative extension of $\mathcal{T}_0$; and
3) $\mathcal{T}_i$ proves that each $\Delta_0(\mathcal{L}_i)$ formula is equivalent to some $\Delta_0(\mathcal{L}_0)$ formula, where $\mathcal{L}_i$ is the vocabulary of $\mathcal{T}_i$.

The induction step follows from the Conservative Extension Lemma.

It follows from the induction arguments above that $\mathcal{T}$ is a polynomial-bounded theory, and that $\mathcal{T}$ proves the equivalence of each $\Delta_0(\mathcal{L})$ formula with some $\Delta_0(\mathcal{L}_0)$ formula, and $T \vdash \Delta_0(\mathcal{L})$-***IND***. It follows from Corollary III.3.6 that T is a conservative extension of T_0.   □

III.3.2. $\overline{I\Delta_0}$: A Universal Conservative Extension of $I\Delta_0$. (This subsection is not needed for the remainder of this chapter, but it is needed for later chapters.)

We begin by introducing terminology that allows us to restate the Herbrand Theorem (see Section II.5).

A *universal formula* is a formula in prenex form (Definition II.5.11) in which all quantifiers are universal. A *universal theory* is a theory which can be axiomatized by universal formulas. Note that by definition (III.1.1), a universal theory can be equivalently axiomatized by a set of quantifier-free formulas, or by a set of ∀ sentences (Definition II.5.1). We can now restate Form 2 of the Herbrand Theorem II.5.5 as follows.

THEOREM III.3.13 (Herbrand Theorem, Form 2). *Let T be a universal theory, and let $\varphi(x_1, \ldots, x_m, y)$ be a quantifier-free formula with all free variables indicated such that*

$$T \vdash \forall x_1 \ldots \forall x_m \exists y \varphi(\vec{x}, y). \tag{26}$$

Then there exist finitely many terms $t_1(\vec{x}), \ldots, t_n(\vec{x})$ such that

$$T \vdash \forall x_1 \ldots \forall x_m \left(\varphi(\vec{x}, t_1(\vec{x})) \vee \cdots \vee \varphi(\vec{x}, t_n(\vec{x})) \right).$$

Note that the theorem easily extends to the case where

$$T \vdash \forall x_1 \ldots \forall x_m \exists y_1 \ldots \exists y_k \varphi(\vec{x}, \vec{y})$$

instead of (26), where $\varphi(\vec{x}, \vec{y})$ is a quantifier-free formula.

PROOF. As we have remarked earlier, T can be axiomatized by a set Γ of ∀ sentences. From (26) it follows that

$$\Gamma \cup \{\exists x_1 \ldots \exists x_m \forall y \neg \varphi(\vec{x}, y)\} \tag{27}$$

is unsatisfiable. Let $c_1, \ldots, c_m$ be new constant symbols. Then it is easy to check that (27) is unsatisfiable if and only if

$$\Gamma \cup \{\forall y \neg \varphi(\vec{c}, y)\}$$

is unsatisfiable. (We will need only the ONLY IF ($\Longrightarrow$) direction.)

Now by Form 1 (Theorem II.5.4), there are terms $t_1(\vec{c}), \ldots, t_n(\vec{c})$ such that

$$\Gamma \cup \{\neg \varphi(\vec{c}, t_1(\vec{c})), \ldots, \neg \varphi(\vec{c}, t_n(\vec{c}))\}$$

is unsatisfiable. (We can assume that $n \geq 1$, since $n = 0$ implies that Γ is itself unsatisfiable, and in that case the theorem is vacuously true.) Then it follows easily that

$$T \vdash \forall x_1 \ldots \forall x_m \left(\varphi(\vec{x}, t_1(\vec{x})) \vee \cdots \vee \varphi(\vec{x}, t_n(\vec{x})) \right). \qquad □$$

As stated, the Herbrand Theorem applies only to universal theories. However every theory has a universal conservative extension, which can be obtained by introducing "Skolem functions". The idea is that these functions explicitly *witness* the existence of existentially quantified variables. Thus we can replace each axiom (which contains $\exists$) of a theory T by a universal axiom.

LEMMA III.3.14. *Suppose that $\psi(\vec{x}) \equiv \exists y \varphi(\vec{x}, y)$ is an axiom of a theory T. Let f be a new function symbol, and let T' be the theory over the extended vocabulary with the same set of axioms as T except that $\psi(\vec{x})$ is replaced by*

$$\varphi(\vec{x}, f(\vec{x})).$$

Then T' is a conservative extension of T.

The new function f is called a *Skolem function*.

EXERCISE III.3.15. Prove the above lemma by a simple model-theoretic argument showing that every model of T can be expanded to a model of T'. It may be helpful to assume that the vocabulary of T is countable, so by the Löwenheim/Skolem Theorem (Theorem II.4.1) we may restrict attention to countable models.

By the lemma, for each axiom of T we can successively eliminate the existential quantifiers, starting from the outermost quantifier, using the Skolem functions. It follows that every theory has a universal conservative extension. For example, we can obtain a universal conservative extension of $I\Delta_0$ by introducing Skolem functions for every instance of the Δ_0-*IND* axiom scheme. Let $\varphi(z)$ be a Δ_0 formula (possibly with other free variables $\vec{x}$). Then the induction scheme for $\varphi(z)$ can be written as

$$\forall \vec{x} \forall z \big(\varphi(z) \vee \neg\varphi(0) \vee \exists y (\varphi(y) \wedge \neg\varphi(y+1))\big).$$

Consider the simple case where φ is an open formula. The single Skolem function (as a function of $\vec{x}, z$) for the above formula is required to "witness" the existence of y (in case such a y exists).

Although the Skolem functions witness the existence of existentially quantified variables, it is not specified which values they take (and in general there may be many different values). Here we can construct a universal conservative extension of $I\Delta_0$ by explicitly taking the smallest values of the witnesses if they exist. Using the least number principle (Definition III.1.13), these functions are indeed definable in $I\Delta_0$.

Let $\varphi(z)$ be an open formula (possibly with other free variables), and t a term. Let $\vec{x}$ be the list of all variables of t and other free variables of $\varphi(z)$ (thus $\vec{x}$ may contain z if t does). Let $f_{\varphi(z),t}(\vec{x})$ be the least $y < t$ such that $\varphi(y)$ holds, or t if no such y exists. Then $f_{\varphi(z),t}$ is total and can be defined as follows (we assume that y, v do not appear in $\vec{x}$):

$$y = f_{\varphi(z),t}(\vec{x}) \leftrightarrow \big(y \leq t \wedge (y < t \supset \varphi(y)) \wedge \forall v < y \neg\varphi(v)\big). \qquad (28)$$

Note that (28) contains an implicit existential quantifier $\exists v$ (consider the direction $\leftarrow$). Our universal theory will contain the following equivalent axiom instead:

$$f(\vec{x}) \le t \wedge \left(f(\vec{x}) < t \supset \varphi(f(\vec{x}))\right) \wedge \left(v < f(\vec{x}) \supset \neg\varphi(v)\right) \qquad (29)$$

(here $f = f_{\varphi(z),t}$).

A consequence of (29) is

$$\exists z \le t\varphi(z) \leftrightarrow \varphi(f_{\varphi(z),t}(\vec{x}))$$

so introduction of the function symbols $f_{\varphi(z),t}$ allows us to eliminate bounded quantifiers (Lemma III.3.19).

Although the predecessor function $pd(x)$ can be defined by a formula of the form (29), we will use the following two recursive defining axioms instead.

D1′. $pd(0) = 0$;
D1″. $x \ne 0 \supset pd(x) + 1 = x$.

Note that **D1″** implies **D1** (see Example III.1.9), and **D1′** is needed to define $pd(0)$.

We are now ready to define the vocabulary $\mathcal{L}_{\Delta_0}$ of the universal theory $\overline{I\Delta_0}$. This vocabulary has a function symbol for every Δ_0-definable function in $I\Delta_0$.

DEFINITION III.3.16 ($\mathcal{L}_{\Delta_0}$). Let $\mathcal{L}_{\Delta_0}$ be the smallest set that satisfies

1) $\mathcal{L}_{\Delta_0}$ includes $\mathcal{L}_A \cup \{pd\}$;
2) For each open $\mathcal{L}_{\Delta_0}$-formula $\varphi(z)$ and $\mathcal{L}_A$-term t there is a function $f_{\varphi(z),t}$ in $\mathcal{L}_{\Delta_0}$.

Note that $\mathcal{L}_{\Delta_0}$ can be alternatively defined as follows. Let

$$\mathcal{L}_0 = \mathcal{L}_A \cup \{pd\},$$

$$\text{for } n \ge 0: \mathcal{L}_{n+1} = \mathcal{L}_n \cup \{f_{\varphi(z),t} : \varphi(z) \text{ is an open } \mathcal{L}_n\text{-formula},$$

$$t \text{ is an } \mathcal{L}_A\text{-term}\}.$$

Then

$$\mathcal{L}_{\Delta_0} = \bigcup_{n \ge 0} \mathcal{L}_n.$$

Our universal theory $\overline{I\Delta_0}$ requires two more axioms in the style of 1-***BASIC***.

B8′. $0 \le x$;
B8″. $x < x + 1$.

DEFINITION III.3.17 ($\overline{I\Delta_0}$). Let $\overline{I\Delta_0}$ be the theory over $\mathcal{L}_{\Delta_0}$ with the following set of axioms: **B1**, ... , **B8**, **B8′**, **B8″**, **D1′**, **D1″** and (29) for each function $f_{\varphi(z),t}$ of $\mathcal{L}_{\Delta_0}$.

Thus $\overline{I\Delta_0}$ is a universal theory. Note that there is no induction scheme among its axioms. Nevertheless we show below that $\overline{I\Delta_0}$ proves the Δ_0-*IND* axiom scheme, and hence $\overline{I\Delta_0}$ extends $I\Delta_0$. From this it is easy to verify that $\overline{I\Delta_0}$ is a polynomial-bounded theory.

THEOREM III.3.18. $\overline{I\Delta_0}$ *is a conservative extension of* $I\Delta_0$.

To show that $\overline{I\Delta_0}$ extends $I\Delta_0$ we show that it proves the Δ_0-*IND* axiom scheme. Note that if the functions of $\mathcal{L}_{\Delta_0}$ receive their intended meaning, then every bounded $\mathcal{L}_A$-formula is equivalent to an open $\mathcal{L}_{\Delta_0}$-formula. Therefore, roughly speaking, the Δ_0-*MIN* (and thus Δ_0-*IND*) axiom scheme is satisfied by considering the appropriate functions of $\mathcal{L}_{\Delta_0}$.

LEMMA III.3.19. *For each* $\Delta_0(\mathcal{L}_A)$ *formula* φ, *there is an open* $\mathcal{L}_{\Delta_0}$-*formula* φ' *such that* $\overline{I\Delta_0} \vdash \varphi \leftrightarrow \varphi'$.

PROOF. We use structural induction on φ. The only interesting cases are for bounded quantifiers. It suffices to consider the case when φ is $\exists y \le t\psi(y)$. Then take φ' to be $\psi'(f_{\psi',t}(\vec{x}))$. It is easy to check that $\overline{I\Delta_0} \vdash \varphi \leftrightarrow \varphi'$ using (29). No properties of $\le$ and $<$ are needed for this implication except the definition $y < f(\vec{x})$ stands for $(y \le f(\vec{x}) \wedge y \ne f(\vec{x}))$. $\qquad\square$

PROOF OF THEOREM III.3.18. First we show that $\overline{I\Delta_0}$ is an extension of $I\Delta_0$, i.e., Δ_0-*IND* is provable in $\overline{I\Delta_0}$.

By the above lemma, it suffices to show that $\overline{I\Delta_0}$ proves the Induction axiom scheme for open $\mathcal{L}_{\Delta_0}$-formulas. Let $\varphi(\vec{x}, z)$ be any open $\mathcal{L}_{\Delta_0}$-formula. We need to show that (omitting $\vec{x}$)

$$\overline{I\Delta_0} \vdash (\varphi(0) \wedge \neg\varphi(z)) \supset \exists y(\varphi(y) \wedge \neg\varphi(y+1)).$$

Assuming $(\varphi(0) \wedge \neg\varphi(z))$, we show in $\overline{I\Delta_0}$ that $(\varphi(y) \wedge \neg\varphi(y+1))$ holds for $y = pd(f_{\neg\varphi,z}(\vec{x}, z))$, using (29). We need to be careful when arguing about $\le$, because the properties **O1**–**O9** and **D1**–**D10** which we have been using for reasoning in $I\Delta_0$ require induction to prove.

First we rewrite (29) for the case f is $f_{\neg\varphi,z}$.

$$f(\vec{x}, z) \le z \wedge (f(\vec{x}, z) < z \supset \neg\varphi(f(\vec{x}, z))) \wedge (v < f(\vec{x}, z) \supset \varphi(v)). \tag{30}$$

Now $0 < z$ by **B8'** and our assumptions $\varphi(0)$ and $\neg\varphi(z)$, so $f(\vec{x}, z) \ne 0$ by (30). Hence $y + 1 = pd(f(\vec{x}, z)) + 1 = f(\vec{x}, z)$ by **D1''**. Therefore $\neg\varphi(y+1)$ by (30) and the assumption $\neg\varphi(z)$.

To establish $\varphi(y)$ it suffices by (30) to show $y < f(\vec{x}, z)$. This holds because $f(\vec{x}, z) = y + 1$ as shown above, and $y < y + 1$ by **B8''**.

This completes the proof that $\overline{I\Delta_0}$ extends $I\Delta_0$. Next, we show that $\overline{I\Delta_0}$ is conservative over $I\Delta_0$. Let $f_1 = pd, f_2, f_3, \ldots$ be an enumeration of $\mathcal{L}_{\Delta_0} \setminus \mathcal{L}_A$ such that for $n \ge 1$, f_{n+1} is defined using some $\mathcal{L}_A$-term t and $(\mathcal{L}_A \cup \{f_1, \ldots, f_n\})$-formula φ as in (29).

For $n \geq 0$ let $\mathcal{L}_n$ denote $\mathcal{L}_A \cup \{f_1, \ldots, f_n\}$. Let $T_0 = I\Delta_0$, and for $n \geq 0$ let T_{n+1} be the theory over $\mathcal{L}_{n+1}$ which is obtained from T_n by adding the defining axiom for f_{n+1} (in particular, T_1 is axiomatized by $I\Delta_0$ and **D1′**, **D1″**). Then

$$T_0 = I\Delta_0 \subset T_1 \subset T_2 \subset \cdots \qquad \text{and} \qquad \overline{I\Delta_0} = \bigcup_{n \geq 0} T_n.$$

By Corollary III.3.6, it suffices to show that for each $n \geq 0$, f_{n+1} is definable in T_n. In fact, we prove the following by induction on $n \geq 0$:

1) T_n proves the $\Delta_0(\mathcal{L}_n)$-**IND** axiom scheme;
2) f_{n+1} is $\Delta_0(\mathcal{L}_n)$-definable in T_n.

Consider the induction step. Suppose that the hypothesis is true for n ($n \geq 0$). We prove it for $n + 1$. By the induction hypothesis, T_n proves the $\Delta_0(\mathcal{L}_n)$-**IND** axiom scheme and $\Delta_0(\mathcal{L}_n)$-defines f_{n+1}. Therefore by Corollary III.3.11, T_{n+1} proves the $\Delta_0(\mathcal{L}_{n+1})$-**IND** axiom scheme. Consequently, T_{n+1} also proves the $\Delta_0(\mathcal{L}_{n+1})$-**MIN** axiom scheme. The defining equation for f_{n+2} has the form (29), and hence T_{n+1} proves (28) where f is f_{n+2}. Thus (28) is a defining axiom which shows that f_{n+2} is $\Delta_0(\mathcal{L}_{n+1})$-definable in T_{n+1}. Here we use the $\Delta_0(\mathcal{L}_{n+1})$-**MIN** axiom scheme to prove $\exists y$ in (24). □

III.3.2.1. *An alternative proof of Parikh's Theorem for $I\Delta_0$.* Now we will present an alternative proof of Parikh's Theorem for $I\Delta_0$ from Herbrand Theorem applied to $\overline{I\Delta_0}$, using the fact that $\overline{I\Delta_0}$ is a conservative extension of $I\Delta_0$.

In proving that $\overline{I\Delta_0}$ is conservative over $I\Delta_0$ (see the proof of Theorem III.3.18), in the induction step we have used Corollary III.3.11 (the case of adding Σ_1-definable function) to show that T_n proves the $\Delta_0(\mathcal{L}_n)$-**IND** axiom scheme. The proof of Corollary III.3.11 (and of the Conservative Extension Lemma) in turn relies on the Bounded Definability Theorem III.3.8, which is proved using Parikh's Theorem. However, for $\overline{I\Delta_0}$, the function f_{n+1} in the induction step in the proof of Theorem III.3.18 is already Δ_0-definable in T_n and comes with a bounding term t. Therefore we have actually used only a simple case of Corollary III.3.11 (i.e., adding Δ_0-definable functions with bounding terms). Thus in fact Parikh's Theorem is not necessary in proving Theorem III.3.18.

PROOF OF PARIKH'S THEOREM. Suppose that $\forall \vec{x} \exists y \varphi(\vec{x}, y)$ is a theorem of $I\Delta_0$, where φ is a bounded formula. We will show that there is an $\mathcal{L}_A$-term s such that

$$I\Delta_0 \vdash \forall \vec{x} \exists y \leq s \varphi(\vec{x}, y).$$

By Lemma III.3.19, there is an open $\mathcal{L}_{\Delta_0}$-formula $\varphi'(\vec{x}, y)$ such that

$$\overline{I\Delta_0} \vdash \forall \vec{x} \forall y (\varphi(\vec{x}, y) \leftrightarrow \varphi'(\vec{x}, y)).$$

Then since $\overline{I\Delta_0}$ extends $I\Delta_0$, it follows that

$$\overline{I\Delta_0} \vdash \forall \vec{x} \exists y \varphi'(\vec{x}, y).$$

Now since $\overline{I\Delta_0}$ is a universal theory, by Form 2 of the Herbrand Theorem III.3.13 there are $\mathcal{L}_{\Delta_0}$-terms $t_1, \ldots, t_n$ such that

$$\overline{I\Delta_0} \vdash \forall \vec{x} \big(\varphi'(\vec{x}, t_1(\vec{x})) \vee \cdots \vee \varphi'(\vec{x}, t_n(\vec{x})) \big). \tag{31}$$

Also since $\overline{I\Delta_0}$ is a polynomial-bounded theory, there is an $\mathcal{L}_A$-term s such that

$$\overline{I\Delta_0} \vdash t_i(\vec{x}) < s(\vec{x}) \qquad \text{for all } i, \, 1 \leq i \leq n.$$

Consequently,

$$\overline{I\Delta_0} \vdash \forall \vec{x} \exists y < s \varphi'(\vec{x}, y).$$

Hence

$$\overline{I\Delta_0} \vdash \forall \vec{x} \exists y < s \varphi(\vec{x}, y).$$

By the fact that $\overline{I\Delta_0}$ is conservative over $I\Delta_0$ we have

$$I\Delta_0 \vdash \forall \vec{x} \exists y < s \varphi(\vec{x}, y). \qquad \square$$

Note that we have proved more than a bound on the existential quantifier $\exists y$. In fact, (31) allows us to explicitly define a Skolem function $y = f(\vec{x})$, using definition by cases. This idea will serve as a method for proving witnessing theorems in future chapters.

III.3.3. Defining $y = 2^x$ and $BIT(i, x)$ in $I\Delta_0$. In this subsection we show that the relation $BIT(i, x)$ is Δ_0-definable in $I\Delta_0$, where $BIT(i, x)$ holds iff the i-th bit in the binary notation for x is 1. This is useful particularly in Section III.4 where we show that $I\Delta_0$ characterizes the Linear Time Hierarchy.

In order to define BIT we will show that the relation $y = 2^x$ is Δ_0-definable in $I\Delta_0$. On the other hand, by Parikh's Theorem III.2.3, the *function* $f(x) = 2^x$ is not Σ_1-definable in $I\Delta_0$, because it grows faster than any polynomial.

Our method is to introduce a sequence of new function and predicate symbols, and show that each can be Δ_0-defined in $I\Delta_0$ extended by the previous symbols. These new symbols together with their defining axioms determine a sequence of conservative extensions of $I\Delta_0$, and according to the Conservative Extension Theorem III.3.12, bounded formulas using the new symbols are provably equivalent to bounded formulas in the vocabulary $\mathcal{L}_A$ of $I\Delta_0$, and hence the induction scheme is available on bounded formulas with the new symbols. Finally the bounded formula $\varphi_{exp}(x, y)$ given in (34) defines $(y = 2^x)$, and the bounded formula $BIT(i, x)$ given in (35) defines the BIT predicate. These formulas are provably equivalent to bounded formulas in $I\Delta_0$, and $I\Delta_0$ proves the properties of their translations, such as those in Exercise III.3.28.

We start by Δ_0-defining the following functions in $I\Delta_0$: $x \dot{-} y$, $\lfloor x/y \rfloor$, $x \bmod y$ and $\lfloor \sqrt{x} \rfloor$. We will show in detail that $x \dot{-} y$ is Δ_0-definable in $I\Delta_0$. A detailed proof for other functions is left as an exercise. It might be helpful to revisit the basic properties **O1**, ... , **O10**, **D1**, ... , **D10** of $I\Delta_0$ in Examples III.1.8, III.1.9.

1) *Limited subtraction.* The function $x \dot{-} y = max\{0, x - y\}$ can be defined by

$$z = x \dot{-} y \leftrightarrow \big((y + z = x) \vee (x \leq y \wedge z = 0)\big).$$

In order to show that $I\Delta_0$ can Δ_0-define this function we must show that

$$I\Delta_0 \vdash \forall x \forall y \exists! z \varphi(x, y, z)$$

where φ is the RHS of the above equivalence (see Definition III.3.2 (b)).

For the existence of z, by **D2** we know that there is some z' such that

$$x + z' = y \vee y + z' = x.$$

If $y + z' = x$ then simply take $z = z'$. Otherwise $x + z' = y$, then by **B8**, $x \leq x + z'$, hence $x \leq y$, and thus we can take $z = 0$.

For the uniqueness of z, first suppose that $x \leq y$. Then we have to show that $y + z = x \supset z = 0$. Assume $y + z = x$. By **B8**, $y \leq y + z$, hence $y \leq x$. Therefore $x = y$ by **B7**. Now from $x + 0 = x$ (**B3**) and $x + z = x$ we have $z = 0$, by **O2** (Commutativity of $+$) and **O6** (Cancellation law for $+$).

Next, suppose that $\neg(x \leq y)$. Then $y + z = x$, and by **O2** and **O6**, $y + z = x \wedge y + z' = x \supset z = z'$.

2) *Division.* The function $x \operatorname{div} y = \lfloor x/y \rfloor$ can be defined by

$$z = \lfloor x/y \rfloor \leftrightarrow \big((y \cdot z \leq x \wedge x < y \cdot (z + 1)) \vee (y = 0 \wedge z = 0)\big).$$

The existence of z is proved by induction on x. The uniqueness of z follows from transitivity of $\leq$ (**D4**), Total Order (**D5**), and **O5**, **D7**.

3) *Remainder.* The function $x \bmod y$ can be defined by

$$x \bmod y = x \dot{-} (y \cdot \lfloor x/y \rfloor).$$

Since $x \bmod y$ is a composition of Σ_1-definable functions, it is Σ_1-definable by Exercise III.3.3. Hence it is Δ_0-definable by Corollary III.3.9.

4) *Square root.*

$$y = \lfloor \sqrt{x} \rfloor \leftrightarrow \big(y \cdot y \leq x \wedge x < (y + 1)(y + 1)\big).$$

The existence of y follows from the least number principle. The uniqueness of y follows from Transitivity of $\leq$ (**D4**), Total Order (**D5**), and **O5**, **D7**.

EXERCISE III.3.20. Show carefully that the functions $\lfloor x/y \rfloor$ and $\lfloor \sqrt{x} \rfloor$ are Δ_0-definable in $I\Delta_0$.

Next we define the following relations $x|y$, $Pow2(x)$, $Pow4(x)$ and $LenBit(y, x)$:

5) *Divisibility.* This relation is defined by
$$x|y \leftrightarrow \exists z \leq y (x \cdot z = y).$$

6) *Powers of 2 and 4.*
   x is a power of 2:
   $$Pow2(x) \leftrightarrow (x \neq 0 \wedge \forall y \leq x((1 < y \wedge y|x) \supset 2|y)),$$
   x is a power of 4: $Pow4(x) \leftrightarrow (Pow2(x) \wedge x \bmod 3 = 1)$.

7) *LenBit.* We want the relation $LenBit(2^i, x)$ to hold iff the i-th bit in the binary expansion of x is 1, where the least significant bit is bit 0. Although we cannot yet define $y = 2^i$, we can define
   $$LenBit(y, x) \leftrightarrow (\lfloor x/y \rfloor \bmod 2 = 1).$$

Note that we intend to use $LenBit(y, x)$ only when y is a power of 2, but it is defined for all values of y.

NOTATION. $(\forall 2^i)$ stands for "for all powers of 2", i.e.,

$(\forall 2^i)\, A(2^i)$ stands for $\forall x\, (Pow2(x) \supset A(x))$,

$(\forall 2^i \leq t)\, A(2^i)$ stands for $\forall x\, ((Pow2(x) \wedge x \leq t) \supset A(x))$.

Same for $(\exists 2^i)$ and $(\exists 2^i \leq t)$.

EXERCISE III.3.21. Show that the following are theorems of $I\Delta_0$:
(a) $Pow2(x) \leftrightarrow Pow2(2x)$.
(b) $(\forall 2^i)(\forall 2^j)(2^i < 2^j \supset 2^i|2^j)$. (Hint: using strong induction (12).)
(c) $(\forall 2^i)(\forall 2^j \leq 2^i)\, Pow2(\lfloor 2^i/2^j \rfloor)$.
(d) $(\forall 2^i)(\forall 2^j)(2^i < 2^j \supset 2 \cdot 2^i \leq 2^j)$.
(e) $(\forall 2^i)(\forall 2^j)\, Pow2(2^i \cdot 2^j)$.
(f) $(\forall 2^i)(\exists 2^j \leq 2^i)\, ((2^j)^2 = 2^i \vee 2(2^j)^2 = 2^i))$.

We also need the following function:

8) *Greatest power of 2 less than or equal to x.*

$y = gp(x) \leftrightarrow$
$$((x = 0 \wedge y = 0) \vee (Pow2(y) \wedge y \leq x \wedge (\forall 2^i \leq x)\, 2^i \leq y)).$$

EXERCISE III.3.22. Show that $I\Delta_0$ can Δ_0-define $gp(x)$. (Hint: Use induction on x.)

EXERCISE III.3.23. Prove the following in $I\Delta_0$:
(a) $x > 0 \supset (gp(x) \leq x < 2gp(x))$.
(b) $x > 0 \supset LenBit(gp(x), x)$.

(c) $y = x \mathbin{\dot{-}} gp(x) \supset (\forall 2^i \leq y)\,(LenBit(2^i, y) \leftrightarrow LenBit(2^i, x))$.

It is a theorem of $I\Delta_0$ that the binary representation of a number uniquely determines the number. This theorem can be proved in $I\Delta_0$ by using strong induction (12) and part (c) of the above exercise. Details are left as an exercise.

THEOREM III.3.24.

$$I\Delta_0 \vdash \forall y \forall x < y (\exists 2^i \leq y)(LenBit(2^i, y) \wedge \neg LenBit(2^i, x)).$$

EXERCISE III.3.25. Prove the above theorem.

III.3.3.1. *Defining the Relation $y = 2^x$.* This is much more difficult to Δ_0-define than any of the previous relations and functions. A first attempt to define $y = 2^x$ might be to assert the existence of a number s coding the sequence $\langle 2^0, 2^1, \ldots, 2^x \rangle$. The main difficulty in this attempt is that the number of bits in s is $\Omega(|y|^2)$ (where $|y|$ is the number of bits in y), and so s cannot be bounded by any $I\Delta_0$ term in x and y.

We get around this by coding a much shorter sequence, of length $|x|$ instead of length x, of numbers of the form 2^z. Suppose that $x > 0$, and $(x_{k-1} \ldots x_0)_2$ is the binary representation of x (where $x_{k-1} = 1$), i.e.,

$$x = \sum_{i=0}^{k-1} x_i 2^i \qquad (\text{and } x_{k-1} = 1).$$

We start by coding the sequence $\langle a_1, a_2, \ldots, a_k \rangle$, where a_i consists of the first i high-order bits of x, so $a_k = x$. Then we code the sequence $\langle b_1, \ldots, b_k \rangle$, where $b_i = 2^{a_i}$, so $y = b_k$.

We have (note that $x_{k-1} = 1$):

$$a_1 = 1, \qquad b_1 = 2.$$
$$\text{For } 1 \leq i < k: a_{i+1} = x_{k-i-1} + 2a_i, \qquad b_{i+1} = 2^{x_{k-i-1}} b_i^2. \tag{32}$$

Note that $a_i < 2^i$ and $b_i < 2^{2^i}$ for $1 \leq i \leq k$.

We will code the sequences $\langle a_1, \ldots, a_k \rangle$ and $\langle b_1, \ldots, b_k \rangle$ by the numbers a and b, respectively, such that a_i and b_i are represented by the bits 2^i to $2^{i+1} - 1$ of a and b, respectively. In order to extract a_i and b_i from a and b we use the function

$$ext(u, z) = \lfloor z/u \rfloor \bmod u. \tag{33}$$

Thus if $u = 2^{2^i}$ then $a_i = ext(u, a)$ and $b_i = ext(u, b)$. It is easy to see that the function ext is Δ_0-definable in $I\Delta_0$.

Note that $a, b < 2^{2^{k+1}}$, and $y \geq 2^{2^{k-1}}$. Hence the numbers a and b can be bounded by $a, b < y^4$. Below we will explain how to express the condition that a number has the form 2^{2^i}. Once this is done, we can express

$$y = 2^x \leftrightarrow \varphi_{exp}(x, y) \quad \text{where} \quad \varphi_{exp} \equiv (x = 0 \wedge y = 1) \vee$$
$$\exists a, b < y^4 \psi_{exp}(x, y, a, b) \quad (34)$$

and $\psi_{exp}(x, y, a, b)$ is the formula stating that the following conditions (expressing the above recurrences) hold, for $x > 0$, $y > 1$:

1) $ext(2^{2^1}, a) = 1$, and $ext(2^{2^1}, b) = 2$.
2) For all u, $2^{2^1} \leq u \leq y$ of the form 2^{2^i}, either
   (a) $ext(u^2, a) = 2ext(u, a)$ and $ext(u^2, b) = (ext(u, b))^2$, or
   (b) $ext(u^2, a) = 1 + 2ext(u, a)$ and $ext(u^2, b) = 2(ext(u, b))^2$.
3) There is $u \leq y^2$ of the form 2^{2^i} such that $ext(u, a) = x$ and $ext(u, b) = y$.

Note that condition (2)(a) holds if $x_{k-i} = 0$, and condition (2)(b) holds if $x_{k-i} = 1$. The conditions do not need to mention x_{k-i} explicitly, because condition (3) ensures that $a_i = x$ for some i, so all bits of x must have been chosen correctly up to this point.

It remains to express "x has the form 2^{2^i}". First, the set of numbers of the form

$$m_\ell = \sum_{i=0}^{\ell} 2^{2^i}$$

can be Δ_0-defined by the formula

$$\varphi_p(x) \equiv \neg LenBit(1, x) \wedge LenBit(2, x) \wedge \forall 2^i \leq x, \ 2 < 2^i \supset$$
$$(LenBit(2^i, x) \leftrightarrow (Pow4(2^i) \wedge LenBit(\lfloor \sqrt{2^i} \rfloor, x))).$$

From this we can Δ_0-define numbers of the form $x = 2^{2^i}$ as the powers of 2 for which $LenBit(x, m_\ell)$ holds for some $m_\ell < 2x$:

$$x \text{ is of form } 2^{2^i}: PPow2(x) \leftrightarrow Pow2(x) \wedge \exists m < 2x \ (\varphi_p(m) \wedge$$
$$LenBit(x, m)).$$

This completes our description of the defining axiom $\varphi_{exp}(x, y)$ for the relation $y = 2^x$. It remains to show that $I\Delta_0$ proves some properties of this relation. First we need to verify in $I\Delta_0$ the properties of $PPow2$.

EXERCISE III.3.26. The following are theorems of $I\Delta_0$:

(a) $PPow2(z) \leftrightarrow PPow2(z^2)$.
(b) $(PPow2(z) \wedge PPow2(z') \wedge z < z') \supset z^2 \leq z'$.
(c) $(PPow2(x) \wedge 4 \leq x) \supset \lfloor \sqrt{x} \rfloor^2 = x$.

We have noted earlier that $a_i < 2^i$ and $b_i < 2^{2^i}$. Here we need to show that these are indeed provable in $I\Delta_0$. We will need this fact in order to prove (in $I\Delta_0$) the correctness of our defining axiom φ_{exp} for the relation $y = 2^x$ (e.g., Exercise III.3.28 (c) and (d)).

EXERCISE III.3.27. Assuming $(y > 1 \wedge \psi_{exp}(x, y, a, b))$, show in $I\Delta_0$ that
(a) $\forall u \leq y^2,\ (PPow2(u) \wedge 4 \leq u) \supset 1 + ext(u, a) < u.$
(b) $\forall u \leq y^2,\ (PPow2(u) \wedge 4 \leq u) \supset 2ext(u, b) \leq u.$

EXERCISE III.3.28. Show that $I\Delta_0$ proves the following:
(a) $\varphi_{exp}(x, y) \supset Pow2(y).$
(b) $Pow2(y) \supset \exists x < y\ \varphi_{exp}(x, y).$ (Hint: strong induction on y, using Exercise III.3.21 (f).)
(c) $\varphi_{exp}(x, y_1) \wedge \varphi_{exp}(x, y_2) \supset y_1 = y_2.$
(d) $\varphi_{exp}(x_1, y) \wedge \varphi_{exp}(x_2, y) \supset x_1 = x_2.$
(e) $\varphi_{exp}(x + 1, 2y) \leftrightarrow \varphi_{exp}(x, y).$ (Hint: Look at the least significant 0 bit of x.)
(f) $\varphi_{exp}(x_1, y_1) \wedge \varphi_{exp}(x_2, y_2) \supset \varphi_{exp}(x_1 + x_2, y_1 \cdot y_2).$ (Hint: Induction on y_2.)

Although the function 2^x is not Δ_0-definable in $I\Delta_0$, it is easy to see using φ_{exp} (and useful to know) that the function

$$Exp(x, y) = \min(2^x, y)$$

is Δ_0-definable in $I\Delta_0$.

EXERCISE III.3.29. The relation $y = z^x$ can be defined using the same techniques that have been used to define the relation $y = 2^x$. Here the sequence $\langle b_1, \ldots, b_k \rangle$ needs to be modified.
(a) Modify the recurrence in (32).
Each b_i now may not fit in the bits 2^i to $2^{i+1} - 1$ of b, but it fits in a bigger segment of b. Let ℓ be the least number such that

$$z \leq 2^{2^\ell}.$$

(b) Show that for $1 \leq i \leq k$, $zb_i \leq 2^{2^{\ell+i}}.$
(c) Show that the function $lpp(z)$, which is the least number of the form 2^{2^i} that is $\geq z$, is Δ_0-definable in $I\Delta_0$.
(d) Show that $I\Delta_0 \vdash z > 1 \supset (z \leq lpp(z) < z^2).$
(e) What are the bounds on the values of the numbers a and b that respectively code the sequences $\langle a_1, \ldots, a_k \rangle$ and $\langle b_1, \ldots, b_k \rangle$?
(f) Give a formula that defines the relation $y = z^x$ by modifying the conditions 1–3.

III.3.3.2. *The BIT and NUMONES Relations.* The relation $BIT(i, x)$ can be defined as follows, where $BIT(i, x)$ holds iff the i-th bit (i.e., coefficient of 2^i) of the binary notation for x is 1:

$$BIT(i, x) \leftrightarrow \exists z \leq x(z = 2^i \wedge LenBit(z, x)). \qquad (35)$$

EXERCISE III.3.30. Show that the length function, $|x| = \lceil \log_2(x + 1) \rceil$, is Δ_0-definable in $I\Delta_0$.

LEMMA III.3.31. *The relation $NUMONES(x, y)$, asserting that y is the number of one-bits in the binary notation for x, is Δ_0-definable.*

PROOF SKETCH. We code a sequence $\langle s_0, s_1, \ldots, s_n \rangle$ of numbers s_i of at most ℓ bits each using a number s such that bits $i\ell$ to $i\ell + \ell - 1$ of s are the bits of s_i. Then we can extract s_i from s using the equation

$$s_i = \lfloor s/2^{i\ell} \rfloor \bmod 2^{\ell}.$$

Our first attempt to define $NUMONES(x, y)$ might be to state the existence of a sequence $\langle s_0, s_1, \ldots, s_n \rangle$, where $n = |x|$, s_i is the number of ones in the first i bits of x, and $\ell = ||x||$. However the number coding this sequence has $n \log n$ bits, which is too many.

We get around this problem using "Bennett's Trick" [15], which is to state the existence of a sparse subsequence of $\langle s_0, s_1, \ldots, s_n \rangle$ and assert that adjacent pairs in the subsequence can be filled in. Thus

$$NUMONES(x, y) \leftrightarrow \exists m \leq |x|\big(|x| \leq m^2 \wedge$$

$$\exists \langle t_0, \ldots, t_m \rangle \big(t_0 = 0 \wedge t_m = y \wedge \forall i < m \exists \langle u_0, \ldots, u_m \rangle (u_0 = t_i \wedge$$

$$u_m = t_{i+1} \wedge \forall j < m(u_{j+1} = u_j + FBIT(im + j, x)))\big)\big)$$

where the function $FBIT(i, x)$ is bit i of x. $\square$

III.4. $I\Delta_0$ and the Linear Time Hierarchy

III.4.1. The Polynomial and Linear Time Hierarchies.
An element of a complexity class such as **P** (polynomial time) is often taken to be a *vocabulary* L, where L is a set of finite strings over some fixed finite alphabet Σ. In the context of bounded arithmetic, it is convenient to consider elements of **P** to be subsets of $\mathbb{N}$, or more generally relations over $\mathbb{N}$, and in this case it is assumed that numbers are presented in binary notation to the accepting machine. In this context, the notation Σ_0^p is sometimes used for polynomial time. Thus $\Sigma_0^p = \boldsymbol{P}$ is the set of all relations $R(x_1, \ldots, x_k), k \geq 1$ over $\mathbb{N}$ such that some polynomial time Turing machine M_R, given input $x_1, \ldots, x_k$ (k numbers in binary notation separated by blanks) determines whether $R(x_1, \ldots, x_k)$ holds.

The class Σ_0^p has a generalization to $\Sigma_i^p, i \geq 0$, which is the i-th level of the polynomial-time hierarchy. This can be defined inductively by the recurrence

$$\Sigma_{i+1}^p = NP^{\Sigma_i^p}$$

where $NP^{\Sigma_i^p}$ is the set of relations accepted by a nondeterministic polynomial time Turing machine which has access to an oracle in Σ_i^p.

For $i \geq 1$, Σ_i^p can be characterized as the set of relations accepted by some alternating Turing machine (ATM) in polynomial time, making at most i alternations, beginning with an existential state. In any case,

$$\Sigma_1^p = NP.$$

We define the polynomial time hierarchy by

$$PH = \bigcup_{i=0}^{\infty} \Sigma_i^p.$$

In the context of $I\Delta_0$, we are interested in the Linear Time Hierarchy (LTH), which is defined analogously to PH. We use $NLinTime$ to denote time $O(n)$ on a nondeterministic multi-tape Turing machine. Then

$$\Sigma_1^{lin} = NLinTime \tag{36}$$

and for $i \geq 1$

$$\Sigma_{i+1}^{lin} = NLinTime^{\Sigma_i^{lin}}. \tag{37}$$

Alternatively, we can define Σ_i^{lin} to be the relations accepted in linear time on an ATM with i alternations, beginning with an existential state. In either case,[2]

$$LTH = \bigcup_{i=1}^{\infty} \Sigma_i^{lin}.$$

$LinTime$ is not as robust a class as polynomial time; for example it is plausible that a $k + 1$-tape deterministic linear time Turing machine can accept sets not accepted by any k tape such machine, and linear time Random Access Machines may accept sets not in $LinTime$. However it is not hard to see that $NLinTime$ is more robust, in the sense that every set in this class can be accepted by a two tape nondeterministic linear time Turing machine.

III.4.2. Representability of LTH Relations. Recall the definition of definable predicates and functions (Definition III.3.2). If Φ is a class of $\mathcal{L}$-formulas, $\mathcal{T}$ a theory over $\mathcal{L}$, and R a Φ-definable relation (over the natural numbers) in $\mathcal{T}$, then we simply say that R is Φ-definable (or Φ-representable).

Thus when Φ is a class of $\mathcal{L}_A$-formulas, a k-ary relation R over the natural numbers is Φ-definable if there is a formula $\varphi(x_1, \ldots, x_k) \in \Phi$ such that for all $(n_1, \ldots, n_k) \in \mathbb{N}^k$,

$$(n_1, \ldots, n_k) \in R \quad \text{iff} \quad \underline{\mathbb{N}} \models \varphi(n_1, \ldots, n_k). \tag{38}$$

More generally, if Φ is a class of $\mathcal{L}$-formulas for some vocabulary $\mathcal{L}$ extending $\mathcal{L}_A$, then instead of $\underline{\mathbb{N}}$ we will take the expansion of $\underline{\mathbb{N}}$ where the extra symbols in $\mathcal{L}$ have their intended meaning.

---

[2] LTH is different from LH, the *logtime-hierarchy* discussed in Section IV.1.

(Note that a relation $R(\vec{x})$ is sometimes called *representable* (or *weakly representable*) *in a theory* $\mathcal{T}$ if there is some formula $\varphi(\vec{x})$ so that for all $\vec{n} \in \mathbb{N}$,

$$R(\vec{n}) \qquad \text{iff} \qquad \mathcal{T} \vdash \varphi(\underline{\vec{n}}).$$

Our notation here is the special case where $\mathcal{T} = \boldsymbol{TA}$.)

For example, the class of Σ_1-representable sets (i.e., unary relations) is precisely the class of r.e. sets. In the context of Buss's $\boldsymbol{S_2^i}$ hierarchy (Section III.5), $\boldsymbol{NP}$ relations are precisely the Σ_1^b-representable relations. (Σ_1^b is defined for the vocabulary $\mathcal{L}_{S_2}$ of $\boldsymbol{S_2}$.) Here we show that the $\boldsymbol{LTH}$ relations are exactly the Δ_0-representable relations.

DEFINITION III.4.1. $\Delta_0^{\mathbb{N}}$ is the class of Δ_0-representable relations.

For instance, we have shown that the relations *BIT* and *NUMONES* are in $\Delta_0^{\mathbb{N}}$. So is the relation $Prime(x)$ (x is a prime number), because

$$Prime(x) \equiv 1 < x \wedge \forall y < x \forall z < x (y \cdot z \neq x).$$

THEOREM III.4.2 (*LTH* Theorem). $\boldsymbol{LTH} = \Delta_0^{\mathbb{N}}$.

PROOF SKETCH. First consider the inclusion $\boldsymbol{LTH} \subseteq \Delta_0^{\mathbb{N}}$. This can be done using the recurrence (36), (37). The hard part here is the base case, showing $\boldsymbol{NLinTime} \subseteq \Delta_0^{\mathbb{N}}$. Once this is done we can show the induction step by, given a nondeterministic linear time oracle Turing machine M, defining the relation $R_M(x, y, b)$ to assert "M accepts input x, assuming that it makes the sequence of oracle queries coded by y, and the answers to those queries are coded by b." This relation R_M is accepted by some nondeterministic linear time Turing machine (with no oracle), and hence it is in $\Delta_0^{\mathbb{N}}$ by the base case.

To show $\boldsymbol{NLinTime} \subseteq \Delta_0^{\mathbb{N}}$ we need to represent the computation of a nondeterministic linear time Turing machine by a constant number k of strings $x_1, \ldots, x_k$ of linear length. One string will code the sequence of states of the computation, and for each tape there is a string coding the sequence of symbols printed and another string coding the head moves. In order to check that the computation is correctly encoded it is necessary to deduce the position of each tape head at each step of the computation, from the sequence of head moves. This can be done by counting the number of left shifts and of right shifts, using the relation $NUMONES(x, y)$, and subtracting. It is also necessary to determine the symbol appearing on a given tape square at a given step, and this can be done by determining the last time that the head printed a symbol on that square.

We prove the inclusion $\Delta_0^{\mathbb{N}} \subseteq \boldsymbol{LTH}$ by structural induction on Δ_0 formulas. The induction step is easy, since bounded quantifiers correspond to $\exists$ and $\forall$ states in an ATM. The only interesting case is one of the base cases: the atomic formula $x \cdot y = z$. To show that this relation $R(x, y, z)$ is in $\boldsymbol{LTH}$ we use Corollary III.4.5 below which shows that $\boldsymbol{L} \subseteq \boldsymbol{LTH}$. ($\boldsymbol{L}$ is the class of relations computable in logarithmic space using Turing machines. See Appendix A.1.1.) It is not hard to see that using the school algorithm for multiplication the relation $x \cdot y = z$ can be checked in space $O(\log n)$, and thus it is in $\textbf{\textit{L}}$.   □

EXERCISE III.4.3. Give more details of the proof showing $\textbf{\textit{LTH}} \subseteq \Delta_0^{\mathbb{N}}$.

THEOREM III.4.4 (Nepomnjaščij's Theorem). *Let ε be a rational number, $0 < \varepsilon < 1$, and let a be a positive integer. Then*

$$\textbf{\textit{NTimeSpace}}(n^a, n^\varepsilon) \subseteq \textbf{\textit{LTH}}.$$

In the above, $\textbf{\textit{NTimeSpace}}(f(n), g(n))$ consists of all relations accepted simultaneously in time $O(f(n))$ and space $O(g(n))$ on a *nondeterministic multi-tape* Turing machine.

PROOF IDEA. We use Bennett's Trick, as in the proof of Lemma III.3.31. Suppose we want to show

$$\textbf{\textit{NTimeSpace}}(n^2, n^{0.6}) \subseteq \textbf{\textit{LTH}}.$$

Let M be a nondeterministic TM running in time n^2 and space $n^{0.6}$. Then M accepts an input x iff

$$\exists \vec{y}(\vec{y} \text{ represents an accepting computation for } x).$$

Here $\vec{y} = y_1, \ldots, y_{n^2}$, where each y_i is a string of length $n^{0.6}$ representing a configuration of M. The total length of $\vec{y}$ is $|\vec{y}| = n^{2.6}$, which is too long for an ATM to guess in linear time.

So we guess a vector $\vec{z} = z_1, \ldots, z_n$ representing every n-th string in $\vec{y}$, so now M accepts x iff

$$\exists \vec{z} \forall i < n \exists \vec{u}(\vec{u} \text{ shows } z_{i+1} \text{ follows from } z_i \text{ in } n \text{ steps and } z_n \text{ is accepting}).$$

Now the lengths of $\vec{z}$ and $\vec{u}$ are only $n^{1.6}$, and we have made progress. Two more iterations of this idea (one for the $\exists \vec{y}$, one for the $\exists \vec{u}$; increasing the nesting depth of quantifiers) will get the lengths of the quantified strings below linear.   □

For the following corollary, $\textbf{\textit{NL}}$ is the class of relations computable by *nondeterministic* Turing machines in logarithmic space. See Appendix A.2.

COROLLARY III.4.5. $\textbf{\textit{NL}} \subseteq \textbf{\textit{LTH}}$.

PROOF. We use the fact that $\textbf{\textit{NL}} \subseteq \textbf{\textit{NTimeSpace}}(n^{O(1)}, \log n)$.   □

REMARK. We know

$$\textbf{\textit{L}} \subseteq \textbf{\textit{LTH}} \subseteq \textbf{\textit{PH}} \subseteq \textbf{\textit{PSPACE}}$$

where no two adjacent inclusions are known to be proper, although we know $\textbf{\textit{L}} \subset \textbf{\textit{PSPACE}}$ by a simple diagonal argument.

Also $\textbf{\textit{LTH}} \subseteq \textbf{\textit{LinSpace}} \subset \textbf{\textit{PSPACE}}$, where the first inclusion is not known to be proper. Finally $\textbf{\textit{P}}$ and $\textbf{\textit{LTH}}$ are thought to be incomparable, but no proof is known. In fact it is difficult to find a natural example of a problem in $\textbf{\textit{P}}$ which seems not to be in $\textbf{\textit{LTH}}$.

III.4.3. Characterizing the *LTH* by $I\Delta_0$. First note that *LTH* is a class of *relations*. The corresponding class of functions is defined in terms of *function graphs*. Given a function $f(\vec{x})$, its graph $G_f(\vec{x}, y)$ is the relation

$$G_f(\vec{x}, y) \leftrightarrow (y = f(\vec{x})).$$

DEFINITION III.4.6 (*FLTH*). A function $f : \mathbb{N}^k \to \mathbb{N}$ is in *FLTH* if its graph $G_f(\vec{x}, y)$ is in *LTH* and its length has at most linear growth, i.e.,

$$f(\vec{x}) = (x_1 + \cdots + x_k)^{O(1)}.$$

EXERCISE III.4.7. In future chapters we will define the class of functions associated with a class of relations using the bit graph $B_f(i, \vec{x}, y)$ of f instead of the graph $G_f(\vec{x}, y)$, where

$$B_f(i, \vec{x}) \leftrightarrow BIT(i, f(\vec{x})).$$

Show that the class *FLTH* remains the same if B_f replaces G_f in the above definition.

In general, in order to associate a theory with a complexity class we should show that the functions in the class coincide with the Σ_1-definable functions in the theory. The next result justifies associating the theory $I\Delta_0$ with the complexity class *LTH*.

THEOREM III.4.8 ($I\Delta_0$-Definability). *A function is Σ_1-definable in $I\Delta_0$ iff it is in FLTH.*

PROOF. The $\Longrightarrow$ direction follows from the Bounded Definability Theorem III.3.8, the above definition of *LTH* functions and the *LTH* Theorem III.4.2.

For the $\Longleftarrow$ direction, suppose $f(\vec{x})$ is an *LTH* function. By definition the graph $(y = f(\vec{x}))$ is an *LTH* relation, and hence by the *LTH* Theorem III.4.2 there is a Δ_0 formula $\varphi(\vec{x}, y)$ such that

$$y = f(\vec{x}) \leftrightarrow \varphi(\vec{x}, y).$$

Further, by definition, $|f(\vec{x})|$ is linear bounded, so there is an $\mathcal{L}_A$-term $t(\vec{x})$ such that

$$f(\vec{x}) \le t(\vec{x}). \tag{39}$$

The sentence $\forall \vec{x} \exists! y \varphi(\vec{x}, y)$ is true, but unfortunately there is no reason to believe that it is provable in $I\Delta_0$. We can solve the problem of proving uniqueness by taking the least y satisfying $\varphi(\vec{x}, y)$. In general, for any formula $A(y)$, we define $Min_y[A(y)](y)$ to mean that y is the least number satisfying $A(y)$. Thus

$$Min_y[A(y)](y) \equiv_{def} A(y) \land \forall z < y(\neg A(z)).$$

If $A(y)$ is bounded, then we can apply the least number principle to $A(y)$ to obtain

$$I\Delta_0 \vdash \exists y A(y) \supset \exists! y Min_y[A(y)](y). \tag{40}$$

This solves the problem of proving uniqueness. To prove existence, we modify φ and define

$$\psi(\vec{x}, y) \equiv_{def} (\varphi(\vec{x}, y) \vee y = t(\vec{x}) + 1)$$

where $t(\vec{x})$ is the bounding term from (39). Now define

$$\varphi'(\vec{x}, y) \equiv Min_y[\psi(\vec{x}, y)](\vec{x}, y).$$

Then $\varphi'(\vec{x}, y)$ also represents the relation $(y = f(\vec{x}))$, and since trivially $I\Delta_0$ proves $\exists y \psi(\vec{x}, y)$ we have by (40)

$$I\Delta_0 \vdash \forall \vec{x} \exists ! y \varphi'(\vec{x}, y). \qquad \square$$

III.5. Buss's S_2^i Hierarchy: The Road Not Taken

Buss's PhD thesis *Bounded Arithmetic* (published as a book in 1986, [20]) introduced the hierarchies of bounded theories

$$S_2^1 \subseteq T_2^1 \subseteq S_2^2 \subseteq T_2^2 \subseteq \cdots \subseteq S_2^i \subseteq T_2^i \subseteq \cdots.$$

These theories, whose definable functions are those in the polynomial hierarchy, are of central importance in the area of bounded arithmetic.

We present these theories in detail in Section VIII.8. Here we present a brief overview of the theories S_2^i and T_2^i, and their union $S_2 = T_2 = \bigcup_{i=1}^{\infty} S_2^i$. The idea is to modify the theory $I\Delta_0$ so that the definable functions are those in the polynomial hierarchy as opposed to the Linear Time Hierarchy, and more importantly to introduce the theory S_2^1 whose definable functions are precisely the polynomial time functions. In order to do this, the underlying vocabulary is augmented to include the function symbol #, whose intended interpretation is $x \# y = 2^{|x| \cdot |y|}$. Thus terms in S_2 represent functions which grow at the rate of polynomial time functions, as opposed to the linear-time growth rate of $I\Delta_0$ terms. The full vocabulary for S_2 is

$$\mathcal{L}_{S_2} = [0, S, +, \cdot, \#, |x|, \lfloor \tfrac{1}{2} x \rfloor; =, \leq].$$

(S is the *Successor* function, $|x|$ is the length (of the binary representation) of x).

Sharply bounded quantifiers have the form $\forall x \leq |t|$ or $\exists x \leq |t|$ (where x does not occur in t). These are important because sharply bounded (as opposed to just bounded) formulas represent polynomial time relations (and in fact TC^0 relations). The syntactic class Σ_i^b (b for "bounded") consists essentially of those formulas with at most i blocks of bounded quantifiers beginning with $\exists$, with any number of sharply bounded quantifiers of both kinds mixed in. The formulas in Σ_1^b represent precisely the *NP* relations, and more generally formulas in Σ_i^b represent precisely the relations in the level Σ_i^p in the polynomial hierarchy. In summary, bounded formulas in the vocabulary of S_2 represent precisely the relations in the polynomial hierarchy.

The axioms for T_2^i consist of 32 $\forall$-sentences called **BASIC** which define the symbols of $\mathcal{L}_{S_2}$, together with the Σ_i^b-**IND** scheme. The axioms for S_2^i are the same as those of T_2^i, except for Σ_i^b-**IND** is replaced by the Σ_i^b-**PIND** scheme:

$$\left(\varphi(0) \wedge \forall x (\varphi(\lfloor \tfrac{1}{2} x \rfloor) \supset \varphi(x)) \right) \supset \forall x \varphi(x)$$

where $\varphi(x)$ is any Σ_i^b formula. Note that this axiom scheme is true in $\underline{\mathbb{N}}$. Also for $i \geq 1$, T_2^i proves the Σ_i^b-**PIND** axiom scheme, and S_2^{i+1} proves the Σ_i^b-**IND** axiom scheme. (Thus for $i \geq 1$, $S_2^i \subseteq T_2^i \subseteq S_2^{i+1}$.)

For $i \geq 1$, the functions Σ_i^b-definable in S_2^i are precisely those polytime reducible to relations in Σ_{i-1}^p (level $i - 1$ of the polynomial hierarchy). In particular, the functions Σ_1^b-definable in S_2^1 are precisely the polynomial time functions.

Since S_2 is a polynomial-bounded theory, Parikh's Theorem III.2.3 can be applied to show that all Σ_1-definable functions in S_2 are polynomial time reducible to **PH**. To show that the Σ_1^b-definable functions in S_2^1 are polynomial-time computable requires a more sophisticated "witnessing" argument introduced by Buss. We shall present this argument later in the context of the two-sorted first-order theory V^1.

In Chapters VI and VIII we present two-sorted versions V^i of S_2^i and TV^i of T_2^i. In Section VIII.8 we show that two-sorted versions are essentially equivalent to the originals.

III.6. Notes

The main references for this chapter are [27, 28] and [54, pp. 277–293].

Parikh's Theorem originally appears in [88], and the proof there is based in the Herbrand Theorem, and resembles our "Alternative Proof" given at the end of Section III.3.2. Buss [20] gives a proof based on cut elimination which is closer to our first proof.

James Bennett [15] was the first to show that the relation $y = z^x$ can be defined by Δ_0 formulas. Hájek and Pudlák [54] give a different definition and show how to prove its basic properties in $I\Delta_0$, and give a history of such definitions and proofs. Our treatment of the relations $y = 2^x$ and $BIT(i, x)$ in Section III.3.3 follows that of Buss in [28], simplified with an idea from earlier proofs.

Bennett's Trick, described in the proof of Lemma III.3.31, is due to Bennett [15] Section 1.7, where it is used to show that the rudimentary functions are closed under a form of bounded recursion on notation.

Theorem III.4.2, stating $LTH = \Delta_0^{\mathbb{N}}$, is due to Wrathall [111]. Nepomnjaščij's Theorem III.4.4 appears in [81].

Chapter IV

TWO-SORTED LOGIC AND COMPLEXITY CLASSES

In this chapter we introduce two-sorted first-order logic (sometimes called second-order logic), an extension of the (single-sorted) first-order logic that we use in the previous chapters. The reason for using two-sorted logic is that our theories capture complexity classes defined in terms of Turing machines or Boolean circuits. The inputs to these devices are bit strings, whereas the objects in the universe of discourse in our single-sorted theories are numbers. Although we can code numbers by bit strings using binary notation, this indirection is sometimes awkward, especially for low-level complexity classes. In particular, our single-sorted theories all include multiplication as a primitive operation, but binary multiplication is not in the complexity class AC^0, whose theory V^0 serves as the basis for all our two-sorted theories. Our complexity reductions and completeness notions are generally defined using AC^0 functions.

The two-sorted theories retain the natural numbers as the first sort, and the objects in the second sort are bit strings (precisely, finite sets of natural numbers, whose characteristic vectors are bit strings). We need the first sort (numbers) in order to reason about the second sort. The numbers involved for this reasoning are small; they are used to index bit positions in the second sort (strings). In defining two-sorted complexity classes, the number inputs to the devices are coded in unary notation, and are treated as auxiliary to the main (second) sort, whose elements are coded by binary strings. In particular we use these conventions to define the two-sorted complexity class AC^0. We prove the Σ_0^B Representation Theorem IV.3.6, which states that the set Σ_0^B of two-sorted formulas represent precisely the AC^0 relations.

In Chapters VII and X we show how to translate bounded theorems in our theories into families of propositional proofs. This translation is made especially simple and elegant by using two-sorted theories.

The historical basis for using two-sorted logic to represent complexity classes is descriptive complexity theory, where each object (a language or a relation) in a complexity class is described by a logical formula whose set of finite models corresponds to the object. In the two-sorted logic setting, each object corresponds to the set of interpretations of a variable in the formula satisfying the formula in the standard model.

In the first part of this chapter we present a brief introduction to descriptive complexity theory. (A comprehensive treatment can be found in [59].) Then we introduce two-sorted first-order logic, describe two-sorted complexity classes, and explain how relations in these classes are represented by certain classes of formulas. We revisit the **LTH** theorem for two-sorted logic. We present the sequent calculus **LK²**, the two-sorted version of **LK**. Finally we show how to interpret two-sorted logic into single-sorted logic.

IV.1. Basic Descriptive Complexity Theory

In descriptive complexity theory, an object (e.g. a set of graphs) in a complexity class is specified as the set of all finite models of a given formula. Here we consider the case in which the object is a language $L \subseteq \{0,1\}^*$, and the formula is a formula of the first-order predicate calculus. We assume that the underlying vocabulary consists of

$$\mathcal{L}_{FO} = [0, max; X, BIT, \leq, =] \tag{41}$$

where 0, max are constants, X is a unary predicate symbol, and BIT, $\leq$, $=$ are binary predicate symbols. We consider finite $\mathcal{L}_{FO}$-structures $\mathcal{M}$ in which the universe $M = \{0, \ldots, n-1\}$ for some natural number $n \geq 1$, and max is interpreted by $n-1$. The symbols 0, $=$, $\leq$, and BIT receive their standard interpretations. (Recall that $BIT(i, x)$ holds iff the i-th bit in the binary representation of x is 1. In the previous chapter we showed how to define BIT in $I\Delta_0$, but note that here it is a primitive symbol in $\mathcal{L}_{FO}$.)

Thus the only symbol without a fixed interpretation is the unary predicate symbol X, and to specify a structure it suffices to specify the tuple of truth values $\langle X(0), X(1), \ldots, X(n-1) \rangle$. By identifying $\top$ with 1 and $\bot$ with 0, we see that there is a natural bijection between the set of structures and the set $\{0,1\}^+$ of nonempty binary strings.

The class **FO** (First-Order) of languages describable by $\mathcal{L}_{FO}$ formulas is defined as follows. For each binary string X we denote by $\mathcal{M}[X]$ the structure which is specified by X as above. Then the language $L(\varphi)$ associated with an $\mathcal{L}_{FO}$ sentence φ is the set of strings whose associated structures satisfy φ:

$$L(\varphi) = \{X \in \{0,1\}^+ : \mathcal{M}[X] \models \varphi\}.$$

DEFINITION IV.1.1 (The Class **FO**).

$$FO = \{L(\varphi) : \varphi \text{ is an } \mathcal{L}_{FO}\text{-sentence}\}.$$

For example, let L_{even} be the set of strings whose even positions (starting from the right at position 0) have 1. Then $L_{even} \in \boldsymbol{FO}$, since $L_{even} = L(\varphi)$, where

$$\varphi \equiv \forall y(\neg BIT(0, y) \supset X(y)).$$

To give a more interesting example, we use the fact [59, page 14] that the relation $x + y = z$ can be expressed by a first-order formula $\varphi_+(x, y, z)$ in the vocabulary $\mathcal{L}_{FO}$. Then the set PAL of binary palindromes is represented by the sentence

$$\forall x \forall y, \; \varphi_+(x, y, max) \supset (X(x) \leftrightarrow X(y)).$$

Thus $PAL \in \boldsymbol{FO}$.

Immerman showed that the class $\boldsymbol{FO}$ is the same as a uniform version of $\boldsymbol{AC^0}$. Originally $\boldsymbol{AC^0}$ was defined in its nonuniform version, which we shall refer to as $\boldsymbol{AC^0/poly}$. A language in $\boldsymbol{AC^0/poly}$ is specified by a polynomial size bounded depth family $\langle C_n \rangle$ of Boolean circuits, where each circuit C_n has n input bits, and is allowed to have $\neg$-gates, as well as unbounded fan-in $\wedge$-gates and $\vee$-gates. In the uniform version, the circuit C_n must be specified in a uniform way; for example one could require that $\langle C_n \rangle$ is in $\boldsymbol{FO}$. (See also Appendix A.5.)

Immerman showed that this definition of uniform $\boldsymbol{AC^0}$ is robust, in the sense that it has several quite different characterizations. For example, the logtime hierarchy $\boldsymbol{LH}$ consists of all languages recognizable by an ATM (Alternating Turing Machine) in time $O(\log n)$ with a constant number of alternations. Also $\boldsymbol{CRAM}[1]$ consists of all languages recognizable in constant time on a so-called Concurrent Random Access Machine. The following theorem is from [59, Corollary 5.32].

THEOREM IV.1.2. $\boldsymbol{FO} = \boldsymbol{AC^0} = \boldsymbol{CRAM}[1] = \boldsymbol{LH}$.

Of course the nonuniform class $\boldsymbol{AC^0/poly}$ contains non-computable sets, and hence it properly contains the uniform class $\boldsymbol{AC^0}$. Nevertheless in 1983 Ajtai [3] (and independently Furst, Saxe, and Sipser [52]) proved that even such a simple set as $PARITY$ (the set of all strings with an odd number of 1's) is not in $\boldsymbol{AC^0/poly}$ (and hence not in $\boldsymbol{FO}$).

On the positive side, we pointed out that the set PAL of palindromes is in $\boldsymbol{FO}$, and hence in $\boldsymbol{AC^0}$. If we code a triple $\langle U, V, W \rangle$ of strings as a single string in some reasonable way then it is easy to see using a carry look-ahead adder that binary addition (the set $\langle U, V, U + V \rangle$) is in $\boldsymbol{AC^0}$ (see page 85). Do not confuse this result with the result of [59, page 14] mentioned above that some first-order formula $\phi_+(x, y, z)$ represents $x + y = z$, since here x, y, z represent elements in the model $\mathcal{M}$, which have nothing much to do with the input string X.

In fact $PARITY$ is efficiently reducible to binary multiplication, so Ajtai's result implies that the set $\langle U, V, U \cdot V \rangle$ is *not* in $\boldsymbol{AC^0}$. In contrast, there is a first-order formula in the vocabulary $\mathcal{L}_{FO}$ which represents $x \cdot y = z$ in standard model with universe $M = \{0, \ldots, n-1\}$.

IV.2. Two-Sorted First-Order Logic

IV.2.1. Syntax. Our two-sorted first-order logic is an extension of the (single-sorted) first-order logic introduced in Chapter II. Here there are two kinds of variables: the variables $x, y, z, \ldots$ of the first sort are called *number variables*, and are intended to range over the natural numbers; and the variables $X, Y, Z, \ldots$ of the second sort are called *set* (or also *string*) *variables*, and are intended to range over finite subsets of natural numbers (which represent binary strings). Function and predicate symbols may involve either or both sorts.

DEFINITION IV.2.1 (Two-Sorted First-Order Vocabularies). A two-sorted first-order vocabulary (or just two-sorted vocabulary, or vocabulary, or language) $\mathcal{L}$ is specified by a set of function symbols and predicate symbols, just as in the case of a single-sorted vocabulary (Section II.2.1), except that the functions and predicates now can take arguments of both sorts, and there are two kinds of functions: the *number-valued functions* (or just *number functions*) and the *string-valued functions* (or just *string functions*).

In particular, for each $n, m \in \mathbb{N}$, there is a set of (n, m)-ary number function symbols, a set of (n, m)-ary string function symbols, and a set of (n, m)-ary predicate symbols. A $(0, 0)$-ary function symbol is called a constant symbol, which can be either a number constant or a string constant.

We use $f, g, h, \ldots$ as meta-symbols for number function symbols; $F, G, H, \ldots$ for string function symbols; and $P, Q, R, \ldots$ for predicate symbols.

For example, consider the following two-sorted extension of $\mathcal{L}_A$ (Definition II.2.3):

DEFINITION IV.2.2. $\mathcal{L}_A^2 = [0, 1, +, \cdot, |\ | ;\ =_1, =_2, \leq, \in]$.

Here the symbols $0, 1, +, \cdot, =_1$ and $\leq$ are from $\mathcal{L}_A$; they are function and predicate symbols over the first sort ($=_1$ corresponds to $=$ of $\mathcal{L}_A$). The function $|X|$ (the "length of X") is a number-valued function and is intended to denote the least upper bound of the set X (roughly the length of the corresponding string). The binary predicate $\in$ takes a number and a set as arguments, and is intended to denote set membership. Finally, $=_2$ is the equality predicate for the second-sort objects. We will write $=$ for both $=_1$ and $=_2$, since it will be clear from the context which is intended.

We will use the abbreviation

$$X(t) =_{\text{def}} t \in X$$

where t is a number term (Definition IV.2.3 below). Thus we think of $X(i)$ as the i-th bit of the binary string X.

Note that in $\mathcal{L}_A^2$ the function symbols $+, \cdot$ each has arity $(2, 0)$, while $|\ |$ has arity $(0, 1)$ and the predicate symbol $\in$ has arity $(1, 1)$.

For a two-sorted vocabulary $\mathcal{L}$, the notions of $\mathcal{L}$-terms and $\mathcal{L}$-formulas generalize the corresponding notions in the single-sorted case (Definitions II.2.1 and II.2.2). Here we have two kinds of terms: *number terms* and *string terms*. As before, we will drop mention of $\mathcal{L}$ when it is not important, or clear from the context. Also, we are interested only in vocabularies $\mathcal{L}$ that extend $\mathcal{L}_A^2$, and we may list only the elements of the set $\mathcal{L} - \mathcal{L}_A^2$ (sometimes without the braces $\{,\}$ for set). In such cases, the notations $\mathcal{L}$-terms, $\mathcal{L}$-formulas, $\Sigma_i^B(\mathcal{L})$, etc. refer really to the corresponding notions for $\mathcal{L} \cup \mathcal{L}_A^2$.

DEFINITION IV.2.3 ($\mathcal{L}$-Terms). Let $\mathcal{L}$ be a two-sorted vocabulary:

1) Every number variable is an $\mathcal{L}$-number term.
2) Every string variable is an $\mathcal{L}$-string term.
3) If f is an (n, m)-ary number function symbol of $\mathcal{L}$, $t_1, \ldots, t_n$ are $\mathcal{L}$-number terms, and $T_1, \ldots, T_m$ are $\mathcal{L}$-string terms, then $f t_1 \ldots t_n T_1 \ldots T_m$ is an $\mathcal{L}$-number term.
4) If F is an (n, m)-ary string function symbol of $\mathcal{L}$, and $t_1, \ldots, t_n$ and $T_1, \ldots, T_m$ are as above, then $F t_1 \ldots t_n T_1 \ldots T_m$ is an $\mathcal{L}$-string term.

Note that all constants in $\mathcal{L}$ are $\mathcal{L}$-terms.

We often denote number terms by $r, s, t, \ldots$, and string terms by $S, T, \ldots$.

The formulas over a two-sorted vocabulary $\mathcal{L}$ are defined as in the single-sorted case (Definition II.2.2), with the addition of quantifiers over string variables. These are called *string quantifiers*, and the quantifiers over number variables are called *number quantifiers*. Also note that a predicate symbol in general may have arguments from both sorts.

DEFINITION IV.2.4 ($\mathcal{L}$-Formulas). Let $\mathcal{L}$ be a two-sorted first-order vocabulary. Then a *two-sorted first-order formula in $\mathcal{L}$* (or $\mathcal{L}$-*formula*, or just *formula*) are defined inductively as follows:

1) If P is an (n, m)-ary predicate symbol of $\mathcal{L}$, $t_1, \ldots, t_n$ are $\mathcal{L}$-number terms and $T_1, \ldots, T_m$ are $\mathcal{L}$-string terms, then

$$P t_1 \ldots t_n T_1 \ldots T_m$$

is an atomic $\mathcal{L}$-formula. Also, each of the logical constants $\perp, \top$ is an atomic formula.
2) If φ, ψ are $\mathcal{L}$-formulas, so are $\neg\varphi$, $(\varphi \wedge \psi)$, and $(\varphi \vee \psi)$.
3) If φ is an $\mathcal{L}$-formula, x is a number variable and X is a string variable, then $\forall x \varphi$, $\exists x \varphi$, $\forall X \varphi$ and $\exists X \varphi$ are $\mathcal{L}$-formulas.

We often denote formulas by $\varphi, \psi, \ldots$.

For readability we will usually use commas to separate the arguments of functions and predicates. Thus we write $f(x_1, \ldots, x_n, X_1, \ldots, X_m)$ and $P(x_1, \ldots, x_n, X_1, \ldots, X_m)$ instead of $f x_1 \ldots x_n X_1 \ldots X_m$ and $P x_1 \ldots x_n X_1 \ldots X_m$.

Recall that in $\mathcal{L}_A^2$ we write $X(t)$ for $t \in X$.

EXAMPLE IV.2.5 ($\mathcal{L}_A^2$-Terms and $\mathcal{L}_A^2$-Formulas).

1) The only string terms of $\mathcal{L}_A^2$ are the string variables $X, Y, Z, \ldots$.
2) The number terms of $\mathcal{L}_A^2$ are obtained from the constants 0, 1, number variables $x, y, z, \ldots$, and the lengths of the string variables $|X|, |Y|, |Z|, \ldots$ using the binary function symbols $+, \cdot$.
3) The only atomic formulas of $\mathcal{L}_A^2$ are $\bot, \top$ or those of the form $s = t$, $X = Y, s \leq t$ and $X(t)$ for string variables X, Y and number terms s, t.

IV.2.2. Semantics. As for single-sorted first-order logic, the semantics of a two-sorted vocabulary is given by structures and object assignments. Here the universe of a structure contains two sorts of objects, one for the number variables and one for the string variables. As in the single-sorted case, we also require that the predicate symbols $=_1$ and $=_2$ must be interpreted as the true equality in the respective sort. The following definition generalizes the notion of a (single-sorted) structure given in Definition II.2.6.

DEFINITION IV.2.6 (Two-Sorted Structures). Let $\mathcal{L}$ be a two-sorted vocabulary. Then an $\mathcal{L}$-structure $\mathcal{M}$ consists of the following:

1) A pair of two nonempty sets U_1 and U_2, which together are called the *universe*. Number (resp. string) variables in $\mathcal{L}$-formulas are intended to range over U_1 (resp. U_2).
2) For each (n, m)-ary number function symbol f of $\mathcal{L}$ an associated function $f^{\mathcal{M}} : U_1^n \times U_2^m \to U_1$.
3) For each (n, m)-ary string function symbol F of $\mathcal{L}$ an associated function $F^{\mathcal{M}} : U_1^n \times U_2^m \to U_2$.
4) For each (n, m)-ary predicate symbol P of $\mathcal{L}$ an associated relation $P^{\mathcal{M}} \subseteq U_1^n \times U_2^m$.

Thus for our "base" vocabulary $\mathcal{L}_A^2$, an $\mathcal{L}_A^2$-structure with universe $\langle U_1, U_2 \rangle$ contains the following interpretations of $\mathcal{L}_A^2$:

- Elements $0^{\mathcal{M}}, 1^{\mathcal{M}} \in U_1$ to interpret 0 and 1, respectively;
- Binary functions $+^{\mathcal{M}}, \cdot^{\mathcal{M}} : U_1 \times U_1 \to U_1$ to interpret $+$ and $\cdot$, respectively;
- A binary predicate $\leq^{\mathcal{M}} \subseteq U_1^2$ interpreting $\leq$;
- A function $|\ |^{\mathcal{M}} : U_2 \to U_1$;
- A binary relation $\in^{\mathcal{M}} \subseteq U_1 \times U_2$.

In this book all two-sorted vocabularies $\mathcal{L}$ that we consider contain $\mathcal{L}_A^2$, so in particular they contain the length function $|\ |$ and the element-of predicate $\in$. Our intention is that an element $\alpha \in U_2$ can be specified by a pair $(|\alpha|, S_\alpha)$, where $|\alpha| = |\alpha|^{\mathcal{M}} \in U_1$, and $S_\alpha = \{u \in U_1 : u \in^{\mathcal{M}} \alpha\}$. Thus we want two elements α_1 and α_2 in U_2 to be equal iff $|\alpha_1|^{\mathcal{M}} = |\alpha_2|^{\mathcal{M}}$ and the subsets of U_1 specified by interpreting $\in^{\mathcal{M}}$ at α_1 and α_2 are the same. In fact all two-sorted theories that we consider include the extensionality axiom **SE** (see Figure 2 on page 96). Thus we may assume that in any model of such a theory, elements α of U_2 are specified by the pair $(|\alpha|, S_\alpha)$ as above.

EXAMPLE IV.2.7 (The Standard Two-Sorted Model $\underline{\mathbb{N}}_2$). The *standard model* $\underline{\mathbb{N}}_2$ has $U_1 = \mathbb{N}$ and U_2 the set of finite subsets of $\mathbb{N}$. The number part of the structure is the standard single-sorted first-order structure $\underline{\mathbb{N}}$. The relation $\in$ gets its usual interpretation (membership), and for each finite subset $S \subseteq \mathbb{N}$, $|S|$ is interpreted as one plus the largest element in S, or 0 if S is empty.

As in the single-sorted case, the truth value of a formula in a structure is defined based on the interpretations of free variables occurring in it. Here we need to generalize the notion of an object assignment (Definition II.2.7):

DEFINITION IV.2.8 (Two-Sorted Object Assignment). A two-sorted object assignment (or just an object assignment) σ for a two-sorted structure $\mathcal{M}$ is a mapping from the number variables to U_1 together with a mapping from the string variables to U_2.

NOTATION. We will write $\sigma(x)$ for the first-sort object assigned to the number variable x by σ, and $\sigma(X)$ for the second-sort object assigned to the string variable X by σ. Also as in the single-sorted case, if x is a variable and $m \in U_1$, then the object assignment $\sigma(m/x)$ is the same as σ except it maps x to m, and if X is a variable and $M \in U_2$, then the object assignment $\sigma(M/X)$ is the same as σ except it maps X to M.

Now the Basic Semantic Definition II.2.8 generalizes in the obvious way.

DEFINITION IV.2.9 (Basic Semantic Definition, Two-Sorted Case). Let $\mathcal{L}$ be a two-sorted first-order vocabulary, let $\mathcal{M}$ be an $\mathcal{L}$-structure with universe $\langle U_1, U_2 \rangle$, and let σ be an object assignment for $\mathcal{M}$. Each $\mathcal{L}$-number term t is assigned an element $t^{\mathcal{M}}[\sigma]$ in U_1, and each $\mathcal{L}$-string term T is assigned an element $T^{\mathcal{M}}[\sigma]$ in U_2, defined by structural induction on terms t and T, as follows (refer to Definition IV.2.3 for the definition of $\mathcal{L}$-term):

(a) $x^{\mathcal{M}}[\sigma]$ is $\sigma(x)$, for each number variable x;
(b) $X^{\mathcal{M}}[\sigma]$ is $\sigma(X)$, for each string variable X;
(c) $(f t_1 \cdots t_n T_1 \ldots T_m)^{\mathcal{M}}[\sigma] = f^{\mathcal{M}}(t_1^{\mathcal{M}}[\sigma], \ldots, t_n^{\mathcal{M}}[\sigma],$
$$T_1^{\mathcal{M}}[\sigma], \ldots, T_m^{\mathcal{M}}[\sigma]);$$

(d) $(Ft_1 \cdots t_n T_1 \ldots T_m)^{\mathcal{M}}[\sigma] = F^{\mathcal{M}}(t_1^{\mathcal{M}}[\sigma], \ldots, t_n^{\mathcal{M}}[\sigma],$
$$T_1^{\mathcal{M}}[\sigma], \ldots, T_m^{\mathcal{M}}[\sigma]).$$

DEFINITION IV.2.10. For φ an $\mathcal{L}$-formula, the notion $\mathcal{M} \models \varphi[\sigma]$ ($\mathcal{M}$ satisfies φ under σ) is defined by structural induction on formulas φ as follows (refer to Definition IV.2.4 for the definition of a formula):

(a) $\mathcal{M} \models \top$ and $\mathcal{M} \not\models \bot$.
(b) $\mathcal{M} \models (Pt_1 \cdots t_n T_1 \ldots T_m)[\sigma]$ iff

$$\langle t_1^{\mathcal{M}}[\sigma], \ldots, t_n^{\mathcal{M}}[\sigma], T_1^{\mathcal{M}}[\sigma], \ldots, T_m^{\mathcal{M}}[\sigma] \rangle \in P^{\mathcal{M}}.$$

(c1) If $\mathcal{L}$ contains $=_1$, then $\mathcal{M} \models (s = t)[\sigma]$ iff $s^{\mathcal{M}}[\sigma] = t^{\mathcal{M}}[\sigma]$.
(c2) If $\mathcal{L}$ contains $=_2$, then $\mathcal{M} \models (S = T)[\sigma]$ iff $S^{\mathcal{M}}[\sigma] = T^{\mathcal{M}}[\sigma]$.
(d) $\mathcal{M} \models \neg\varphi[\sigma]$ iff $\mathcal{M} \not\models \varphi[\sigma]$.
(e) $\mathcal{M} \models (\varphi \vee \psi)[\sigma]$ iff $\mathcal{M} \models \varphi[\sigma]$ or $\mathcal{M} \models \psi[\sigma]$.
(f) $\mathcal{M} \models (\varphi \wedge \psi)[\sigma]$ iff $\mathcal{M} \models \varphi[\sigma]$ and $\mathcal{M} \models \psi[\sigma]$.
(g1) $\mathcal{M} \models (\forall x\varphi)[\sigma]$ iff $\mathcal{M} \models \varphi[\sigma(m/x)]$ for all $m \in U_1$.
(g2) $\mathcal{M} \models (\forall X\varphi)[\sigma]$ iff $\mathcal{M} \models \varphi[\sigma(M/X)]$ for all $M \in U_2$.
(h1) $\mathcal{M} \models (\exists x\varphi)[\sigma]$ iff $\mathcal{M} \models \varphi[\sigma(m/x)]$ for some $m \in U_1$.
(h2) $\mathcal{M} \models (\exists X\varphi)[\sigma]$ iff $\mathcal{M} \models \varphi[\sigma(M/X)]$ for some $M \in U_2$.

Note that items (c1) and (c2) in the definition of $\mathcal{M} \models A[\sigma]$ follow from (b) and the fact that $=_1^{\mathcal{M}}$ and $=_2^{\mathcal{M}}$ are always the equality relations in the respective sorts.

The notions of "$\mathcal{M} \models \varphi$", "logical consequence", "validity", etc., are defined as before (Definition II.2.10), and we do not repeat them here. Also, the Substitution Theorem (II.2.14) generalizes to the current context, and the Formula Replacement Theorem (II.2.15) continues to hold, and we will not restate them.

IV.3. Two-Sorted Complexity Classes

IV.3.1. Notation for Numbers and Finite Sets. In Section III.4 we explained how to interpret an element of a complexity class such as $\boldsymbol{P}$ (polynomial time) and $\boldsymbol{LTH}$ (Linear Time Hierarchy) as a relation over $\mathbb{N}$. In this context the numerical inputs $x_1, \ldots, x_k$ of a relation $R(x_1, \ldots, x_k)$ are presented in binary to the accepting machine. In the two-sorted context, however, the relations $R(x_1, \ldots, x_k, X_1, \ldots, X_m)$ in question have arguments of both sorts, and now the numbers x_i are presented to the accepting machines using unary notation (n is represented by a string of n 1's) instead of binary. The elements X_i of the second sort are finite subsets of $\mathbb{N}$, and we represent them as binary strings (see below) for the purpose of presenting them as inputs to the accepting machine. The intuitive reason that we represent the numerical arguments in unary is that now they play an auxiliary role as indices to the string arguments, and hence their values are comparable in size to the length of the string arguments.

Thus a numerical relation $R(x)$ with no string argument is in two-sorted polynomial time iff it is computed in time $2^{O(n)}$ on some Turing machine, where n is the binary length of the input x. In particular, the relation $Prime(x)$ is easily seen to be in this class, using a "brute force" algorithm that tries all possible divisors between 1 and x.

The binary string representation of a finite subset of $\mathbb{N}$ is defined as follows. Recall that we write $S(i)$ for $i \in S$ (for $i \in \mathbb{N}$ and $S \subseteq \mathbb{N}$). Thus if we write 0 for $\bot$ and 1 for $\top$, then we can use the binary string

$$w(S) = S(n)S(n-1) \ldots S(1)S(0) \qquad (42)$$

to interpret the finite nonempty subset S of $\mathbb{N}$, where n is the largest member of S. We define $w(\varnothing)$ to be the empty string. For example,

$$w(\{0, 2, 3\}) = 1101.$$

Notice that the intended interpretation of $|S|$ (one plus the largest element of S, or 0 if $S = \varnothing$) is precisely the length of the associated string $w(S)$.

Thus w is an injective map from finite subsets of $\mathbb{N}$ to $\{0, 1\}^*$, but it is not surjective, since the string $w(S)$ begins with 1 for all nonempty S. Nevertheless $w(S)$ is a useful way to represent S as an input to a Turing machine or circuit.

Using the method just described of representing numbers and strings, we can define two-sorted complexity classes as sets of relations. For example two-sorted P consists of the set of all relations $R(\vec{x}, \vec{X})$ which are accepted in polynomial time by some deterministic Turing machine, where each numerical argument x_i is represented in unary as an input, and each subset argument X_i is represented by the string $w(X_i)$ as an input. Similar definitions specify the two-sorted polynomial hierarchy PH, and the two-sorted complexity classes AC^0 and LTH.

IV.3.2. Representation Theorems.

NOTATION. If $\vec{T} = T_1, \ldots T_n$, is a sequence of string terms, then $|\vec{T}|$ denotes the sequence $|T_1|, \ldots, |T_n|$ of number terms.

Bounded number quantifiers are defined as in the single-sorted case (Definition III.1.6). To define bounded string quantifiers, we need the length function $|X|$ of $\mathcal{L}^2_A$.

NOTATION. A two-sorted vocabulary $\mathcal{L}$ is always assumed to be an extension of $\mathcal{L}^2_A$.

DEFINITION IV.3.1 (Bounded Formulas). Let $\mathcal{L}$ be a two-sorted vocabulary. If x is a number variable and X a string variable that do not occur in the $\mathcal{L}$-number term t, then $\exists x \leq t\varphi$ stands for $\exists x(x \leq t \wedge \varphi)$, $\forall x \leq t\varphi$ stands for $\forall x(x \leq t \supset \varphi)$, $\exists X \leq t\varphi$ stands for $\exists X(|X| \leq t \wedge \varphi)$, and

$\forall X \leq t\varphi$ stands for $\forall X(|X| \leq t \supset \varphi)$. Quantifiers that occur in this form are said to be *bounded*, and a *bounded formula* is one in which every quantifier is bounded.

Notation. $\exists \vec{x} \leq \vec{t}\varphi$ stands for $\exists x_1 \leq t_1 \ldots \exists x_k \leq t_k\varphi$ for some k, where no x_i occurs in any t_j (even if $i < j$). Similarly for $\forall \vec{x} \leq \vec{t}, \exists \vec{X} \leq \vec{t}$, and $\forall \vec{X} \leq \vec{t}$.

If the above convention is violated in the sense that x_i occurs in t_j for $i < j$, and the terms $\vec{t}$ are $\mathcal{L}_A^2$-terms, then new bounding terms $\vec{t'}$ in $\mathcal{L}_A^2$ can be found which satisfy the convention. For example $\exists x_1 \leq t_1 \exists x_2 \leq t_2(x_1)\varphi$ is equivalent to

$$\exists x_1 \leq t_1 \exists x_2 \leq t_2(t_1)(x_2 \leq t_2(x_1) \wedge \varphi).$$

We will now define the following important classes of formulas.

Definition IV.3.2 (The $\Sigma_1^1(\mathcal{L})$, $\Sigma_i^B(\mathcal{L})$ and $\Pi_i^B(\mathcal{L})$ Formulas). Let $\mathcal{L} \supseteq \mathcal{L}_A^2$ be a two-sorted vocabulary. Then $\Sigma_0^B(\mathcal{L}) = \Pi_0^B(\mathcal{L})$ is the set of $\mathcal{L}$-formulas whose only quantifiers are bounded number quantifiers (there can be free string variables). For $i \geq 0$, $\Sigma_{i+1}^B(\mathcal{L})$ (resp. $\Pi_{i+1}^B(\mathcal{L})$) is the set of formulas of the form $\exists \vec{X} \leq \vec{t}\varphi(\vec{X})$ (resp. $\forall \vec{X} \leq \vec{t}\varphi(\vec{X})$), where φ is a $\Pi_i^B(\mathcal{L})$ formula (resp. a $\Sigma_i^B(\mathcal{L})$ formula), and $\vec{t}$ is a sequence of $\mathcal{L}_A^2$-terms not involving any variable in $\vec{X}$. Also, a $\Sigma_1^1(\mathcal{L})$ formula is one of the form $\exists \vec{X}\varphi$, where $\vec{X}$ is a vector of zero or more string variables, and φ is a $\Sigma_0^B(\mathcal{L})$ formula.

We usually write Σ_i^B for $\Sigma_i^B(\mathcal{L}_A^2)$ and Π_i^B for $\Pi_i^B(\mathcal{L}_A^2)$.
We have

$$\Sigma_0^B(\mathcal{L}) \subseteq \Sigma_1^B(\mathcal{L}) \subseteq \Sigma_2^B(\mathcal{L}) \subseteq \cdots,$$

$$\Sigma_0^B(\mathcal{L}) \subseteq \Pi_1^B(\mathcal{L}) \subseteq \Pi_2^B(\mathcal{L}) \subseteq \cdots$$

and for $i \geq 0$

$$\Sigma_i^B(\mathcal{L}) \subseteq \Pi_{i+1}^B(\mathcal{L}) \quad \text{and} \quad \Pi_i^B(\mathcal{L}) \subseteq \Sigma_{i+1}^B(\mathcal{L}).$$

Notice the "strict" requirements on $\Sigma_i^B(\mathcal{L})$ and $\Pi_i^B(\mathcal{L})$: all string quantifiers must occur in front. For example, $\Sigma_1^B(\mathcal{L}_A^2)$ is sometimes called *strict* $\Sigma_1^{1,b}$ in the literature. (Also notice that the bounding terms $\vec{t}$ must be in the basic vocabulary $\mathcal{L}_A^2$.) We will show that some theories prove *replacement* theorems, which assert the equivalence of a non-strict Σ_i^B formula (for certain values if i) with its strict counterpart.

In Section III.3.1 we discussed the definability of predicates (i.e., relations) and functions in a single-sorted theory. In the case of relations, the notion is purely semantic, and does not depend on the theory, but only the underlying vocabulary and the standard model. The situation is the same for the two-sorted case, and so we will define the notion of a relation $R(\vec{x}, \vec{X})$ *represented* by a formula, without reference to a theory. As in the single-sorted case, we assume that each relation symbol has a standard interpretation in an expansion of the standard model, in this case $\underline{\mathbb{N}}_2$, and formulas in the following definition are interpreted in the same model.

DEFINITION IV.3.3 (Representable/Definable Relations). Let $\mathcal{L} \supseteq \mathcal{L}^2_A$ be a two-sorted vocabulary, and let φ be an $\mathcal{L}$-formula. Then we say that $\varphi(\vec{x}, \vec{X})$ *represents* (or *defines*) a relation $R(\vec{x}, \vec{X})$ if

$$R(\vec{x}, \vec{X}) \leftrightarrow \varphi(\vec{x}, \vec{X}). \tag{43}$$

If Φ is a set of $\mathcal{L}$-formulas, then we say that $R(\vec{x}, \vec{X})$ is Φ-*representable* (or Φ-*definable*) if it is represented by some $\varphi \in \Phi$.

If we want to precisely represent a language $L \subseteq \{0, 1\}^*$, then we need to consider strings that do not necessarily begin with 1. Thus the relation $R_L(X)$ corresponding to L is defined by

$$R_L(X) \leftrightarrow w'(X) \in L$$

where the string $w'(X)$ is obtained from $w(X)$ (42) by deleting the initial 1 (and $w'(\varnothing)$ and $w'(\{1\})$ both are the empty string).

EXAMPLE IV.3.4. The language PAL (page 75) of binary palindromes is represented by the formula

$$\varphi_{PAL}(X) \leftrightarrow |X| \leq 1 \vee \forall x, y < |X|, x + y + 2 = |X| \supset (X(x) \leftrightarrow X(y)).$$

Despite this example, we emphasize that the objects of the second sort in our complexity classes are finite sets of natural numbers, and we will not be much concerned by the fact that the corresponding strings (for nonempty sets) all begin with 1.

We define two-sorted AC^0 using the log time hierarchy LH. We could define LH using alternating Turing machines (those relations accepted in log time with a constant number of alternations), but we choose instead to define the levels of the hierarchy using a recurrence analogous to our definition of LTH in Section III.4.1. Thus we define $NLogTime$ to be the class of relations $R(\vec{x}, \vec{X})$ accepted by a nondeterministic index Turing machine M in time $\mathcal{O}(\log n)$. (See also Appendix A.2.)

As explained before, normally inputs $\vec{x}$ are presented in unary and $\vec{X}$ are presented in binary. However in defining LH it is convenient to change this convention and assume that the number inputs $\vec{x}$ are presented in binary (string inputs $\vec{X}$ are also presented in binary as before). To keep the meaning of "log time" unchanged, we define the length of a number input x_i to be x_i, even though the actual length of the binary notation is $|x_i|$. The reason for using binary notation is that in time $O(\log x_i)$ a Turing machine M can read the entire binary notation for x_i.

The machine M accesses its string inputs using index tapes; one such tape for each string argument X_i of $R(\vec{x}, \vec{X})$. When M enters the query state for an input X_i, if the index tape contains the number j written in binary, then j-th bit of X_i is returned. The index tape is not erased between input queries. Since M runs in log time, only $O(\log|X_i|)$ bits of X_i can be accessed during any one computation.

Now define

$$\Sigma_1^{log} = NLogTime \tag{44}$$

and for $i \geq 1$

$$\Sigma_{i+1}^{log} = NLogTime^{\Sigma_i^{log}}. \tag{45}$$

Then

$$LH = \bigcup_i \Sigma_i^{log}.$$

DEFINITION IV.3.5 (Two-Sorted AC^0). $AC^0 = LH$.

The notation $NLogTime^{\Sigma_i^{log}}$ in (45) refers to a nondeterministic log time Turing machine M as above, except now M has access to an oracle for a relation $S(\vec{y}, \vec{Y})$ in Σ_i^{log}. In order to explain how M in log time accesses an arbitrary input $(\vec{y}, \vec{Y})$ to S, we simplify things by requiring that $\vec{X} = \vec{Y}$; that is the string inputs to S are the same as the string inputs to M. However the number inputs $\vec{y}$ to S are arbitrary: M has time to write them in binary on a special query tape. (See Appendix A.3 for oracle Turing machines).

Two-sorted AC^0 restricted to numerical relations $R(\vec{x})$ is exactly the same as single-sorted LTH as defined in Section III.4.1. The amount of time allotted for the Turing machines under the two definitions for an input $\vec{x}$ is the same, namely $O(\log(\Sigma x_i))$.

Thus for numerical relations, the following representation theorem is the same as the LTH Theorem III.4.2 ($LTH = \Delta_0^{\mathbb{N}}$). For string relations, it can be considered a restatement of Theorem IV.1.2 ($FO = AC^0$).

THEOREM IV.3.6 (Σ_0^B Representation). *A relation $R(\vec{x}, \vec{X})$ is in AC^0 iff it is represented by some Σ_0^B formula $\varphi(\vec{x}, \vec{X})$.*

PROOF SKETCH. In light of the above discussion, the proof is essentially the same as for Theorem III.4.2. To show that every relation $R(\vec{x}, \vec{X})$ in AC^0 (i.e. LH) is representable by a Σ_0^B formula $\varphi(\vec{x}, \vec{X})$ we use the recurrence (44), (45). The proof is almost the same as showing $LTH \subseteq \Delta_0^{\mathbb{N}}$. There is an extra consideration in the base case, showing how the formula $\varphi(\vec{x}, \vec{X})$ represents the computation of a log time nondeterministic Turing machine M that now accesses its string inputs using index tapes. The computation is represented as before, except now $\varphi(\vec{x}, \vec{X})$ uses an extra number variable j_i for each string input variable X_i. Here j_i holds the current numerical value of the index tape for X_i.

The proof of the converse, that every relation representable by a Σ_0^B formula is in $\boldsymbol{LH}$, is straightforward and similar to the proof that $\Delta_0^{\mathbb{N}} \subseteq \boldsymbol{LTH}$. □

NOTATION. For X a finite subset of $\mathbb{N}$, let $bin(X)$ be the number whose binary notation is $w(X)$ (see (42)). Thus

$$bin(X) = \sum_i X(i)2^i \tag{46}$$

where here we treat the predicate $X(i)$ as a 0-1-valued function. For example, $bin(\{0, 2, 3\}) = 2^2 + 2^3 = 12$.

Define the relations R_+ and $R_\times$ by

$$R_+(X, Y, Z) \leftrightarrow bin(X) + bin(Y) = bin(Z),$$
$$R_\times(X, Y, Z) \leftrightarrow bin(X) \cdot bin(Y) = bin(Z).$$

As mentioned earlier, $PARITY$ is efficiently reducible to $R_\times$, and hence $R_\times$ is not in $\boldsymbol{AC^0}$, and cannot be represented by any Σ_0^B formula. However R_+ is in $\boldsymbol{AC^0}$. To represent it as a Σ_0^B formula, we first define the relation $Carry(i, X, Y)$ to mean that there is a carry into bit position i when computing $bin(X) + bin(Y)$. Then (using the idea behind a carry-lookahead adder)

$$Carry(i, X, Y) \leftrightarrow \exists k < i \big(X(k) \wedge Y(k) \wedge$$
$$\forall j < i(k < j \supset (X(j) \vee Y(j)))\big). \tag{47}$$

Thus

$$R_+(X, Y, Z) \leftrightarrow \big(|Z| \leq |X| + |Y| \wedge$$
$$\forall i < |X| + |Y|(Z(i) \leftrightarrow (X(i) \oplus Y(i) \oplus Carry(i, X, Y)))\big)$$

where $\oplus$ represents exclusive or.

Note that the Σ_0^B Representation Theorem can be alternatively proved by using the characterization $\boldsymbol{AC^0} = \boldsymbol{FO}$. Here we need the fact that

$$\boldsymbol{FO[BIT]} = \boldsymbol{FO[PLUS, TIMES]}$$

i.e., the vocabulary $\mathcal{L}_{FO}$ in (41) can be equivalently defined as

$$[0, max, +, \cdot; X, \leq, =].$$

Note also that in $\mathcal{L}_{FO}$ we have only one "free" unary predicate symbol X, so technically speaking, $\mathcal{L}_{FO}$ formulas can describe only unary relations (i.e., languages). In order to describe a k-ary relation, one way is to extend the vocabulary $\mathcal{L}_{FO}$ to include additional "free" unary predicates. Then Theorem IV.1.2 continues to hold. Now the Σ_0^B Representation Theorem can be proved by translating any Σ_0^B formula φ into an $\boldsymbol{FO}$ formula φ' that *describes* the relation *represented* by φ, and vice versa.

We use Σ_i^P to denote level $i \geq 1$ of the two-sorted polynomial hierarchy. In particular, Σ_1^P denotes two-sorted **NP**. Thus a relation $R(\vec{x}, \vec{X})$ is in Σ_i^P iff it is accepted by some polynomial time ATM with at most i alternations, starting with existential, using the input conventions described in Section IV.3.1. (See also Appendices A.2 and A.3.)

THEOREM IV.3.7 (Σ_i^B and Σ_1^1 Representation). *For $i \geq 1$, a relation $R(\vec{x}, \vec{X})$ is in Σ_i^P iff it is represented by some Σ_i^B formula. The relation is recursively enumerable iff it is represented by some Σ_1^1 formula.*

PROOF. We show that a relation $R(\vec{x}, \vec{X})$ is in **NP** iff it is represented by a Σ_1^B formula. (The other cases are proved similarly.) First suppose that $R(\vec{x}, \vec{X})$ is accepted by a nondeterministic polytime Turing machine M. Then the Σ_1^B formula that represents R has the form

$$\exists Y \leq t(\vec{x}, \vec{X}) \; \varphi(\vec{x}, \vec{X}, Y)$$

where Y codes an accepting computation of M on input $\langle \vec{x}, \vec{X} \rangle$, t represents the upper bound on the length of such computation, and φ is a Σ_0^B formula that verifies the correctness of Y. Here the bounding term t exists by the assumption that M works in polynomial time, and the formula φ can be easily constructed given the transition function of M.

On the other hand, suppose that $R(\vec{x}, \vec{X})$ is represented by the Σ_1^B formula

$$\exists \vec{Y} \leq \vec{t}(\vec{x}, \vec{X}) \; \varphi(\vec{x}, \vec{X}, \vec{Y}).$$

Then the polytime NTM M that accepts R works as follows. On input $\langle \vec{x}, \vec{X} \rangle$ M simply guesses the values of $\vec{Y}$, and then verifies that $\varphi(\vec{x}, \vec{X}, \vec{Y})$ holds. The verification can be easily done in polytime (it is in fact in AC^0 as shown by the Σ_0^B Representation Theorem). □

IV.3.3. The *LTH* Revisited. Consider **LTH** (Linear Time Hierarchy, Section III.4) as a two-sorted complexity class. Here we can define the relations in this class by *linearly bounded formulas*, a concept defined below.

DEFINITION IV.3.8. A formula φ over $\mathcal{L}_A^2$ is called a *linearly bounded formula* if all of its quantifiers are bounded by terms not involving $\cdot$.

THEOREM IV.3.9 (Two-Sorted **LTH**). *A relation is in **LTH** if and only if it is represented by some linearly bounded formula.*

The proof of this theorem is similar to the proof of Theorem III.4.2. Here the ($\Longleftarrow$) direction is simpler: For the base case, we need to calculate the number terms $t(x_1, \ldots, x_k, |X_1|, \ldots, |X_m|)$ in time *linear in* ($\sum x_i + \sum |X_j|$), and this is straightforward.

For the other direction, as in the proof of the single-sorted **LTH** Theorem, the interesting part is to show that relations in **NLinTime** can be represented by linearly bounded formulas. Here we do not need to define the relation $y = 2^x$ as in the single-sorted case, since the relation $X(i)$ (which stands for $i \in X$) is already in our vocabulary. We still need to "count" the number of 1-bits in a string, i.e., we need to define the two-sorted version of *Numones*: $Numones_2(a, i, X)$ is true iff a is the number of 1-bits in the first i low-order bits of X. Again, $Numones_2$ can be defined using Bennett's Trick.

EXERCISE IV.3.10. (a) Define using linearly bounded formula the relation $m = \lceil \sqrt{i} \rceil$.
  (b) Define using linearly bounded formula the relation "$k =$ the number of 1-bits in the substring $X(im) \ldots X(im + m - 1)$".
  (c) Now define $Numones_2(a, i, X)$ using linearly bounded formula.

EXERCISE IV.3.11. Complete the proof of the Two-Sorted LTH Theorem.

In [113], Zambella considers the subset of $\mathcal{L}^2_A$ without the number function $\cdot$, denoted here by $\mathcal{L}^{2-}_A$, and introduces the notion of *linear formulas*, which are the bounded formulas in the vocabulary $\mathcal{L}^{2-}_A$. Then LTH is also characterized as the class of relations representable by linear formulas. In order to prove this claim from the Two-Sorted LTH Theorem above, we need to show that the relation $x \cdot y = z$ is definable by some *linear formula*.

EXERCISE IV.3.12. Define the relation $x \cdot y = z$ using a linear formula. (Hint: First define the relation "z is a multiple of y".)

We have shown how to define the relation $y = 2^x$ using Δ_0 formula in Section III.3.3. Here it is much easier to define this relation using linearly bounded formulas.

EXERCISE IV.3.13. Show how to express $y = 2^x$ using linearly bounded formula. (Hint: Use $Numones_2$ from Exercise IV.3.10.)

IV.4. The Proof System LK^2

Now we extend the sequent system LK (Section II.2.3) to a system LK^2 for a two-sorted vocabulary $\mathcal{L}^2$. As for LK, here we introduce the *free string variables* denoted by $\alpha, \beta, \gamma, \ldots$, and the *bound string variables* $X, Y, Z, \ldots$ in addition to the *free number variables* denoted by $a, b, c, \ldots$, and the *bound number variables* denoted by $x, y, z, \ldots$.

Also, in LK^2 the terms (of both sorts) do not involve any bound variable, and the formulas do not have any free occurrence of any bound variable.

The system LK^2 includes all axioms and rules for LK as described in Section II.2.3, where the term t is a number term respecting our convention for free and bound variables above. In addition LK^2 has the following four rules introducing string quantifiers, here T is any string term that does not contain any bound string variable $X, Y, Z, \ldots$:

String $\forall$ introduction rules

$$\text{left: } \frac{\varphi(T), \Gamma \longrightarrow \Delta}{\forall X \varphi(X), \Gamma \longrightarrow \Delta} \qquad \text{right: } \frac{\Gamma \longrightarrow \Delta, \varphi(\beta)}{\Gamma \longrightarrow \Delta, \forall X \varphi(X)}$$

String $\exists$ introduction rules

$$\text{left: } \frac{\varphi(\beta), \Gamma \longrightarrow \Delta}{\exists X \varphi(X), \Gamma \longrightarrow \Delta} \qquad \text{right: } \frac{\Gamma \longrightarrow \Delta, \varphi(T)}{\Gamma \longrightarrow \Delta, \exists X \varphi(X)}$$

Restriction. The free variable β must not occur in the conclusion of $\forall$-right and $\exists$-left.

The notion of $\boldsymbol{LK}^2$ *proofs* generalizes the notion of $\boldsymbol{LK}$ proofs and anchored $\boldsymbol{LK}$ proofs. The Derivational Soundness, the Completeness Theorem (II.2.23), and the Anchored Completeness Theorem (II.2.28) continue to hold for $\boldsymbol{LK}^2$ (without equality).

In general, when the vocabulary $\mathcal{L}$ does not contain either of the equality predicate symbols, then the notion of $\boldsymbol{LK}^2$-Φ proof is defined as in Definition II.2.21. In the sequel our two-sorted vocabularies will all contain both of the equality predicates, so we will restrict our attention to this case. Here we need to generalize the Equality Axioms given in Definition II.3.6. Recall that we write $=$ for both $=_1$ and $=_2$.

DEFINITION IV.4.1 ($\boldsymbol{LK}^2$ Equality Axioms for $\mathcal{L}$). Suppose that $\mathcal{L}$ is a two-sorted vocabulary containing both $=_1$ and $=_2$. The $\boldsymbol{LK}^2$ *Equality Axioms for* $\mathcal{L}$ consists of the following axioms. (We let Λ stand for

$$t_1 = u_1, \ldots, t_n = u_n, T_1 = U_1, \ldots, T_m = U_m$$

in E4′, E4″ and E5′.) Here t, u, t_i, u_i are number terms, and T, U, T_i, U_i are string terms.

E1′. $\longrightarrow t = t$;
E1″. $\longrightarrow T = T$;
E2′. $t = u \longrightarrow u = t$;
E2″. $T = U \longrightarrow U = T$;
E3′. $t = u, u = v \longrightarrow t = v$;
E3″. $T = U, U = V \longrightarrow T = V$;
E4′. $\Lambda \longrightarrow f t_1 \ldots t_n T_1 \ldots T_m = f u_1 \ldots u_n U_1 \ldots U_m$ for each f in $\mathcal{L}$;
E4″. $\Lambda \longrightarrow F t_1 \ldots t_n T_1 \ldots T_m = F u_1 \ldots u_n U_1 \ldots U_m$ for each F in $\mathcal{L}$;
E5′. $\Lambda, P t_1 \ldots t_n T_1 \ldots T_m \longrightarrow P u_1 \ldots u_n U_1 \ldots U_m$ for each P in $\mathcal{L}$ (here P is not $=_1$ or $=_2$).

DEFINITION IV.4.2 ($\boldsymbol{LK}^2$-Φ Proofs). Suppose that $\mathcal{L}$ is a two-sorted vocabulary containing both $=_1$ and $=_2$, and Φ is a set of $\mathcal{L}$-formulas. Then an $\boldsymbol{LK}^2$-Φ proof (or a Φ-proof) is an $\boldsymbol{LK}^2$-Ψ proof in the sense of Definition II.2.21, where Ψ is Φ together with all instances of the $\boldsymbol{LK}^2$ Equality

Axioms E1′, E1″, ..., E4′, E4″, E5′ for $\mathcal{L}$. If Φ is empty, we simply refer to an LK^2-proof (but allow E1′, ..., E5′ as axioms).

Recall that if φ is a formula with free variables $a_1, \ldots, a_n, \alpha_1, \ldots, \alpha_m$, then $\forall\varphi$, the universal closure of φ, is the sentence

$$\forall x_1 \ldots \forall x_n \forall X_1 \ldots \forall X_m \varphi(x_1/a_1, \ldots, x_n/a_n, X_1/\alpha_1, \ldots, X_m/\alpha_m)$$

where $x_1, \ldots, x_n, X_1, \ldots, X_m$ is a list of new bound variables. Also recall that if Φ is a set of formulas, then $\forall\Phi$ is the set of all sentences $\forall\varphi$, for $\varphi \in \Phi$.

The following Soundness and Completeness Theorem for the two-sorted system LK^2 is the analogue of Theorem II.3.8, and is proved in the same way.

THEOREM IV.4.3 (Soundness and Completeness of LK^2). *For any set Φ of formulas and sequent S,*

$$\forall\Phi \models S \text{ iff } S \text{ has an } LK^2\text{-}\Phi \text{ proof.}$$

Below we will state the two-sorted analogue of the Anchored LK Completeness Theorem and the Subformula Property of Anchored LK Proofs (Theorems II.3.10 and II.3.11). They can be proved just as in the case of LK.

DEFINITION IV.4.4 (Anchored LK^2 Proof). An LK^2-Φ proof π is *anchored* provided every cut formula in π is a formula in some non-logical axiom of π (including possibly E1′, E1″, ..., E5′).

THEOREM IV.4.5 (Anchored LK^2 Completeness). *Suppose that Φ is a set of formulas closed under substitution of terms for variables and that the sequent S is a logical consequence of $\forall\Phi$. Then there is an anchored LK^2-Φ proof of S.*

THEOREM IV.4.6 (Subformula Property of Anchored LK^2 Proofs). *If π is an anchored LK^2-Φ proof of a sequent S, then every formula in every sequent of π is a term substitution instance of a sub-formula of a formula either in S or in a non-logical axiom of π (including E1′, ..., E4″, E5′).*

As in the case for LK where the Anchored LK Completeness Theorem is used to prove the Compactness Theorem (Theorem II.4.2), the above Anchored LK^2 Completeness Theorem can be used to prove the following (two-sorted) Compactness Theorem.

THEOREM IV.4.7 (Compactness). *If Φ is an unsatisfiable set of (two-sorted) formulas, then some finite subset of Φ is unsatisfiable.*

(See also the three alternative forms in Theorem II.1.16.)

Form 1 of the Herbrand Theorem (Theorem II.5.4) can also be extended to the two-sorted logic, with the set of (single-sorted) equality axioms $\mathcal{E}_{\mathcal{L}}$ now replaced by the set of two-sorted equality axioms E1′, E1″, ..., E4″, E5′ above. Below we will state only Form 2 of the Herbrand Theorem for the two-sorted logics. Note that it also follows from Form 1, just as in the single-sorted case.

A *two-sorted theory* (or just *theory*, when it is clear) is defined as in Definition III.1.1, where now it is understood that the underlying vocabulary $\mathcal{L}$ is a two-sorted vocabulary. Also, a universal theory is a theory which can be axiomatized by universal formulas, (i.e., formulas in prenex form, in which all quantifiers are universal).

THEOREM IV.4.8 (Herbrand Theorem for Two-Sorted Logic).   (a) *Let T be a universal (two-sorted) theory, and let $\varphi(x_1, \ldots, x_k, X_1, \ldots, X_m, Z)$ be a quantifier-free formula with all free variables displayed such that*

$$T \vdash \forall x_1 \ldots \forall x_k \forall X_1 \ldots \forall X_m \exists Z \varphi(\vec{x}, \vec{X}, Z).$$

Then there exist finitely many string terms $T_1(\vec{x}, \vec{X}), \ldots, T_n(\vec{x}, \vec{X})$ such that

$$T \vdash \forall x_1 \ldots \forall x_k \forall X_1 \ldots \forall X_m \big(\varphi(\vec{x}, \vec{X}, T_1(\vec{x}, \vec{X})) \vee \cdots \vee$$
$$\varphi(\vec{x}, \vec{X}, T_n(\vec{x}, \vec{X}))\big).$$

(b) *Similarly, let the theory T be as above, and let*

$$\varphi(x_1, \ldots, x_k, z, X_1, \ldots, X_m)$$

be a quantifier-free formula with all free variables displayed such that

$$T \vdash \forall x_1 \ldots \forall x_k \forall X_1 \ldots \forall X_m \exists z \varphi(\vec{x}, z, \vec{X}).$$

Then there exist finitely many number terms $t_1(\vec{x}, \vec{X}), \ldots, t_n(\vec{x}, \vec{X})$ such that

$$T \vdash \forall x_1 \ldots \forall x_k \forall X_1 \ldots \forall X_m \big(\varphi(\vec{x}, t_1(\vec{x}, \vec{X}), \vec{X}) \vee \cdots \vee$$
$$\varphi(\vec{x}, t_n(\vec{x}, \vec{X}), \vec{X})\big).$$

The theorem easily extends to the cases where

$$T \vdash \forall \vec{x} \forall \vec{X} \exists z_1 \ldots \exists z_m \exists Z_1 \ldots \exists Z_n \varphi(\vec{x}, \vec{z}, \vec{X}, \vec{Z}).$$

IV.4.1. Two-Sorted Free Variable Normal Form. The notion of free variable normal form (Section II.2.4) generalizes naturally to LK^2 proofs, where now the term *free variable* refers to free variables of both sorts. Again there is a simple procedure for putting any LK^2 proof into free variable normal form (with the same endsequent), provided that the underlying vocabulary has constant symbols of both sorts. This procedure preserves the size and shape of the proof, and takes an anchored LK^2-Φ proof to an anchored LK^2-Φ proof, provided that the set Φ of formulas is closed under substitution of terms for free variables.

In the case of $\mathcal{L}_A^2$, there is no string constant symbol, so we expand the notion of a LK^2-Φ proof over $\mathcal{L}_A^2$ by allowing the constant symbol $\varnothing$ (for the empty string) and assume that Φ contains the following axiom:

E. $|\varnothing| = 0$.

Adding this symbol and axiom to any theory $\mathcal{T}$ over $\mathcal{L}_A^2$ we consider will result in a conservative extension of $\mathcal{T}$, since every model for $\mathcal{T}$ can trivially be expanded to a model of $\mathcal{T} \cup \{\mathbf{E}\}$. Now any LK^2 proof over $\mathcal{L}_A^2$ can be transformed to one in free variable normal form with the same endsequent, and similarly for LK^2-Φ for suitable Φ.

IV.5. Single-Sorted Logic Interpretation

In this section we will briefly discuss how the Compactness Theorem and Herbrand Theorem in the two-sorted logic follow from the analogous results for the single-sorted logic that we have seen in Chapter II. This section is independent with the rest of the book, and it is the approach that we follow to prove the above theorems in Section IV.4 that will be useful in later chapters, not the approach that we present here.

Although a two-sorted logic is a generalization of a single-sorted logic by having one more sort, it can be interpreted as a single-sorted logic by merging both sorts and using 2 extra unary predicate symbols to identify elements of the 2 sorts.

More precisely, for each two-sorted vocabulary $\mathcal{L}$, w.l.o.g., we can assume that it does not contain the unary predicate symbols FS (for first sort) and SS (for second sort). Let $\mathcal{L}^1 = \{FS, SS\} \cup \mathcal{L}$, where it is understood that the functions and predicates in $\mathcal{L}_1$ take arguments from a single sort.

In addition, let $\Phi_{\mathcal{L}}$ be the set of $\mathcal{L}^1$-formulas which consists of

1) $(\forall x, FS(x) \lor SS(x)) \land (\exists x \exists y, FS(x) \land SS(y))$.
2) For each function symbol f of $\mathcal{L}^1$ (where f has arity (n, m) in $\mathcal{L}$) the formula

$$\forall \vec{x} \forall \vec{y}, (FS(x_1) \land \cdots \land FS(x_n) \land SS(y_1) \land \cdots \land SS(y_m)) \supset FS(f(\vec{x}, \vec{y})).$$

(If f is a number constant c, the above formula is just $FS(c)$.)
3) For each function symbol F of $\mathcal{L}^1$ (where F has arity (n, m) in $\mathcal{L}$) the formula

$$\forall \vec{x} \forall \vec{y}, (FS(x_1) \land \cdots \land FS(x_n) \land SS(y_1) \land \cdots \land SS(y_m)) \supset SS(F(\vec{x}, \vec{y})).$$

(If F is a string constant α, the above formula is just $SS(\alpha)$.)
4) For each predicate symbol P of $\mathcal{L}^1$ (where P has arity (n, m) in $\mathcal{L}$) the formula

$$\forall \vec{x} \forall \vec{y}, P(\vec{x}, \vec{y}) \supset (FS(x_1) \land \cdots \land FS(x_n) \land SS(y_1) \land \cdots \land SS(y_m)).$$

LEMMA IV.5.1. *For each nonempty two-sorted vocabulary $\mathcal{L}$, the set $\Phi_{\mathcal{L}}$ is satisfiable.*

PROOF. The proof is straightforward: For an arbitrary (two-sorted) $\mathcal{L}$-structure $\mathcal{M}$ with universe $\langle U_1, U_2 \rangle$, we construct a (single-sorted) $\mathcal{L}_1$-structure $\mathcal{M}_1$ that has universe $\langle U_1, U_2 \rangle$, $\mathsf{FS}^{\mathcal{M}_1} = U_1$, $\mathsf{SS}^{\mathcal{M}_1} = U_2$, and the same interpretation as in $\mathcal{M}$ for each symbol of $\mathcal{L}$. It is easy to verify that $\mathcal{M}_1 \models \Phi_{\mathcal{L}}$. $\qquad\square$

It is also evident from the above proof that any model $\mathcal{M}_1$ of $\Phi_{\mathcal{L}}$ can be interpreted as a two-sorted $\mathcal{L}$-structure $\mathcal{M}$.

Now we construct for each $\mathcal{L}$-formula φ an $\mathcal{L}^1$-formula φ^1 inductively as follows.

1) If φ is an atomic sentence, then $\varphi^1 =_{\text{def}} \varphi$.
2) If $\varphi \equiv \varphi_1 \wedge \varphi_2$ (or $\varphi \equiv \varphi_1 \vee \varphi_2$, or $\varphi \equiv \neg\psi$), then $\varphi^1 =_{\text{def}} \varphi_1^1 \wedge \varphi_2^1$ (or $\varphi^1 \equiv \varphi_1^1 \vee \varphi_2^1$, or $\varphi^1 \equiv \neg\psi^1$, respectively).
3) If $\varphi \equiv \exists x \psi(x)$, then $\varphi^1 =_{\text{def}} \exists x(\mathsf{FS}(x) \wedge \psi^1(x))$.
4) If $\varphi \equiv \forall x \psi(x)$, then $\varphi^1 =_{\text{def}} \forall x(\mathsf{FS}(x) \supset \psi^1(x))$.
5) If $\varphi \equiv \exists X \psi(X)$, then $\varphi^1 =_{\text{def}} \exists x(\mathsf{SS}(x) \wedge \psi^1(x))$.
6) If $\varphi \equiv \forall X \psi(X)$, then $\varphi^1 =_{\text{def}} \forall x(\mathsf{SS}(x) \supset \psi^1(x))$.

Note that when φ is a sentence, then φ^1 is also a sentence.

For a set Ψ of $\mathcal{L}$-formulas, let Ψ^1 denote the set $\{\varphi^1 : \varphi \in \Psi\}$. The lemma above can strengthened as follows.

THEOREM IV.5.2. *A set Ψ of $\mathcal{L}$-sentences φ is satisfiable iff the set of $\Phi_{\mathcal{L}} \cup \Psi^1$ of $\mathcal{L}^1$-sentences is satisfiable.*

Notice that in the statement of the theorem, Ψ is a set of *sentences*. In general, the theorem may not be true if Ψ is a set of formulas.

PROOF. For simplicity, we will prove the theorem when Ψ is the set of a single sentence φ. The proof for the general case is similar.

For the ONLY IF direction, for any model $\mathcal{M}$ of φ we construct a $\mathcal{L}_1$-structure $\mathcal{M}_1$ as in the proof of Lemma IV.5.1. It can be proved by structural induction on φ that $\mathcal{M}_1 \models \varphi^1$. By the lemma, $\mathcal{M}_1 \models \Phi_{\mathcal{L}}$. Hence $\mathcal{M}_1 \models \Phi_{\mathcal{L}} \cup \{\varphi^1\}$.

For the other direction, suppose that $\mathcal{M}_1$ is a model for $\Phi_{\mathcal{L}} \cup \{\varphi^1\}$. Construct the two-sorted $\mathcal{L}$-structure $\mathcal{M}$ from $\mathcal{M}_1$ as in the remark following the proof of Lemma IV.5.1. Now we can prove by structural induction on φ that $\mathcal{M}$ is a model for φ. Therefore φ is also satisfiable. $\qquad\square$

EXERCISE IV.5.3. Prove the Compactness Theorem for the two-sorted logic (IV.4.7) from the Compactness Theorem for single-sorted logic (II.4.2).

EXERCISE IV.5.4. Prove the Herbrand Theorem for the two-sorted logic (IV.4.8) from Form 2 of the Herbrand Theorem for single-sorted logic (III.3.13).

IV.6. Notes

Historically, Buss [20] was the first to use multi-sorted theories to capture complexity classes such as polynomial space and exponential time.
The main reference for Section IV.1 is [59] Sections 1.1, 1.2, 5.5. Our two-sorted vocabulary $\mathcal{L}_A^2$ is from Zambella [112, 113]. Zambella [112] states the representation theorems IV.3.6 and IV.3.7, although Theorem IV.3.7 essentially goes back to [111], [50], and [105].

Chapter V

THE THEORY V^0 AND AC^0

In this chapter we introduce the family of two-sorted theories $V^0 \subset V^1 \subseteq V^2 \subseteq \cdots$ over the vocabulary $\mathcal{L}^2_A$. For $i \geq 1$, V^i corresponds to Buss's single-sorted theory S^i_2 (Section III.5). The theory V^0 characterizes AC^0 in the same way that $I\Delta_0$ characterizes LTH. Similarly V^1 characterizes P, and in general for $i > 1$, V^i is related to the i-th level of the polynomial time hierarchy.

Here we concentrate on the theory V^0, which will serve as the base theory: all two-sorted theories introduced in this book are extensions of V^0. It is axiomatized by the set 2-*BASIC* of the defining axioms for the symbols in $\mathcal{L}^2_A$, together with Σ^B_0-*COMP* (the comprehension axiom scheme for Σ^B_0 formulas). For $i \geq 1$, V^i is the same as V^0 except that Σ^B_0-*COMP* is replaced by Σ^B_i-*COMP*. We show that for $i \geq 0$, V^i proves the Σ^B_i induction scheme, even though it is not explicitly postulated as a set of axioms. We generalize Parikh's Theorem, and show that it applies to each of the theories V^i.

The main result of this chapter is that V^0 characterizes AC^0: The provably total functions in V^0 are precisely the AC^0 functions. The proof of this characterization is somewhat more involved than the proof of the analogous characterization of LTH by $I\Delta_0$ (Theorem III.4.8). The hard part here is the *Witnessing Theorem for* V^0, which is proved by analyzing anchored LK^2-V^0 proofs. We also give an alternative proof of the witnessing theorem based on the universal conservative extension $\overline{V}^0$ of V^0, using the Herbrand Theorem.

V.1. Definition and Basic Properties of V^i

The set 2-*BASIC* of axioms is given in Figure 2. Recall that $t < u$ stands for $(t \leq u \wedge t \neq u)$.

Axioms **B1**, ... , **B8** are taken from the axioms in 1-*BASIC* for $I\Delta_0$, and **B9**, ... , **B12** are theorems of $I\Delta_0$ (see Examples III.1.8 and III.1.9). Axioms **L1** and **L2** characterize $|X|$ to be one more than the largest element of X, or 0 if X is empty. Axiom **SE** (extensionality) specifies that

B1. $x + 1 \neq 0$	B7. $(x \leq y \wedge y \leq x) \supset x = y$
B2. $x + 1 = y + 1 \supset x = y$	B8. $x \leq x + y$
B3. $x + 0 = x$	B9. $0 \leq x$
B4. $x + (y + 1) = (x + y) + 1$	B10. $x \leq y \vee y \leq x$
B5. $x \cdot 0 = 0$	B11. $x \leq y \leftrightarrow x < y + 1$
B6. $x \cdot (y + 1) = (x \cdot y) + x$	B12. $x \neq 0 \supset \exists y \leq x(y + 1 = x)$
L1. $X(y) \supset y < |X|$	L2. $y + 1 = |X| \supset X(y)$
SE. $\big(|X| = |Y| \wedge \forall i < |X|(X(i) \leftrightarrow Y(i))\big) \supset X = Y$	

FIGURE 2. 2-**BASIC**.

sets X and Y are the same if they have the same elements. Note that the converse

$$X = Y \supset (|X| = |Y| \wedge \forall i < |X|(X(i) \leftrightarrow Y(i)))$$

is valid because in every $\mathcal{L}_A^2$-structure, $=_2$ must be interpreted as true equality over the strings.

EXERCISE V.1.1. Show that the following formulas are provable from 2-**BASIC**.

(a) $\neg x < 0$.
(b) $x < x + 1$.
(c) $0 < x + 1$.
(d) $x < y \supset x + 1 \leq y$. (Use **B10**, **B11**, **B7**.)
(e) $x < y \supset x + 1 < y + 1$.

DEFINITION V.1.2 (Comprehension Axiom). If Φ is a set of formulas, then the *comprehension axiom scheme* for Φ, denoted by Φ-**COMP**, is the set of all formulas

$$\exists X \leq y \forall z < y(X(z) \leftrightarrow \varphi(z)), \tag{48}$$

where $\varphi(z)$ is any formula in Φ, and X does not occur free in $\varphi(z)$.

In the above definition $\varphi(z)$ may have free variables of both sorts, in addition to z. We are mainly interested in the cases in which Φ is one of the formula classes Σ_i^B.

NOTATION. Since (48) states the existence of a finite set X of numbers, we will sometimes use standard set-theoretic notation in defining X:

$$X = \{z : z < y \wedge \varphi(z)\}. \tag{49}$$

DEFINITION V.1.3 (V^i). For $i \geq 0$, the theory V^i has the vocabulary $\mathcal{L}_A^2$ and is axiomatized by 2-**BASIC** and Σ_i^B-**COMP**.

There are no explicit induction axioms for V^i, but nevertheless induction is provable (See Corollary V.1.8).

NOTATION. Since now there are two sorts of variables, there are two different types of *induction axioms*: One is on numbers, and is defined as in Definition III.1.4 (where now Φ is a set of two-sorted formulas), and one is on strings, which we will discuss later. For this reason, we will speak of *number induction axioms* and *string induction axioms*. Similarly, we will use the notion of *number minimization axioms*, which is different from the *string minimization axioms* (to be introduced later). For convenience we repeat the definitions of the axiom schemes for numbers below.

DEFINITION V.1.4 (Number Induction Axiom). If Φ is a set of two-sorted formulas, then Φ-**IND** axioms are the formulas

$$\bigl(\varphi(0) \wedge \forall x(\varphi(x) \supset \varphi(x+1))\bigr) \supset \forall z \varphi(z)$$

where φ is a formula in Φ.

DEFINITION V.1.5 (Number Minimization and Maximization Axioms). The number minimization axioms (or *least number principle* axioms) for a set Φ of two-sorted formulas are denoted Φ-**MIN** and consist of the formulas

$$\varphi(y) \supset \exists x \leq y\bigl(\varphi(x) \wedge \neg \exists z < x \varphi(z)\bigr)$$

where φ is a formula in Φ. Similarly the number maximization axioms for Φ are denoted Φ-**MAX** and consist of the formulas

$$\varphi(0) \supset \exists x \leq y\bigl(\varphi(x) \wedge \neg \exists z \leq y(x < z \wedge \varphi(z))\bigr)$$

where φ is a formula in Φ.

In the above definitions, $\varphi(x)$ is permitted to have free variables of both sorts, in addition to x.

Notice that all axioms of V^0 hold in the standard model $\underline{\mathbb{N}}_2$ (page 79). In particular, all theorems of V^0 about numbers are true in $\underline{\mathbb{N}}$. Indeed we will show that V^0 is a conservative extension of $I\Delta_0$: all theorems of $I\Delta_0$ are theorems of V^0, and all theorems of V^0 over $\mathcal{L}_A$ are theorems of $I\Delta_0$.

For the first direction, note that the above axiomatization of V^0 contains no explicit induction axioms, so we need to show that it proves the number induction axioms for the Δ_0 formulas. In fact, we will show that it proves Σ_0^B-**IND** by showing first that it proves the X-**MIN** axiom, where

$$X\text{-}MIN \equiv 0 < |X| \supset \exists x < |X|(X(x) \wedge \forall y < x \neg X(y)).$$

LEMMA V.1.6. $V^0 \vdash X\text{-}MIN$.

PROOF. We reason in V^0: By Σ_0^B-**COMP** there is a set Y such that $|Y| \leq |X|$ and for all $z < |X|$

$$Y(z) \leftrightarrow \forall y \leq z \neg X(y). \tag{50}$$

Thus the set Y consists of the numbers smaller than every element in X. Assuming $0 < |X|$, we will show that $|Y|$ is the least member of X. Intuitively, this is because $|Y|$ is the least number that is larger than any member of Y. Formally, we need to show: (i) $X(|Y|)$, and (ii) $\forall y < |Y| \neg X(y)$. Details are as follows.

First suppose that Y is empty. Then $|Y| = 0$ by **B12** and **L2**, hence (ii) holds vacuously by Exercise V.1.1 (a). Also, $X(0)$ holds, since otherwise $Y(0)$ holds by **B7** and **B9**. Thus we have proved (i).

Now suppose that Y is not empty, i.e., $Y(y)$ holds for some y. Then $y < |Y|$ by **L1**, and thus $|Y| \neq 0$ by Exercise V.1.1 (a). By **B12**, $|Y| = z+1$ for some z and hence $(Y(z) \land \neg Y(z+1))$ by **L1** and **L2**. Hence by (50) we have

$$\forall y \leq z \neg X(y) \land \exists i \leq z + 1 X(i).$$

It follows that $i = z + 1$ in the second conjunct, since if $i < z + 1$ then $i \leq z$ by **B11**, which contradicts the first conjunct. This establishes (i) and (ii), since $i = z + 1 = |Y|$. □

Consider the following instance of $\Sigma_0^B\text{-}IND$:

$$X\text{-}IND \equiv \big(X(0) \land \forall y < z(X(y) \supset X(y + 1)) \big) \supset X(z).$$

Corollary V.1.7. $V^0 \vdash X\text{-}IND$.

Proof. We prove by contradiction. Assume $\neg X\text{-}IND$, then we have for some z:

$$X(0) \land \neg X(z) \land \forall y < z(X(y) \supset X(y + 1)).$$

By $\Sigma_0^B\text{-}COMP$, there is a set Y with $|Y| \leq z + 1$ such that

$$\forall y < z + 1(Y(y) \leftrightarrow \neg X(y)).$$

Then $Y(z)$ holds by Exercise V.1.1 (b), so $0 < |Y|$ by (a) and **L1**. By $Y\text{-}MIN$, Y has a least element y_0. Then $y_0 \neq 0$ because $X(0)$, hence $y_0 = x_0 + 1$ for some x_0, by **B12**. But then we must have $X(x_0)$ and $\neg X(x_0 + 1)$, which contradicts our assumption. □

Corollary V.1.8. *Let* T *be an extension of* V^0 *and* Φ *be a set of formulas in* T. *Suppose that* T *proves the* $\Phi\text{-}COMP$ *axiom scheme. Then* T *also proves the* $\Phi\text{-}IND$ *axiom scheme, the* $\Phi\text{-}MIN$ *axiom scheme, and the* $\Phi\text{-}MAX$ *axiom scheme.*

Proof. We show that T proves the $\Phi\text{-}IND$ axiom scheme. This will show that V^0 proves $\Sigma_0^B\text{-}IND$, and hence extends $I\Delta_0$ and proves the arithmetic properties in Examples III.1.8 and III.1.9. The proof for the $\Phi\text{-}MIN$ and $\Phi\text{-}MAX$ axiom schemes is similar to that for $\Phi\text{-}IND$, but easier since these properties are now available.

Let $\varphi(x) \in \Phi$. We need to show that

$$T \vdash \big(\varphi(0) \land \forall y(\varphi(y) \supset \varphi(y + 1)) \big) \supset \varphi(z).$$

Reasoning in V^0, assume

$$\varphi(0) \land \forall y(\varphi(y) \supset \varphi(y + 1)). \tag{51}$$

By Φ-*COMP*, there exists X such that $|X| \leq z + 1$ and

$$\forall y < z + 1 \ (X(y) \leftrightarrow \varphi(y)). \tag{52}$$

By **B**11, Exercise V.1.1 (c,e) and (51) we conclude from this

$$X(0) \wedge \forall y < z(X(y) \supset X(y + 1)).$$

Finally $X(z)$ follows from this and X-*IND*, and so $\varphi(z)$ follows from (52) and Exercise V.1.1 (b).   □

It follows from the corollary that for all $i \geq 0$, V^i proves Σ_i^B-*IND*, Σ_i^B-*MIN*, and Σ_i^B-*MAX*.

THEOREM V.1.9. *V^0 is a conservative extension of $I\Delta_0$.*

PROOF. The axioms for $I\Delta_0$ consist of **B**1, ... , **B**8 and the Δ_0-*IND* axioms. Since **B**1, ... , **B**8 are also axioms of V^0, and we have just shown that V^0 proves the Σ_0^B-*IND* axioms (which include the Δ_0-*IND* axioms), it follows that V^0 extends $I\Delta_0$. To show that V^0 is conservative over $I\Delta_0$ (i.e. theorems of V^0 in the vocabulary of $I\Delta_0$ are also theorems of $I\Delta_0$), we prove the following lemma.

LEMMA V.1.10. *Any model $\mathcal{M}$ for $I\Delta_0$ can be expanded to a model $\mathcal{M}'$ for V^0, where the "number" part of $\mathcal{M}'$ is $\mathcal{M}$.*

Note that Theorem V.1.9 follows immediately from the above lemma, because if φ is in the vocabulary of $I\Delta_0$, then the truth of φ in $\mathcal{M}'$ depends only on the truth of φ in $\mathcal{M}$. (See the proof of the Extension by Definition Theorem III.3.5.)   □

PROOF OF LEMMA V.1.10. Suppose that $\mathcal{M}$ is a model of $I\Delta_0$ with universe $M = U_1$. Recall that $I\Delta_0$ proves **B**1, ... , **B**12, so $\mathcal{M}$ satisfies these axioms. According to the semantics for $\mathcal{L}_A^2$ (Section IV.2.2), to expand $\mathcal{M}$ to a model $\mathcal{M}'$ for V^0 we must construct a suitable universe U_2 whose elements are determined by pairs (m, S), where $S \subseteq M$ and $m = |S|$. In order to satisfy axioms L1 and L2, if $S \in U_2$ is empty, then $|S| = 0$, and if S is nonempty, then S must have a largest element s and $|S| = s + 1$. Since $S \subseteq M$ and $|S|$ is determined by S, it follows that the extensionality axiom SE is satisfied.

The other requirement for U_2 is that the Σ_0^B-*COMP* axioms must be satisfied. We will construct U_2 to consist of all bounded subsets of M defined by Δ_0-formulas with parameters in M. We use the following conventional notation: If $\varphi(x)$ is a formula and c is an element in M, then $\varphi(c)$ represents $\varphi(x)$ with a constant symbol (also denoted c) substituted for x in φ, where it is understood that the symbol c is interpreted as the element c in M. If $\varphi(x, \vec{y})$ is a formula and $c, \vec{d}$ are elements of M, we use the notation

$$S(c, \varphi(x, \vec{d})) = \{e \in M | e < c \text{ and } \mathcal{M} \text{ satisfies } \varphi(e, \vec{d})\}.$$

Then we define

$$U_2 = \{S(c, \varphi(x, \vec{d})) : c, d_1, \ldots, d_k \in M \text{ and}$$
$$\varphi(x, \vec{y}) \text{ is a } \Delta_0(\mathcal{L}_A) \text{ formula}\}. \quad (53)$$

We must show that every nonempty element S of U_2 has a largest element, so that $|S|$ can be defined to satisfy L1 and L2. The largest element exists because the differences between the upper bound c for S and elements of S have a minimum element, by Δ_0-*MIN*. Specifically, if $S = S(c, \varphi(x, \vec{d}))$ is nonempty and m is the least z satisfying $\varphi(c \dot{-} 1 \dot{-} z, \vec{d})$, then define $|S| = \ell_\varphi(c, \vec{d})$ where $\ell_\varphi(c, \vec{d}) = c \dot{-} m$. Then

$$\ell_\varphi(c, \vec{d}) = \begin{cases} \sup(S(c, \varphi(x, \vec{d}))) + 1 & \text{if } S \neq \varnothing, \\ 0 & \text{otherwise.} \end{cases}$$

The preceding argument shows that the function $\ell_\varphi(z, \vec{y})$ is provably total in $I\Delta_0$.

It remains to show that Σ_0^B-*COMP* holds in $\mathcal{M}'$. This means that for every Σ_0^B formula $\psi(z, \vec{x}, \vec{Y})$ (with all free variables indicated) and for every vector $\vec{d}$ of elements of M interpreting $\vec{x}$ and every vector $\vec{S}$ of elements in U_2 interpreting $\vec{Y}$ and for every $c \in M$, the set

$$T = \{e \in M : e < c \text{ and } \mathcal{M}' \models \psi(e, \vec{d}, \vec{S})\} \quad (54)$$

must be in U_2. Suppose that

$$S_i = S(c_i, \varphi_i(u, \vec{d_i}))$$

for some Δ_0 formulas $\varphi_i(x, \vec{y_i})$. Let $\theta(z, \vec{x}, \vec{y_1}, \vec{y_2}, \ldots, w_1, w_2, \ldots)$ be the result of replacing every sub-formula of the form $Y_i(t)$ in $\psi(z, \vec{x}, \vec{Y})$ by $(\varphi_i(t, \vec{y_i}) \wedge t < w_i)$ and every occurrence of $|Y_i|$ by $\ell_{\varphi_i}(w_i, \vec{y_i})$. (We may assume that ψ has no occurrence of $=_2$ by replacing every equation $X =_2 Z$ by a Σ_0^B formula using the extensionality axiom **SE**.) Finally let

$$T = S(c, \theta(z, \vec{d}, \vec{d_1}, \vec{d_2}, \ldots, c_1, c_2, \ldots)).$$

Then T satisfies (54). Since the functions ℓ_{φ_i} are Σ_1-definable in $I\Delta_0$, by the Conservative Extension Lemma III.3.10, θ can be transformed into an equivalent $\Delta_0(\mathcal{L}_A)$ formula. Thus $T \in U_2$. □

EXERCISE V.1.11. Suppose that instead of defining U_2 according to (53), we defined U_2 to consist of all subsets of M which have a largest element, together with $\varnothing$. Then for each set $S \subset U_1$ in U_2 we define $|S|$ in the obvious way to satisfy axioms L1 and L2. Prove that if $\mathcal{M}$ is a nonstandard model of $I\Delta_0$, then the resulting two-sorted structure (U_1, U_2) is not a model of V^0.

EXERCISE V.1.12. Suppose that we want to prove that V^0 is conservative over $I\Delta_0$ by considering an anchored LK^2 proof instead of the above model-theoretic argument. Here we consider a small part of such an argument. Suppose that φ is a formula in the vocabulary of $I\Delta_0$ and π is an anchored $\boldsymbol{LK^2}\text{-}\boldsymbol{V^0}$ proof of $\longrightarrow \varphi$. Suppose (to make things easy) that no formula in π contains a string quantifier. Show explicitly how to convert π to an $\boldsymbol{LK}\text{-}I\Delta_0$ proof π' of $\longrightarrow \varphi$.

Since according to Theorem V.1.9 V^0 extends $I\Delta_0$, we will freely use the results in Chapter III when reasoning in V^0 in the sequel.

V.2. Two-Sorted Functions

Complexity classes of two-sorted relations were discussed in Section IV.3. Now we associate with each two-sorted complexity class $\mathbf{C}$ of relations a two-sorted function class $\boldsymbol{FC}$. Two-sorted functions are either *number functions* or *string functions*. A number function $f(\vec{x}, \vec{Y})$ takes values in $\mathbb{N}$, and a string function $F(\vec{x}, \vec{Y})$ takes finite subsets of $\mathbb{N}$ as values.

DEFINITION V.2.1. A function f or F is *polynomially bounded* (or *p-bounded*) if there is a polynomial $p(\vec{x}, \vec{y})$ such that $f(\vec{x}, \vec{Y}) \leq p(\vec{x}, |\vec{Y}|)$ or $|F(\vec{x}, \vec{Y})| \leq p(\vec{x}, |\vec{Y}|)$.

All function complexity classes we consider here contain only p-bounded functions.

In defining the functions associated with a complexity class of relations the natural relation to use for a number function is its graph. However this does not work well for string functions. For example the function $F(X)$ which gives the prime factorization of X (considered as a binary number) is not known to be polynomial time computable, but its graph is a polynomial time relation. It turns out that the right relation to associate with a string function is its *bit graph*.

DEFINITION V.2.2 (Graph, Bit Graph). The *graph* G_f of a number function $f(\vec{x}, \vec{Y})$ is defined by

$$G_f(z, \vec{x}, \vec{Y}) \leftrightarrow z = f(\vec{x}, \vec{Y}).$$

The *bit graph* B_F of a string function $F(\vec{x}, \vec{Y})$ is defined by

$$B_F(i, \vec{x}, \vec{Y}) \leftrightarrow F(\vec{x}, \vec{Y})(i).$$

DEFINITION V.2.3 (Function Class). If $\mathbf{C}$ is a two-sorted complexity class of relations, then the corresponding function class $\boldsymbol{FC}$ consists of all p-bounded number functions whose graphs are in $\mathbf{C}$, together with all p-bounded string functions whose bit graphs are in $\mathbf{C}$.

In particular, the string functions in $\boldsymbol{FAC^0}$ are those p-bounded functions whose bit graphs are in AC^0. The nonuniform version $\boldsymbol{FAC^0}/poly$ has a nice circuit characterization like that of $AC^0/poly$ (see page 75).

Thus a string function $F(X)$ is in $\boldsymbol{FAC^0}/poly$ iff there is a polynomial size bounded depth family $\langle C_n \rangle$ of Boolean circuits (with unbounded fan-in $\land$-gates and $\lor$-gates) such that each C_n has n input bits specifying the input string X, and the output bits of C_n specify the string $F(X)$.

The following characterization of $\boldsymbol{FAC^0}$ follows from the above definitions and the Σ_0^B Representation Theorem (Theorem IV.3.6).

COROLLARY V.2.4. *A string function is in* $\boldsymbol{FAC^0}$ *if and only if it is p-bounded, and its bit graph is represented by a* Σ_0^B *formula. The same holds for a number function, with graph replacing bit graph.*

An interesting example of a string function in $\boldsymbol{FAC^0}$ is binary addition. Note that as in (46) we can treat a finite subset $X \subset \mathbb{N}$ as the natural number

$$bin(X) = \sum_i X(i)2^i$$

where we write 0 for $\perp$ and 1 for $\top$. We will write $X + Y$ for the string function "binary addition", so $X + Y = bin(X) + bin(Y)$. Let $Carry(i, X, Y)$ hold iff there is a carry into bit position i when computing $X + Y$. Then $Carry(i, X, Y)$ is represented by the Σ_0^B formula given in (47).

The bit graph of $X + Y$ can be defined as follows.

EXAMPLE V.2.5 (Bit Graph of String Addition). The bit graph of $X + Y$ is

$$(X + Y)(i) \leftrightarrow \left(i < |X| + |Y| \land (X(i) \oplus Y(i) \oplus Carry(i, X, Y)) \right) \quad (55)$$

where $p \oplus q \equiv ((p \land \neg q) \lor (\neg p \land q))$.

In general, the graph $G_F(\vec{x}, \vec{Y}, Z) \equiv (Z = F(\vec{x}, \vec{Y}))$ of a string function $F(\vec{x}, \vec{Y})$ can be defined from its bit graph as follows:

$$G_F(\vec{x}, \vec{Y}, Z) \leftrightarrow \forall i \, (Z(i) \leftrightarrow B_F(i, \vec{x}, \vec{Y})).$$

So if F is polynomially bounded and its bit graph is in $\boldsymbol{AC^0}$, then its graph is also in $\boldsymbol{AC^0}$, because

$$G_F(\vec{x}, \vec{Y}, Z) \leftrightarrow \left(|Z| \leq t \land \forall i < t \, (Z(i) \leftrightarrow B_F(i, \vec{x}, \vec{Y})) \right) \quad (56)$$

where t is the bound on the length of F.

As we noted earlier (Section IV.1), the relation $R_\times$ is not in $\boldsymbol{AC^0}$, where

$$R_\times(X, Y, Z) \leftrightarrow bin(X) \cdot bin(Y) = bin(Z)$$

(because PARITY, which is not in $\boldsymbol{AC^0}$, is reducible to it). As a result, the bit graph of $(X \times Y)(i)$ is not representable by any Σ_0^B formula, where $X \times Y = bin(X) \cdot bin(Y)$ is the string function "binary multiplication".

If a string function $F(X)$ is polynomially bounded, it is not enough to say that its graph is an $\boldsymbol{AC^0}$ relation in order to ensure that $F \in \boldsymbol{FAC^0}$. For example, let M be a fixed polynomial-time Turing machine, and define $F(X)$ to be a string coding the computation of M on input X. If the computation is nicely encoded then $F(X)$ is polynomially bounded and the graph $Y = F(X)$ is an AC^0 relation, but if the Turing machine computes a function not in AC^0 (such as the number of ones in X) then $F \notin FAC^0$.

For the same reason that the numerical AC^0 relations in the two-sorted setting are precisely the LTH relations in the single-sorted setting (see the proof of the Σ_0^B Representation Theorem, IV.3.6), number functions with no string arguments are AC^0 functions iff they are single-sorted LTH functions.

The nonuniform version of FAC^0 consists of functions computable by bounded-depth polynomial-size circuits, and it is clear from this definition that the class is closed under composition. It is also clear that nonuniform AC^0 is closed under substitution of (nonuniform) AC^0 functions for parameters. These are some of the natural properties that also hold for uniform AC^0 and FAC^0.

EXERCISE V.2.6. Show that a number function $f(\vec{x}, \vec{X})$ is in FAC^0 if and only if

$$f(\vec{x}, \vec{X}) = |F(\vec{x}, \vec{X})|$$

for some string function $F(\vec{x}, \vec{X})$ in FAC^0.

THEOREM V.2.7. (a) *The AC^0 relations are closed under substitution of AC^0 functions for variables.*
(b) *The AC^0 functions are closed under composition.*
(c) *The AC^0 functions are closed under definition by cases, i.e., if φ is an AC^0 relation, g, h and G, H are functions in FAC^0, then the functions f and F defined by*

$$f = \begin{cases} g & \text{if } \varphi, \\ h & \text{otherwise} \end{cases} \qquad F = \begin{cases} G & \text{if } \varphi, \\ H & \text{otherwise} \end{cases}$$

are also in FAC^0.

PROOF. We will prove (a) for the case of substituting a string function for a string variable. The case of substituting a number function for a number variable is left as an easy exercise. Part (b) follows easily from part (a). We leave part (c) as an exercise.

Suppose that $R(\vec{x}, \vec{X}, Y)$ is an AC^0 relation and $F(\vec{x}, \vec{X})$ an AC^0 function. We need to show that the relation $Q(\vec{x}, \vec{X}) \equiv R(\vec{x}, \vec{X}, F(\vec{x}, \vec{X}))$ is also an AC^0 relation, i.e., it is representable by some Σ_0^B formula.

By the Σ_0^B Representation Theorem (IV.3.6) there is a Σ_0^B formula $\varphi(\vec{x}, \vec{X}, Y)$ that represents R:

$$R(\vec{x}, \vec{X}, Y) \leftrightarrow \varphi(\vec{x}, \vec{X}, Y).$$

By Corollary V.2.4 there is a Σ_0^B formula $\theta(i, \vec{x}, \vec{X})$ and a number term $t(\vec{x}, \vec{X})$ such that

$$F(\vec{x}, \vec{X})(i) \leftrightarrow i < t(\vec{x}, \vec{X}) \wedge \theta(i, \vec{x}, \vec{X}). \qquad (57)$$

It follows from Exercise V.2.6 that the relation $z = |F(\vec{x}, \vec{X})|$ is represented by a Σ_0^B formula η, so

$$z = |F(\vec{x}, \vec{X})| \leftrightarrow \eta(z, \vec{x}, \vec{X}). \qquad (58)$$

The Σ_0^B formula that represents the relation $Q(\vec{x}, \vec{X})$ is obtained from $\varphi(\vec{x}, \vec{X}, Y)$ by successively eliminating each occurrence of Y using (57) and (58) as follows.

First eliminate all atomic formulas of the form $Y = Z$ (or $Z = Y$) in φ by replacing them with equivalent formulas using the extensionality axiom **SE**. Thus

$$Y = Z \leftrightarrow (|Y| = |Z|) \wedge \forall i < |Y|(Y(i) \leftrightarrow Z(i)).$$

Now Y can only occur in the form $|Y|$ or $Y(r)$, for some term r. Any occurrence of $|Y|$ in $\varphi(\vec{x}, \vec{X}, Y)$ must be in the context of an atomic formula $\psi(\vec{x}, \vec{X}, |Y|)$, which we replace with

$$\exists z \leq t(\vec{x}, \vec{X})\,(\eta(z, \vec{x}, \vec{X}) \wedge \psi(\vec{x}, \vec{X}, z)).$$

Finally we replace each occurrence of $Y(r)$ in $\varphi(\vec{x}, \vec{X}, Y)$ by

$$r < t(\vec{x}, \vec{X}) \wedge \theta(r, \vec{x}, \vec{X}).$$

The result is a Σ_0^B formula which represents $Q(\vec{x}, \vec{X})$. $\square$

EXERCISE V.2.8. Prove part (a) of Theorem V.2.7 for the case of substitution of number functions for variables. Also prove parts (b) and (c) of the theorem.

V.3. Parikh's Theorem for Two-Sorted Logic

Recall (Section III.2) that a term $t(\vec{x})$ is a bounding term for a function symbol f in a single-sorted theory $\mathcal{T}$ if

$$\mathcal{T} \vdash \forall \vec{x}\, f(\vec{x}) \leq t(\vec{x}).$$

For a two-sorted theory $\mathcal{T}$ whose vocabulary is an extension of $\mathcal{L}_A^2$, we say that a number term $t(\vec{x}, \vec{X})$ is a bounding term for a number function f in $\mathcal{T}$ if

$$\mathcal{T} \vdash \forall \vec{x} \forall \vec{X}\, f(\vec{x}, \vec{X}) \leq t(\vec{x}, \vec{X}).$$

Also, $t(\vec{x}, \vec{X})$ is a bounding term for a string function F in $\mathcal{T}$ if

$$\mathcal{T} \vdash \forall \vec{x} \forall \vec{X}\, |F(\vec{x}, \vec{X})| \leq t(\vec{x}, \vec{X}).$$

DEFINITION V.3.1. A number function or a string function is *polynomially bounded* in T if it has a bounding term in the vocabulary $\mathcal{L}_A^2$.

EXERCISE V.3.2. Let T be a two-sorted theory over the vocabulary $\mathcal{L} \supseteq \mathcal{L}_A^2$. Suppose that T extends $I\Delta_0$. Show that if the functions of $\mathcal{L}$ are polynomially bounded in T, then for each number term $s(\vec{x}, \vec{X})$ and string term $T(\vec{x}, \vec{X})$ of $\mathcal{L}$, there is an $\mathcal{L}_A^2$-number term $t(\vec{x}, \vec{X})$ such that

$$T \vdash \forall \vec{x} \forall \vec{X}\, s(\vec{x}, \vec{X}) \leq t(\vec{x}, \vec{X}) \quad \text{and} \quad T \vdash \forall \vec{x} \forall \vec{X}\, |T(\vec{x}, \vec{X})| \leq t(\vec{x}, \vec{X}).$$

Note that a bounded formula is one in which every quantifier (both string and number quantifiers) is bounded. Recall the definition of a polynomial-bounded single-sorted theory (Definition III.2.2).

In two-sorted logic, a polynomial-bounded theory is required to extend V^0. The formal definition follows.

DEFINITION V.3.3 (Polynomial-Bounded Two-Sorted Theory). Let T be a two-sorted theory over the vocabulary $\mathcal{L}$. Then T is a *polynomial-bounded theory* if (i) it extends V^0; (ii) it can be axiomatized by a set of bounded formulas; and (iii) each function f or F in $\mathcal{L}$ is polynomially bounded in T.

Note that each theory $V^i, i \geq 0$, is a polynomial-bounded theory. In fact, all two-sorted theories considered in this book are polynomial-bounded.

THEOREM V.3.4 (Parikh's Theorem, Two-Sorted Case). *Suppose that T is a polynomial-bounded theory and $\varphi(\vec{x}, \vec{y}, \vec{X}, \vec{Y})$ is a bounded formula with all free variables indicated such that*

$$T \vdash \forall \vec{x} \forall \vec{X} \exists \vec{y} \exists \vec{Y} \varphi(\vec{x}, \vec{y}, \vec{X}, \vec{Y}). \tag{59}$$

Then

$$T \vdash \forall \vec{x} \forall \vec{X} \exists \vec{y} \leq t \exists \vec{Y} \leq t \varphi(\vec{x}, \vec{y}, \vec{X}, \vec{Y}) \tag{60}$$

for some $\mathcal{L}_A^2$-term $t = t(\vec{x}, \vec{X})$ containing only the variables $(\vec{x}, \vec{X})$.

It follows from Exercise V.3.2 that the bounding term t can be taken to be a term in $\mathcal{L}_A^2$.

It suffices to prove the following simple form of the above theorem.

LEMMA V.3.5. *Suppose that T is a polynomial-bounded theory, and $\varphi(z, \vec{x}, \vec{X})$ is a bounded formula with all free variables indicated such that*

$$T \vdash \forall \vec{x} \forall \vec{X} \exists z \varphi(z, \vec{x}, \vec{X}).$$

Then

$$T \vdash \forall \vec{x} \forall \vec{X} \exists z \leq t(\vec{x}, \vec{X}) \varphi(z, \vec{x}, \vec{X})$$

for some term $t(\vec{x}, \vec{X})$ with all variables indicated.

Proof of Parikh's Theorem from Lemma V.3.5. Define (omitting $\vec{x}$ and $\vec{X}$)

$$\psi(z) \equiv \exists \vec{y} \leq z \exists \vec{Y} \leq z \varphi(\vec{y}, \vec{Y}).$$

From the assumption (59) we conclude that $\mathcal{T} \vdash \exists z \psi(z)$, since we can take

$$z = y_1 + \cdots + y_k + |Y_1| + \cdots + |Y_\ell|.$$

Since φ is a bounded formula, ψ is also a bounded formula. By the lemma, we conclude that $\mathcal{T}$ proves $\exists z \leq t \psi(z)$, where the variables in t satisfy Parikh's Theorem. Thus (60) follows.   □

Proof of Lemma V.3.5. The proof is the same as the proof of Parikh's Theorem in the single-sorted logic (page 46), with minor modifications. Refer to Section IV.4 for the system LK^2. Here we consider an anchored LK^2-$\mathcal{T}$ proof π of $\exists z \varphi(z, \vec{a}, \vec{\alpha})$, where $\mathcal{T}$ is the set of all term substitution instances of axioms of $\mathcal{T}$ (note that now we have both the substitution of number terms for number variables and string terms for string variables). We assume that π is in free variable normal form (see Section IV.4.1).

We convert π to a proof π' by converting each sequent S in π into a sequent S' and providing an associated derivation $D(S)$, where S' and $D(S)$ are defined by induction on the depth of S in π so that the following is satisfied:

Induction Hypothesis. If S has no occurrence of $\exists y \varphi$, then $S' = S$. If S has one or more occurrences of $\exists y \varphi$, then S' is a sequent which is the same as S except all occurrences of $\exists y \varphi$ are replaced by a single occurrence of $\exists y \leq t \varphi$, where t is an $\mathcal{L}_A^2$-number term that depends on S and the placement of S in π. Further every variable in t occurs free in the original sequent S.

As discussed in Section IV.4.1, if the underlying vocabulary has no string constant symbol (for example $\mathcal{L}_A^2$), then we allow the string constant $\varnothing$ to occur in π, in order to assume that it is in free variable normal form. Thus the bounding term t in the endsequent $\longrightarrow \exists y \leq t \varphi$ may contain $\varnothing$. Since t is an $\mathcal{L}_A^2(\varnothing)$-term, each occurrence of $\varnothing$ is in the context $|\varnothing|$, and hence can be replaced by 0 using the axiom **E**: $|\varnothing| = 0$.

The *Cases I–V* are supplemented to consider the four string quantifier rules, which are treated in the same way as their LK counterparts.   □

V.4. Definability in V^0

Recall the notion of Φ-definable single-sorted function (Definition III.3.2). For a two-sorted theory $\mathcal{T}$, this notion is defined in the same way for functions of each sort, and in particular $\mathcal{T}$ must be able to prove existence and uniqueness of function values.

DEFINITION V.4.1 (Two-Sorted Definability). Let $\mathcal{T}$ be a theory with vocabulary $\mathcal{L} \supseteq \mathcal{L}_A^2$, and let Φ be a set of $\mathcal{L}$-formulas. A number function f is Φ-*definable in* $\mathcal{T}$ if there is a formula $\varphi(\vec{x}, y, \vec{X})$ in Φ such that

$$\mathcal{T} \vdash \forall \vec{x} \forall \vec{X} \exists! y \varphi(\vec{x}, y, \vec{X}) \tag{61}$$

and

$$y = f(\vec{x}, \vec{X}) \leftrightarrow \varphi(\vec{x}, y, \vec{X}). \tag{62}$$

A string function F is Φ-*definable in* $\mathcal{T}$ if there is a formula $\varphi(\vec{x}, \vec{X}, Y)$ in Φ such that

$$\mathcal{T} \vdash \forall \vec{x} \forall \vec{X} \exists! Y \varphi(\vec{x}, \vec{X}, Y) \tag{63}$$

and

$$Y = F(\vec{x}, \vec{X}) \leftrightarrow \varphi(\vec{x}, \vec{X}, Y). \tag{64}$$

Then (62) is a *defining axiom* for f and (64) is a *defining axiom* for F, and we write $\mathcal{T}(f)$ or $\mathcal{T}(F)$ for the theory extending $\mathcal{T}$ by adding f or F and its corresponding defining axiom to $\mathcal{T}$. We say that f or F is *definable in* $\mathcal{T}$ if it is Φ-definable in $\mathcal{T}$ for some Φ.

Note that if f (or F) is in $\mathcal{L}$ then it can be trivially defined in any theory over $\mathcal{L}$ by the formula $y = f(\vec{x}, \vec{X})$ (or $Y = F(\vec{x}, \vec{X})$).

THEOREM V.4.2 (Two-Sorted Extension by Definition). $\mathcal{T}(f)$ *and* $\mathcal{T}(F)$ (*as defined above*) *are conservative extensions of* $\mathcal{T}$.

PROOF. This is proved in the same way as its single-sorted version Theorem III.3.5. □

If Φ is the set of all $\mathcal{L}_A^2$-formulas, then every arithmetical function (that is, every function whose graph is represented by an $\mathcal{L}_A^2$-formula) is Φ-definable in V^0. To see this, suppose that $F(\vec{x}, \vec{X})$ has defining axiom (64). Then the graph of F is also defined by the following formula $\varphi'(\vec{x}, \vec{X}, Y)$:

$$(\exists! Z \varphi(\vec{x}, \vec{X}, Z) \wedge \varphi(\vec{x}, \vec{X}, Y)) \vee (\neg \exists! Z \varphi(\vec{x}, \vec{X}, Z) \wedge Y = \varnothing).$$

Then (63) with φ' for φ is trivially provable in V^0.

We want to choose a standard class Φ of formulas such that the class of Φ-definable functions in a theory $\mathcal{T}$ depends nicely on the proving power of $\mathcal{T}$, so that various complexity classes can be characterized by fixing Φ and varying $\mathcal{T}$. In single-sorted logic, our choice for Φ was Σ_1, and we defined the provably total functions of $\mathcal{T}$ to be the Σ_1-definable functions in $\mathcal{T}$. Here our choice for Φ is Σ_1^1 (recall (Definition IV.3.2) that a Σ_1^1 formula is a formula of the form $\exists \vec{X} \varphi$, where φ is a Σ_0^B formula). The notion of a *provably total function* in two-sorted logic is defined as follows.

DEFINITION V.4.3 (Provably Total Function). A function (which can be either a number function or a string function) is said to be *provably total* in a theory T iff it is Σ_1^1-definable in T.

If T consists of all formulas of $\mathcal{L}_A^2$ which are true in the standard model $\underline{\mathbb{N}}_2$, then the functions provably total in T are precisely all total functions computable on a Turing machine. The idea here is that the existential string quantifiers in a Σ_1^1 formula can be used to code the computation of a Turing machine computing the function. If T is a polynomially bounded theory, then both the function values and the computation must be polynomially bounded. In fact, the following result in a corollary of Parikh's Theorem.

COROLLARY V.4.4. *Let T be a polynomial-bounded theory. Then all provably total functions in T are polynomially bounded. A function is provably total in T iff it is Σ_1^B-definable in T.*

We will show that the provably total functions in V^0 are precisely the functions in FAC^0, and in the next chapter we will show that the provably total functions in V^1 are precisely the polynomial time functions. Later we will give similar characterizations of other complexity classes.

EXERCISE V.4.5. Show that for any theory T whose vocabulary includes $\mathcal{L}_A^2$, the set of provably total functions of T is closed under composition.

In two-sorted logic, for string functions we have the notion of a *bit-definable function* in addition to that of a definable function.

DEFINITION V.4.6 (Bit-definable Function). Let Φ be a set of $\mathcal{L}$ formulas where $\mathcal{L} \supseteq \mathcal{L}_A^2$. We say that a string function symbol $F(\vec{x}, \vec{Y})$ not in $\mathcal{L}$ is Φ-*bit-definable* from $\mathcal{L}$ if there is a formula $\varphi(i, \vec{x}, \vec{Y})$ in Φ and an $\mathcal{L}_A^2$-number term $t(\vec{x}, \vec{Y})$ such that the bit graph of F satisfies

$$F(\vec{x}, \vec{Y})(i) \leftrightarrow (i < t(\vec{x}, \vec{Y}) \wedge \varphi(i, \vec{x}, \vec{Y})). \qquad (65)$$

We say that the formula on the RHS of (65) is a *bit-defining axiom*, or *bit definition*, of F.

The choice of φ and t in the above definition is not uniquely determined by F. However we will assume that a specific formula φ and a specific number term t has been chosen, so we will speak of *the bit-defining axiom*, or *the bit definition*, of F. Note also that such a F is polynomially bounded in T, and t is a bounding term for F.

The following proposition follows easily from the above definition and Corollary V.2.4.

PROPOSITION V.4.7. *A string function is Σ_0^B-bit-definable iff it is in FAC^0.*

EXERCISE V.4.8. Let T be a theory which extends V^0 and proves the bit-defining axiom (65) for a string function F, where φ is a Σ_0^B formula.

Show that there is a Σ_0^B formula $\eta(z, \vec{x}, \vec{Y})$ such that $\mathcal{T}$ proves

$$z = |F(\vec{x}, \vec{Y})| \leftrightarrow \eta(z, \vec{x}, \vec{Y}).$$

It is important to distinguish between a "definable function" and a "bit-definable function". In particular, if a theory $\mathcal{T}_2$ is obtained from a theory $\mathcal{T}_1$ by adding a Φ-bit-definable function F together with its bit-defining axiom (65), then in general we cannot conclude that $\mathcal{T}_2$ is a conservative extension of $\mathcal{T}_1$. For example, it is easy to show that the string multiplication function $X \times Y$ has a Σ_1^B bit definition. However, as we noted earlier, this function is not Σ_1^B-definable in V^0. The theory that results from adding this function together with its Σ_1^B-bit-definition to V^0 is *not* a conservative extension of V^0.

To get definability, and hence conservativity, it suffices to assume that $\mathcal{T}_1$ proves a comprehension axiom scheme. The following definition is useful here and in Chapter VI.

DEFINITION V.4.9 (Σ_0^B-Closure). Let Φ be a set of formulas over a vocabulary $\mathcal{L}$ which extends $\mathcal{L}_A^2$. Then $\Sigma_0^B(\Phi)$ is the closure of Φ under the operations $\neg, \wedge, \vee$ and bounded number quantification. That is, if φ and ψ are formulas in $\Sigma_0^B(\Phi)$ and t is an $\mathcal{L}_A^2$-term not containing x, then the following formulas are also in $\Sigma_0^B(\Phi)$: $\neg\varphi$, $(\varphi \wedge \psi)$, $(\varphi \vee \psi)$, $\forall x \leq t\varphi$, and $\exists x \leq t\varphi$.

LEMMA V.4.10 (Extension by Bit Definition). *Let $\mathcal{T}$ be a theory over $\mathcal{L}$ that contains V^0, and Φ be a set of $\mathcal{L}$-formulas such that $\Phi \supseteq \Sigma_0^B$. Suppose that $\mathcal{T}$ proves the Φ-COMP axiom scheme. Then any polynomially bounded number function whose graph is Φ-representable, or a polynomially bounded string function whose bit graph is Φ-representable, is $\Sigma_0^B(\Phi)$-definable in $\mathcal{T}$.*

PROOF. Consider the case of a string function. Suppose that F is a polynomially bounded string function with bit graph in Φ, so there are an $\mathcal{L}_A^2$-number term t and a formula $\varphi \in \Phi$ such that

$$F(\vec{x}, \vec{Y})(i) \leftrightarrow (i < t(\vec{x}, \vec{Y}) \wedge \varphi(i, \vec{x}, \vec{Y})).$$

As in (56), the graph G_F of F can be defined as follows:

$$G_F(\vec{x}, \vec{Y}, Z) \equiv |Z| \leq t \wedge \forall i < t\, (Z(i) \leftrightarrow \varphi(i, \vec{x}, \vec{Y})). \qquad (66)$$

Now since $\mathcal{T}$ proves the Φ-COMP, we have

$$\mathcal{T} \vdash \forall \vec{x} \forall \vec{Y} \exists Z\, G_F(\vec{x}, \vec{Y}, Z). \qquad (67)$$

Also $\mathcal{T}$ proves that such Z is unique, by the extensionality axiom SE in 2-BASIC. Since the formula $G_F(\vec{x}, \vec{Y}, Z)$ is in $\Sigma_0^B(\Phi)$, it follows that F is $\Sigma_0^B(\Phi)$-definable in $\mathcal{T}$.

Next consider the case of a number function. Let f be a polynomially bounded number function whose graph is in Φ, so there are an $\mathcal{L}_A^2$-number term t and a formula $\varphi \in \Phi$ such that

$$y = f(\vec{x}, \vec{X}) \leftrightarrow (y < t(\vec{x}, \vec{X}) \wedge \varphi(y, \vec{x}, \vec{X})).$$

By Corollary V.1.8, $\mathcal{T}$ proves the Φ-**MIN** axiom scheme. Therefore f is definable in $\mathcal{T}$ by using the following $\Sigma_0^B(\Phi)$ formula for its graph:

$$G_f(y, \vec{x}, \vec{X}) \equiv (\forall z < y \neg \varphi(z, \vec{x}, \vec{X})) \wedge (y < t \supset \varphi(y, \vec{x}, \vec{X})) \qquad (68)$$

(i.e., y is the least number $< t$ such that $\varphi(y)$ holds, or t if no such y exists).   □

In this lemma, if we take $\mathcal{T} = V^0$ and $\Phi = \Sigma_0^B$, then (since $\Sigma_0^B(\Sigma_0^B) = \Sigma_0^B$) we can apply Corollary V.2.4 and Proposition V.4.7 to obtain the following:

COROLLARY V.4.11. *Every function in* $\boldsymbol{FAC}^0$ *is* Σ_0^B*-definable in* V^0.

This result can be generalized, using the following definition.

DEFINITION V.4.12. [3] *A string function is* Σ_0^B*-definable from a collection* $\mathcal{L}$ *of two-sorted functions and relations if it is p-bounded and its bit graph is represented by a* $\Sigma_0^B(\mathcal{L})$ *formula. Similarly, a number function is* Σ_0^B*-definable from* $\mathcal{L}$ *if it is p-bounded and its graph is represented by a* $\Sigma_0^B(\mathcal{L})$ *formula.*

This "semantic" notion of Σ_0^B-definability should not be confused with Σ_0^B-definability in a theory (Definition V.4.1), which involves provabililty. The next result connects the two notions.

COROLLARY V.4.13. *Let* $\mathcal{T}$ *be a theory over* $\mathcal{L}$ *that contains* V^0, *and suppose that* $\mathcal{T}$ *proves the* $\Sigma_0^B(\mathcal{L})$*-**COMP** axiom scheme. Then a function which is* Σ_0^B*-definable from* $\mathcal{L}$ *is* $\Sigma_0^B(\mathcal{L})$*-definable in* $\mathcal{T}$.

In Section V.5 we will prove the Witnessing Theorem for V^0, which says that any Σ_1^1-definable function of V^0 is in $\boldsymbol{FAC}^0$. This will complete our characterization of $\boldsymbol{FAC}^0$ by V^0. (Compare this with Proposition V.4.7, which characterizes $\boldsymbol{FAC}^0$ in terms of bit-definability, independent of any theory.)

COROLLARY V.4.14. *Suppose that the theory* $\mathcal{T}$ *proves* $\Sigma_0^B(\mathcal{L})$*-**COMP**, where* $\mathcal{L}$ *is the vocabulary of* $\mathcal{T}$. *Then the theory resulting from* $\mathcal{T}$ *by adding the* $\Sigma_0^B(\mathcal{L})$*-defining axioms or the* $\Sigma_0^B(\mathcal{L})$*-bit-defining axioms for a collection of number functions and string functions is a conservative extension of* $\mathcal{T}$.

The following result shows in particular that if we extend V^0 by a sequence of Σ_0^B defining axioms and bit-defining axioms, the resulting theory is not only conservative over V^0, it also proves the $\Sigma_0^B(\mathcal{L})$-**COMP** and $\Sigma_0^B(\mathcal{L})$-**IND** axioms, where $\mathcal{L}$ is the resulting vocabulary. We state it generally for $\Sigma_i^B(\mathcal{L})$ formulas.

---

[3]This notion is important for our definition of AC^0 reduction, Definition IX.1.1.

Lemma V.4.15 (Σ_0^B-Transformation). *Let $\mathcal{T}$ be a polynomial-bounded theory which extends V^0, and assume that the vocabulary $\mathcal{L}$ of $\mathcal{T}$ has the same predicate symbols as $\mathcal{L}_A^2$. Suppose that for every number function f in $\mathcal{L}$, $\mathcal{T}$ proves a $\Sigma_0^B(\mathcal{L}_A^2)$ defining axiom for f, and for every string function F in $\mathcal{L}$, $\mathcal{T}$ proves a $\Sigma_0^B(\mathcal{L}_A^2)$ bit-defining axiom for F. Then for every $i \geq 0$ and every $\Sigma_i^B(\mathcal{L})$ formula φ^+ there is a $\Sigma_i^B(\mathcal{L}_A^2)$ formula φ such that*

$$\mathcal{T} \vdash \varphi^+ \leftrightarrow \varphi.$$

Proof. We prove the conclusion for the case $i = 0$. The case $i > 0$ follows immediately from this case. We may assume by the axiom **SE** that φ^+ does not contain $=_2$. We proceed by induction on the maximum nesting depth of any function symbol in φ^+, where in defining nesting depth we only count functions which are in $\mathcal{L}$ but not in $\mathcal{L}_A^2$. The base case is nesting depth 0, so φ^+ is already a $\Sigma_0^B(\mathcal{L}_A^2)$ formula, and there is nothing to prove.

For the induction step, assume that φ^+ has at least one occurrence of a function not in $\mathcal{L}_A^2$. It suffices to consider the case in which φ^+ is an atomic formula. Since by assumption the only predicate symbols in $\mathcal{L}$ are those in $\mathcal{L}_A^2$, the only predicate symbols we need consider are $\varepsilon, =, \leq$. First consider the case $\in$, so φ^+ has the form $F(\vec{t}, \vec{T})(s)$. Then by assumption $\mathcal{T}$ proves a bit definition of the form

$$F(\vec{x}, \vec{X})(i) \leftrightarrow (i < r(\vec{x}, \vec{X}) \wedge \psi(i, \vec{x}, \vec{X}))$$

where r is an $\mathcal{L}_A^2$ term and ψ is a $\Sigma_0^B(\mathcal{L}_A^2)$ formula. Then $\mathcal{T}$ proves

$$\varphi^+ \leftrightarrow (s < r(\vec{t}, \vec{T}) \wedge \psi(s, \vec{t}, \vec{T})).$$

The RHS has nesting depth at most that of φ^+ and $\vec{t}, \vec{T}$ have smaller nesting depth, and hence we have reduced the induction step to the case that φ^+ has the form $\rho(\vec{s})$ where $\rho(\vec{x})$ is an atomic formula over $\mathcal{L}_A^2$ and each term s_i has one of the forms $f(\vec{t}, \vec{T})$, for f not in $\mathcal{L}_A^2$, or $|F(\vec{t}, \vec{T})|$. In either case, using the defining axiom for f or Exercise V.4.8, for each term s_i there is a $\Sigma_0^B(\mathcal{L}_A^2)$ formula $\eta_i(z, \vec{x}, \vec{X})$ and a bounding term $r_i(\vec{x}, \vec{X})$ of $\mathcal{L}_A^2$ such that $\mathcal{T}$ proves

$$z = s_i \leftrightarrow (z < r_i(\vec{t}, \vec{T}) \wedge \eta_i(z, \vec{t}, \vec{T})).$$

Hence (since φ^+ is $\rho(\vec{s})$), $\mathcal{T}$ proves

$$\varphi^+ \leftrightarrow \exists \vec{z} < \vec{r}(\vec{t}, \vec{T})\big(\rho(\vec{z}) \wedge \bigwedge_i \eta_i(z_i, \vec{t}, \vec{T})\big).$$

Thus we have reduced the nesting depth of φ^+, and we can apply the induction hypothesis. □

The following result is immediate from the preceding lemma, Definitions V.4.12 and V.2.3, and the Σ_0^B Representation Theorem IV.3.6.

COROLLARY V.4.16 (***FAC***^{0} Closed under Σ_0^B-Definability). *Every function Σ_0^B-definable from a collection of **FAC**^{0} functions is in **FAC**^{0}.*

Below we give Σ_0^B-bit-definitions of the string functions $\varnothing$ (zero, or empty string), $S(X)$ (successor), $X + Y$ and several other useful AC^0 functions: *Row, seq, left* and *right*. Each of these functions is Σ_0^B-definable in V^0, and the above lemmas and corollaries apply.

EXAMPLE V.4.17 ($\varnothing, S, +$). The string constant $\varnothing$ has bit defining axiom

$$\varnothing(z) \leftrightarrow z < 0.$$

Binary successor $S(X)$ has bit-defining axiom

$$S(X)(i) \leftrightarrow \big(i \leq |X| \wedge ((X(i) \wedge \exists j < i \neg X(j)) \vee (\neg X(i) \wedge \forall j < i X(j)))\big).$$

Recall from (55) that binary addition $X \mid Y$ has the following bit-defining axiom:

$$(X + Y)(i) \leftrightarrow \big(i < |X| + |Y| \wedge (X(i) \oplus Y(i) \oplus Carry(i, X, Y))\big)$$

where $\oplus$ is exclusive OR, and

$$Carry(i, X, Y) \equiv \exists k < i \big(X(k) \wedge Y(k) \wedge \forall j < i(k < j \supset (X(j) \vee Y(j)))\big).$$

EXERCISE V.4.18. Show that V^0 proves

$$\neg Carry(0, X, Y) \wedge$$
$$Carry(i + 1, X, Y) \leftrightarrow MAJ(Carry(i, X, Y), X(i), Y(i))$$

where the Boolean function $MAJ(P, Q, R)$ holds iff at least two of P, Q, R are true. This formula gives a recursive definition of Carry which is the binary analog to the school method for computing carries in decimal addition.

EXERCISE V.4.19. Let $V^0(\varnothing, S, +)$ be V^0 extended by $\varnothing, S, +$ and their bit-defining axioms. Show that the following are theorems of $V^0(\varnothing, S, +)$:
(a) $X + \varnothing = X$.
(b) $X + S(Y) = S(X + Y)$ Use the previous exercise, and the fact that in computing the successor of a binary number the lowest order 0 turns to 1, the 1's to the right turn to 0's, and the other bits remain the same. Compare the positions of this lowest order 0 in X and in $X + Y$.
(c) $X + Y = Y + X$ (Commutativity).
(d) $(X + Y) + Z = X + (Y + Z)$ (Associativity).
  For Associativity, first show in $V^0(+)$ that

$$Carry(i, Y, Z) \oplus Carry(i, X, Y + Z) \leftrightarrow$$
$$Carry(i, X, Y) \oplus Carry(i, X + Y, Z).$$

Derive a stronger statement than this, and prove it by induction on i.

Example V.4.20 (The Pairing Function). We define the pairing function $\langle x, y \rangle$ as the following term of $I\Delta_0$:

$$\langle x, y \rangle =_{\text{def}} (x + y)(x + y + 1) + 2y. \tag{69}$$

Exercise V.4.21. Show using results in Section III.1 that $I\Delta_0$ proves $\langle x, y \rangle$ is a one-one function. That is

$$I\Delta_0 \vdash \langle x_1, y_1 \rangle = \langle x_2, y_2 \rangle \supset x_1 = x_2 \wedge y_1 = y_2. \tag{70}$$

(First show that the LHS implies $x_1 + y_1 = x_2 + y_2$.)

In general we can "pair" more than 2 numbers, e.g., define

$$\langle x_1, \ldots, x_{k+1} \rangle = \langle \langle x_1, \ldots, x_k \rangle, x_{k+1} \rangle.$$

We will refer to the term $\langle x_1, \ldots, x_{k+1} \rangle$ as a *tupling function*.

For any constant $k \in \mathbb{N}$, $k \geq 2$, we can use the tupling function to code a k-dimensional bit array by a single string Z by defining

Notation.

$$Z(x_1, \ldots, x_k) =_{\text{def}} Z(\langle x_1, \ldots, x_k \rangle). \tag{71}$$

Example V.4.22 (The Projection Functions). Consider the (partial) projection functions:

$$y = \mathit{left}(x) \leftrightarrow \exists z \leq x \, (x = \langle y, z \rangle),$$
$$z = \mathit{right}(x) \leftrightarrow \exists y \leq x \, (x = \langle y, z \rangle).$$

To make these functions total, we define

$$\mathit{left}(x) = \mathit{right}(x) = 0 \qquad \text{if } \neg \mathit{Pair}(x)$$

where

$$\mathit{Pair}(x) \equiv \exists y \leq x \exists z \leq x \, (x = \langle y, z \rangle).$$

For constants n and $k \leq n$, if x codes an n-tuple, then the k-th component $\langle x \rangle_k^n$ of x can be extracted using *left* and *right*, e.g.,

$$\langle x \rangle_2^3 = \mathit{right}(\mathit{left}(x)).$$

Exercise V.4.23. Use Exercise V.4.21 to show that $\mathit{left}(x)$ and $\mathit{right}(x)$ are Σ_0^B-definable in $I\Delta_0$. Show that $I\Delta_0(\mathit{left,right})$ proves the following properties of *Pair* and the projection functions:

(a) $\forall y \forall z \, \mathit{Pair}(\langle y, z \rangle)$.
(b) $\forall x (\mathit{Pair}(x) \supset x = \langle \mathit{left}(x), \mathit{right}(x) \rangle)$.
(c) $x = \langle x_1, x_2 \rangle \supset (x_1 = \mathit{left}(x) \wedge x_2 = \mathit{right}(x))$.

Now we can generalize the Σ_0^B-comprehension axiom scheme to multiple dimensions.

DEFINITION V.4.24 (Multiple Comprehension Axiom). If Φ is a set of formulas, then the *multiple comprehension axiom scheme* for Φ, denoted by Φ-***MULTICOMP***, is the set of all formulas

$$\exists X \le \langle y_1, \ldots, y_k \rangle \forall z_1 < y_1 \ldots \forall z_k < y_k (X(z_1, \ldots, z_k) \leftrightarrow \varphi(z_1, \ldots, z_k))$$
(72)

where $k \ge 2$ and $\varphi(z)$ is any formula in Φ which may contain other free variables, but not X.

LEMMA V.4.25 (Multiple Comprehension). *Suppose that $\mathcal{T} \supseteq V^0$ is a theory with vocabulary $\mathcal{L}$ which proves the $\Sigma_0^B(\mathcal{L})$-**COMP** axioms. Then $\mathcal{T}$ proves the $\Sigma_0^B(\mathcal{L})$-**MULTICOMP** axioms.*

PROOF. For the case $\mathcal{L} = \mathcal{L}_A^2$ we could work in the conservative extension $\mathcal{T}(left, right)$ and apply Lemma V.4.15 to prove this. However for general $\mathcal{L}$ we use another method.

For simplicity we prove the case $k = 2$. Define $\psi(z)$ by

$$\psi(z) \equiv \exists z_1 \le z \exists z_2 \le z (z = \langle z_1, z_2 \rangle \wedge \varphi(z_1, z_2)).$$

Now by Σ_0^B-***COMP***,

$$\mathcal{T} \vdash \exists X \le \langle y_1, y_2 \rangle \forall z < \langle y_1, y_2 \rangle (X(z) \leftrightarrow \psi(z)).$$

By Exercise V.4.21, $\mathcal{T}$ proves that such X satisfies (72). □

Notice that the string X in (72) is not unique, because there are numbers $z < \langle y_1, \ldots, y_k \rangle$ which are not of the form $\langle z_1, \ldots, z_k \rangle$ (the pairing function (69) is not surjective). This, however, is not important, since we will be using only the truth values of $X(z)$ where $z = \langle z_1, \ldots, z_k \rangle$ for $z_i < y_i$, $1 \le i \le k$. (A unique such X can be defined as in the proof above.)

Now we introduce the string function $Row(x, Z)$ (or $Z^{[x]}$) in ***FAC***^{0} to represent row x of the binary array Z.

DEFINITION V.4.26 (*Row* and $V^0(Row)$). The function $Row(x, Z)$ (also denoted $Z^{[x]}$) has the bit-defining axiom

$$Row(x, Z)(i) \leftrightarrow (i < |Z| \wedge Z(x, i)). \tag{73}$$

$V^0(Row)$ is the extension of V^0 obtained from V^0 by adding to it the string function Row and its Σ_0^B-bit-definition (73).

Note that by Corollary V.4.14, $V^0(Row)$ is a conservative extension of V^0.

The next result follows immediately from Lemma V.4.15.

LEMMA V.4.27 (*Row* Elimination). *For every $\Sigma_0^B(Row)$ formula φ, there is Σ_0^B formula φ' such that $V^0(Row) \vdash \varphi \leftrightarrow \varphi'$. Hence $V^0(Row)$ proves the $\Sigma_0^B(Row)$-**COMP** axiom scheme.*

We can use *Row* to represent a tuple $X_1, \ldots, X_k$ of strings by a single string Z, where $X_i = Z^{[i]}$. The following result follows immediately from the Multiple Comprehension Lemma.

LEMMA V.4.28. $V^0(Row)$ *proves*

$$\forall X_1 \ldots \forall X_k \exists Z \leq t(X_1 = Z^{[1]} \wedge \cdots \wedge X_k = Z^{[k]}) \tag{74}$$

where $t = \langle k, |X_1| + \cdots + |X_k| \rangle$. $\qquad\qquad\qquad\qquad\qquad\qquad\qquad$ □

DEFINITION V.4.29. A *single*-$\Sigma_1^B(\mathcal{L})$ formula is one of the form $\exists X \leq t\varphi$, where φ is $\Sigma_0^B(\mathcal{L})$.

EXERCISE V.4.30. Let $\mathcal{T}$ be a polynomial-bounded theory with vocabulary $\mathcal{L}$ such that $\mathcal{T}$ extends $V^0(Row)$. Prove that for every $\Sigma_1^B(\mathcal{L})$ formula φ there is a *single*-$\Sigma_1^B(\mathcal{L})$ formula φ' such that $\mathcal{T} \vdash \varphi \leftrightarrow \varphi'$.
Now use Lemma V.4.27 to show that the same is true when $\mathcal{T}$ is V^0 and $\mathcal{L}$ is $\mathcal{L}_A^2$.

Just as we use a "two-dimensional" string $Z(x, y)$ to code a sequence $Z^{[0]}, Z^{[1]}, \ldots$ of strings, we use a similar idea to allow Z to code a sequence $y_0, y_1, \ldots$ of numbers. Now y_i is the smallest element of $Z^{[i]}$, or $|Z|$ if $Z^{[i]}$ is empty. We define an AC^0 function $seq(i, Z)$ (also denoted $(Z)^i$) to extract y_i.

DEFINITION V.4.31 (Coding a Bounded Sequence of Numbers). The number function $seq(x, Z)$ (also denoted $(Z)^x$) has the defining axiom:

$$y = seq(x, Z) \leftrightarrow (y < |Z| \wedge Z(x, y) \wedge$$
$$\forall z < y \neg Z(x, z)) \vee (\forall z < |Z| \neg Z(x, z) \wedge y = |Z|).$$

It is easy to check that V^0 proves the existence and uniqueness of y satisfying the RHS of the above formula, and hence seq is Σ_0^B-definable in V^0. As in the case of *Row*, it follows from Lemma V.4.15 that any $\Sigma_0^B(seq)$ formula is provably equivalent in $V^0(seq)$ to a $\Sigma_0^B(\mathcal{L}_A^2)$ formula. (See also the AC^0 Elimination Lemma V.6.7 for a more general result.)

V.4.1. Δ_1^1-Definable Predicates. Recall the notion of a Φ-definable (or Φ-representable) predicate symbol, where Φ is a class of formulas (Definition III.3.2). Recall also that we obtain a conservative extension of a theory $\mathcal{T}$ by adding to it a definable predicate symbol P and its defining axiom. Below we define the notions of a "$\Delta_1^1(\mathcal{L})$-definable predicate symbol" and a "$\Delta_1^B(\mathcal{L})$-definable predicate symbol". Note that here $\Delta_1^1(\mathcal{L})$ and $\Delta_1^B(\mathcal{L})$ depend on the theory $\mathcal{T}$, in contrast to Definition III.3.2.

DEFINITION V.4.32 ($\Delta_1^1(\mathcal{L})$ and $\Delta_1^B(\mathcal{L})$ Definable Predicate). Let $\mathcal{T}$ be a theory over the vocabulary $\mathcal{L}$ and P a predicate symbol not in $\mathcal{L}$. We say that P is $\Delta_1^1(\mathcal{L})$-definable (or simply Δ_1^1-definable) in $\mathcal{T}$ if there are $\Sigma_1^1(\mathcal{L})$

formulas $\varphi(\vec{x}, \vec{Y})$ and $\psi(\vec{x}, \vec{Y})$ such that

$$R(\vec{x}, \vec{Y}) \leftrightarrow \varphi(\vec{x}, \vec{Y}), \quad \text{and} \quad \mathcal{T} \vdash \varphi(\vec{x}, \vec{Y}) \leftrightarrow \neg \psi(\vec{x}, \vec{Y}). \quad (75)$$

We say that P is $\Delta_1^B(\mathcal{L})$-definable (or simply Δ_1^B-definable) in $\mathcal{T}$ if the formulas φ and ψ above are Σ_1^B formulas.

The following exercise can be proved using Parikh's Theorem.

EXERCISE V.4.33. Show that if $\mathcal{T}$ is a polynomial-bounded theory, then a predicate is Δ_1^1-definable in $\mathcal{T}$ iff it is Δ_1^B-definable in $\mathcal{T}$.

DEFINITION V.4.34 (Characteristic Function). The *characteristic function* of a relation $R(\vec{x}, \vec{X})$, denoted by $f_R(\vec{x}, \vec{X})$, is defined as follows:

$$f_R(\vec{x}, \vec{X}) = \begin{cases} 1 & \text{if } R(\vec{x}, \vec{X}), \\ 0 & \text{otherwise.} \end{cases}$$

We will show that $\boldsymbol{FAC^0}$ coincides with the class of provably total functions in V^0. It follows that AC^0 relations are precisely the Δ_1^1 definable relations in V^0. More generally we have the following theorem.

THEOREM V.4.35. *If the vocabulary of a theory $\mathcal{T}$ includes $\mathcal{L}_A^2$, and a complexity class C has the property that for all relations R, $R \in C$ iff $f_R \in \boldsymbol{FC}$, and the class of Σ_1^1-definable functions in $\mathcal{T}$ coincides with $\boldsymbol{FC}$, then the class of Δ_1^1-definable relations in $\mathcal{T}$ coincides with C.*

PROOF. Assume the hypotheses of the theorem, and suppose that the relation $R(\vec{x}, \vec{X})$ is Δ_1^1-definable in $\mathcal{T}$. Then there are Σ_0^B formulas φ and ψ such that

$$R(\vec{x}, \vec{X}) \leftrightarrow \exists \vec{Y} \, \varphi(\vec{x}, \vec{X}, \vec{Y})$$

and

$$\mathcal{T} \vdash \exists \vec{Y} \, \varphi(\vec{x}, \vec{X}, \vec{Y}) \leftrightarrow \neg \exists \vec{Y} \, \psi(\vec{x}, \vec{X}, \vec{Y}). \quad (76)$$

Thus the characteristic function $f_R(\vec{x}, \vec{X})$ of R satisfies

$$y = f_R(\vec{x}, \vec{X}) \leftrightarrow \theta(y, \vec{x}, \vec{X}) \quad (77)$$

where

$$\theta(y, \vec{x}, \vec{X}) \equiv \exists \vec{Y}((y = 1 \wedge \varphi(\vec{x}, \vec{X}, \vec{Y})) \vee (y = 0 \wedge \psi(\vec{x}, \vec{X}, \vec{Y}))).$$

Then $\mathcal{T}$ proves $\exists! y \theta(y, \vec{x}, \vec{X})$, where the existence of y and $\vec{Y}$ follows from the $\leftarrow$ direction of (76) and the uniqueness of y follows from the $\rightarrow$ direction of (76). Thus f_R is Σ_1^1-definable in $\mathcal{T}$, so f_R is in $\boldsymbol{FC}$, and therefore R is in C.

Conversely, suppose that $R(\vec{x}, \vec{X})$ is in C, so f_R is in $\boldsymbol{FC}$. Then f_R is Σ_1^1-definable in $\mathcal{T}$, so there is a Σ_1^1 formula $\theta(y, \vec{x}, \vec{X})$ such that (77) holds and

$$\mathcal{T} \vdash \exists! y \theta(y, \vec{x}, \vec{X}).$$

Then $R(\vec{x}, \vec{X}) \leftrightarrow \exists y (y \neq 0 \wedge \theta(y, \vec{x}, \vec{X}))$ and

$$\mathcal{T} \vdash \exists y (y \neq 0 \wedge \theta(y, \vec{x}, \vec{X})) \leftrightarrow \neg\theta(0, \vec{x}, \vec{X}).$$

Since $\exists y (y \neq 0 \wedge \theta(y, \vec{x}, \vec{X}))$ is equivalent to a Σ_1^1 formula, it follows that R is Δ_1^1-definable in $\mathcal{T}$.   □

V.5. The Witnessing Theorem for V^0

NOTATION. For a theory $\mathcal{T}$ and a list $\mathcal{L}$ of functions that are definable/bit-definable in $\mathcal{T}$, we denote by $\mathcal{T}(\mathcal{L})$ the theory $\mathcal{T}$ extended by the defining/bit-defining axioms for the symbols in $\mathcal{L}$.

Recall that number functions in FAC^0 are Σ_0^B-definable in V^0, and string functions in FAC^0 are Σ_0^B-bit-definable in V^0 (see Proposition V.4.7 and Corollary V.4.11). It follows from Corollary V.4.14 that $V^0(\mathcal{L})$ is a conservative extension of V^0, for any collection $\mathcal{L}$ of FAC^0 functions.

Our goal now is to prove the following theorem.

THEOREM V.5.1 (Witnessing for V^0). *Suppose that* $\varphi(\vec{x}, \vec{y}, \vec{X}, \vec{Y})$ *is a* Σ_0^B *formula such that*

$$V^0 \vdash \forall \vec{x} \forall \vec{X} \exists \vec{y} \exists \vec{Y} \; \varphi(\vec{x}, \vec{y}, \vec{X}, \vec{Y}).$$

Then there are FAC^0 *functions* $f_1, \dots, f_k, F_1, \dots, F_m$ *so that*

$$V^0(f_1, \dots, f_k, F_1, \dots, F_m) \vdash \forall \vec{x} \forall \vec{X} \varphi(\vec{x}, \vec{f}(\vec{x}, \vec{X}), \vec{X}, \vec{F}(\vec{x}, \vec{X})).$$

The functions f_i and F_j are called the *witnessing functions*, for y_i and Y_j, respectively.

We will prove the Witnessing Theorem for V^0 in the next section. First, we list some of its corollaries.

The next corollary follows from the above theorem and Corollary V.4.11.

COROLLARY V.5.2 (Σ_1^1-Definability Theorem for V^0). *A function is in* FAC^0 *iff it is* Σ_1^1-*definable in* V^0 *iff it is* Σ_1^B-*definable in* V^0 *iff it is* Σ_0^B-*definable in* V^0.

COROLLARY V.5.3. *A relation is in* AC^0 *iff it is* Δ_1^1 *definable in* V^0 *iff it is* Δ_1^B *definable in* V^0.

It follows from the Σ_0^B-Representation Theorem IV.3.6 that a relation is in AC^0 iff its characteristic function is in AC^0. Therefore Corollary V.5.3 follows from the Σ_1^1-Definability Theorem for V^0 and Theorem V.4.35. Alternatively, it can be proved using the Witnessing Theorem for V^0 as follows.

PROOF. Since each AC^0 relation R is represented by a Σ_0^B formula θ, it is obvious that they are Δ_1^B (and hence Δ_1^1) definable in V^0: In (75) simply let φ be θ, and ψ be $\neg\theta$.

On the other hand, suppose that R is a Δ_1^1-definable relation of V^0. In other words, there are Σ_0^B formulas $\varphi(\vec{x}, \vec{X}, \vec{Y})$ and $\psi(\vec{x}, \vec{X}, \vec{Y})$ so that

$$R(\vec{x}, \vec{X}) \leftrightarrow \exists \vec{Y} \varphi(\vec{x}, \vec{X}, \vec{Y})$$

and

$$V^0 \vdash \exists \vec{Y} \varphi(\vec{x}, \vec{X}, \vec{Y}) \leftrightarrow \neg \exists \vec{Y} \psi(\vec{x}, \vec{X}, \vec{Y}). \tag{78}$$

In particular,

$$V^0 \vdash \exists \vec{Y}(\varphi(\vec{x}, \vec{X}, \vec{Y}) \vee \psi(\vec{x}, \vec{X}, \vec{Y})).$$

By the Witnessing Theorem for V^0, there are AC^0 functions $F_1, \ldots, F_k$ so that

$$V^0(F_1, \ldots, F_k) \vdash \forall \vec{x} \forall \vec{X}(\varphi(\vec{x}, \vec{X}, \vec{F}(\vec{x}, \vec{X})) \vee \psi(\vec{x}, \vec{X}, \vec{F}(\vec{x}, \vec{X}))). \tag{79}$$

We claim that $V^0(F_1, \ldots, F_k)$ proves

$$\forall \vec{x} \forall \vec{X}(\exists \vec{Y} \varphi(\vec{x}, \vec{X}, \vec{Y}) \leftrightarrow \varphi(\vec{x}, \vec{X}, \vec{F}(\vec{x}, \vec{X}))).$$

The $\Leftarrow$ direction is trivial. The other direction follows from (78) and (79).

Consequently $\varphi(\vec{x}, \vec{X}, \vec{F}(\vec{x}, \vec{X}))$ also represents $R(\vec{x}, \vec{X})$. Here R is obtained from the relation represented by $\varphi(\vec{x}, \vec{X}, \vec{Y})$ by substituting the AC^0 functions $\vec{F}$ for $\vec{Y}$. By Theorem V.2.7 (a), R is also an AC^0 relation.
□

V.5.1. Independence Follows from the Witnessing Theorem for V^0.

We can use the Witnessing Theorem to show the unprovability in V^0 of $\exists Z\ \varphi(Z)$ by showing that no AC^0 function can witness the quantifier $\exists Z$. Recall that the relation $PARITY(X)$ is defined by

$$PARITY(X) \leftrightarrow \text{the set } X \text{ has an odd number of elements.}$$

Then a well known result in complexity theory states:

PROPOSITION V.5.4. $PARITY \notin AC^0$.

First, it follows that the characteristic function $parity(X)$ of $PARITY(X)$ is not in FAC^0. Therefore $parity$ is not Σ_1^1-definable in V^0. In the next chapter we will show that $parity$ is Σ_1^1-definable in the theory V^1. This will show that V^0 is a proper sub-theory of V^1.

Now consider the Σ_0^B formula $\varphi_{parity}(X, Y)$:

$$\neg Y(0) \wedge \forall i < |X|(Y(i+1) \leftrightarrow (X(i) \oplus Y(i))) \tag{80}$$

where $\oplus$ is exclusive OR. Thus $\varphi_{parity}(X, Y)$ asserts that for $0 \leq i < |X|$, bit $Y(i+1)$ is 1 iff the number of 1's among bits $X(0), \ldots, X(i)$ is odd. Define

$$\varphi(X) \equiv \exists Y \leq (|X| + 1)\ \varphi_{parity}(X, Y).$$

Then $\forall X \varphi(X)$ is true in the standard model $\underline{\mathbb{N}}_2$, but by the above proposition, no function $F(X)$ satisfying $\forall X \varphi_{parity}(X, F(X))$ can be in $\textbf{FAC}^0$. Hence by the Witnessing Theorem for V^0,

$$V^0 \nvdash \forall X \exists Y \leq (|X| + 1)\ \varphi_{parity}(X, Y).$$

Note that this independence result does not follow from Parikh's Theorem.

V.5.2. Proof of the Witnessing Theorem for V^0. Recall the analogous statement in single-sorted logic for $I\Delta_0$ (i.e., that a Σ_1 theorem of $I\Delta_0$ can be "witnessed" by a single-sorted $\textbf{LTH}$ function) which is proved in Theorem III.4.8. There we use the Bounded Definability Theorem III.3.8 (which follows from Parikh's Theorem) to show that the graph of any Σ_1-definable function of $I\Delta_0$ is actually definable by a Δ_0 formula, and hence an $\textbf{LTH}$ relation.

Unfortunately, a similar method does not work here. We can also use Parikh's Theorem to show that the graph of a Σ_1^1-definable function of V^0 is representable by a Σ_1^B formula. However this does not suffice, since there are string functions whose graphs are in $\textbf{AC}^0$ (i.e., representable by Σ_0^B formulas), but which do not belong to $\textbf{FAC}^0$. An example is the counting function whose graph is given by the Σ_0^B formula $\delta_{NUM}(x, X, Y)$ (227).

Our first proof is by the Anchored $\textbf{LK}^2$ Completeness Theorem IV.4.5. This proof is important because the same method can be used to prove the witnessing theorem for V^1 (Theorem VI.4.1). Our second proof method (see Section V.6.1) is based on the Herbrand Theorem and does not work for V^1.

We will prove the following simple form of the theorem, since it implies the general form.

LEMMA V.5.5. *Suppose that* $\varphi(\vec{x}, \vec{X}, Y)$ *is a* Σ_0^B *formula such that*

$$V^0 \vdash \forall \vec{x} \forall \vec{X} \exists Z \varphi(\vec{x}, \vec{X}, Z).$$

Then there is an $\textbf{FAC}^0$ *function* F *so that*

$$V^0(F) \vdash \forall \vec{x} \forall \vec{X} \varphi(\vec{x}, \vec{X}, F(\vec{x}, \vec{X})).$$

PROOF OF THEOREM V.5.1 FROM LEMMA V.5.5. The idea is to use the function Row to encode the tuple $\langle \vec{y}, \vec{Y} \rangle$ by a single string variable Z, as in Lemma V.4.28. Then by the above lemma, Z is witnessed by an $\textbf{AC}^0$ function F. The witnessing functions for $y_1, \ldots, y_k, Y_1, \ldots, Y_m$ will then be extracted from F using the function Row. Details are as follows.

Assume the hypothesis of the Witnessing Theorem for V^0, i.e.,

$$V^0 \vdash \forall \vec{x} \forall \vec{X} \exists \vec{y} \exists \vec{Y}\ \varphi(\vec{x}, \vec{y}, \vec{X}, \vec{Y})$$

for a Σ_0^B formula $\varphi(\vec{x}, \vec{y}, \vec{X}, \vec{Y})$. Then since $V^0(Row)$ extends V^0, we have also

$$V^0(Row) \vdash \forall \vec{x} \forall \vec{X} \exists \vec{y} \exists \vec{Y}\ \varphi(\vec{x}, \vec{y}, \vec{X}, \vec{Y}).$$

Note that

$$V^0(Row) \vdash \forall y_1 \ldots \forall y_k \forall Y_1 \ldots \forall Y_m \exists Z \left(\bigwedge_{i=1}^{k} |Z^{[i]}| = y_i \wedge \bigwedge_{j=1}^{m} Z^{[k+j]} = Y_j \right).$$

(See also Lemma V.4.28.) Thus

$$V^0(Row) \vdash \forall \vec{x} \forall \vec{X} \exists Z \; \varphi(\vec{x}, |Z^{[1]}|, \ldots, |Z^{[k]}|, \vec{X}, Z^{[k+1]}, \ldots, Z^{[k+m]})$$

i.e.,

$$V^0(Row) \vdash \forall \vec{x} \forall \vec{X} \exists Z \psi(\vec{x}, \vec{X}, Z)$$

where

$$\psi(\vec{x}, \vec{X}, Z) \equiv \varphi(\vec{x}, |Z^{[1]}|, \ldots, |Z^{[k]}|, \vec{X}, Z^{[k+1]}, \ldots, Z^{[k+m]})$$

is a $\Sigma_0^B(\mathcal{L}_A^2 \cup \{Row\})$ formula.

Now by Lemma V.4.27, there is a $\Sigma_0^B(\mathcal{L}_A^2)$ formula $\psi'(\vec{x}, \vec{X}, Z)$ so that

$$V^0(Row) \vdash \forall \vec{x} \forall \vec{X} \forall Z(\psi(\vec{x}, \vec{X}, Z) \leftrightarrow \psi'(\vec{x}, \vec{X}, Z)).$$

As a result, since $V^0(Row)$ is conservative over V^0, we also have

$$V^0 \vdash \forall \vec{x} \forall \vec{X} \exists Z \psi'(\vec{x}, \vec{X}, Z).$$

Applying Lemma V.5.5, there is an AC^0 function F so that

$$V^0(F) \vdash \forall \vec{x} \forall \vec{X} \psi'(\vec{x}, \vec{X}, F(\vec{x}, \vec{X})).$$

Therefore

$$V^0(Row, F) \vdash \forall \vec{x} \forall \vec{X} \psi(\vec{x}, \vec{X}, F(\vec{x}, \vec{X}))$$

i.e.,

$$V^0(Row, F) \vdash \forall \vec{x} \forall \vec{X} \; \varphi(\vec{x}, |F^{[1]}|, \ldots, |F^{[k]}|, \vec{X}, F^{[k+1]}, \ldots, F^{[k+m]})$$

where we write F for $F(\vec{x}, \vec{X})$.

Let $f_i(\vec{x}, \vec{X}) = |(F(\vec{x}, \vec{X}))^{[i]}|$ for $1 \leq i \leq k$ and $F_j(\vec{x}, \vec{X}) = (F(\vec{x}, \vec{X}))^{[k+j]}$ for $1 \leq j \leq m$ and denote $\{f_1, \ldots, f_k, F_1, \ldots, F_m\}$ by $\mathcal{L}$, we have

$$V^0(\{Row, F\} \cup \mathcal{L}) \vdash \forall \vec{x} \forall \vec{X} \varphi(\vec{x}, \vec{f}, \vec{X}, \vec{F}).$$

By Corollary V.4.14, $V^0(\{Row, F\} \cup \mathcal{L})$ is a conservative extension of $V^0(\mathcal{L})$. Consequently,

$$V^0(\mathcal{L}) \vdash \forall \vec{x} \forall \vec{X} \; \varphi(\vec{x}, \vec{f}, \vec{X}, \vec{F}). \qquad \square$$

The rest of this section is devoted to the proof of Lemma V.5.5.

PROOF OF LEMMA V.5.5. The proof method is similar to that of Lemma V.3.5 (for Parikh's Theorem). Suppose that $\exists Z \varphi(\vec{a}, \vec{\alpha}, Z)$ is a theorem of V^0. By the Anchored LK^2 Completeness Theorem, there is an anchored LK^2-T proof π of

$$\longrightarrow \exists Z \varphi(\vec{a}, \vec{\alpha}, Z)$$

where T is the set of all term substitution instances of the axioms for V^0. We assume that π is in free variable normal form (see Section IV.4.1).

Note that all instances of the $\Sigma_0^B\text{-}COMP$ axioms (48) are Σ_1^1 formulas (they are in fact Σ_1^B formulas). Since the endsequent of π is also a Σ_1^1 formula, by the Subformula Property (Theorem IV.4.6), all formulas in π are Σ_1^1 formulas, and in fact they contain at most one string quantifier $\exists X$ in front. In particular, every sequent in π has the form

$$\exists X_1\theta_1(X_1),\ldots,\exists X_m\theta_m(X_m),\Gamma \longrightarrow \Delta,\exists Y_1\psi_1(Y_1),\ldots,\exists Y_n\psi_n(Y_n) \tag{81}$$

for $m,n \geq 0$, where θ_i and ψ_j and all formulas in Γ and Δ are Σ_0^B.

We will prove by induction on the depth in π of a sequent $\mathcal{S}$ of the form (81) that there are Σ_0^B-bit-definable string functions $F_1,\ldots,F_n$ (i.e., the witnessing functions) such that there is a collection of Σ_0^B-bit-definable functions $\mathcal{L}$ including $F_1,\ldots,F_n$ and an $LK^2\text{-}V^0(\mathcal{L})$ proof of

$$\mathcal{S}' =_{\text{def}} \theta_1(\beta_1),\ldots,\theta_m(\beta_m),\Gamma \longrightarrow \Delta,\psi_1(F_1),\ldots,\psi_n(F_n) \tag{82}$$

where F_i stands for $F_i(\vec{a},\vec{\alpha},\vec{\beta})$, and $\vec{a},\vec{\alpha}$ is a list of exactly those variables with free occurrences in $\mathcal{S}$. (This list may be different for different sequents.) Here $\beta_1,\ldots,\beta_m$ are distinct new free variables corresponding to the bound variables $X_1,\ldots,X_m$, although the latter variables may not be distinct.

It follows that for the endsequent $\longrightarrow \exists Z\varphi(\vec{a},\vec{\alpha},Z)$ of π, there is a finite collection $\mathcal{L}$ of FAC^0 functions, and an $F \in \mathcal{L}$ so that

$$V^0(\mathcal{L}) \vdash \varphi(\vec{a},\vec{\alpha},F(\vec{a},\vec{\alpha})).$$

Note that by Corollary V.4.14, $V^0(\mathcal{L})$ is a conservative extension of $V^0(F)$. Consequently we have

$$V^0(F) \vdash \varphi(\vec{a},\vec{\alpha},F(\vec{a},\vec{\alpha}))$$

and we are done.

Our inductive proof has several cases, depending on whether $\mathcal{S}$ is a V^0 axiom, or which rule is used to generate $\mathcal{S}$. In each case we will introduce suitable witnessing functions when required, and it is an easy exercise to check that in each of the functions introduced has a $\Sigma_0^B(\mathcal{L}_A^2)$-bit-definition.

To show that the arguments $\vec{a},\vec{\alpha}$ of previously-introduced witnessing functions continue to include only those variables with free occurrences in the sequent $\mathcal{S}$, we use the fact that the proof π is in free variable normal form, and hence no free variable is eliminated by any rule in the proof except $\forall$-right and $\exists$-left. (We made a similar argument concerning the free variables in the bounding terms t in the proof of Lemma V.3.5).

In general we will show that $\mathcal{S}'$ has an $LK^2\text{-}V^0(\mathcal{L})$ proof not by constructing the proof, but rather by arguing that the formula giving the semantics of $\mathcal{S}'$ (Definition II.2.16) is provable in V^0 from the bit-defining axioms of the functions $\mathcal{L}$, and invoking the LK^2 Completeness Theorem. However in each case the $LK^2\text{-}V^0(\mathcal{L})$ proof is not hard to find.

Specifically, if we write (82) in the form

$$S' = \quad A_1, \ldots, A_k \longrightarrow B_1, \ldots, B_\ell$$

then we assert

$$V^0(\mathcal{L}) \vdash \forall \vec{x} \forall \vec{X} \forall \vec{Y} \big((A_1 \wedge \cdots \wedge A_k) \supset (B_1 \vee \cdots \vee B_\ell) \big). \quad (83)$$

Case I. S is an axiom of V^0. If the axiom only involves Σ_0^B formulas, then no witnessing functions are needed. Otherwise S comes from a Σ_0^B-*COMP* axiom, i.e.,

$$S =_{\text{def}} \longrightarrow \exists X \leq b \forall z < b(X(z) \leftrightarrow \psi(z, b, \vec{a}, \vec{\alpha})).$$

Then a function witnessing X has bit-defining axiom

$$F(b, \vec{a}, \vec{\alpha})(z) \leftrightarrow z < b \wedge \psi(z, b, \vec{a}, \vec{\alpha}).$$

Case II. S is obtained by an application of the rule string $\exists$-right. Then S is the bottom of the inference

$$\frac{S_1}{S} = \frac{\Lambda \longrightarrow \Pi, \psi(T)}{\Lambda \longrightarrow \Pi, \exists X \psi(X)}$$

where the string term T is either a variable γ or the constant $\varnothing$ introduced when putting π in free variable normal form. In the former case, γ must have a free occurrence in S, and we may witness the new quantifier $\exists X$ by the function F with bit-defining axiom

$$F(\vec{a}, \gamma, \vec{\alpha}, \vec{\beta})(z) \leftrightarrow z < |\gamma| \wedge \gamma(z)$$

In the latter case T is $\varnothing$, and we define

$$F(\vec{a}, \vec{\alpha}, \vec{\beta})(z) \leftrightarrow z < 0.$$

Case III. S is obtained by an application of the rule string $\exists$-left. Then S is the bottom of the inference

$$\frac{S_1}{S} = \frac{\theta(\gamma), \Lambda \longrightarrow \Pi}{\exists X \theta(X), \Lambda \longrightarrow \Pi}$$

Note that γ cannot occur in S, by the restriction for this rule, but S' has a new variable β' available corresponding to $\exists X$ (see (82)). No new witnessing function is required. Each witnessing function $F_j(\vec{a}, \gamma, \vec{\alpha}, \vec{\beta})$ for the top sequent is replaced by the witnessing function

$$F_j'(\vec{a}, \vec{\alpha}, \beta', \vec{\beta}) = F_j(\vec{a}, \beta', \vec{\alpha}, \vec{\beta})$$

for S'.

Case IV. S is obtained by an application of the rule number $\exists$-right or number $\forall$-left. No new witnessing functions are required.

Case V. S follows from an application of rule number ∃-left or number ∀-right. We consider number ∃-left, since number ∀-right is similar. Then S is the bottom sequent in the inference

$$\frac{S_1}{S} = \frac{b \leq t \wedge \theta(b), \Lambda \longrightarrow \Pi}{\exists x \leq t\theta(x), \Lambda \longrightarrow \Pi}$$

No new witnessing function is needed, but the free variable b is eliminated as an argument to the existing witnessing functions, and it must be given a value. We give it a value which satisfies the new existential quantifier, if one exists. Thus define the FAC^0 number function

$$g(\vec{a}, \vec{\alpha}) = \min b \leq t\ \theta(b).$$

For each witnessing function $F_j(b, \vec{a}, \vec{\alpha}, \vec{\beta})$ for the top sequent define the corresponding witnessing function for the bottom sequent by

$$F_j'(\vec{a}, \vec{\alpha}, \vec{\beta}) = F_j(g(\vec{a}, \vec{\alpha}), \vec{a}, \vec{\alpha}, \vec{\beta}).$$

Case VI. S is obtained by the cut rule. Then S is the bottom of the inference

$$\frac{S_1 \qquad S_2}{S} = \frac{\Lambda \longrightarrow \Pi, \psi \qquad \psi, \Lambda \longrightarrow \Pi}{\Lambda \longrightarrow \Pi}$$

Assume first that ψ is Σ_0^B. For $i = 1, 2$, let $F_1^i(\vec{a}, \vec{\alpha}), \ldots, F_n^i(\vec{a}, \vec{\alpha})$ be the witnessing functions for Π in S_i'. Then we define witnessing functions $F_1, \ldots, F_n$ for these formulas in the conclusion S' by the bit-defining axioms

$$F_j(\vec{a}, \vec{\alpha})(z) \leftrightarrow ((\neg\psi \wedge F_j^1(\vec{a}, \vec{\alpha})(z)) \vee (\psi \wedge F_j^2(\vec{a}, \vec{\alpha})(z))).$$

Now assume that ψ is not Σ_0^B, so ψ has the form

$$\psi \equiv \exists X \theta(X) \tag{84}$$

where $\theta(X)$ is Σ_0^B. Let $G(\vec{a}, \vec{\alpha})$ be the witnessing function for $\exists X$ in S_1' and let β be the variable in S_2' corresponding to X. Let $F_1^1(\vec{a}, \vec{\alpha}), \ldots, F_n^1(\vec{a}, \vec{\alpha})$ be the other witnessing functions for Π in S_1', and $F_1^2(\vec{a}, \vec{\alpha}, \beta), \ldots, F_n^2(\vec{a}, \vec{\alpha}, \beta)$ be the witnessing functions for Π in S_2'. The corresponding witnessing function F_j in S' has defining axiom (replace $\ldots$ by $\vec{a}, \vec{\alpha}$)

$$F_j(\ldots)(z) \leftrightarrow$$
$$(\neg\theta(G(\ldots)) \wedge F_j^1(\ldots)(z)) \vee (\theta(G(\ldots)) \wedge F_j^2(\ldots, G(\ldots))(z)). \tag{85}$$

EXERCISE V.5.6. Show correctness of this definition of F in the special case where the cut formula ψ has the form (84), and Π has only one Σ_1^1 formula, by arguing that $V^0(\mathcal{L})$ can prove the semantic translation (83) of S' from the semantic translations of S_1' and S_2'.

Case VII. $\mathcal{S}$ is obtained from an instance of the rule $\wedge$-left or $\vee$-right. These are both handled in the same manner. Consider $\wedge$-right.

$$\frac{\mathcal{S}_1 \quad \mathcal{S}_2}{\mathcal{S}} = \frac{\Lambda \longrightarrow \Pi, A \quad \Lambda \longrightarrow \Pi, B}{\Lambda \to \Pi, (A \wedge B)}$$

Here, as in (81),

$$\Lambda =_{\mathrm{def}} \exists X_1 \theta_1(X_1), \dots, \exists X_m \theta_m(X_m), \Gamma$$
$$\text{and} \quad \Pi =_{\mathrm{def}} \Delta, \exists Y_1 \psi_1(Y_1), \dots, \exists Y_n \psi_n(Y_n)$$

for $m, n \geq 0$, where θ_i and ψ_j and all formulas in Γ and Δ are Σ_0^B. Also, A and B are Σ_0^B formulas.

Let $F_j^1(\vec{a}, \vec{\alpha})$ and $F_j^2(\vec{a}, \vec{\alpha})$ witness Y_j in $\mathcal{S}_1'$ and $\mathcal{S}_2'$, respectively. Then we define the witness $F_j(\vec{a}, \vec{\alpha})$ for Y_j in $\mathcal{S}'$ to be $F_j^1(\vec{a}, \vec{\alpha})$ or $F_j^2(\vec{a}, \vec{\alpha})$, depending on whether $F_j^1(\vec{a}, \vec{\alpha})$ works as a witness. In particular (replace $\dots$ by $\vec{a}, \vec{\alpha}$):

$$F_j(\dots)(z) \leftrightarrow \left((\psi_j(F_j(\dots)) \wedge F_j^1(\dots)(z)) \vee (\neg \psi_j(F_j(\dots)) \wedge F_j^2(\dots)(z)) \right).$$

Case VIII. $\mathcal{S}$ is obtained by any of the other rules. Weakening is easy. There is nothing to do for exchange and $\neg$ introduction. The contraction rules can be derived from cut and exchanges.   □

EXERCISE V.5.7. Show that in the *Cases V, VI,* and *VII* above, the new functions introduced have $\Sigma_0^B(\mathcal{L}_A^2)$-bit-definitions.

V.6. $\overline{V}^0$: Universal Conservative Extension of V^0

Recall that a universal formula is a formula in prenex form in which all quantifiers are universal, and a universal theory is a theory which can be axiomatized by universal formulas. Recall also the universal single-sorted theory $\overline{IA_0}$ introduced in Section III.3.2.

The universal theory $\overline{V}^0$ extends $\overline{IA_0}$, and is defined in the same way as $\overline{IA_0}$. Here we show that $\overline{V}^0$ is a conservative extension of V^0, and that this gives us an alternative proof of the Witnessing Theorem for V^0 by applying the Herbrand Theorem IV.4.8 for $\overline{V}^0$.

The idea is to introduce number functions with universal defining axioms, and string functions with universal bit-defining axioms, which are provably total in V^0. Thus we obtain a conservative extension of V^0. Furthermore, the new functions are defined in such a way that the axioms of V^0 with existential quantifiers (namely Σ_0^B-*COMP* and **B**12, **SE**) can be proved from other axioms, and hence can be deduced from our set of universal axioms for $\overline{V}^0$.

We use the following notation. For any formula $\varphi(z, \vec{x}, \vec{X})$ and $\mathcal{L}_A^2$-term $t(\vec{x}, \vec{X})$, let $F_{\varphi(z),t}(\vec{x}, \vec{X})$ be the string function with bit definition

$$F_{\varphi(z),t}(\vec{x}, \vec{X})(z) \leftrightarrow \left(z < t(\vec{x}, \vec{X}) \wedge \varphi(z, \vec{x}, \vec{X}) \right). \tag{86}$$

Also, let $f_{\varphi(z),t}(\vec{x}, \vec{X})$ be the number function defined as in (28) to be the least $y < t$ such that $\varphi(y, \vec{x}, \vec{X})$ holds, or t if no such y exists. Then $f_{\varphi(z),t}$ has defining axiom (we write f for $f_{\varphi(z),t}(\vec{x}, \vec{X})$ and also omit the arguments $\vec{x}, \vec{X}$ from φ and t)

$$f \leq t \wedge (f < t \supset \varphi(f)) \wedge (v < f \supset \neg\varphi(v)). \tag{87}$$

As in Section III.3.2 we can use the functions $f_{\varphi(z),t}$ to eliminate bounded number quantifiers from a formula, since the above defining axiom implies

$$\exists z \leq t \varphi(z) \leftrightarrow \varphi(f_{\varphi(z),t}(\vec{x}, \vec{X})).$$

Recall that the predecessor function pd has the defining axioms:

B12′. $pd(0) = 0$,   **B12″.** $x \neq 0 \supset pd(x) + 1 = x.$ (88)

(**B12′** and **B12″** are called respectively **D1′** and **D2″** in Section III.3.2.)

In two-sorted logic, the extensionality axiom **SE** contains an implicit existential quantifier $\exists i < |X|$. Therefore we introduce the function f_{SE} with the defining axiom (87), where $\varphi(z, X, Y) \equiv X(z) \not\leftrightarrow Y(z)$, and $t(X, Y) = |X|$. Intuitively, $f_{SE}(X, Y)$ is the smallest number $< |X|$ that distinguishes X and Y, and $|X|$ if no such number exists.

$$f_{SE}(X, Y) \leq |X| \wedge f_{SE}(X, Y) < |X| \supset \neg(X(f_{SE}(X, Y)) \leftrightarrow$$
$$Y(f_{SE}(X, Y))) \wedge z < f_{SE}(X, Y) \supset (X(z) \leftrightarrow Y(z)). \tag{89}$$

Let **SE′** be the following axiom

$$(|X| = |Y| \wedge f_{SE}(X, Y) = |X|) \supset X = Y. \tag{90}$$

The vocabulary $\mathcal{L}_{FAC^0}$ is defined below. It contains a function symbol for every AC^0 function. Note that it extends $\mathcal{L}_{\Lambda_0}$ (Definition III.3.16).

Definition V.6.1. $\mathcal{L}_{FAC^0}$ is the smallest set that satisfies

1) $\mathcal{L}_{FAC^0}$ includes $\mathcal{L}_A^2 \cup \{pd, f_{SE}\}$.
2) For each open formula $\varphi(z, \vec{x}, \vec{X})$ over $\mathcal{L}_{FAC^0}$ and term $t = t(\vec{x}, \vec{X})$ of $\mathcal{L}_A^2$ there is a string function $F_{\varphi(z),t}$ and a number function $f_{\varphi(z),t}$ in $\mathcal{L}_{FAC^0}$.

Definition V.6.2. $\overline{V}^0$ is the theory over $\mathcal{L}_{FAC^0}$ with the following set of axioms: **B1–B11**, **L1**, **L2** (Figure 2), **B12′** and **B12″** (88), (89), **SE′** (90), and (86) for each function $F_{\varphi(z),t}$ and (87) for each function $f_{\varphi(z),t}$ of $\mathcal{L}_{FAC^0}$.

Thus $\overline{V}^0$ extends $\overline{I\Delta}_0$. Also, the axioms for $\overline{V}^0$ do not include any comprehension axiom. However, we will show that $\overline{V}^0$ proves the Σ_0^B-*COMP* axiom scheme, and hence $\overline{V}^0$ extends V^0.

Recall that an open formula is a formula without quantifier. The following lemma can be proved by structural induction on φ in the same way as Lemma III.3.19.

LEMMA V.6.3. *For every $\Sigma_0^B(\mathcal{L}_{FAC^0})$ formula φ there is an open $\mathcal{L}_{FAC^0}$-formula φ^+ such that $\overline{V}^0 \vdash \varphi \leftrightarrow \varphi^+$.*

LEMMA V.6.4. $\overline{V}^0$ *proves the $\Sigma_0^B(\mathcal{L}_{FAC^0})$-COMP, $\Sigma_0^B(\mathcal{L}_{FAC^0})$-IND, and $\Sigma_0^B(\mathcal{L}_{FAC^0})$-MIN axiom schemes.*

PROOF. For comprehension, we need to show, for each $\Sigma_0^B(\mathcal{L}_{FAC^0})$ formula $\varphi(z, \vec{x}, \vec{X})$,

$$\overline{V}^0 \vdash \exists Z \leq y \forall z < y (Z(z) \leftrightarrow \varphi(z, \vec{x}, \vec{X})).$$

By Lemma V.6.3 we may assume that φ is open. Thus we can take $Z = F_{\varphi,y}(\vec{x}, \vec{X})$ and apply (86). For induction and minimization we use Corollary V.1.8.   □

THEOREM V.6.5. *The theory $\overline{V}^0$ is a conservative extension of V^0.*

PROOF. To show that $\overline{V}^0$ extends V^0, we need to verify that $\overline{V}^0$ proves B12, SE and Σ_0^B-*COMP*. First, B12 follows from B12″. We prove *SE* in $\overline{V}^0$ as follows. Assume that

$$|X| = |Y| \wedge \forall z < |X|(X(z) \leftrightarrow Y(z)).$$

Then from (89) we have $f_{SE}(X, Y) = |X|$. Hence by (90) we obtain $X = Y$.

That $\overline{V}^0$ proves Σ_0^B-*COMP* follows from Lemma V.6.4.

Now we show that $\overline{V}^0$ is conservative over V^0. Let

$$pd, f_{SE}, \ldots \tag{91}$$

be an enumeration of $\mathcal{L}_{FAC^0}$ such that the n-th function is defined or bit-defined by an open formula using only the first $(n-1)$ functions. Let $\mathcal{L}_n$ denote the union of $\mathcal{L}_A^2$ and the set of the first n functions in the enumeration, and $V^0(\mathcal{L}_n)$ denote V^0 together with the defining axioms or bit-defining axioms for the functions of $\mathcal{L}_n$ ($n \geq 0$). Then

$$\overline{V}^0 = \bigcup_{n \geq 0} V^0(\mathcal{L}_n).$$

First we prove:

CLAIM. For $n \geq 1$, $V^0(\mathcal{L}_n)$ satisfies the hypothesis of Lemma V.4.15.

From Lemma V.4.15 and the claim we have

$$V^0(\mathcal{L}_n) \vdash \Sigma_0^B(\mathcal{L}_n)\text{-}COMP.$$

Therefore by Corollary V.4.14 $V^0(\mathcal{L}_{n+1})$ is conservative over $V^0(\mathcal{L}_n)$. Then by Compactness Theorem, it follows that $\overline{V}^0$ is also conservative over V^0. (See also Corollary III.3.6.)

It remains to prove the claim.

First note that $V^0(\mathcal{L}_n)$ extends V^0 for all $n \geq 1$. Also $\mathcal{L}_{FAC^0}$ has the same predicates as $\mathcal{L}_A^2$. We will prove by induction on n that each string function in $\mathcal{L}_n$ has a $\Sigma_0^B(\mathcal{L}_A^2)$-bit-defining axiom in $V^0(\mathcal{L}_n)$, and each number function in $\mathcal{L}_n$ has a $\Sigma_0^B(\mathcal{L}_A^2)$-defining axiom in $V^0(\mathcal{L}_n)$, and thus establishing the claim.

For the base case, $n = 1$, by **B**12′ and **B**12″ pd has a $\Sigma_0^B(\mathcal{L}_A^2)$-defining axiom in V^0, therefore $V^0(\mathcal{L}_1)$ (which is $V^0(pd)$) satisfies the hypothesis of Lemma V.4.15.

For the induction step we need to show that the $(n+1)$-st function f_{n+1} or F_{n+1} in (91) has a $\Sigma_0^B(\mathcal{L}_A^2)$-defining axiom or a $\Sigma_0^B(\mathcal{L}_A^2)$-bit-defining axiom in $V^0(\mathcal{L}_{n+1})$. By definition, the function f_{n+1}/F_{n+1} already has an open defining/bit-defining axiom in the vocabulary $\mathcal{L}_n$. From the induction hypothesis, $V^0(\mathcal{L}_n)$ satisfies the hypothesis of Lemma V.4.15. Consequently the defining/bit-defining axiom for f_{n+1}/F_{n+1} is provably equivalent in $V^0(\mathcal{L}_n)$ to a $\Sigma_0^B(\mathcal{L}_A^2)$ formula. Hence $V^0(\mathcal{L}_{n+1})$ proves that f_{n+1}/F_{n+1} has a $\Sigma_0^B(\mathcal{L}_A^2)$ defining/bit-defining axiom, and this completes the proof of the claim. □

Inspection of the above proof shows that each number function of $\mathcal{L}_{FAC^0}$ has a $\Sigma_0^B(\mathcal{L}_A^2)$-defining axiom, and each string function of $\mathcal{L}_{FAC^0}$ has a $\Sigma_0^B(\mathcal{L}_A^2)$-bit-defining axiom.

COROLLARY V.6.6. *The $\mathcal{L}_{FAC^0}$ functions are precisely the functions of* **FAC**^{0}.

PROOF. By the above remark and the Σ_0^B-Representation Theorem IV.3.6, the $\mathcal{L}_{FAC^0}$ functions are in **FAC**^{0}. The other inclusion follows from the Σ_0^B-Representation Theorem IV.3.6 and Lemma V.6.3. □

The next lemma follows from Lemma V.4.15 and the claim in the above proof of Theorem V.6.5. It generalizes the *Row* Elimination Lemma V.4.27.

LEMMA V.6.7 (**FAC**^{0} Elimination). *Suppose that $\mathcal{L} \subseteq \mathcal{L}_{FAC^0}$. Then for every $i \geq 0$ and every $\Sigma_i^B(\mathcal{L})$ formula φ^+ there is a $\Sigma_i^B(\mathcal{L}_A^2)$ formula φ so that $V^0(\mathcal{L}) \vdash \varphi^+ \leftrightarrow \varphi$.*

V.6.1. Alternative Proof of the Witnessing Theorem for $\overline{V}^0$. Here we show how to apply the Herbrand Theorem to $\overline{V}^0$ to obtain a simple proof of Theorem V.5.1. For notational simplicity, we consider the case of a single existential string quantifier, and prove Lemma V.5.5.

Suppose that $\varphi(\vec{x}, \vec{X}, Z)$ is a Σ_0^B formula such that

$$V^0 \vdash \forall \vec{x} \forall \vec{X} \exists Z \varphi(\vec{x}, \vec{X}, Z).$$

By Lemma V.6.3 there is an open formula φ' over $\mathcal{L}_{FAC^0}$ such that $\overline{V}^0 \vdash \varphi \leftrightarrow \varphi'$. Since $\overline{V}^0$ extends V^0, we have

$$\overline{V}^0 \vdash \forall \vec{x} \forall \vec{X} \exists Z \varphi'(\vec{x}, \vec{X}, Z).$$

Now $\overline{V}^0$ is a universal theory, so by the Herbrand Theorem IV.4.8, there are terms $T_1(\vec{x}, \vec{X}), \ldots, T_n(\vec{x}, \vec{X})$ of $\overline{V}^0$ such that

$$\overline{V}^0 \vdash \forall \vec{x} \forall \vec{X} \big(\varphi'(\vec{x}, \vec{X}, T_1(\vec{x}, \vec{X})) \vee \cdots \vee \varphi'(\vec{x}, \vec{X}, T_n(\vec{x}, \vec{X})) \big).$$

Define $F(\vec{x}, \vec{X})$ by cases as follows:

$$F(\vec{x}, \vec{X}) = \begin{cases} T_1(\vec{x}, \vec{X}) & \text{if } \varphi'(\vec{x}, \vec{X}, T_1(\vec{x}, \vec{X})), \\ \vdots \\ T_{n-1}(\vec{x}, \vec{X}) & \text{if } \varphi'(\vec{x}, \vec{X}, T_{n-1}(\vec{x}, \vec{X})), \\ T_n(\vec{x}, \vec{X}) & \text{otherwise.} \end{cases}$$

It is easy to see that $F(\vec{x}, \vec{X})$ has a bit definition (86), and hence is a function in $\mathcal{L}_{FAC^0}$, and

$$\overline{V}^0 \vdash \forall \vec{x} \forall \vec{X} \varphi'(\vec{x}, \vec{X}, F(\vec{x}, \vec{X})).$$

Now $\overline{V}^0 \vdash \varphi \leftrightarrow \varphi'$, and also the proof of Theorem V.6.5 shows that $\overline{V}^0$ is conservative over $V^0(F)$ (the extension of V^0 resulting by adding the defining axioms for F). Hence

$$V^0(F) \vdash \forall \vec{x} \forall \vec{X} \varphi(\vec{x}, \vec{X}, F(\vec{x}, \vec{X}))$$

as required.   □

The above proof shows that adding true Σ_0^B axioms to a theory does not increase the set of provably total functions in the theory. For example, let $True\Sigma_0^B$ be the set of all Σ_0^B formulas which are true in the standard model $\underline{\mathbb{N}}_2$. Let $V^0(True\Sigma_0^B)$ be the result of adding $True\Sigma_0^B$ as axioms to V^0, and let $\overline{V}^0(True\Sigma_0^B)$ be the result of adding $True\Sigma_0^B$ as axioms to $\overline{V}^0$. Then $\overline{V}^0(True\Sigma_0^B)$ is a conservative extension of $V^0(True\Sigma_0^B)$, and the above proof goes through to show that the same class FAC^0 of functions serve to witness the Σ_1^1 theorems of $V^0(True\Sigma_0^B)$. Thus we have shown

COROLLARY V.6.8. *The provably total functions in* $V^0(True\Sigma_0^B)$ *are precisely the functions in* FAC^0.

V.7. Finite Axiomatizability

THEOREM V.7.1. V^0 *is finitely axiomatizable.*

PROOF. It suffices to show that all Σ_0^B-***COMP*** axioms follow from finitely many theorems of V^0. Let 2-***BASIC***$^+$ (or simply B^+) denote the 2-***BASIC*** axioms (Fig. 2) along with the finitely many theorems of $I\Delta_0$ (and hence of V^0) given in Examples III.1.8 and III.1.9 asserting that $+, \cdot, \leq$ satisfy the properties of a commutative discretely-ordered semiring.

We show more generally that both Σ_0^B-***COMP*** and the multiple comprehension axioms (72) for all Σ_0^B formulas follow from B^+ and finitely many such comprehension instances. We use the notation $\varphi[\vec{a}, \vec{Q}](\vec{x})$ to indicate that the Σ_0^B formula φ can contain the free variables $\vec{a}, \vec{Q}$ in addition to $\vec{x} = x_1, \ldots, x_k$. Then for $k \geq 1$, ***COMP***$_\varphi(\vec{a}, \vec{Q}, \vec{b})$ denotes the comprehension formula

$$\exists Y \leq \langle b_1, \ldots, b_k \rangle \forall x_1 < b_1 \ldots \forall x_k < b_k (Y(\vec{x}) \leftrightarrow \varphi(\vec{x})). \tag{92}$$

We will show that ***COMP***$_\varphi$ for the following 12 formulas φ will suffice.

$$
\begin{aligned}
\varphi_1(x_1, x_2) &\equiv x_1 = x_2, \\
\varphi_2(x_1, x_2, x_3) &\equiv x_3 = x_1, \\
\varphi_3(x_1, x_2, x_3) &\equiv x_3 = x_2, \\
\varphi_4[Q_1, Q_2](x_1, x_2) &\equiv \exists y \leq x_1 (Q_1(x_1, y) \wedge Q_2(y, x_2)), \\
\varphi_5[a](x, y) &\equiv y = a, \\
\varphi_6[Q_1, Q_2](x, y) &\equiv \exists z_1 \leq y \exists z_2 \leq y (Q_1(x, z_1) \wedge Q_2(x, z_2) \wedge y = z_1 + z_2), \\
\varphi_7[Q_1, Q_2](x, y) &\equiv \exists z_1 \leq y \exists z_2 \leq y (Q_1(x, z_1) \wedge Q_2(x, z_2) \wedge y = z_1 \cdot z_2), \\
\varphi_8[Q_1, Q_2, c](x) &\equiv \exists y_1 \leq c \exists y_2 \leq c (Q_1(x, y_1) \wedge Q_2(x, y_2) \wedge y_1 \leq y_2), \\
\varphi_9[X, Q, c](x) &\equiv \exists y \leq c (Q(x, y) \wedge X(y)), \\
\varphi_{10}[Q](x) &\equiv \neg Q(x), \\
\varphi_{11}[Q_1, Q_2](x) &\equiv Q_1(x) \wedge Q_2(x), \\
\varphi_{12}[Q, c](x) &\equiv \forall y \leq c Q(x, y).
\end{aligned}
$$

In the following lemmas, we abbreviate ***COMP***$_{\varphi_i}(\ldots)$ by C_i.

LEMMA V.7.2. *For each $k \geq 1$ and $1 \leq i \leq k$ let*

$$\psi_{ik}(x_1, \ldots, x_k, y) \equiv y = x_i.$$

Then $B^+, C_1, C_2, C_3, C_4 \vdash$ ***COMP***$_{\psi_{ik}}$.

PROOF. We proceed by induction on k. For $k = 1$ we have $\psi_{1,1} \leftrightarrow \varphi_1(x_1, y)$ and for $k = 2$ we have $\psi_{2,1} \leftrightarrow \varphi_2(x_1, x_2, y)$ and $\psi_{2,2} \leftrightarrow \varphi_3(x_1, x_2, y)$. For $k > 2$, recall $\langle x_1, \ldots, x_k \rangle = \langle \langle x_1, \ldots, x_{k-1} \rangle, x_k \rangle$. Hence

$$B^+, C_3 \vdash \text{***COMP***}_{\psi_{kk}}.$$

For $1 \leq i < k$ use C_4 with Q_1 defined by C_2 and Q_2 defined by ***COMP***$_{\psi_{i,k-1}}$. □

LEMMA V.7.3. *Let* $\vec{x} = x_1, \cdots, x_k$, $k \geq 1$, *be a list of variables and let* $t(\vec{x})$ *be a term which in addition to possibly involving variables from $\vec{x}$ may involve other variables* $\vec{a}, \vec{Q}$. *Let* $\psi_t[\vec{a}, \vec{Q}](\vec{x}, y) \equiv y = t(\vec{x})$. *Then*

$$B^+, C_1, \ldots, C_7 \vdash \textbf{\textit{COMP}}_{\psi_t}(\vec{a}, \vec{Q}, \vec{b}, d).$$

PROOF. By using algebraic theorems in B^+ we may suppose that $t(\vec{x})$ is a sum of monomials in $x_1, \ldots, x_k$, where the coefficients are terms involving $\vec{a}, \vec{Q}$. The case $t \equiv u$, where u does not involve any x_i is obtained from C_5 with $a \leftarrow u$. The cases $t \equiv x_i$ are obtained from Lemma V.7.2. We then build monomials using C_7 repeatedly, and build the general case by repeated use of C_6.   □

LEMMA V.7.4. *Let* $t_1(\vec{x}), t_2(\vec{x})$ *be terms with variables among* $\vec{x}, \vec{a}, \vec{Q}$. *Suppose*

$$\psi_1[\vec{a}, \vec{Q}](\vec{x}) \quad \equiv t_1(\vec{x}) \leq t_2(\vec{x}),$$
$$\psi_2[\vec{a}, \vec{Q}, X](\vec{x}) \equiv X(t_1(\vec{x})).$$

Then $B^+, C_1, \ldots, C_9 \vdash \textbf{\textit{COMP}}_{\psi_i}$, *for* $i = 1, 2$.

PROOF. $\textbf{\textit{COMP}}_{\psi_1}(\vec{a}, \vec{Q}, \vec{b})$ follows from $\textbf{\textit{COMP}}_{\varphi_8}(Q_1, Q_2, c, b)$ with for $i = 1, 2$, Q_i defined from $\textbf{\textit{COMP}}_{\psi_{t_i}}$ in Lemma V.7.3 with $d \leftarrow t_1(\vec{b}) + t_2(\vec{b}) + 1$, so

$$\forall \vec{x} < \vec{b} \forall y < (t_1(\vec{b}) + t_2(\vec{b}) + 1) \, (Q_i(\vec{x}, y) \leftrightarrow y = t_i(\vec{x})).$$

In $\textbf{\textit{COMP}}_{\varphi_8}$ we take $c \leftarrow t_1(\vec{b}) + t_2(\vec{b})$ and $b \leftarrow \langle b_1, \ldots, b_k \rangle$.

For $\textbf{\textit{COMP}}_{\psi_2}(\vec{a}, \vec{Q}, X, \vec{b})$ we use $\textbf{\textit{COMP}}_{\varphi_9}(X, P, c, b)$ with $c \leftarrow t_1(\vec{b})$ and $b \leftarrow \langle b_1, \ldots, b_k \rangle$ and P defined from Lemma V.7.3 similarly to Q_1 above.   □

Now we can complete the proof of the theorem. Lemma V.7.4 takes care of the case when φ is an atomic formula, since equations $t_1(\vec{x}) = t_2(\vec{x})$ can be initially replaced by $t_1(\vec{x}) \leq t_2(\vec{x}) \wedge t_2(\vec{x}) \leq t_1(\vec{x})$. Then by repeated applications of $\textbf{\textit{COMP}}_{\varphi_{10}}$ and $\textbf{\textit{COMP}}_{\varphi_{11}}$ we handle the case in which φ is quantifier-free.

Now suppose $\varphi(\vec{x}) \equiv \forall y \leq t(\vec{x}) \psi(\vec{x}, y)$. We assume as an induction hypothesis that we can define Q satisfying

$$\forall \vec{x} < \vec{b} \forall y < t(\vec{b}) + 1 \big(Q(\vec{x}, y) \leftrightarrow (y \leq t(\vec{x}) \supset \psi(\vec{x}, y)) \big).$$

Then $\textbf{\textit{COMP}}_{\varphi}(\vec{b})$ follows from $\textbf{\textit{COMP}}_{\varphi_{12}}(Q, c, b)$ with $c \leftarrow t(\vec{b})$ and $b \leftarrow \langle b_1, \ldots, b_k \rangle$.   □

V.8. Notes

The system V^0 we introduce in this chapter is essentially Σ_0^p-*comp* in [112], and $I\Sigma_0^{1,b}$ (without #) in [72]. Zambella [112] used $\mathcal{R}$ for $\textbf{\textit{FAC}}^0$ and called it the class of *rudimentary* functions. However there is danger here of confusion with Smullyan's rudimentary relations [103].

The set 2-***BASIC*** is similar to the axioms for Zambella's theory Θ in [112], and forms the two-sorted analog of Buss's single-sorted axioms ***BASIC*** [20]. It is slightly different from that which are presented in [43] and [42].

The statement and proof of Theorem V.5.1 (witnessing) are inspired by [20], although our treatment here is simplified because we only witness formulas in which all string quantifiers are in front.

The universal theory $\overline{V}^0$ is taken from [42].

Theorem V.7.1 (finite axiomatizability) is taken from Section 7 of [43].

Chapter VI

THE THEORY V^1 AND POLYNOMIAL TIME

In this chapter we show that the theory V^1 (the two-sorted version of Buss's theory S_2^1) characterizes P in the same way that V^0 characterizes AC^0. This is stated in the Σ_1^1-Definability Theorem VI.2.2 for V^1: A function is Σ_1^1-definable (equivalently Σ_1^B-definable) in V^1 if and only if it is in FP. The "only if" direction follows from the *Witnessing Theorem for V^1*.

The theory of algorithms can be viewed, to a large extent, as the study of polynomial time functions. All polytime algorithms can be described in V^1, and experience has shown that proofs of their important properties can usually be formalized in V^1. (See Example VI.4.3, prime recognition, for an apparent exception.) Razborov [96] has shown how to formalize lower bound proofs for Boolean complexity in V^1. Standard theorems from graph theory, including Kuratowski's Theorem, Hall's Theorem, and Menger's Theorem can be formalized in V^1.

In Chapter VIII we will introduce the (apparently) weaker theories TV^0 and VPV for polynomial time, and prove that they have the same Σ_1^B-theorems (and hence the same Σ_1^1-theorems) as V^1.

VI.1. Induction Schemes in V^i

Recall (Definition V.1.3) that V^i is axiomatized by 2-*BASIC* and Σ_i^B-*COMP*, where Σ_i^B-*COMP* consists of all formulas of the form

$$\exists X \leq y \forall z < y (X(z) \leftrightarrow \varphi(z)) \tag{93}$$

where $\varphi(z)$ is a Σ_i^B formula, and X does not occur free in $\varphi(z)$.
The next result follows from Corollary V.1.8.

COROLLARY VI.1.1. *For $i \geq 0$, V^i proves the Σ_i^B-IND, Σ_i^B-MIN, and Σ_i^B-MAX axiom schemes.*

It turns out that V^i proves these schemes for a wider class of formulas than just Σ_i^B. To show this, we start with a partial generalization of the Multiple Comprehension Lemma V.4.25. Recall the projection functions *left* and *right* (Example V.4.22).

Lemma VI.1.2 (Multiple Comprehension Revisited). *Let $\mathcal{T}$ be a theory which extends V^0 and has vocabulary $\mathcal{L}$, and suppose that either $\mathcal{L} = \mathcal{L}_A^2$ or $\mathcal{L}$ includes the projection functions left and right. For each $i \geq 0$, if $\mathcal{T}$ proves the $\Sigma_i^B(\mathcal{L})$-COMP axioms, then $\mathcal{T}$ proves the multiple comprehension axiom* (72):

$$\exists X \leq \langle y_1, \ldots, y_k \rangle \forall z_1 < y_1 \ldots \forall z_k < y_k (X(z_1, \ldots, z_k) \leftrightarrow \varphi(z_1, \ldots, z_k))$$
(94)

for any $k \geq 2$ and any $\varphi \in \Sigma_i^B(\mathcal{L})$. In particular, for all $i \geq 0$, V^i proves Σ_i^B-MULTICOMP.

Proof. The method used to prove the earlier version, Lemma V.4.25, does not work here, because for $i \geq 1$ the $\Sigma_i^B(\mathcal{L})$-formulas are not closed under bounded number quantification.

For notational simplicity we prove the case $k = 2$. First we consider the case that $\mathcal{L}$ includes *left* and *right*. Assuming that $\varphi(z_1, z_2)$ is in $\Sigma_i^B(\mathcal{L})$ and $\mathcal{T}$ proves the $\Sigma_i^B(\mathcal{L})$-COMP axioms, it follows that $\mathcal{T}$ proves

$$\exists X \leq \langle y_1, y_2 \rangle \forall z < \langle y_1, y_2 \rangle (X(z) \leftrightarrow \varphi(left(z), right(z))).$$

Now (94) follows by the properties of *left,right* (Exercise V.4.23) and the notation (71) stating that $X(z_1, z_2) \equiv X(\langle z_1, z_2 \rangle)$.

For the case $\mathcal{L} = \mathcal{L}_A^2$, we work in the conservative extension $\mathcal{T}(left, right)$ of $\mathcal{T}$. By the FAC^0 Elimination Lemma V.6.7, if $\mathcal{T}$ proves the Σ_i^B-COMP axioms, it follows that $\mathcal{T}(left, right)$ proves the $\Sigma_i^B(left, right)$-COMP axioms and hence also (94) by the previous case. Hence also $\mathcal{T}$ proves (94) by conservativity.   □

The next result refers to the Σ_0^B-closure of a set of formulas (Definition V.4.9).

Theorem VI.1.3. *Let $\mathcal{T}$ be a theory over a vocabulary $\mathcal{L}$ which extends V^0 and proves the multiple comprehension axioms* (94) *for every $k \geq 1$ and every φ in some class Φ of $\mathcal{L}$-formulas. Then $\mathcal{T}$ proves the $\Sigma_0^B(\Phi)$-COMP axioms.*

The following result is an immediate consequence of this theorem, Lemma VI.1.2, and Corollary V.1.8, since every Π_i^B formula is equivalent to a negated Σ_i^B formula.

Corollary VI.1.4. *For $i \geq 0$ let Φ_i be $\Sigma_0^B(\Sigma_i^B \cup \Pi_i^B)$. Then V^i proves the Φ_i-COMP, Φ_i-IND, Φ_i-MIN, and Φ_i-MAX axiom schemes.*

Proof of Theorem VI.1.3. We prove the stronger assertion that $\mathcal{T}$ proves the multiple comprehension axioms (94) for $\varphi \in \Sigma_0^B(\Phi)$, by structural induction on φ relative to Φ. We use the fact that $\mathcal{T}$ extends V^0 and hence by Lemma VI.1.2 proves the multiple comprehension axioms for Σ_0^B-formulas.

The base case, $\varphi \in \Phi$, holds by hypothesis. For the induction step, consider the case that φ has the form $\neg \psi$. By the induction hypothesis $\mathcal{T}$

proves

$$\exists Y \leq \langle \vec{y} \rangle \forall \vec{z} < \vec{y}(Y(\vec{z}) \leftrightarrow \psi(\vec{z}))$$

and by Lemma VI.1.2, $\mathcal{T}$ proves

$$\exists X \leq \langle \vec{y} \rangle \forall \vec{z} < \vec{y}(X(\vec{z}) \leftrightarrow \neg Y(\vec{z})).$$

Thus $\mathcal{T}$ proves (94).

The cases $\wedge$ and $\vee$ are similar. Finally we consider the case that $\varphi(\vec{z})$ has the form $\forall x \leq t\psi(x, \vec{z})$. By the induction hypothesis $\mathcal{T}$ proves

$$\exists Y \leq \langle t+1, \vec{y} \rangle \forall x \leq t \forall \vec{z} < \vec{y}(Y(x, \vec{z}) \leftrightarrow \psi(x, \vec{z})).$$

By Lemma V.4.25 V^0 proves

$$\exists X \leq \langle \vec{y} \rangle \forall \vec{z} < \vec{y}(X(\vec{z}) \leftrightarrow \forall x \leq t\, Y(x, \vec{z})).$$

Now (94) follows from these two formulas. $\square$

VI.2. Characterizing P by V^1

The class (two-sorted) P consists of relations computable in polynomial time by a deterministic Turing machine (i.e., *polytime relations*), and *FP* is the class of functions computable in polynomial time by a deterministic Turing machine (i.e., *polytime functions*). Alternatively (Definition V.2.3) *FP* is the class of the polynomially bounded number functions whose graphs are in P, and the polynomially bounded string functions whose bit graphs are in P. (See also Appendix A.1.)

Recall that a number input to the accepting machine is represented as a unary string, and a set input is represented as a binary string (page 81). (Thus a purely numerical function $f(\vec{x})$ is in *FP* iff it is computed in time $2^{O(n)}$, where n is the length of the *binary* notation for its arguments.)

The following proposition follows easily from the definitions involved.

PROPOSITION VI.2.1.   (a) *A number function $f(\vec{x}, \vec{X})$ is in FP iff there is a string function $F(\vec{x}, \vec{X})$ in FP so that $f(\vec{x}, \vec{X}) = |F(\vec{x}, \vec{X})|$.*

(b) *A relation is in P iff its characteristic function is in FP.*

We will prove that the theory V^1 characterizes P in the same way that V^0 characterizes AC^0:

THEOREM VI.2.2 (Σ_1^1-Definability for V^1). *A function is Σ_1^1-definable in V^1 iff it is in FP.*

The "if" direction is proved in Section VI.2.1. The "only-if" direction follows immediately from the Witnessing Theorem for V^1 (Theorem VI.4.1).

Note that V^1 is a polynomial-bounded theory (Definition V.3.3). The following corollary follows from the Σ_1^1-Definability Theorem for V^1 above, and Parikh's Theorem (see Corollary V.4.4).

COROLLARY VI.2.3. *A function is in FP iff it is Σ_1^B-definable in V^1.*

The next corollary follows from the results above and Theorem V.4.35.

COROLLARY VI.2.4. *A relation is in* P *iff it is is* Δ_1^1*-definable in* V^1 *iff it is* Δ_1^B*-definable in* V^1.

Recall (Theorem IV.3.7) that the Σ_1^B formulas represent precisely the *NP* relations, and hence by Definition V.4.32 a relation is Δ_1^B definable in a theory $\mathcal{T}$ iff $\mathcal{T}$ proves that the relation is in both *NP* and *co-NP*. Thus the above corollary says that a relation is in P iff V^1 proves that it is in $NP \cap co\text{-}NP$.

COROLLARY VI.2.5. V^1 *is a proper extension of* V^0.

PROOF. There are relations (such as $PARITY(X)$, page 118) which are in P but not in AC^0.   □

EXERCISE VI.2.6 (*parity*(X) in V^1). Recall the formula $\varphi_{parity}(X, Y)$ ((80) on page 118). Show that the function *parity*(X), which is the characteristic function of *PARITY* (page 118), is Σ_1^1-definable in V^1 by showing that

$$V^1 \vdash \forall X \exists! Y \varphi_{parity}(X, Y).$$

EXERCISE VI.2.7 (String Multiplication in V^1). Consider the string multiplication function $X \times Y$ where

$$X \times Y = Z \leftrightarrow bin(Z) = bin(X) \cdot bin(Y)$$

and $bin(X)$ is the integer value of the binary string X (see (46) on page 85). Consider the the Σ_1^1 defining axiom for $X \times Y$ in V^1 that is based on the "school" algorithm for multiplying two integers written in binary notation. First, we construct the *table* $X \otimes Y$ that has $|Y|$ rows and whose ith row is either 0, if $Y(i) = 0$ (i.e., $\neg Y(i)$), or a copy of X shifted left by i bits, if $Y(i) = 1$. Thus, $X \otimes Y$ can be defined by (see Definition V.4.26 for row notation)

$$X \otimes Y = Z \leftrightarrow |Z| \leq \langle |Y|, |X| + |Y| \rangle \wedge$$

$$\forall i < |Y| \forall z < i + |X| \left(Z^{[i]}(z) \leftrightarrow (Y(i) \wedge \exists u \leq z \, (u + i = z \wedge X(u))) \right).$$

(a) Let $Z = X \otimes Y$. Show that V^0 proves the existence and uniqueness of Z.

(b) Show that V^1 proves the existence and uniqueness of W, where

$$|W| \leq 1 + \langle |Y|, |X| + |Y| \rangle \wedge |W^{[0]}| = 0 \wedge$$

$$\forall i < |Y| (W^{[i+1]} = W^{[i]} + Z^{[i]}).$$

(Hint: Use Σ_1^B-*IND*. For the bound on $|W|$, show that $|W^{[i]}| \leq |X| + i$.)

(c) Define $X \times Y$ in terms of $X \otimes Y$. Conclude that the string multiplication function is provably total in V^1.

(d) Recall string functions $\varnothing$, S and $X + Y$ from Example V.4.17. Show that the following are theorems of $V^1(\varnothing, S, +, \times)$:

   (i) $X \times \varnothing = \varnothing$.

   (ii) $X \times S(Y) = (X \times Y) + X$.

Now we argue that the subtheory *IOPEN* of Peano Arithmetic (Definition III.1.7) can be interpreted in V^1 by interpreting each number x by the unique string X such that $bin(X) = x$. Then $+$ and $\times$ in *IOPEN* are interpreted by Example V.4.17 and Exercise VI.2.7 respectively, 0 is interpreted by $\varnothing$, and 1 by the string constant $1 = \{0\}$. It is easy to give a Σ_0^B formula defining the relation $X \le Y$ to interpret $\le$. Then one can check that $V^0 \vdash S(X) = X + 1$, and with the help of Exercises V.4.19 and VI.2.7 it is not hard to show that $V^1(\varnothing, 1, +, \times)$ proves the string interpretations of axioms **B1**, ..., **B8** for *PA*.

It remains to show that $V^1(\varnothing, 1, +, \times)$ proves the string interpretations of the induction axiom scheme (10) (on page 41) for open formulas $\varphi(x)$. In fact this will follow from our discussion in Section VIII.3 (see Corollary VIII.3.20 and Theorem VIII.2.11).

Consequently V^1 proves the string interpretations of the formulas given in Example III.1.8 (commutativity and associativity of $+$, $\times$ etc). In fact it is not hard to show that V^1 also proves the formulas (involving $+$, $\times$, $\le$) given in Example III.1.9, even though some of the proofs given there involve induction on Σ_1^b-formulas rather than on just open formulas.

EXERCISE VI.2.8 (String Division and Remainder in V^1). Consider the string division function $X \div Y = \lfloor X/Y \rfloor$ and the string remainder function $Rem(X, Y) = X - Y \times (X \div Y)$. These functions can be Σ_1^1-defined in V^1 by the following steps. Suppose that $Y \le X$, and let z be such that $z + |Y| = |X|$.

(a) Give a Σ_0^B-bit-definition for the table U, where the row $U^{[i]}$ of U is Y shifted left by i bits, for $0 \le i \le z$.

(b) Prove in V^1 the existence and uniqueness of a table W such that

$$W^{[z]} = X \wedge \forall i < z\big((W^{[i+1]} < U^{[i+1]} \supset W^{[i]} = W^{[i+1]}) \wedge$$
$$(U^{[i+1]} \le W^{[i+1]} \supset W^{[i]} + U^{[i+1]} = W^{[i+1]})\big).$$

(c) Define $X \div Y$ and $Rem(X, Y)$ using W.

(d) Show in $V^1(+, \times, \div, Rem)$ that

$$X = (Y \times (X \div Y)) + Rem(X, Y).$$

VI.2.1. The "If" Direction of Theorem VI.2.2. We will give two proofs of the fact that every polynomial time function is Σ_1^1-definable in V^1. The first is based directly on Turing machine computations, and the second is based on Cobham's characterization of *FP*. We give the second proof in more detail, since it provides the basis for the universal theory *VPV* described in Chapter VIII.

The key idea for the first proof is that the computation of a polytime Turing machine M on a given input $\vec{x}, \vec{X}$ can be encoded as a string of configurations (see Definition V.4.26 for notation)

$$Z = \langle Z^{[0]}, Z^{[1]}, \ldots, Z^{[m]} \rangle$$

whose length is bounded by some polynomial in $\vec{x}, |\vec{X}|$, and whose existence we need to prove in V^1. The output of M can then be extracted from Z easily. The defining axiom for the polytime function computed by M is the formula that states the existence of such Z.

EXERCISE VI.2.9. Describe a method of coding Turing machine configurations by strings, and show that for each Turing machine M working on input $\vec{x}, \vec{X}$ there are Σ_0^B-definable string functions in V^0: $Init_M(\vec{x}, \vec{X})$, $Next_M(Z)$ and $Out_M(Z)$ such that

- $Init_M(\vec{x}, \vec{X})$ is the initial configuration of M on input $(\vec{x}, \vec{X})$;
- $Z' = Next_M(Z)$ if Z and Z' code two consecutive configurations of M, or $Z' = Z$ if Z codes a final configuration of M, or $Z' = \varnothing$ if Z does not code a configuration of M.
- $Out_M(Z)$ is the tape contents of a configuration Z of M, or $\varnothing$ if Z does not code a configuration of M.

Below we will use all three functions in the above exercise, as well as the string function $Row(z, Y)$ (Definition V.4.26). Because these functions are Σ_0^B-definable in V^0, it follows from the FAC^0 Elimination Lemma V.6.7 that any $\Sigma_0^B(\mathcal{L}_A^2 \cup \{Init, Next, Out, Row\})$ formula can be transformed into a provably equivalent $\Sigma_0^B(\mathcal{L}_A^2)$ formula. Formally we will work in the conservative extension of V^1 consisting of V^1 together with the defining axioms for these functions, although we will continue to refer to this theory as simply V^1. Thus each Σ_0^B (resp. Σ_1^B) formula below with the new functions is provably equivalent to a Σ_0^B (resp. Σ_1^B) formula in the vocabulary of V^1.

FIRST PROOF OF THE $\Longleftarrow$ DIRECTION OF THEOREM VI.2.2. Consider the case of string functions. (The case of number functions is similar.) Suppose that $F(\vec{x}, \vec{X})$ is a polytime function. Let M be a Turing machine which computes $F(\vec{x}, \vec{X})$ in time polynomial of $\vec{x}, |\vec{X}|$, and let $t(\vec{x}, |\vec{X}|)$ be a bound on the running time of M on input $\vec{x}, \vec{X}$. We may assume that M halts with $F(\vec{x}, \vec{X})$ equal to the contents of its tape, so that $Out_M(Z) = F(\vec{x}, \vec{X})$ if Z codes the final configuration. Then

$$Y = F(\vec{x}, \vec{X}) \leftrightarrow \exists Z \le \langle t, t \rangle (\varphi_M(\vec{x}, \vec{X}, Z) \wedge Y = Out_M(Z^{[t]})) \quad (95)$$

where $\varphi_M(\vec{x}, \vec{X}, Z)$ is the formula

$$Z^{[0]} = Init_M(\vec{x}, \vec{X}) \wedge \forall z < t \ (Z^{[z+1]} = Next_M(Z^{[z]})).$$

We will show that the RHS of (95) is a defining axiom for F in V^1, i.e.,

$$V^1 \vdash \forall \vec{x} \forall \vec{X} \exists ! Y \exists Z \leq \langle t, t \rangle (\varphi_M(\vec{x}, \vec{X}, Z) \wedge Y = Out_M(Z^{[t]})).$$

For the uniqueness of Y, it suffices to verify that if Z_1 and Z_2 are two strings satisfying

$$|Z_k| \leq \langle t, t \rangle \wedge \varphi_M(\vec{x}, \vec{X}, Z_k)$$

(for $k = 1, 2$), then for all z,

$$z \leq t \supset Z_1^{[z]} = Z_2^{[z]}. \tag{96}$$

This follows in V^1 using $\Sigma_0^B\text{-}IND$ on the formula (96) with induction on z. For the existence of Y, we need to show that V^1 proves

$$\forall \vec{x} \forall \vec{X} \exists Z \leq \langle t, t \rangle \; \varphi_M(\vec{x}, \vec{X}, Z).$$

This formula can be proved in V^1 by using number induction axiom (Corollary VI.1.1) on b for the Σ_1^B formula

$$\exists W \leq \langle b, t \rangle (W^{[0]} = Init_M(\vec{x}, \vec{X}) \wedge \forall z < b \, W^{[z+1]} = Next_M(W^{[z]})). \quad \square$$

Exercise VI.2.10. Carry out details of the induction step in the proof of the above formula.

An alternative proof for the above direction of Theorem VI.2.2 can be obtained by using Cobham's characterization of FP. To explain this, we need the notion of *limited recursion*. First we introduce the AC^0 string function $Cut(x, X)$, which is the initial segment of X and contains all elements of X that are $< x$. It has the Σ_0^B-bit-defining axiom

$$Cut(x, X)(z) \leftrightarrow z < x \wedge X(z). \tag{97}$$

Notation. We will sometimes write $X^{<x}$ for $Cut(x, X)$.

Definition VI.2.11 (Limited Recursion). A string function $F(y, \vec{x}, \vec{X})$ is defined by *limited recursion* from $G(\vec{x}, \vec{X})$ and $H(y, \vec{x}, \vec{X}, Z)$ iff

$$F(0, \vec{x}, \vec{X}) = G(\vec{x}, \vec{X}), \tag{98}$$

$$F(y + 1, \vec{x}, \vec{X}) = (H(y, \vec{x}, \vec{X}, F(y, \vec{x}, \vec{X})))^{<t(y, \vec{x}, \vec{X})} \tag{99}$$

for some $\mathcal{L}_A^2$-term t representing a polynomial in $y, \vec{x}, |\vec{X}|$.

For two-sorted function classes, we can also define the notion of limited recursion for a number function. However here we can just appeal to Proposition VI.2.1 (a) when we have to deal with number functions. A version of Cobham's characterization of FP is as follows.

Theorem VI.2.12 (Cobham's Characterization of FP). *A string function is in FP iff it can be obtained from AC^0 functions by finitely many applications of composition and limited recursion.*

PROOF SKETCH. The $\Longleftarrow$ direction follows from the fact that AC^0 functions are in FP, and that applying the operations composition and limited recursion to functions in FP results in functions in FP.

For the $\Longrightarrow$ direction, the function F computed by a polytime Turing machine M can be defined from the AC^0 functions $Init_M$, $Next_M$ and Out_M by limited recursion and composition. In more detail, we can define a string function $Conf_M(y, \vec{x}, \vec{X})$ to be the string coding the configuration of M on input $(\vec{x}, \vec{X})$ at time y. Then $Conf_M$ satisfies the recursion

$$Conf_M(0, \vec{x}, \vec{X}) = Init_M(\vec{x}, \vec{X}),$$

$$Conf_M(y + 1, \vec{x}, \vec{X}) = Next_M(Conf_M(y, \vec{x}, \vec{X})).$$

To turn this recursion into one fitting Definition VI.2.11 we apply

$$Cut(t(y, \vec{x}, \vec{X}), \dots)$$

to the RHS of the second equation, for a suitable $\mathcal{L}^2_A$-term t bounding the run time of M. Then

$$F(\vec{x}, \vec{X}) = Out_M(Conf_M(t(\vec{x}, \vec{X}), \vec{x}, \vec{X})). \qquad (100) \qquad \square$$

VI.2.2. Application of Cobham's Theorem.

SECOND PROOF OF THE $\Longleftarrow$ DIRECTION OF THEOREM VI.2.2. We use Cobham's characterization of FP to show that the polytime string functions are Σ^1_1-definable in V^1. It follows from Proposition VI.2.1 that the polytime number functions are also Σ^1_1-definable in V^1.

We proceed by induction on the number of applications of composition and limited recursion needed to obtain F from AC^0 functions. For the base case, the AC^0 functions are Σ^1_1-definable in V^0 (Corollary V.5.2), hence also in V^1. For the induction step, we need to show that the Σ^1_1-definable functions of V^1 are closed under composition and limited recursion. The case of composition is easily seen to hold for any theory $\mathcal{T}$ (see exercise V.4.5). Hence it suffices to prove the case of limited recursion.

Suppose that $G(\vec{x}, \vec{X})$ and $H(y, \vec{x}, \vec{X}, Z)$ are Σ^1_1-definable functions in V^1, and $F(y, \vec{x}, \vec{X})$ is defined by limited recursion from G and H as in (98) and (99) for some polynomial p. Then we can Σ^1_1-define F by coding the sequence of values $F(0), F(1), \dots, F(y)$ as the rows $W^{[0]}, W^{[1]}, \dots, W^{[y]}$ of a single array W. Thus (omitting $\vec{x}, \vec{X}$):

$$Y = F(y) \leftrightarrow \left(\exists W \ W^{[0]} = G() \wedge \right.$$

$$\left. (\forall z < y \ W^{[z+1]} = (H(z, W^{[z]}))^{<t(z)}) \wedge Y = W^{[y]} \right).$$

The RHS is not immediately equivalent to a Σ^1_1 formula when the equations involving G and H are replaced by Σ^1_1 formulas using the defining axioms for G and H. This is because of the number quantifier $\forall z < y$ of the middle conjunct, which is mixed in between the existential string quantifiers. We obtain a Σ^1_1-defining axiom for F from the RHS as follows:

By assumption, G and H have Σ_1^1-defining axioms. Therefore there are Σ_0^B formulas φ_G and φ_H so that

$$W = G() \leftrightarrow \exists \vec{U} \varphi_G(\vec{U}, W), \qquad W = H(y, Z) \leftrightarrow \exists \vec{V} \varphi_H(y, Z, \vec{V}, W)$$

and

$$V^1 \vdash \exists! W \exists \vec{U} \varphi_G(\vec{U}, W), \tag{101}$$

$$V^1 \vdash \forall y \forall Z \exists! W \exists \vec{V} \varphi_H(y, Z, \vec{V}, W). \tag{102}$$

The Σ_1^1-defining axiom for F is obtained by using arrays $\vec{V}$ for which $\vec{V}^{[z]}$ (row z in the arrays $\vec{V}$) codes the values of $\vec{V}$ needed to satisfy (102) when evaluating $H(z, W^{[z]})$.

$$Y = F(y) \leftrightarrow \left(\exists W \exists \vec{U} \exists \vec{V} \, \varphi_G(\vec{U}, W^{[0]}) \wedge \right.$$
$$\left. (\forall z < y \varphi_H(z, W^{[z]}, \vec{V}^{[z]}, (W^{[z+1]})^{<t(z)})) \wedge Y = W^{[y]} \right). \tag{103}$$

Since the terms such as $(W^{[z+1]})^{<t(z)}$ are easily seen to be Σ_0^B-bit-definable, it follows from Lemma V.6.7 that this defining axiom can be replaced by an equivalent Σ_1^1-formula (see the discussion following Exercise VI.2.9).

It is easy to see that V^1 proves the uniqueness of Y by proving that if W_1 and W_2 satisfy (103), then for $z \le y$ we have $W_1^{[z]} = W_2^{[z]}$. This is by number induction on $z \le y$, and follows from the uniqueness of W in (101) and (102).

Now we show that V^1 proves the existence of Y satisfying the RHS of (103). We start by noting that all of the initial string quantifiers can be bounded. This follows from Parikh's Theorem, using (101) and (102). Let $\psi(y)$ be the Σ_1^B-formula obtained from this bounded form of the RHS of (103), with the final conjunct $Y = W^{[y]}$ deleted. Thus $\psi(y)$ asserts the existence of an array

$$W = (W^{[0]}, W^{[1]}, \dots, W^{[y]})$$

whose rows are the successive values

$$F(0), F(1), \dots, F(y).$$

We show that V^1 proves $\psi(y)$ by induction on y. The base case follows from (101): If W' satisfies the existential quantifier $\exists W$ in (101), then W satisfying $\psi(y)$ can be defined using multiple comprehension (Lemma VI.1.2):

$$W(0, i) \leftrightarrow W'(i).$$

For the induction step, the new values of W and $\vec{V}$ for $y + 1$ are obtained by pasting together the previous values for y, together with values from (102) with (y, Z) in φ_H replaced by $(y, W^{[y]})$. The pasting is again defined using multiple comprehension.

Hence $V^1 \vdash \psi(y)$. From this it follows that V^1 proves the existence of Y satisfying the RHS of (103): just set $Y = W^{[y]}$. Hence $F(y)$ is Σ_1^1-definable in V^1.   □

VI.3. The Replacement Axiom Scheme

Recall that the classes Σ_i^B and Π_i^B consist of bounded formulas in which all string quantifiers occur in front. We now define more general classes which allow mixing bounded number quantifiers with bounded string quantifiers.

DEFINITION VI.3.1 ($g\Sigma_i^B(\mathcal{L})$ and $g\Pi_i^B(\mathcal{L})$). For a vocabulary $\mathcal{L}$ extending $\mathcal{L}_A^2$, define

$$g\Sigma_0^B(\mathcal{L}) = g\Pi_0^B(\mathcal{L}) = \Sigma_0^B(\mathcal{L}).$$

For $i \geq 0$, $g\Sigma_{i+1}^B(\mathcal{L})$ is the closure of $g\Pi_i^B(\mathcal{L})$ under $\wedge, \vee, \forall x \leq t, \exists x \leq t$ and $\exists X \leq t$. Similarly, $g\Pi_{i+1}^B(\mathcal{L})$ is the closure of $g\Sigma_i^B(\mathcal{L})$ under $\wedge, \vee, \forall x \leq t, \exists x \leq t$ and $\forall X \leq t$.

We usually write $g\Sigma_i^B$ for $g\Sigma_i^B(\mathcal{L}_A^2)$ and $g\Pi_i^B$ for $g\Pi_i^B(\mathcal{L}_A^2)$. It is easy to see that

$$\Sigma_0^B(\mathcal{L}) \subset g\Sigma_1^B(\mathcal{L}) \subset g\Sigma_2^B(\mathcal{L}) \subset \cdots,$$
$$\Sigma_0^B(\mathcal{L}) \subset g\Pi_1^B(\mathcal{L}) \subset g\Pi_2^B(\mathcal{L}) \subset \cdots.$$

Although for $i \geq 1$ the syntactic classes $g\Sigma_i^B(\mathcal{L})$ and $g\Pi_i^B(\mathcal{L})$ do not allow negations in front of string quantifiers, note that a negated $g\Sigma_i^B(\mathcal{L})$ formula is logically equivalent to a $g\Pi_i^B(\mathcal{L})$ formula, and *vice versa*. Hence every formula in $\Sigma_0^B(\Sigma_i^B(\mathcal{L}) \cup \Pi_i^B(\mathcal{L}))$ is equivalent to one in $g\Sigma_{i+1}^B(\mathcal{L})$ and one in $g\Pi_{i+1}^B(\mathcal{L})$

For any formula φ^+ in $g\Sigma_i^B$ there is a formula φ in Σ_i^B so that in the standard model $\underline{\mathbb{N}}_2$ we have $\varphi^+ \leftrightarrow \varphi$. In particular, when φ^+ is a $g\Sigma_1^B$ formula of the form

$$\forall x \leq t \exists X \leq t \psi(x, X)$$

where ψ is a Σ_0^B formula, then we can collect the values of X for $x = 0, 1, \ldots, t$ into a single array Y whose rows $Y^{[0]}, Y^{[1]}, \ldots, Y^{[t]}$ are these successive values of X. Thus we can take φ to be

$$\exists Y \leq \langle t, t \rangle \forall x \leq t (|Y^{[x]}| \leq t \wedge \psi(x, Y^{[x]})).$$

In this case $\varphi \supset \varphi^+$ is logically valid, and $\varphi^+ \supset \varphi$ is true in the standard model $\underline{\mathbb{N}}_2$, but may not be valid. In this section we are concerned with the provability of formulas of the type $\varphi^+ \supset \varphi$ in our theories. Consider the following axiom scheme.

DEFINITION VI.3.2 (Replacement Axiom). For a set Φ of formulas over the vocabulary $\mathcal{L}$, the *replacement axiom scheme* for Φ, denoted by Φ-*REPL*, is the set of all formulas (over $\mathcal{L} \cup \{Row\}$):

$$(\forall x \leq b \exists X \leq c\varphi(x, X)) \supset$$
$$\exists Z \leq \langle b, c \rangle \forall x \leq b(|Z^{[x]}| \leq c \wedge \varphi(x, Z^{[x]})) \quad (104)$$

where φ is in Φ.

Note that in (104) the LHS is a logical consequence of the RHS. Also (104) is true in the expansion of the standard model $\underline{N}_2$, for any formula φ.

The function *Row* occurs on the RHS of (104), but by the *Row* Elimination Lemma V.4.27 (or more generally the FAC^0 Elimination Lemma V.6.7), any $\Sigma_0^B(Row)$ formula is equivalent to a $\Sigma_0^B(\mathcal{L}_A^2)$ formula. So in the context of the theories with underlying vocabulary $\mathcal{L}_A^2$ (such as V^i, or $\widetilde{V}^1$ below), we define (104) to be the equivalent $\mathcal{L}_A^2$ formula which is obtained by transforming every atomic sub-formula containing *Row* into a $\Sigma_0^B(\mathcal{L}_A^2)$ formula.

NOTATION. When we say that a theory $\mathcal{T}$ with vocabulary $\mathcal{L}$ proves a *REPL* axiom scheme (e.g., $\Sigma_0^B(\mathcal{L})$-*REPL*), then either $\mathcal{L}_A^2 \cup \{Row\} \subseteq \mathcal{L}$, or $\mathcal{L} = \mathcal{L}_A^2$ and (104) is as above.

Recall that a *single*-Σ_1^B formula has the form $\exists X \leq t\psi(X)$, where ψ is a Σ_0^B formula.

LEMMA VI.3.3. *Suppose that $\mathcal{T}$ is a polynomial–bounded theory which proves the $\Sigma_0^B(\mathcal{L})$-REPL axiom scheme, where $\mathcal{L}$ is the vocabulary of $\mathcal{T}$ (so either $\mathcal{L} = \mathcal{L}_A^2$, or $\mathcal{L}_A^2 \cup \{Row\} \subseteq \mathcal{L}$). Then for each $g\Sigma_1^B(\mathcal{L})$ formula φ there is a single-$\Sigma_1^B(\mathcal{L})$ formula φ' so that $\mathcal{T} \vdash \varphi \leftrightarrow \varphi'$.*

PROOF. We prove by structural induction on the formula φ. For the base case, if φ is a $\Sigma_0^B(\mathcal{L})$ formula, then we can simply take $\varphi' \equiv \varphi$.

For the induction step, consider the interesting case where φ has the form $\forall x \leq s\theta(x)$, where θ is a $g\Sigma_1^B(\mathcal{L})$ formula but not a $\Sigma_0^B(\mathcal{L})$ formula. By the induction hypothesis, $\theta(x)$ is equivalent in $\mathcal{T}$ to a *single*-$\Sigma_1^B(\mathcal{L})$ formula $\exists X \leq t\psi(x, X)$, where ψ is a $\Sigma_0^B(\mathcal{L})$ formula. In other words,

$$\mathcal{T} \vdash \varphi \leftrightarrow \forall x \leq s \exists X \leq t\psi(x, X).$$

Now $\mathcal{T}$ proves φ is equivalent to a *single*-$\Sigma_1^B(\mathcal{L})$ formula by $\Sigma_0^B(\mathcal{L})$-*REPL*.

The other cases for the induction step follow easily with the help of exercise V.4.30, which shows that a prefix of several bounded string quantifiers can be collapsed into a single one. $\square$

In the next lemma we generalize the previous lemma. Part (b) follows easily from (a), and (a) can be proved by induction on i. The base case is proved in Lemma VI.3.3. The induction step is similar to the base case.

LEMMA VI.3.4. *Let $\mathcal{T}$ be a polynomial–bounded theory with vocabulary $\mathcal{L}$ which proves the $\Pi_i^B(\mathcal{L})$-**REPL** axiom scheme, for some $i \geq 0$ (so either $\mathcal{L} = \mathcal{L}_A^2$, or $\mathcal{L}_A^2 \cup \{Row\} \subseteq \mathcal{L}$). Then*

   (a) *For each $g\Sigma_{i+1}^B(\mathcal{L})$ formula φ there is a $\Sigma_{i+1}^B(\mathcal{L})$ formula φ' so that $\mathcal{T} \vdash \varphi \leftrightarrow \varphi'$.*

   (b) *For each $g\Pi_{i+1}^B(\mathcal{L})$ formula φ there is a $\Pi_{i+1}^B(\mathcal{L})$ formula φ' so that $\mathcal{T} \vdash \varphi \leftrightarrow \varphi'$.*

EXERCISE VI.3.5. Prove the above lemma.

EXERCISE VI.3.6. Let $\mathcal{T}$, $\mathcal{L}$ and i be as in Lemma VI.3.4 above. Show that $\mathcal{T}$ proves the $\Sigma_{i+1}^B(\mathcal{L})$-**REPL** axiom scheme.

The next lemma shows that V^1 proves the Σ_1^B-**REPL** axiom scheme. It is important to note that the analogous statement does not hold for V^0: we will prove later (see Section VIII.6) that V^0 *does not* prove the Σ_0^B-**REPL** axiom scheme. (It follows from Exercise VI.3.6 that over V^0 Σ_1^B-**REPL** follows from Σ_0^B-**REPL**, i.e., the two axioms schemes are equivalent over V^0.) Also, we will introduce the universal theory VPV which characterizes P in the same way that V^1 characterizes P, and we will show that it is unlikely that VPV proves Σ_1^B-**REPL**.

LEMMA VI.3.7. *Let $\mathcal{T}$ be an extension of V^0, where the vocabulary $\mathcal{L}$ of $\mathcal{T}$ is either $\mathcal{L}_A^2$ or $\mathcal{L}_A^2 \cup \{Row\} \subseteq \mathcal{L}$). Suppose that $\mathcal{T}$ proves the $\Sigma_{i+1}^B(\mathcal{L})$-**IND** axiom scheme, for some $i \geq 0$. Then $\mathcal{T}$ also proves the $\Pi_i^B(\mathcal{L})$-**REPL** axiom scheme.*

PROOF. Let φ be a $\Pi_i^B(\mathcal{L})$ formula. We will show that $\mathcal{T}$ proves (104). Intuitively, the RHS of (104) is the formula which states the existence of an array Z having b rows, whose x-th row $Z^{[x]}$ satisfies $\varphi(x, Z^{[x]})$. We will prove by number induction the existence of the initial segments of Z, and hence derive the existence of Z.

Formally we need to make sure that the RHS of (104) is equivalent to a $\Sigma_{i+1}^B(\mathcal{L})$ formula. First consider the case where $i = 0$, so φ is a $\Sigma_0^B(\mathcal{L})$ formula. Let

$$\psi(z) \equiv \exists Z \leq \langle z, c \rangle \forall x \leq z (|Z^{[x]}| \leq c \wedge \varphi(x, Z^{[x]})).$$

Then $\psi(z)$ is a $\Sigma_1^B(\mathcal{L})$ formula and the RHS of (104) is just $\psi(b)$. Our task is to show in $\mathcal{T}$ that $\psi(z)$ holds for $z \leq b$, assuming the LHS of (104). This is proved in $\mathcal{T}$ by induction on $z \leq b$. For the base case, $\psi(0)$ follows from the LHS of (104) by putting $x = 0$. The induction step follows from the induction hypothesis and the LHS of (104), using Σ_0^B-**COMP**.

For the case where $i \geq 1$, note that when φ is a $\Pi_i^B(\mathcal{L})$ formula, the RHS of (104) is not really a $\Sigma_{i+1}^B(\mathcal{L})$ formula. But it is equivalent (in $\mathcal{T}$) to:

$$\exists Z \leq \langle b, c \rangle \forall Y \leq b \, (|Z^{[|Y|]}| \leq c \wedge \varphi(x, Z^{[|Y|]}))$$

which is equivalent to a $\Sigma_{i+1}^B(\mathcal{L})$ formula. Let ψ be the equivalent $\Sigma_{i+1}^B(\mathcal{L})$ formula, then we can use the same arguments as for the case $i = 0$.   □

From Exercise VI.3.6, Lemma VI.3.7, Corollary VI.1.1, Corollary VI.1.4, and Lemma VI.3.3 we have:

COROLLARY VI.3.8. *For $i \geq 1$, the theory V^i proves the $g\Sigma_i^B$-REPL axiom scheme. For each $g\Sigma_i^B$ (resp. $g\Pi_i^B$) formula φ, there is a single-Σ_i^B (resp. single-Π_i^B) formula φ' such that $V^i \vdash \varphi \leftrightarrow \varphi'$. Also V^i proves Φ_i'-COMP, Φ_i'-IND, Φ_i'-MIN and Φ_i'-MAX, where $\Phi_i' = \Sigma_0^B(g\Sigma_i^B \cup g\Pi_i^B)$.*

VI.3.1. Extending V^1 by Polytime Functions. By the Extension by Definition Theorem III.3.5, if we extend V^1 by a collection $\mathcal{L}$ of its Σ_1^1-definable functions (i.e., polytime functions), Δ_1^1-definable predicates (i.e., polytime predicates), and their defining axioms, then we obtain a conservative extension $V^1(\mathcal{L})$ of V^1. Here we want to show further that $V^1(\mathcal{L})$ proves the $\Sigma_1^B(\mathcal{L})$-COMP axiom scheme. This is similar to the situation for V^0, where it follows from Corollary V.4.14 and Lemma V.4.15 that $V^0(\mathcal{L})$ is conservative over V^0, and it proves the $\Sigma_0^B(\mathcal{L})$-COMP axiom scheme for a collection $\mathcal{L}$ of AC^0 functions. Note that for the case of V^0, the AC^0 string functions are Σ_0^B-bit-definable in V^0.

Here it suffices to show that any $\Sigma_1^B(\mathcal{L})$ formula is provably equivalent in $V^1(\mathcal{L})$ to a $\Sigma_1^B(\mathcal{L}_A^2)$ formula. We will prove this by structural induction on the $\Sigma_1^B(\mathcal{L})$ formula. For the induction step, we use Corollary VI.3.8 above. More generally, we prove:

LEMMA VI.3.9 (Σ_1^B-Transformation). *Let $\mathcal{T}$ be a polynomial-bounded theory over the vocabulary $\mathcal{L} \supseteq \mathcal{L}_A^2 \cup \{Row\}$. Suppose that $\mathcal{T}$ proves $\Sigma_0^B(\mathcal{L})$-REPL. Let $\mathcal{T}'$ be the extension of $\mathcal{T}$ which is obtained by adding to $\mathcal{T}$ a $\Sigma_1^1(\mathcal{L})$-definable function or a $\Delta_1^1(\mathcal{L})$-definable predicate, and its defining axiom, and $\mathcal{L}'$ be the vocabulary of $\mathcal{T}'$. Then*

(a) *$\mathcal{T}'$ is conservative over $\mathcal{T}$, and $\mathcal{T}'$ is polynomial-bounded;*
(b) *For any $\Sigma_1^B(\mathcal{L}')$ formula φ^+, there is a $\Sigma_1^B(\mathcal{L})$ formula φ so that $\mathcal{T}' \vdash \varphi^+ \leftrightarrow \varphi$;*
(c) *For any $\Sigma_0^B(\mathcal{L}')$ formula φ^+, there are a $\Sigma_1^B(\mathcal{L})$ formula φ_1 and a $\Pi_1^B(\mathcal{L})$ formula φ_2 so that $\mathcal{T}' \vdash \varphi^+ \leftrightarrow \varphi_1$, and $\mathcal{T} \vdash \varphi_1 \leftrightarrow \varphi_2$;*
(d) *$\mathcal{T}'$ proves the $\Sigma_1^B(\mathcal{L}')$-REPL axiom scheme.*

Indeed, by Exercise V.4.30, the formulas φ and φ_1 can be taken to be single-$\Sigma_1^B(\mathcal{L})$ formulas, and φ_2 can be taken to be a single-$\Pi_1^B(\mathcal{L})$ formula.

PROOF. For (a) the conservativity of $\mathcal{T}'$ over $\mathcal{T}$ follows from the Extension by Definition Theorem III.3.5. Also, $\mathcal{T}'$ is polynomial-bounded because $\mathcal{T}$ is, and the Σ_1^1-definable functions of $\mathcal{T}$ are polynomially bounded (Corollary V.4.4).

Part (b) follows from (c), and (d) follows from (c) and Exercise VI.3.6 (for the case $i = 0$). We prove (c) for the case of extending $\mathcal{T}$ by a

Σ_1^1-definable string function. The case of adding a Σ_1^1-definable number function or a Δ_1^1-definable predicate is similar, and is left as an exercise.

Let F be the $\Sigma_1^1(\mathcal{L})$-definable function in $\mathcal{T}$. Since $\mathcal{T}$ is a polynomial-bounded theory, F is polynomially bounded in $\mathcal{T}$, and is $\Sigma_1^B(\mathcal{L})$-definable in $\mathcal{T}$ (Corollary V.4.4). So there is a $\Sigma_1^B(\mathcal{L})$ formula $\varphi_F(\vec{x}, \vec{X}, Y)$ such that

$$Y = F(\vec{x}, \vec{X}) \leftrightarrow \varphi_F(\vec{x}, \vec{X}, Y) \tag{105}$$

and

$$\mathcal{T} \vdash \forall \vec{x} \forall \vec{X} \exists! Y \leq t \varphi_F(\vec{x}, \vec{X}, Y). \tag{106}$$

By Lemma VI.3.3, it suffices to prove a simpler statement, i.e., that there exist a $g\Sigma_1^B(\mathcal{L})$ formula φ_1 and a $g\Pi_1^B(\mathcal{L})$ formula φ_2 such that $\mathcal{T}' \vdash \varphi^+ \leftrightarrow \varphi_1$ and $\mathcal{T} \vdash \varphi_1 \leftrightarrow \varphi_2$. We prove this by induction on the nesting depth of F in φ^+. For the base case, F does not occur in φ^+, and there is nothing to prove. For the induction step, first we prove:

CLAIM. Suppose that for each atomic sub-formula ψ of φ^+, there are a $g\Sigma_1^B(\mathcal{L})$ formula ψ_1 and a $g\Pi_1^B(\mathcal{L})$ formula ψ_2 so that $\mathcal{T}' \vdash \psi^+ \leftrightarrow \psi_1$ and $\mathcal{T} \vdash \psi_1 \leftrightarrow \psi_2$. Then there are a $g\Sigma_1^B(\mathcal{L})$ formula φ_1 and a $g\Pi_1^B(\mathcal{L})$ formula φ_2 so that $\mathcal{T}' \vdash \varphi^+ \leftrightarrow \varphi_1$ and $\mathcal{T} \vdash \varphi_1 \leftrightarrow \varphi_2$.

We prove the claim by structural induction on φ^+. The base case holds trivially. The induction step is immediate from definition of $g\Sigma_1^B(\mathcal{L})$ formulas and De Morgan's laws.

Now we return to the proof of the induction step for (c). By the claim, it suffices to consider the atomic formulas over $\mathcal{L}'$. We can reduce the nesting depth of F as follows. The maximum nesting depth of F is the depth of F in (different) terms of the form $F(\vec{s}, \vec{T})$, where $\vec{s}, \vec{T}$ are terms with less nesting depth of F. We will show how to eliminate one such term from φ^+. In the general case all such terms can be eliminated using the same method. Write φ^+ as $\varphi^+(F(\vec{s}, \vec{T}))$. Then using (105) and (106) it is easy to see that (writing t for $t(\vec{s}, \vec{T})$):

$$\mathcal{T}' \vdash \varphi^+(F(\vec{s}, \vec{T})) \leftrightarrow \exists Y \leq t (\varphi_F(\vec{s}, \vec{T}, Y) \wedge \varphi^+(Y))$$

and

$$\mathcal{T}' \vdash \exists Y \leq t (\varphi_F(\vec{s}, \vec{T}, Y) \wedge \varphi^+(Y)) \leftrightarrow \forall Y \leq t (\varphi_F(\vec{s}, \vec{T}, Y) \supset \varphi^+(Y)).$$

The last line has the form $\mathcal{T}' \vdash \varphi_1' \leftrightarrow \varphi_2'$, where φ_1' is equivalent to a $\Sigma_1^B(\mathcal{L}')$ formula and φ_2' is equivalent to a $\Pi_1^B(\mathcal{L}')$ formula. Further φ_1' and φ_2' have less nesting depth of F than $\varphi^+(F(\vec{s}, \vec{T}))$. By applying the induction hypothesis to the atomic sub-formulas, we obtain a $g\Sigma_1^B(\mathcal{L})$ formula φ_1 and a $g\Pi_1^B(\mathcal{L})$ formula φ_2 that satisfy the induction step. $\quad\square$

EXERCISE VI.3.10. Prove Lemma VI.3.9 (c) for the cases of extending $\mathcal{T}$ by a Σ_1^1-definable number function and a Δ_1^1-definable predicate.

COROLLARY VI.3.11. *Suppose that T_0 is a polynomial-bounded theory with vocabulary $\mathcal{L}_0 \supseteq \mathcal{L}_A^2 \cup \{Row\}$, and that T_0 proves the $\Sigma_0^B(\mathcal{L}_0)$-REPL axiom scheme. Let $T_0 \subset T_1 \subset T_2 \subset \cdots$ be a sequence of extensions of T_0 where each T_i has vocabulary $\mathcal{L}_i$ and each T_{i+1} is obtained from T_i by adding the defining axiom for a $\Sigma_1^1(\mathcal{L}_i)$-definable function or a $\Delta_1^1(\mathcal{L}_i)$-definable predicate. Let*

$$T = \bigcup_{i \geq 0} T_i.$$

Then T is a polynomial-bounded theory which is conservative over T_0 and proves the $\Sigma_1^B(\mathcal{L})$-REPL axiom scheme, where $\mathcal{L}$ is the vocabulary of T. Furthermore, each function in $\mathcal{L}$ is $\Sigma_1^1(\mathcal{L}_0)$-definable in T_0, and each predicate in $\mathcal{L}$ is $\Delta_1^1(\mathcal{L}_0)$-definable in T_0. Finally each $\Sigma_1^B(\mathcal{L})$ formula is provably equivalent in T to a $\Sigma_1^B(\mathcal{L}_0)$ formula.

The corollary is proved using Lemma VI.3.9 by proving by induction on i that the analogous statement holds for each theory T_i. The conservativity of T follows from the conservativity of each T_i by compactness.

The corollary can be applied to the case in which $T_0 = V^1$, since by Corollary VI.3.8, V^1 proves Σ_1^B-REPL, and we may assume that T_1 is $V^1(Row)$. We will use Corollary VI.3.11 for $T_0 = V^1(Row)$ in Subsection VI.4.2 when we prove the Witnessing Theorem for V^1.

VI.4. The Witnessing Theorem for V^1

To prove the $\Longrightarrow$ direction of Theorem VI.2.2, i.e., every Σ_1^1-definable function in V^1 is a polytime function, we prove the Witnessing Theorem for V^1 below. Recall that by the $\Longleftarrow$ direction, each polytime function has a Σ_1^1-defining axiom in V^1.

THEOREM VI.4.1 (Witnessing Theorem for V^1). *Suppose that $\varphi(\vec{x}, \vec{y}, \vec{X}, \vec{Y})$ is a Σ_0^B formula, and that*

$$V^1 \vdash \forall \vec{x} \forall \vec{X} \exists \vec{y} \exists \vec{Y} \varphi(\vec{x}, \vec{y}, \vec{X}, \vec{Y}).$$

Then there are polytime functions $f_1, \ldots, f_k, F_1, \ldots, F_m$ so that

$$V^1(f_1, \ldots, f_k, F_1, \ldots, F_m) \vdash \forall \vec{x} \forall \vec{X} \varphi(\vec{x}, \vec{f}(\vec{x}, \vec{X}), \vec{X}, \vec{F}(\vec{x}, \vec{X})).$$

A more general witnessing statement follows from this theorem and Corollary VI.3.11 and Lemma VI.3.3.

COROLLARY VI.4.2. *Let T be a theory with vocabulary $\mathcal{L}$ which results from V^1 by a sequence of extensions by Σ_1^1-definable functions and Δ_1^1-definable predicates. If*

$$T \vdash \forall \vec{x} \forall \vec{X} \exists Y \varphi(\vec{x}, \vec{X}, Y)$$

where φ is in $g\Sigma_1^B(\mathcal{L})$ then there is a polytime function F such that

$$\mathcal{T}(F) \vdash \forall \vec{x} \forall \vec{X} \varphi(\vec{x}, \vec{X}, F(\vec{x}, \vec{X})).$$

EXAMPLE VI.4.3 (Prime Recognition). Any polynomial time prime recognition algorithm (such as the one by Agrawal et al [2]) gives a predicate $Prime(X)$ which according to Corollary VI.2.4 is Δ_1^B definable in V^1. It follows by the Witnessing Theorem that if V^1 proves the correctness of the algorithm, then binary integers can be factored in polynomial time. Here correctness means

$$Prime(X) \leftrightarrow \left(2 \leq |X| \wedge \forall Y \forall Z(Y \times Z = X \supset (X = Y \vee X = Z))\right).$$

(Recall that $Y \times Z$ is Σ_1^1 definable in V^1, by Exercise VI.2.7). In fact, the right-to-left direction of this correctness statement implies

$$\forall X \exists Y \exists Z\left((Y \times Z = X \wedge X \neq Y \wedge X \neq Z) \vee Prime(X) \vee |X| < 2\right).$$

Thus if $V^1(Prime, \times)$ proves correctness then polynomial time witnessing functions for Y and Z would provide proper factors for each nonprime X with $|X| \geq 2$.

EXERCISE VI.4.4 (Prime Factorization). Show that V^1 proves that every binary integer X greater than 1 can be represented as a product of primes. Use the fact that V^1 proves the Σ_1^B-$\textbf{MAX}$ axioms (Corollary V.1.8), where we are trying to maximize k such that for some string $Y = \langle Z_1, \ldots, Z_k \rangle$ with each Z_i a binary number ≥ 2, $\prod Z_i = X$. Explain why it does not follow from the Witnessing Theorem for V^1 that binary integers can be factored into primes in polynomial time.

As in the proof of the Witnessing Theorem for V^0 (Subsection V.5.2), the Witnessing Theorem for V^1 follows from the following special case.

LEMMA VI.4.5. *Suppose that $\varphi(\vec{x}, \vec{X}, Y)$ is a Σ_0^B formula such that*

$$V^1 \vdash \forall \vec{x} \forall \vec{X} \exists Y \varphi(\vec{x}, \vec{X}, Y).$$

Then there is a polytime function F so that

$$V^1(F) \vdash \forall \vec{x} \forall \vec{X} \varphi(\vec{x}, \vec{X}, F(\vec{x}, \vec{X})).$$

Our first attempt to prove the lemma would be to consider an anchored LK^2-V^1 proof π of $\exists Y \leq t\ \varphi(\vec{x}, \vec{X}, Y)$, and proceed as in the proof of Lemma V.5.5. In this case, however, a Σ_1^B-$\textbf{COMP}$ axiom

$$\exists X \leq y \forall z < y(X(z) \leftrightarrow \varphi(z)) \tag{107}$$

is not in general provably equivalent to a Σ_1^B formula, because of the clause $\varphi(z) \supset X(z)$. So the LK^2-V^1 proof π could contain formulas which are not Σ_1^1. To get around this difficulty, we begin by showing that V^1 can be axiomatized by Σ_1^B-$\textbf{IND}$ and Σ_0^B-$\textbf{COMP}$ instead of Σ_1^B-$\textbf{COMP}$. Consider the theory $\widetilde{V}^1$:

Definition VI.4.6. The theory $\widetilde{V}^1$ has vocabulary $\mathcal{L}_A^2$ and has the axioms of V^0 and the Σ_1^B-*IND* axiom scheme.

By Exercise V.4.30, $\widetilde{V}^1$ can be axiomatized by V^0 and the *single*-Σ_1^B-*IND* axiom scheme.

Lemma VI.4.7. $\widetilde{V}^1$ *proves the* Σ_1^B-*REPL axioms.*

Proof. Corollary VI.3.8 states this for V^1, and the only properties of V^1 used in the proof are that V^1 extends V^0 and proves the Σ_1^B-*IND* axioms. Hence the same proof works for $\widetilde{V}^1$. □

Theorem VI.4.8. *The theories* V^1 *and* $\widetilde{V}^1$ *are the same.*

Proof. By Corollary VI.1.1, V^1 proves the Σ_1^B-*IND* axiom scheme. Therefore $\widetilde{V}^1 \subseteq V^1$. It remains to prove the other direction. As noted earlier, (107) is not in general equivalent to a Σ_1^B formula, so we cannot use Σ_1^B-*IND* directly on (107) to prove the existence of X. We introduce the number function *numones*(y, X), which is the number of elements of X that are $< y$. Recall that $seq(u, Z) = (Z)^u$ is the AC^0 function used for coding a finite sequence of numbers (Definition V.4.31). The function *numones* has the defining axiom:

numones$(y, X) = z \leftrightarrow$

$$z \leq y \wedge \exists Z \leq 1 + \langle y, y \rangle ((Z)^0 = 0 \wedge (Z)^y = z \wedge \forall u < y((X(u) \supset$$
$$(Z)^{u+1} = (Z)^u + 1) \wedge (\neg X(u) \supset (Z)^{u+1} = (Z)^u))). \quad (108)$$

Here Z codes a sequence of $(y+1)$ numbers so that $(Z)^u = $ *numones*(u, X), for $u \leq y$.

Exercise VI.4.9. (a) Show that (108) is a Σ_1^B definition of *numones* in $\widetilde{V}^1$, i.e., show that $\widetilde{V}^1 \vdash \forall y \forall X \exists! z \varphi_{numones}(y, z, X)$, where $\varphi_{numones}(y, z, X)$ is the RHS of (108).
(b) Show that the following is a theorem of $\widetilde{V}^1$(*numones*).

$$\exists x < y (X(x) \wedge \neg Y(x) \wedge \forall u < y (u \neq x \supset (X(u) \leftrightarrow Y(u)))) \supset$$
$$numones(y, X) = numones(y, Y) + 1.$$

Although (107) may not be Σ_1^B, the result of replacing $\leftrightarrow$ by $\supset$ is Σ_1^B. Motivated by this, we define

$$\eta(y, Y) \equiv \forall z < y (Y(z) \supset \varphi(z)).$$

Let X be the set satisfying the existential quantifier in (107). Then $\eta(y, Y)$ asserts $Y \subseteq X$.

Now consider the formula

$$\psi(w, y) \equiv \exists Y \leq y (\eta(y, Y) \wedge w = numones(y, Y)).$$

For any w and Y that satisfy $\psi(w, y)$, we have $w \leq$ *numones*(y, X), and $Y = X$ iff Y satisfies $\psi(w_0, y)$, where w_0 is the maximal value for

w. To formalize this argument, we need the Σ_1^B-MAX axioms, which by Definition V.1.5 have the form

$$\varphi(0) \supset \exists x \leq y\big(\varphi(x) \land \neg \exists z \leq y(x < z \land \varphi(z))\big)$$

where $\varphi(x)$ is Σ_1^B. These are provable in V^1 by Corollary VI.1.1.

EXERCISE VI.4.10. Show that $\widetilde{V}^1$ proves the Σ_1^B-MAX axioms. Hint: Apply Σ_1^B-IND to the formula $\varphi'(x)$ given by

$$\exists z \leq y\big(x \leq z \land \varphi(z)\big).$$

Since *numones* is Σ_1^1-definable in $\widetilde{V}^1$, it follows from Lemmas VI.3.8 and VI.3.9 that $\widetilde{V}^1(numones)$ is a conservative over $\widetilde{V}^1$ and proves that every $\Sigma_1^B(numones)$-formula is equivalent to some Σ_1^B-formula. Hence by Exercise VI.4.10, $\widetilde{V}^1(numones)$ proves the Σ_1^B-$MAX(numones)$ axioms.

Now apply Σ_1^B-MAX for the case $\varphi(w)$ is $\psi(w, y)$. Arguing in $\widetilde{V}^1$, we have $\psi(0, y)$ (take Y to be the empty set), and hence there is a maximum $w_0 \leq y$ satisfying $\psi(w_0, y)$. We argued above that the set Y corresponding to w_0 is the set X satisfying (107), and this argument can be formalized in $\widetilde{V}^1$ using Exercise VI.4.9.   □

VI.4.1. The Sequent System LK^2-$\widetilde{V}^1$. We now convert $\widetilde{V}^1$ into an equivalent sequent system LK^2-$\widetilde{V}^1$, which is defined essentially as in Definition IV.4.2 (for $\Phi = \widetilde{V}^1$), but now we replace the Σ_1^B-IND axiom scheme by the Σ_1^B-IND inference rule. Recall that for LK^2, terms do not contain any bound variables $x, y, z, \ldots, X, Y, Z, \ldots$, and formulas do not contain free occurrence of any bound variable, or bound occurrence of any free variable.

DEFINITION VI.4.11 (The IND Rule). For a set Φ of formulas, the Φ-IND rule consists of the inferences of the form

$$\frac{\Gamma, A(b) \longrightarrow A(b+1), \Delta}{\Gamma, A(0) \longrightarrow A(t), \Delta} \tag{109}$$

where A is a formula in Φ.
Restriction. The variable b is called an *eigenvariable* and does not occur in the bottom sequent.

NOTATION. In general, we refer to an LK^2 proof where the IND rule is allowed as an $LK^2 + IND$ proof.

In this chapter we are mainly interested in this rule for the case where Φ is Σ_1^B.

DEFINITION VI.4.12 (LK^2-$\widetilde{V}^1$). The rules of LK^2-$\widetilde{V}^1$ consist of the rules of LK^2 (Section IV.4), together with the *single-Σ_1^B-IND* rule (109). The non-logical axioms of LK^2-$\widetilde{V}^1$ are sequents of the form $\longrightarrow A$, where A

is any term substitution instance of a Σ_0^B-***COMP*** axiom or a 2-***BASIC***
axiom (Figure 2) or an LK^2 equality axiom (Definition IV.4.1).

Thus the axioms of LK^2-$\widetilde{V}^1$ are the same as those of LK^2-V^0.
The notion of an *anchored* LK^2-$\widetilde{V}^1$ proof generalizes the notion of an
anchored LK^2 proof (Definition IV.4.4) to include the rule Σ_1^B-***IND*** above.
Note that the axioms of LK^2-$\widetilde{V}^1$ are closed under substitution of terms
for free variables. More generally, we have:

DEFINITION VI.4.13 (Anchored LK^2 Proof with the ***IND*** Rule).   An
LK^2 proof π where the rule Φ-***IND*** is allowed, for some set Φ of formulas,
is said to be *anchored* provided that every cut formula in π occurs also
either as a formula in the non-logical axioms of π, or as one of the formulas
$A(0), A(t)$ in an instance of the rule Φ-***IND*** (109).

The following exercise is to show the soundness of LK^2+IND in general.
It follows that LK^2-$\widetilde{V}^1$ is sound, in the sense that the sequents provable in
LK^2-$\widetilde{V}^1$ are also provable in $\widetilde{V}^1$.

EXERCISE VI.4.14 (Soundness of LK^2+IND). Let Ψ and Φ be sets of
formulas. Show that if A has an LK^2-Ψ proof, where the Φ-***IND*** rule is
allowed, then A is a theorem of the theory axiomatized by $\Psi \cup \Phi$-***IND***.

To prove the Witnessing Theorem for V^1, we first prove that every the-
orem of $\widetilde{V}^1$ has an anchored LK^2-$\widetilde{V}^1$ proof. This is stated more generally
as follows.

THEOREM VI.4.15 (Anchored Completeness for LK^2+IND). *Let Ψ
and Φ be two sets of formulas over a vocabulary $\mathcal{L}$, and suppose that Ψ
includes formulas which are the semantic equivalents of the equality axioms
(Definition IV.4.1). Suppose that $\mathcal{T}$ is the theory which is axiomatized by
the set of axioms $\Psi \cup \Phi$-IND. Let Ψ' and Φ' be the closures of Ψ and
Φ respectively under substitution of terms for free variables. Then for any
theorem A of $\mathcal{T}$ there is an anchored LK^2-Ψ' proof of $\longrightarrow A$ where instances
of the Φ'-IND rule are allowed.*

To apply this to $\widetilde{V}^1$ (and hence to V^1, by Theorem VI.4.8) take $\mathcal{T} = \widetilde{V}^1$,
$\Phi = \Sigma_1^B$ and $\Psi =$ 2-***BASIC*** $\cup \Sigma_0^B$-***COMP***.

COROLLARY VI.4.16. *Every theorem of V^1 has an anchored LK^2-$\widetilde{V}^1$
proof.*

PROOF OF THEOREM VI.4.15. We refer to an anchored LK^2+IND proof
of the type stated above simply as an anchored LK^2-Ψ' proof, with the
understanding that the Φ'-***IND*** rule is allowed. We will show that if a
sequent $\Gamma \longrightarrow \Delta$ is a theorem of $\mathcal{T}$ (in the sense that its semantic formula
given in Definition II.2.16 is a theorem of $\mathcal{T}$), then there is an anchored
LK^2-Ψ' proof of $\Gamma \longrightarrow \Delta$.

Recall the proof of the Completeness Lemma II.2.24 and the Anchored **LK** Completeness Theorem II.2.28 (outlined in Exercise II.2.29). Our proof here is by the same method, i.e., for a sequent $\Gamma \longrightarrow \Delta$ purportedly provable in $\mathcal{T}$, we try to find an anchored $LK^2\text{-}\Psi'$ proof of $\Gamma \longrightarrow \Delta$. Our procedure guarantees that in the case where no such proof is found, then we will be able to define a structure that satisfies $\mathcal{T}$ but does not satisfy $\Gamma \longrightarrow \Delta$. Thus we can conclude that $\Gamma \longrightarrow \Delta$ is not provable in $\mathcal{T}$.

We begin by listing all formulas, variables, and terms. In two-sorted logic, there are two sorts of terms: number terms and string terms. So we enumerate all quadruples $\langle A_i, c_j, t_k, T_\ell \rangle$, where A_i is an $\mathcal{L}$-formula, c_j is a free variable, t_k is an $\mathcal{L}$-number term, and T_ℓ is an $\mathcal{L}$-string term. (The term t_k contains only free variables $a, b, \dots, \alpha, \beta, \dots$.) The enumeration is such that each quadruple $\langle A_i, c_j, t_k, T_\ell \rangle$ occurs infinitely many times.

The proof π is constructed in stages. Initially π consists of just the sequent $\Gamma \longrightarrow \Delta$. At each stage we expand π by applying the **IND** rule and the rules of LK^2 in reverse. We follow the 3 steps listed in the proof of the Completeness Lemma, with necessary modifications. The idea is that if this proof-building procedure does not terminate, then the term model $\mathcal{M}$ derived from it satisfies $\mathcal{T}$ but not $\Gamma \longrightarrow \Delta$. In particular, in this case the procedure produces an infinite sequence of sequents $\Gamma_n \longrightarrow \Delta_n$ (starting with $\Gamma \longrightarrow \Delta$), and $\mathcal{M}$ is defined in such a way that it satisfies every formula in the antecedents Γ_n, and falsifies every formula in the succedents Δ_n.

We modify the notion of an *active sequent* as follows.

NOTATION. In the process of constructing π, a sequent is said to be *active* if it is active as defined on page 27, and it cannot be derived from $\longrightarrow B$ for some B in Ψ' using only the exchange and weakening rules.

We use one quadruple $\langle A_i, c_j, t_k, T_\ell \rangle$ of our enumeration in each stage. Here are the details for the next stage in general.

Let $\langle A_i, c_j, t_k, T_\ell \rangle$ be the next quadruple in our enumeration. Call A_i the *active formula* for this stage.

Step 1. If A_i is in Ψ', then expand π at every active sequent $\Gamma' \longrightarrow \Delta'$ as follows:

$$\frac{A_i, \Gamma' \longrightarrow \Delta' \qquad \dfrac{\longrightarrow A_i}{\Gamma' \longrightarrow \Delta', A_i}\text{ (weakening)}}{\Gamma' \longrightarrow \Delta'}\text{ (cut)}$$

Step 2a. If $A_i \in \Phi$ and c_j has one or more free occurrences in A_i, then we incorporate an application of the **IND** rule for A_i. Let b be a new free variable that does not occur in the proof so far, and let $A(b)$ be the result of substituting b for c_j in A_i. For each active sequent $\Gamma' \longrightarrow \Delta'$ we expand π as follows:

$$\cfrac{\cfrac{A(t_j), \Gamma' \longrightarrow \Delta' \qquad \cfrac{\Gamma', A(b) \longrightarrow A(b+1), \Delta'}{\Gamma', A(0) \longrightarrow A(t_j), \Delta'}}{\cfrac{\cfrac{A(t_j), \Gamma', A(0) \longrightarrow \Delta'}{\Gamma' \longrightarrow \Delta', A(0)} \qquad \Gamma', A(0) \longrightarrow \Delta'}{\Gamma' \longrightarrow \Delta'}}}{}$$

Here the top-right inference is by the Φ-*IND* rule, and the three double lines are for the weakening, cut and exchange rules (with cut formulas $A(0)$, $A(t_j)$).

Step 2b. Proceed as in the *Step 2* in the proof of the Anchored *LK* Completeness Lemma II.2.24. Here we use the string term T_k in our enumeration for the string quantifiers, in addition to the number term t_j which is for the number quantifiers, just as in the mentioned proof.

Step 3. If there is no active sequent remaining in π, then exit from the algorithm. Otherwise continue to the next stage.

It is easy to verify that if the above procedure terminates, then the resulting proof π is an anchored LK^2-Ψ' proof of $\Gamma \longrightarrow \Delta$. It remains to show that if the procedure does not halt, then the sequent $\Gamma \longrightarrow \Delta$ is not a logical consequence of $\mathcal{T}$. This is similar as for the Completeness Lemma II.2.24, and is left as an exercise. $\square$

EXERCISE VI.4.17. Complete the proof of the Anchored Completeness Lemma for LK^2+*IND* above by constructing, in the case where the procedure does not terminate, a term model $\mathcal{M}$ (see Definition II.2.26) that satisfies $\mathcal{T}$ but not the sequent $\Gamma \longrightarrow \Delta$. The two equality relations $=_1$ and $=_2$ are not necessarily interpreted as true equality in the term model, but by our assumption on Ψ the equality axioms of Definition IV.4.1 are satisfied, so the equivalence classes of terms form a true model. Also note that the occurrences of $A(0)$ in the antecedent of the construction for *Step 2a* disappear from the sequents above them, so the term model must be defined in such a way that $A(0)$ is not necessarily satisfied. Show nevertheless that the Φ-*IND* axioms are satisfied.

Effectively we have shown that any LK^2 proof with axioms from $\mathcal{T}$ can be transformed into an anchored LK^2+*IND* proof with axioms only from Ψ'. The advantage of the latter type of LK proofs is that the cut formulas are now essentially from $\Phi \cup \Psi'$, instead of the instances of Φ-*IND* $\cup \Psi$. In the case of LK^2-V^1 proofs, the cut formulas are restricted to Σ_1^B formulas (indeed, *single*-Σ_1^B formulas), while normally, an LK^2 proof with axiom from $\widetilde{V}^1$ (Definition II.2.21) contains cut formulas which are in general not Σ_1^B. This property of LK^2-$\widetilde{V}^1$ proofs is important for our proof of the Witnessing Theorem for V^1 that we present in the next section.

PROPOSITION VI.4.18 (Subformula Property of LK^2+*IND*). *Suppose that Ψ and Φ are sets of formulas, both of which are closed under substitution* *of terms for free variables. Suppose that π is an anchored LK^2-Ψ proof of S, where the Φ-IND rule is allowed. Then every formula in every sequent of π is a sub-formula of a formula in S or in $\Psi \cup \Phi$.*

VI.4.2. Proof of the Witnessing Theorem for V^1. Now we prove the Witnessing Theorem for V^1, using the same method as for the proof of the Witnessing Theorem for V^0 (Subsection V.5.2). Here it suffices to prove Lemma VI.4.5.

Suppose that $\exists Z \varphi(\vec{a}, \vec{\alpha}, Z)$ is a Σ_1^1 theorem of V^1, where φ is a Σ_0^B formula. Then by the Anchored LK^2-$\tilde{V}^1$ Completeness Theorem VI.4.15, there is an anchored LK^2-$\tilde{V}^1$ proof π of $\exists Z \varphi(\vec{a}, \vec{\alpha}, Z)$. We may assume that π is in free variable normal form, where now Definition II.2.20 is modified to allow applications of the Σ_1^B-IND rule to eliminate a variable from a sequent (in addition to $\forall$-right and $\exists$-left). By the Subformula Property of LK^2-$\tilde{V}^1$ (Proposition VI.4.18), the formulas in π are Σ_1^1 formulas, and in fact they are Σ_0^B formulas or *single-Σ_1^1* formulas. As a result, every sequent in π has the form (81):

$$\exists X_1 \theta_1(X_1), \ldots, \exists X_m \theta_m(X_m), \Gamma \longrightarrow \Delta, \exists Y_1 \psi_1(Y_1), \ldots, \exists Y_n \psi_n(Y_n)$$
$$(110)$$

for $m, n \geq 0$, where θ_i and ψ_j and all formulas in Γ and Δ are Σ_0^B.

We will prove by induction on the depth in π of a sequent S of the form (110) that there is a finite collection of polytime functions

$$\mathcal{L} = \{F_1, \ldots, F_n, \ldots\}$$

so that $V^1(\mathcal{L})$ proves the (semantic equivalent of the) sequent

$$S' =_{\text{def}} \theta_1(\beta_1), \ldots, \theta_m(\beta_m), \Gamma \longrightarrow \Delta, \psi_1(F_1), \ldots, \psi_n(F_n) \qquad (111)$$

i.e., there is an LK^2-$V^1(\mathcal{L})$ proof of S'. Here F_i stands for $F_i(\vec{a}, \vec{\alpha}, \vec{\beta})$, and $\vec{a}, \vec{\alpha}$ is a list of exactly those variables with free occurrences in S. (This list may be different for different sequents.) Also $\beta_1, \ldots, \beta_m$ are distinct new free variables corresponding to the bound variables $X_1, \ldots, X_m$, although the latter variables may not be distinct.

We proceed as in the proof of the Witnessing Theorem for V^0 in Section V.5.2 by considering the cases where S is an axiom of LK^2-$\tilde{V}^1$ (i.e., an axiom of V^0), or S is generated using inference rules of LK^2-$\tilde{V}^1$. The case of the non-logical axioms or the introduction rules for $\neg, \wedge, \vee$ and bounded number quantifiers are dealt with just as in *Cases I–VIII* in the proof for V^0. Here we will consider the only new case, i.e., the case of the Σ_1^B-IND rule. This is the one that causes the introduction of non-AC^0 witnessing functions.

Case IX. $\mathcal{S}$ is obtained by an application of the Σ_1^B-**IND** rule. Then $\mathcal{S}$ is the bottom sequent of

$$\frac{\mathcal{S}_1}{\mathcal{S}} = \frac{\Lambda, \exists X \leq r(b)\psi(b, X) \longrightarrow \exists X \leq r(b+1)\psi(b+1, X), \Pi}{\Lambda, \exists X \leq r(0)\psi(0, X) \longrightarrow \exists X \leq r(t)\psi(t, X), \Pi}$$

where b does not occur in $\mathcal{S}$, and ψ is Σ_0^B.

By the induction hypothesis for the top sequent $\mathcal{S}_1$, there is a finite collection $\mathcal{L}$ of polytime functions, and a polytime function $G(b, \beta) \in \mathcal{L}$ (suppressing arguments for the other variables present) such that $V^1(\mathcal{L})$ proves the sequent $\mathcal{S}_1'$, which is

$$\Lambda', |\beta| \leq r(b) \wedge \psi(b, \beta) \longrightarrow |G(b, \beta)| \leq r(b+1) \wedge$$
$$\psi(b+1, G(b, \beta)), \Pi'. \quad (112)$$

Note that by the variable restriction, b and β do not occur in Λ', and can only occur in Π' as arguments to witnessing functions $F_i(b, \beta)$.

We define the witness function $\hat{G}(t, \beta)$ for the formula $\exists X \leq r(t)\psi(t, X)$ in the succedent of $\mathcal{S}$ by limited recursion (Definition VI.2.11) as follows:

$$\hat{G}(0, \beta) = \beta, \quad (113)$$
$$\hat{G}(z+1, \beta) = (G(z, \hat{G}(z, \beta)))^{<r(z+1)}. \quad (114)$$

Since G is a polytime function, by Cobham's Theorem VI.2.12, $\hat{G}$ is also a polytime function.

Let $F_1^1(b, \beta), \ldots, F_m^1(b, \beta) \in \mathcal{L}$ be the witnessing functions in Π'. Consider the sequent

$$\Lambda', |\hat{G}(b, \beta)| \leq r(b) \wedge \psi(b, \hat{G}(b, \beta)) \longrightarrow$$
$$|\hat{G}(b+1, \beta)| \leq r(b+1) \wedge \psi(b+1, \hat{G}(b+1, \beta)), \Pi'' \quad (115)$$

which is obtained from (112) by substituting $\hat{G}(b, \beta)$ for β, and writing $\hat{G}(b+1, \beta)$ for $G(b, \hat{G}(b, \beta))$ (using (114)). In particular, Π'' is obtained from Π' by replacing each witnessing function $F_i^1(b, \beta)$ for $\mathcal{S}_1$ by $F_i^2(b, \beta)$, where

$$F_i^2(b, \beta) = F_i^1(b, \hat{G}(b, \beta)) \quad (1 \leq i \leq m).$$

Let $\mathcal{L}' = \mathcal{L} \cup \{\hat{G}, F_1^2, \ldots, F_m^2\}$. Then since (112) is a theorem of $\boldsymbol{LK}^2$-$V^1(\mathcal{L})$, (115) is a theorem of $\boldsymbol{LK}^2$-$V^1(\mathcal{L}')$. Note that (115) is of the form

$$\Lambda', \rho(b, \beta) \longrightarrow \rho(b+1, \beta), \Pi'' \quad (116)$$

where

$$\rho(b, \beta) \equiv |\hat{G}(b, \beta)| \leq r(b) \wedge \psi(b, \hat{G}(b, \beta)).$$

Here ρ is a $\Sigma_0^B(\mathcal{L}')$ formula.

Notice that in Π'', b occurs (only) as an argument to F_i^2. So we cannot apply the ***IND*** rule to (116). Moreover, b should not occur in our desired sequent $\mathcal{S}'$. We remove b from Π'' by introducing the number function h:

$$h(\beta) = \min y < t \, \neg p(y + 1, \beta)$$

i.e., h has the $\Sigma_0^B(\mathcal{L}')$-defining axiom

$$h(\beta) = y \leftrightarrow y \leq t \wedge (y = t \vee \neg p(y + 1, \beta)) \wedge \forall z < y p(z + 1, \beta). \tag{117}$$

Then h is a polytime function, and can be defined from $p(b, \beta)$ using limited recursion. Define for each i, $1 \leq i \leq m$,

$$F_i(\beta) = F_i^2(h(\beta), \beta).$$

Then F_i is a polytime function. Let Π''' be Π'' with each witnessing function $F_i^2(b, \beta)$ replaced by $F_i(\beta)$. Also define (by composition):

$$G^*(\beta) = \hat{G}(t, \beta).$$

Now define $\mathcal{S}'$ to be the sequent:

$$\mathcal{S}' = \Lambda', |\beta| \leq r(0) \wedge \psi(0, \beta) \longrightarrow |G^*(\beta)| \leq r(t) \wedge \psi(t, G^*(\beta)), \Pi'''. \tag{118}$$

Then $\mathcal{S}'$ is of the form (111). It remains to show that $\mathcal{S}'$ is provable in LK^2-$V^1(\mathcal{L}'')$, where $\mathcal{L}''$ is $\mathcal{L}'$ together with the new functions in $\mathcal{S}'$, i.e., $\mathcal{L}'' = \mathcal{L}' \cup \{h, F_1, \ldots, F_m, G^*\}$.

First, by (113) the sequent (118) is equivalent to

$$\Lambda', p(0, \beta) \longrightarrow p(t, \beta), \Pi'''. \tag{119}$$

Then by replacing b in (116) with $h(\beta)$, LK^2-$V^1(\mathcal{L}'')$ proves

$$\Lambda', p(h(\beta), \beta) \longrightarrow p(h(\beta) + 1, \beta), \Pi'''. \tag{120}$$

Next, by the definition of h (117), LK^2-$V^1(\mathcal{L}'')$ proves the sequents

$$p(0, \beta) \longrightarrow p(h(\beta), \beta) \quad \text{and} \quad p(h(\beta) + 1, \beta) \longrightarrow p(t, \beta).$$

From this and (120), it follows that LK^2-$V^1(\mathcal{L}'')$ proves (119), and hence (118). $\qquad\square$

VI.5. Notes

Our theory V^1 is essentially Zambella's Theory Σ_1^p-*comp* in [112], and is a variation of the theory V_1^1 in [72], which in turn is defined in the style of Buss's second-order theories [20]. It is a two-sorted version of Buss's S_2^1. Our Σ_1^B formulas correspond to *strict* Σ_1^b formulas, but this does not really matter, as shown in Section VI.3.

The Σ_1^1 Definability Theorem for V^1 is essentially due to Buss [20] who proved it for his first-order theory S_2^1. Exercise VI.4.4 (V^1 proves the prime factorization theorem) is due to Jeřábek [60]. The interesting part of Theorem VI.4.8, that $\widetilde{V}^1$ proves the Σ_1^B-*COMP* axioms, is essentially Theorem 1 in [23].

Chapter VII

PROPOSITIONAL TRANSLATIONS

In Section II.1 we presented Gentzen's Propositional Calculus **PK** and showed that **PK** is sound and complete; i.e. a propositional formula is valid iff it is provable in **PK**. In this chapter we introduce the general notion of *propositional proof system* (or simply *proof system*) and study its complexity. We are particularly interested in which families of tautologies have polynomial length proofs. In the (apparently unlikely) event that there is a polynomial $p(n)$ such that for every n, every tautology of length n has a proof in the system of length at most $p(n)$, then we say that the system is *polynomially bounded*. The question of existence (or nonexistence) of a polynomially bounded proof system is equivalent to the important complexity theory question of whether $NP = co\text{-}NP$.

Here our main interest is the relationship between bounded arithmetic and propositional proof systems. There is an extensive literature on the complexity of proof systems (see for example [72] and [13]) which we will barely touch.

One of our goals is to associate a proof system with each of our theories, such as V^0, V^1, ... In this chapter we associate the proof system constant-depth *Frege* (AC^0-*Frege*) with V^0 and the system extended Frege (*eFrege*) with V^1. Each Σ_0^B theorem in the theory can be translated into a family of tautologies which have polynomial size proofs in the corresponding proof system (the propositional translation), showing that the proof system is sufficiently powerful. On the other hand, in Chapter X we show that the soundness of a proof system is provable in the associated theory (the Reflection Principle), showing that the proof system is not too powerful.

In order to associate proof systems with other theories, and in order to translate $\Sigma_1^B, \Sigma_2^B, \ldots$ theorems of our theories (and not just Σ_0^B theorems), we need to generalize the propositional calculus to the quantified propositional calculus (QPC). This we do in Section VII.3, and introduce the QPC proof system G and its subsystems $G_0^*, G_1^*, \ldots$ and $G_0, G_1, \ldots$. We show that for $i \geq 1$ each bounded theorem of V^i can be translated into a family of valid QPC formulas with polynomial size G_i^* proofs. In Chapter VIII we introduce the hierarchy of theories TV^i and in Chapter X we show a similar relation between TV^i and G_i. This and other results justify saying that G_i^* is a kind of nonuniform version of V^i when considering Σ_i^B-theorems, but not for theorems in general. Similarly for G_i and TV^i.

VII.1. Propositional Proof Systems

Recall (Section II.1) that a propositional formula is built from the logical constants $\bot$, $\top$ (for False, True), the propositional variables (or atoms) $p_1, p_2, \ldots$, connectives $\neg, \vee, \wedge$ and parentheses (,). Also, a tautology is a valid propositional formula (Definition II.1.1). We assume that tautologies are coded as binary strings (or more properly finite subsets of $\mathbb{N}$) using some efficient encoding.

Definition VII.1.1. $TAUT$ is the set of (strings coding) propositional tautologies.

A propositional proof system is a formal system for proving tautologies. An example is the system PK introduced in Section II.1, where a formal proof of a formula A is a tree of sequents, where the root is $\longrightarrow A$, the leaves are axioms, and the sequent at each internal node follows from its parent sequent(s) by a rule of inference. The soundness and completeness theorems state that $TAUT$ is exactly the set of formulas with formal PK proofs. Below we give a very general definition of proof system, and then explain how to make PK fit this definition.

Definition VII.1.2 (Propositional Proof System). A *propositional proof system* (or simply a *proof system*) is a polytime, surjective (onto) function

$$F : \{0, 1\}^* \longrightarrow TAUT.$$

If $F(X) = A$, then we say that X is a proof of A in the system F.

The length of A is denoted $|A|$, and the length (or size) of the proof X is denoted $|X|$. A proof system F is said to be *polynomially bounded* if there is a polynomial $p(n)$ such that for all tautologies A, there is a proof X of A in F such that $|X| \leq p(|A|)$.

Informally, a proof system F is polynomially bounded if every tautology has a short proof in F.

Example VII.1.3. PK can be treated as a proof system in the sense of Definition VII.1.2, because the function

$$PK(X) = \begin{cases} A & \text{if } X \text{ codes a } PK \text{ proof of } \longrightarrow A, \\ \top \ (\text{True}) & \text{otherwise} \end{cases}$$

is a polytime function.

It is not known whether PK is polynomially bounded. In fact, the existence of a polynomially bounded proof system is equivalent to the assertion that $NP = co\text{-}NP$.

Theorem VII.1.4. *There exists a polynomially bounded proof system iff* $NP = co\text{-}NP$.

Proof. Since $TAUT$ is $co\text{-}NP$-complete, we have $NP = co\text{-}NP$ iff $TAUT \in NP$.

($\Longrightarrow$) Suppose that F is a polynomially bounded proof system. Then by definition, there is a polynomial $p(n)$ such that

$$A \in TAUT \Leftrightarrow \exists X \leq p(|A|)F(X) = A.$$

This shows that $TAUT \in NP$: The witness for the membership of A in $TAUT$ is the proof X.

($\Longleftarrow$) If $TAUT \in NP$, then there is a polytime relation $R(Y, A)$, and a polynomial $p(n)$ such that

$$A \in TAUT \Leftrightarrow \exists Y \leq p(|A|)R(Y, A).$$

Define the proof system F by

$$F(X) = \begin{cases} A & \text{if } X \text{ codes a pair } \langle Y, A \rangle, \text{ and } R(Y, A), \\ \top & \text{otherwise.} \end{cases}$$

Clearly F is a polynomially bounded proof system.   □

The general feeling among complexity theorists is that $NP \neq co\text{-}NP$, so the above theorem suggests that no proof system is polynomially bounded. In fact some weak proof systems, including resolution and bounded depth Frege systems (which is introduced below) have been proved to be not polynomially bounded. However it seems to be very difficult to prove this for the system *PK*. The system *PK* is p-equivalent (defined below) to a large class of proof systems, called *Frege* systems, which includes many standard proof systems described in logic text books. This adds interest to the problem of showing that *PK* is not polynomially bounded.

Also because *PK* is p-equivalent to the *Frege* proof systems, we will continue to work with *PK*, and will not define the *Frege* proof systems. Below we introduce *bPK* (*bounded depth PK*) and *ePK* (*extended PK*). They belong respectively to the families call *bounded depth Frege* and *extended Frege*.

Definition VII.1.5. A proof system F_1 is said to *p-simulate* a proof system F_2 if there is a polytime function G such that $F_2(X) = F_1(G(X))$, for all X. Two proof systems F_1 and F_2 are said to be *p-equivalent* if F_1 p-simulates F_2, and vice versa.

Thus F_1 p-simulates F_2 if any given F_2-proof X of a tautology A can be transformed (by a polytime function G) into an F_1-proof $G(X)$ of A.

Exercise VII.1.6. (a) Show that the relation on proof systems "F_1 p-simulates F_2" is transitive and reflexive.

(b) Show that if F_1 p-simulates F_2, and F_2 is polynomially bounded, then F_1 is also polynomially bounded.

VII.1.1. Treelike vs Daglike Proof Systems. Proofs in the system *PK* are trees. This tree structure is potentially inefficient, since each sequent in the proof can be used only once as a hypothesis for a rule, and if it needs to be used again in another part of the proof, then it must be rederived. This motivates allowing the proof structure to be a dag (directed acyclic graph), since this allows each sequent to be used repeatedly to derive others.

DEFINITION VII.1.7 (Treelike vs Daglike). A proof system is *treelike* if the structure of each proof is required to be a tree. The system is *daglike* if a proof is allowed to have the more general structure of a dag.

In general a proof, whether treelike or daglike, can be represented as a sequence of "lines", where each line is the contents of some node in the proof. Each line is either an axiom or it follows from an earlier line or earlier lines in the proof (its parent or parents), and the line might be annotated to indicate this information. The proof is a tree if each sequent is a parent of at most one line.

The notions treelike and daglike can be used as adjectives to indicate different version of a proof system. For example, *treelike PK* is the same as *PK*, but *daglike PK* has the same axioms and rules as *PK*, but allows a proof to take the form of a dag.

The next result shows that for *PK* the distinction is not important. (But it is important for the system G_1^* defined later in this chapter.)

THEOREM VII.1.8 (Krajíček[68]). *Treelike PK p-simulates daglike PK.*

PROOF. Recall that to each sequent $S = A_1, \ldots, A_k \longrightarrow B_1, \ldots, B_\ell$ we associate the formula A_S which gives the meaning of S:

$$A_S \equiv \neg A_1 \lor \cdots \lor \neg A_k \lor B_1 \lor \cdots \lor B_\ell. \tag{121}$$

Here it is not important how we parenthesize A_S (see Lemma VII.1.15). Also, there is a treelike *PK* derivation, whose size is bounded by a polynomial in the size of S, of S from the sequent $\longrightarrow A_S$.

Suppose that $\pi = S_1, \ldots, S_n$ is a daglike *PK* proof. We show:

CLAIM. The sequence

$$\longrightarrow A_{S_1}; \quad \longrightarrow (A_{S_1} \land A_{S_2}); \quad \ldots; \quad \longrightarrow (A_{S_1} \land \cdots \land A_{S_n}); \quad \longrightarrow A_{S_n}$$

can be augmented to a treelike *PK* proof whose size is bounded by a polynomial in the length of π.

Again it is not important how the conjunctions $A_{S_1} \land \cdots \land A_{S_k}$ are parenthesized. The claim follows easily from the exercise below.   □

EXERCISE VII.1.9.   (a) Show that the following sequents have polynomial size cut-free treelike *PK* proofs:
   (i) $\longrightarrow A_S$, where S is any axiom of *PK*.
   (ii) $A \land B \longrightarrow B$, for any *PK* formulas A, B.
   (iii) $A \land B \longrightarrow A \land B \land B$, for any *PK* formulas A, B.

(b) Suppose that S is derived from S_1 (and S_2) by an inference rule of **PK**. Show that the following sequents have polynomial size cut-free treelike **PK** proofs, for any formula A:

(i) $A \wedge A_{S_1} \longrightarrow A \wedge A_S$.

(ii) $A \wedge A_{S_1} \wedge A_{S_2} \longrightarrow A \wedge A_S$.

The next result will be useful later in the chapter.

LEMMA VII.1.10 (**PK***-Replacement). *Let $A(p)$ and B be propositional formulas, and let $A(B)$ be the result of substituting B for p in $A(p)$. Then for all propositional formulas B_1, B_2, the sequent*

$$(B_1 \leftrightarrow B_2) \longrightarrow (A(B_1) \leftrightarrow A(B_2))$$

*has a cut-free treelike **PK** proof of size bounded by a polynomial in its endsequent.*

EXERCISE VII.1.11. Prove the lemma by giving (using structural induction on $A(p)$) cut-free treelike **PK** proofs of size polynomial in the size of the endsequents for the following sequents:

$$A(B_1), B_1 \leftrightarrow B_2 \longrightarrow A(B_2), \qquad A(B_2), B_1 \leftrightarrow B_2 \longrightarrow A(B_1).$$

VII.1.2. The Pigeonhole Principle and Bounded Depth PK. To show that a proof system F is not polynomially bounded, it suffices to exhibit a family of tautologies that requires F-proofs of super-polynomial size. Similarly, to show that a proof system F_2 does not p-simulate a proof system F_1, it suffices to show the existence of a family of tautologies that has polynomial size F_1-proofs, but requires super-polynomial size F_2-proofs.

There is an important family of tautologies that formalizes the Pigeonhole Principle, which states that if $n+1$ pigeons are placed in n holes, then two pigeons will wind up in the same hole. The principle is formulated using the atoms

$$p_{i,j} \qquad (\text{for } 0 \le i \le n, 0 \le j < n)$$

where $p_{i,j}$ is intended to mean that pigeon i gets placed in hole j. First, the negation of the principle is expressed as an unsatisfiable propositional formula $\neg PHP_n^{n+1}$, which is the conjunction of the following clauses:

$$(p_{i,0} \vee \cdots \vee p_{i,n-1}), \qquad 0 \le i \le n, \tag{122}$$

$$(\neg p_{i,j} \vee \neg p_{k,j}), \qquad 0 \le i < k \le n, \ 0 \le j < n. \tag{123}$$

Here, (122) says that the pigeon i is placed in some hole, and (123) says that two pigeons i and k are not placed in the same hole.

The Pigeonhole Principle itself is equivalent to the negation of $\neg PHP_n^{n+1}$, which by applying De Morgan's laws, can be expressed as follows.

Definition VII.1.12 (PHP_n^{n+1}). The propositional formula PHP_n^{n+1} is defined to be

$$(\bigwedge_{0 \le i \le n} \bigvee_{0 \le j < n} p_{i,j}) \supset \bigvee_{0 \le i < k \le n, 0 \le j < n} (p_{i,j} \wedge p_{k,j}). \qquad (124)$$

Define $PHP = \{PHP_n^{n+1} : n \ge 1\}$.

Thus for each $n \ge 1$, PHP_n^{n+1} is a tautology.

In 1985 Armen Haken proved an exponential lower bound on the length of any Resolution refutation of $\neg PHP_n^{n+1}$, one of the early important results in propositional proof complexity. On the other hand, in 1987 Buss presented polynomial size Frege proofs of PHP_n^{n+1}. (Buss's proofs are based on the fact that there are propositional formulas $A_k(p_1, \ldots, p_n)$ of size polynomial in n which express the condition that at least k of $p_1, \ldots, p_n$ are true.) It follows that Resolution does not p-simulate *Frege*. (While it is easy to show that *Frege* p-simulates Resolution.)

In fact the family *PHP* does not have polynomial size proofs in a stronger proof system called *bounded depth Frege* (also known as AC^0-*Frege*). We will define *bPK*, a representative from these systems. First, we formally define the depth of a formula. Here we think of the connectives $\wedge, \vee$ as having arbitrary fan-in.

Definition VII.1.13 (Depth of a Formula). The *depth* of a formula A is the maximal number of times the connective changes in any path in the tree form of A.

So in particular, the formula $(p_1 \vee \cdots \vee p_n)$ has depth 1, for any n, no matter how the parentheses are inserted. The depth of each clause (122) is 2, and the depth of the conjunction $\neg PHP_n^{n+1}$ is 3.

Definition VII.1.14 (Bounded Depth *PK*). For each constant $d \in \mathbb{N}$ we define a d-*PK* proof to be a *PK* proof in which the *cut* formulas have depth at most d. We define a *bounded depth PK system* (or just *bPK*) to be any system d-*PK* for $d \in \mathbb{N}$.

Sometimes the definition for a d-*PK* proof is taken to be that *all* formulas in the proof have depth $\le d$. Our definition given above is more general: For proving a formula of depth $\le d$, the two definitions are the same, but here we allow d-*PK* proofs of any formula (not just formulas of depth $\le d$). Indeed, since any tautology has a *PK* proof without using the cut rule (the *PK* Completeness Theorem II.1.8), it follows that d-*PK* is complete, for any $d \ge 0$.

In general, we are not interested in the exact length of bounded depth *PK* proofs, but only interested in the length up to the application of a polynomial. Because of this and the next lemma, we will ignore how parentheses are placed in a disjunction $(A_1 \vee \cdots \vee A_n)$.

LEMMA VII.1.15. *If A is some parenthesization of $(B_1 \vee \cdots \vee B_n)$, and A' is another such parenthesization, then there is a cut-free treelike PK proof of the sequent $A \longrightarrow A'$ consisting of $O(n^2)$ sequents, where each sequent has length at most that of the sequent $A \longrightarrow A'$.*

For example, we may have

$$A \equiv (B_1 \vee (B_2 \vee B_3)) \vee B_4), \qquad A' \equiv (B_1 \vee (B_2 \vee (B_3 \vee B_4))).$$

PROOF. By repeated use of the rule $\vee$-left, it is easy to see that there is such a d-PK proof of the sequent

$$A \longrightarrow B_1, \ldots, B_n.$$

Now repeated use of $\vee$-right (with exchanges) gives the desired d-PK proof. □

In 1988 Ajtai proved that PHP_n^{n+1} does not have polynomial size bounded depth *Frege* proofs. (In fact he proved the result for a weaker version of PHP which asserts that there is no *bijection* mapping $(n + 1)$ pigeons to n holes, see Section IX.4.3.) This was strengthened by two groups a few years later to prove the following exponential lower bound, which remains one of the strongest lower bound results in propositional proof complexity.

THEOREM VII.1.16 (Bounded Depth Lower Bound [11]). *For every $d \in \mathbb{N}$, every d-PK proof of PHP_n^{n+1} must have size at least*

$$2^{n^{\varepsilon^d}}$$

where $\varepsilon = 1/6$.

In view of Buss's upper bound for PHP_n^{n+1}, we have

COROLLARY VII.1.17. *No bounded depth $Frege$ system p-simulates any $Frege$ system.*

The lower bound results in propositional proof complexity can be used to obtain independence results in the theories of bounded arithmetic. We will explain this in the next sections.

VII.2. Translating V^0 to bPK

In this section we give evidence that the propositional proof system bPK is a kind of nonuniform version of the Σ_0^B-fragment of V^0 (in Chapter X we give more evidence). Intuitively a V^0 proof of a Σ_0^B formula is able to use concepts from the complexity class AC^0. Recall from Subsection IV.1 that a language in nonuniform AC^0 is specified by polynomial size family of bounded depth formulas. Thus the lines in a polynomial size family of bPK proofs express nonuniform AC^0 concepts.

VII.2.1. Translating Σ_0^B Formulas. We begin by showing how to translate each Σ_0^B formula $\varphi(\vec{x}, \vec{X})$ into a polynomial size bounded depth family

$$\|\varphi(\vec{x}, \vec{X})\| = \{\varphi(\vec{x}, \vec{X})[\vec{m}; \vec{n}] : \vec{m}, \vec{n} \in \mathbb{N}\}$$

of propositional calculus formulas, and then we show how to translate a V^0 proof of a Σ_0^B formula into a polynomial size family of bPK proofs. Later we will show how to translate in general a bounded two-sorted formula into a polynomial size family of *quantified propositional calculus*. Here, the depth of each formula in the family $\|\varphi(\vec{x}, \vec{X})\|$ is bounded by a constant which depends only on φ.

We first explain the translation for a Σ_0^B formula $\varphi(X)$ which has a single free (string) variable X. We introduce propositional variables $p_0^X, p_1^X, \ldots$, where p_i^X is intended to mean $X(i)$. The translation has the property that for each $n \in \mathbb{N}$, $\varphi(X)[n]$ is valid iff the formula $\forall X(|X| = \underline{n} \supset \varphi(X))$ is true in the standard model, where $\underline{n}$ is the n-th numeral. More generally, there is a one-one correspondence between truth assignments satisfying $\varphi(X)[n]$ and strings X that satisfies $\varphi(X)$ and $|X| = n$.

NOTATION. We use $val(t)$ for the numerical value of a term t, where t may have numerical constants substituted for variables.

We define $\varphi(X)[n]$ inductively as follows. For the base case, $\varphi(X)$ is an atomic formula. Consider the following possibilities.

- If $\varphi(X)$ is $X = X$, then $\varphi(X)[n] =_{\text{def}} \top$.
- If $\varphi(X)$ is $\top$ or $\bot$, then $\varphi(X)[n] =_{\text{def}} \varphi(X)$.
- If $\varphi(X)$ is $t(|X|) = u(|X|)$, then

$$\varphi(X)[n] =_{\text{def}} \begin{cases} \top & \text{if } val(t(\underline{n})) = val(u(\underline{n})), \\ \bot & \text{otherwise.} \end{cases}$$

- Similarly if $\varphi(X)$ is $t(|X|) \leq (|X|)$.
- If $\varphi(X)$ is $X(t(|X|))$, then we set $j = val(t(\underline{n}))$. Let

$$\varphi(X)[0] =_{\text{def}} \bot$$

and for $n \geq 1$:

$$\varphi(X)[n] =_{\text{def}} \begin{cases} p_j^X & \text{if } j < n - 1, \\ \top & \text{if } j = n - 1, \\ \bot & \text{if } j > n - 1. \end{cases}$$

For the induction step, $\varphi(X)$ is built from smaller formulas using a propositional connective $\wedge, \vee, \neg$, or a bounded number quantifier. For $\wedge, \vee, \neg$ we make the obvious definitions: If both $\psi(X)[n]$ and $\eta(X)[n]$ are not the logical constants $\bot$ or $\top$, then

$$\big(\psi(X) \wedge \eta(X)\big)[n] =_{\text{def}} \big(\psi(X)[n] \wedge \eta(X)[n]\big),$$
$$\big(\psi(X) \vee \eta(X)\big)[n] =_{\text{def}} \big(\psi(X)[n] \vee \eta(X)[n]\big),$$
$$\big(\neg\psi(X)\big)[n] =_{\text{def}} \neg\psi(X)[n].$$

Otherwise, if either $\psi(X)[n]$ or $\eta(X)[n]$ is a logical constant $\bot$ or $\top$, then we simplify the above definitions in the obvious way. For example,

$$\big(\psi(X) \wedge \eta(X)\big)[n] =_{\text{def}} \begin{cases} \eta(X)[n] & \text{if } \psi(X)[n] \text{ is } \top, \\ \psi(X)[n] & \text{if } \eta(X)[n] \text{ is } \top, \\ \bot & \text{if either } \psi(X)[n] \text{ or } \eta(X)[n] \text{ is } \bot. \end{cases}$$

For the case of bounded number quantifiers, $\varphi(X)$ is $\exists y \le t(|X|)\ \psi(y, X)$ or $\forall y \le t(|X|)\ \psi(y, X)$. We define

$$\big(\exists y \le t(|X|)\ \psi(y, X)\big)[n] =_{\text{def}} \bigvee_{i=0}^{m} \psi(\underline{i}, X)[n],$$
$$\big(\forall y \le t(|X|)\ \psi(y, X)\big)[n] =_{\text{def}} \bigwedge_{i=0}^{m} \psi(\underline{i}, X)[n]$$

where $m = val(t(\underline{n}))$, and recall that $\underline{i}$ is the i-th numeral. Also, if any of the $\psi(\underline{i}, X)[n]$ is translated into $\top$ or $\bot$, we simplify $\varphi(X)[n]$ just as above.

Recall that $s < t$ stands for $s \le t \wedge s \ne t$. For $val(t(\underline{n})) \ge 1$ we have

$$\big(\exists y < t(|X|)\ \psi(y, X)\big)[n] \leftrightarrow \bigvee_{i=0}^{m-1} \psi(\underline{i}, X)[n],$$
$$\big(\forall y < t(|X|)\ \psi(y, X)\big)[n] \leftrightarrow \bigwedge_{i=0}^{m-1} \psi(\underline{i}, X)[n].$$

In addition,

$$\big(\exists y < 0\ \psi(y, X)\big)[n] \leftrightarrow \bot, \qquad \big(\forall y < 0\ \psi(y, X)\big)[n] \leftrightarrow \top.$$

Recall that $\langle x, y \rangle$ is the pairing function, and we write $X(x, y)$ for $X(\langle x, y \rangle)$. We formulate the Pigeonhole Principle using a $\Sigma_0^B(\mathcal{L}_A^2)$ formula $\boldsymbol{PHP}(y, X)$ below. Here y stands for the number of holes, and X is intended to be a 2-dimensional Boolean array, with $X(i, j)$ holds iff pigeon i gets placed in hole j (for $0 \le i \le y$, $0 \le j < y$).

EXAMPLE VII.2.1 (Formulation of $\boldsymbol{PHP}$ in Two-Sorted Logic).

$$\boldsymbol{PHP}(y, X) \equiv \forall i \le y \exists j < y X(i, j) \supset$$
$$\exists i \le y \exists k \le y \exists j < y(i < k \wedge X(i, j) \wedge X(k, j)). \quad (125)$$

Then for all $1 \leq n \in \mathbb{N}$, $\boldsymbol{PHP}(\underline{n}, X)[1 + \langle n, n - 1 \rangle]$ is just $\boldsymbol{PHP}_n^{n+1}$ (Definition VII.1.12).

In general, we can define the translation of a $\Sigma_0^B(\mathcal{L}_A^2)$ formula $\varphi(\vec{x}, \vec{X})$ (i.e., with multiple free variables of both sorts). Then for each string variable X_k we associate a list of propositional variables $p_0^{X_k}, p_1^{X_k}, \ldots$, and we give each free number variable a numerical value. Thus the family $\varphi(\vec{x}, \vec{X})[\vec{m}; \vec{n}]$ is defined so that it is valid iff the formula

$$\forall \vec{x} \forall \vec{X}, \left(\bigwedge |X_k| = \underline{n_k} \right) \supset \varphi(\underline{\vec{m}}, \vec{X})$$

is true in the standard model $\underline{\mathbb{N}}_2$. Here for the base case we have to handle an additional case, i.e., where $\varphi(\vec{x}, \vec{X}) \equiv X_i = X_k$, where $i \neq k$. We reduce this case to other cases by considering φ to be its equivalence given by the LHS of the axiom $\boldsymbol{SE}$ (Figure 2):

$$|X_i| = |X_k| \wedge \forall x < |X_i|(X_i(x) \leftrightarrow X_k(x)).$$

LEMMA VII.2.2. *For every $\Sigma_0^B(\mathcal{L}_A^2)$ formula $\varphi(\vec{x}, \vec{X})$, there is a constant $d \in \mathbb{N}$ and a polynomial $p(\vec{m}, \vec{n})$ such that for all $\vec{m}, \vec{n} \in \mathbb{N}$, the propositional formula $\varphi(\vec{x}, \vec{X})[\vec{m}; \vec{n}]$ has depth at most d and size at most $p(\vec{m}, \vec{n})$.*

PROOF. The proof is by structural induction on φ, and is straightforward. □

Now we come to the main result of this section:

THEOREM VII.2.3 (V^0 Translation). *Suppose that $\varphi(\vec{x}, \vec{X})$ is a Σ_0^B formula such that $V^0 \vdash \forall \vec{x} \forall \vec{X} \varphi(\vec{x}, \vec{X})$. Then the propositional family $\|\varphi(\vec{x}, \vec{X})\|$ has polynomial size bounded depth $\boldsymbol{PK}$ proofs. That is, there are a constant d and a polynomial $p(\vec{m}, \vec{n})$ such that for all $1 \leq \vec{m}, \vec{n} \in \mathbb{N}$, $\varphi(\vec{x}, \vec{X})[\vec{m}; \vec{n}]$ has a d-$\boldsymbol{PK}$ proof of size at most $p(\vec{m}, \vec{n})$. Further there is an algorithm which finds a d-$\boldsymbol{PK}$ proof of $\varphi(\vec{x}, \vec{X})[\vec{m}; \vec{n}]$ in time bounded by a polynomial in $(\vec{m}, \vec{n})$.*

(See Theorem VII.5.6 for a generalization of this result which applies to all bounded theorems of V^0.)

In view of the Bounded Depth Lower Bound Theorem VII.1.16 above, we have:

COROLLARY VII.2.4 (Independence of PHP from V^0). *The true $\forall \Sigma_0^B$ sentence*

$$\forall y \forall X \, \boldsymbol{PHP}(y, X)$$

(*see Example VII.2.1*) *is not a theorem of V^0.*

To prove the V^0 Translation Theorem, the idea is to translate each sequent in an $\boldsymbol{LK}^2$ proof of $\varphi(\vec{a}, \vec{\alpha})$ into a $\boldsymbol{bPK}$ sequent which has a short proof. The issue here is that an $\boldsymbol{LK}^2$-V^0 proof may contain Σ_1^B formulas (i.e., the Σ_0^B-$\boldsymbol{COMP}$ axioms), whose translation we have not discussed. We introduce the theory $\widetilde{V}^0$ which plays the same role for V^0 as $\widetilde{V}^1$ does for V^1. In the next subsection we define $\widetilde{V}^0$ and the associated sequent system $LK^2\text{-}\widetilde{V}^0$ (an analogue of $LK^2\text{-}\widetilde{V}^1$), and use these to prove the V^0 Translation Theorem.

VII.2.2. $\widetilde{V}^0$ and $LK^2\text{-}\widetilde{V}^0$.

DEFINITION VII.2.5. The theory $\widetilde{V}^0$ has vocabulary $\mathcal{L}_A^2$ and is axiomatized by 2-**BASIC** and the Σ_0^B-**IND** axiom scheme.

Thus $\widetilde{V}^0$ is the same as V^0, except the Σ_0^B-**COMP** axioms are replaced by the Σ_0^B-**IND** axioms. By Corollary V.1.8, V^0 proves the Σ_0^B-**IND** axiom scheme, hence $\widetilde{V}^0 \subseteq V^0$.

Unlike the $\widetilde{V}^1$, V^1 case, unfortunately V^0 is not the same as $\widetilde{V}^0$, because $\widetilde{V}^0$ does not prove the Σ_0^B-**COMP** axioms. To see this, expand the standard (single-sorted) model $\underline{\mathbb{N}}$ to a $\mathcal{L}_A^2$ structure $\mathcal{M}$ by letting the string universe be $\{\varnothing\}$, where $|\varnothing| = 0$. Then it is easy to see that $\mathcal{M}$ is a model of $\widetilde{V}^0$, but not of V^0. Nevertheless, we can prove a weaker statement.

DEFINITION VII.2.6 (Φ-Conservative Extension). Let Φ be a set of formulas in the vocabulary $\mathcal{L}$. Suppose that $\mathcal{T}$ is a theory over $\mathcal{L}$, and $\mathcal{T}'$ is an extension of $\mathcal{T}$ (the vocabulary of $\mathcal{T}'$ may contain function or predicate symbols not in $\mathcal{L}$). Then we say that $\mathcal{T}'$ is a Φ-*conservative extension* of $\mathcal{T}$ if for every formula $\varphi \in \Phi$, if $\mathcal{T}' \vdash \varphi$ then $\mathcal{T} \vdash \varphi$.

So if Φ is the set of all $\mathcal{L}$ formulas, then $\mathcal{T}'$ is Φ-conservative over $\mathcal{T}$ precisely when it is conservative over $\mathcal{T}$. For the case of $\widetilde{V}^0$ and V^0, we can take Φ to be Σ_0^B.

LEMMA VII.2.7. V^0 *is* Σ_0^B-*conservative over of* $\widetilde{V}^0$.

By our definition of semantics (Sections IV.2.2 and II.2.2), this is the same as saying that V^0 is $\forall \Sigma_0^B$-conservative over $\widetilde{V}^0$, where $\forall \Sigma_0^B$ is the universal closure of Σ_0^B (Definition II.2.22).

PROOF. We noted earlier that $\widetilde{V}^0 \subseteq V^0$ (by Corollary V.1.8). The proof that every Σ_0^B theorem of V^0 is also provable in $\widetilde{V}^0$ is like the proof that V^0 is conservative over $I\Delta_0$ (Theorem V.1.9). We use the following lemma, which is proved in the same way as Lemma V.1.10 (any model of $I\Delta_0$ can be expanded to a model of V^0). In the present case, U_2' is defined as before in (53), except that now the formula φ is allowed parameters from U_2.

LEMMA VII.2.8. *Every model* $\mathcal{M} = \langle U_1, U_2 \rangle$ *for* $\widetilde{V}^0$ *can be extended to a model* $\mathcal{M}' = \langle U_1', U_2' \rangle$ *of* V^0, *where* $U_1 = U_1'$ *and* $U_2 \subseteq U_2'$.

It follows that if $\varphi(\vec{x}, \vec{X})$ is a Σ_0^B formula with all free variables indicated, and $\vec{a}$ are any elements in U_1 and $\vec{\alpha}$ are any elements in U_2, then

$$\mathcal{M} \models \varphi(\vec{a}, \vec{\alpha}) \qquad \text{iff} \qquad \mathcal{M}' \models \varphi(\vec{a}, \vec{\alpha}).$$

(The proof actually shows that V^0 is Φ-conservative over $\widetilde{V}^0$ for a set Φ larger than Σ_0^B, i.e., Φ contains formulas with unbounded number quantifiers and without string quantifiers. But we do not need this fact here.) □

The sequent system $LK^2\text{-}\widetilde{V}^0$ is analogous to $LK^2\text{-}\widetilde{V}^1$:

Definition VII.2.9 $(LK^2\text{-}\widetilde{V}^0)$. The rules of $LK^2\text{-}\widetilde{V}^0$ consist of the rules of LK^2 (Section IV.4), together with the $\Sigma_0^B\text{-}IND$ rule (Definition VI.4.11). The non-logical axioms of $LK^2\text{-}\widetilde{V}^0$ are sequents of the form $\longrightarrow A$, where A is any term substitution instance of a 2-*BASIC* axiom (Figure 2) or an LK^2 equality axiom (Definition IV.4.1).

Recall the notion of an anchored $LK^2\text{-}\widetilde{V}^0$ proof from Definition VI.4.13, and the Anchored Completeness Lemma for $LK^2\text{+}IND$ VI.4.15. We are now ready to prove the V^0 Translation Theorem.

VII.2.3. Proof of the Translation Theorem for V^0. By assumption, $\varphi(\vec{a}, \vec{\alpha})$ is a Σ_0^B theorem of V^0. By the Anchored Completeness Lemma for $LK^2\text{+}IND$ VI.4.15, there is an anchored $LK^2\text{-}\widetilde{V}^0$ proof π of $\varphi(\vec{a}, \vec{\alpha})$. We may assume that π is in free variable normal form, where (as in Subsection VI.4.2) we modify Definition II.2.20 to allow the rule $\Sigma_0^B\text{-}IND$ to eliminate a variable. By the Subformula Property of $LK^2\text{+}IND$ (Proposition VI.4.18), every formula in every sequent of π is Σ_0^B. So every sequent S in π has the form

$$\psi_1(\vec{b}, \vec{\beta}), \ldots, \psi_k(\vec{b}, \vec{\beta}) \longrightarrow \eta_1(\vec{b}, \vec{\beta}), \ldots, \eta_\ell(\vec{b}, \vec{\beta})$$

where ψ_i, η_j are Σ_0^B formulas, and $(\vec{b}, \vec{\beta})$ are all the free variables in S (which may be different for different sequents). The translation $S[\vec{m}; \vec{n}]$ is obtained from the translations $\psi_i(\vec{b}, \vec{\beta})[\vec{m}; \vec{n}]$ and $\eta_j(\vec{b}, \vec{\beta})[\vec{m}; \vec{n}]$ as follows. First, if any $\psi_i(\vec{b}, \vec{\beta})[\vec{m}; \vec{n}]$ is $\bot$, or any $\eta_j(\vec{b}, \vec{\beta})[\vec{m}; \vec{n}]$ is $\top$, then $S[\vec{m}; \vec{n}]$ is the axiom

$$\longrightarrow \top. \tag{126}$$

Otherwise, $S[\vec{m}; \vec{n}]$ has the form

$$S[\vec{m}; \vec{n}] =_{\text{def}} \ldots, \psi_i(\vec{b}, \vec{\beta})[\vec{m}; \vec{n}], \ldots \longrightarrow \ldots, \eta_j(\vec{b}, \vec{\beta})[\vec{m}; \vec{n}], \ldots$$

where the antecedent consists of all $\psi_i(\vec{b}, \vec{\beta})[\vec{m}; \vec{n}]$ that are not $\top$, and the succedent consists of all $\eta_j(\vec{b}, \vec{\beta})[\vec{m}; \vec{n}]$ that are not $\bot$.

We will prove by induction on the number of lines above this sequent in π that there are a constant d and a polynomial p depending on π, such that the propositional sequent $S[\vec{m}; \vec{n}]$ has a d-*PK* proof of size at most $p(\vec{m}, \vec{n})$, for all $\vec{m}, \vec{n} \in \mathbb{N}$. It is straightforward to verify that the proof can be obtained in time polynomial in $\vec{m}, \vec{n}$.

For the base case, S is a non-logical axiom of $LK^2\text{-}\widetilde{V}^0$. Thus S is of the form $\longrightarrow \eta$, where η is a term substitution instance of the 2-*BASIC* axioms, or S is an instance of the Equality axioms (Definition IV.4.1). First, any string variable X can occur in an instance of **B1**–**B12** only in the context of a number term $|X|$. Since these axioms are true in the standard model $\underline{\mathbb{N}}_2$, they translate into the propositional constant $\top$. Therefore if η is an instance of **B1–B12**, then $\longrightarrow \eta$ translates into the axiom (126) of **PK**.

Instances of **L1** and **L2** translate into (126). Consider, for example, an instance of **L1**:

$$\eta(\vec{b}, \gamma, \vec{\beta}) \equiv \gamma(t) \supset t < |\gamma|$$

where $\vec{b}, \vec{\beta}$ denote all (free) variables occurring in the $\mathcal{L}_A^2$-number term $t = t(\vec{b}, |\gamma|, |\vec{\beta}|)$. By definition, in order to get $\eta(\vec{b}, \gamma, \vec{\beta})[\vec{m}; n, \vec{n}]$, first we obtain the formulas

$$\begin{cases} p_i^\gamma \supset \top & \text{if } i < n - 1, \\ \top \supset \top & \text{if } i = n - 1, \\ \bot \supset \bot & \text{if } i > n - 1 \end{cases}$$

where $i = val(t(\vec{m}, n, \vec{n}))$. Simplifying these formulas results in

$$\eta(\vec{b}, \gamma, \vec{\beta})[\vec{m}; n, \vec{n}] =_{\text{def}} \top.$$

By definition, any instance of the axiom **SE** translates into a formula of the form $A \supset A$, where A is the translation of the LHS of **SE**. This tautology has a short cut-free derivation **PK**.

Similar (and simple) arguments show that if S is an instance of any of the Equality Axioms, then its $S[\vec{m}; n, \vec{n}]$ has a short d-**PK** proof, for some small constant d. (This constant accounts for the fact that we translate $X = Y$ using the LHS of **SE**, which translates into a propositional formula of depth 3.)

For the induction step, we consider the rules of LK^2-$\widetilde{V}^0$. Since all formulas in π are Σ_0^B, the string quantifier rules are never applied. If S is obtained from S_1 (and S_2) by one of the introduction rules for the connectives $\wedge, \vee$ and $\neg$ and the translation(s) of the auxiliary formula(s) are not simplified to Boolean constants then we can apply the same rules to get the **PK** proof of $S[\vec{m}; \vec{n}]$ from the **PK** proof(s) of $S_1[\vec{m}; \vec{n}]$ (and $S_2[\vec{m}; \vec{n}]$). Otherwise, if an auxiliary formula is translated into $\top$ or $\bot$ then it can be seen that $S[\vec{m}; \vec{n}]$ is the same as $S_1[\vec{m}; \vec{n}]$ (or $S_2[\vec{m}; \vec{n}]$). No new cut is needed for this step.

For the case of the cut rule, the cut formula $\psi(\vec{b}, \vec{\beta})$ is Σ_0^B, and since π is in free variable normal form, no variable is eliminated by the rule. Consider the interesting case where the translation of $\psi(\vec{b}, \vec{\beta})$ is not a constant $\top$ or $\bot$. The corresponding **PK** proof also uses the cut rule, where the cut formula is a propositional translation $\psi(\vec{b}, \vec{\beta})[\vec{m}; \vec{n}]$ of this formula, which according to Lemma VII.2.2 has bounded depth d independent of $\vec{m}, \vec{n}$.

Consider the case of the number $\forall$-right. Suppose that the inference is

$$\frac{S_1}{S} = \frac{\Lambda \longrightarrow \Pi, c \leq t(\vec{b}, |\vec{\beta}|) \supset \eta(\vec{b}, c, \vec{\beta})}{\Lambda \longrightarrow \Pi, \forall x \leq t(\vec{b}, |\vec{\beta}|)\, \eta(\vec{b}, x, \vec{\beta})}$$

where c does not occur in S. By the induction hypothesis, there are a constant $d \in \mathbb{N}$ and a polynomial $p(\vec{m}, i, \vec{n})$ so that for each $\langle \vec{m}, i, \vec{n} \rangle$, there is a d-$\textbf{PK}$ proof $\pi[\vec{m}, i; \vec{n}]$ of size $\leq p(\vec{m}, i, \vec{n})$ of the sequent $S_1[\vec{m}, i; \vec{n}]$. Note that if for some $i \leq r, \eta(\vec{b}, c, \vec{\beta})[\vec{m}, i; \vec{n}]$ is $\perp$ then $\forall x \leq t(\vec{b}, |\vec{\beta}|)\, \eta(\vec{b}, x, \vec{\beta})$ translates into $\perp$ and hence $S[\vec{m}; \vec{n}] = S_1[\vec{m}, i; \vec{n}]$ and we are done. Moreover, if all $\eta(\vec{b}, c, \vec{\beta})[\vec{m}, i; \vec{n}]$ (for $i \leq r$) are $\top$ then $S[\vec{m}; \vec{n}]$ is the axiom $\longrightarrow \top$ and we are also done.

Now, if some $\eta(\vec{b}, c, \vec{\beta})[\vec{m}, i; \vec{n}]$ is $\top$ then it will be deleted from the translation of $\forall x \leq t(\vec{b}, |\vec{\beta}|)\, \eta(\vec{b}, x, \vec{\beta})$, and the sequent $S_1[\vec{m}, i; \vec{n}]$ is the axiom $\longrightarrow \top$ and it will not be used in the following derivation. So suppose that for all $i \leq r, \eta(\vec{b}, c, \vec{\beta})[\vec{m}, i; \vec{n}]$ is neither $\top$ nor $\perp$. Then for $i \leq r, S_1[\vec{m}, i; \vec{n}]$ is

$$\Lambda[\vec{m}; \vec{n}] \longrightarrow \Pi[\vec{m}; \vec{n}], \; \eta(\vec{b}, c, \vec{\beta})[\vec{m}, i; \vec{n}].$$

The sequent S translates into

$$S[\vec{m}; \vec{n}] =_{\text{def}} \Lambda[\vec{m}; \vec{n}] \longrightarrow \Pi[\vec{m}; \vec{n}], \; \bigwedge_{i=0}^{r} \eta(\vec{b}, \underline{i}, \vec{\beta})[\vec{m}; \vec{n}].$$

Thus $S[\vec{m}; \vec{n}]$ is obtained from $S_1[\vec{m}, i; \vec{n}]$ (for $i = 0, 1, \ldots, r$) by the $\wedge$-right rule. No new instance of the cut rule is needed. This proof of $S[\vec{m}; \vec{n}]$ has size slightly more than the sum of the $(m+1)$ proofs $\pi[\vec{m}, i; \vec{n}]$, and m is a polynomial in $\vec{m}, \vec{n}$. Hence the resulting proof is bounded in size by a polynomial in $\vec{m}, \vec{n}$.

The case $\exists$-left is similar, and the cases $\forall$-left, $\exists$-right are straightforward. These are left as an exercise.

EXERCISE VII.2.10. Take care of the other number quantifier cases.

Finally we consider the case that S is obtained by the Σ_0^B-$\textbf{IND}$ rule:

$$\frac{S_1}{S} = \frac{\Lambda, \psi(c) \longrightarrow \psi(c+1), \Pi}{\Lambda, \psi(0) \longrightarrow \psi(t), \Pi}$$

where c does not occur in S, and we have suppressed all free variables except c (here t is of the form $t(\vec{b}, |\vec{\beta}|)$). By the induction hypothesis, there are polynomial size d-$\textbf{PK}$ proofs $\pi[\vec{m}, i; \vec{n}]$ of the propositional sequents

$$S_1[\vec{m}, i; \vec{n}] =_{\text{def}} \Lambda[\vec{m}; \vec{n}], \psi(c)[\vec{m}, i; \vec{n}] \longrightarrow \psi(c+1)[\vec{m}, i; \vec{n}], \Pi[\vec{m}; \vec{n}]$$

for some constant $d \in \mathbb{N}$. Let $r = val(t(\vec{m}, \vec{n}))$. The sequent S translates into

$$S[\vec{m}; \vec{n}] =_{\text{def}} \Lambda[\vec{m}; \vec{n}], \psi(0)[\vec{m}; \vec{n}] \longrightarrow \psi(r)[\vec{m}; \vec{n}], \Pi[\vec{m}; \vec{n}].$$

Now if $r = 0$ then $\mathcal{S}[\vec{m}; \vec{n}]$ is derived from the following axiom of $\mathbf{PK}$ simply by weakening:

$$\psi(0)[\vec{m}; \vec{n}] \longrightarrow \psi(0)[\vec{m}; \vec{n}].$$

For $r > 0$, we combine these proofs $\pi[\vec{m}, i; \vec{n}]$ for $i = 0, 1, \ldots, r - 1$ by using repeated cuts, with cut formulas $\psi(i)[\vec{m}; \vec{n}]$, $1 \leq i \leq r - 1$. By Lemma VII.2.2, these formulas have depth bounded by a constant depending only on ψ. Also, given that each $\pi[\vec{m}, i; \vec{n}]$ has a polynomial bounded size, the proof $\pi[\vec{m}; \vec{n}]$ is easily shown to be bounded in size by some polynomial in $\vec{m}, \vec{n}$. This completes the proof of the Translation Theorem for V^0.   □

Note that the Σ_0^B-$\mathbf{IND}$ axioms are Σ_0^B. So in fact we could have defined $\mathbf{LK}^2$-$\widetilde{V}^0$ to include the Σ_0^B-$\mathbf{IND}$ *axiom scheme* instead of the Σ_0^B-$\mathbf{IND}$ *rule*. Here we can use the following version of the Σ_0^B-$\mathbf{IND}$ axiom:

$$(\varphi(0) \wedge \forall x < t(\varphi(x) \supset \varphi(x + 1))) \supset \forall z \leq t\varphi(z) \qquad (127)$$

where t is any term not involving x or z, and φ is a Σ_0^B formula which may contain other free variables.

In this way, the case of the Σ_0^B-$\mathbf{IND}$ rule in the induction step of the proof above is replaced by two cases: One for the base case where the axiom is an Σ_0^B-$\mathbf{IND}$ axiom, and one for the induction step, in the case of the cut rule where the cut formula is an instance of the Σ_0^B-$\mathbf{IND}$ axioms. The latter is dealt with just as any other instance of the cut rule. Handling the former is left as an exercise.

EXERCISE VII.2.11. Show directly (without using Theorem VII.2.3) that the translation of (127) above has polynomial size d-$\mathbf{PK}$ proofs, where d depends only on φ.

VII.3. Quantified Propositional Calculus

Quantified Propositional Calculus (QPC) is an extension of the Propositional Calculus (Section II.1) which allows quantifiers over propositional variables. In this section we will discuss the sequent system $\mathbf{G}$ which extends Gentzen's system $\mathbf{PK}$ by the introduction rules for the propositional quantifiers. There are subsystems of $\mathbf{G}$ that relate to the first-order theories in the same way that $\mathbf{bPK}$ relates to V^0. Here we will show this relationship between V^1 and the subsystem $\mathbf{G}_1^*$ of $\mathbf{G}$.

Formally, QPC formulas (or simply formulas) are built from

- propositional constants $\top, \bot$,
- free variables $p, q, r, \ldots$,
- bound variables $x, y, z, \ldots$,
- connectives $\wedge, \vee, \neg$,
- quantifiers $\exists, \forall$,

- parentheses $(,)$

according to the following rules:

(a) $\top$, $\bot$, and p are *atomic* formulas, for any free variable p;
(b) if φ and ψ are formulas, so are $(\varphi \wedge \psi)$, $(\varphi \vee \psi)$, $\neg\varphi$;
(c) if $\varphi(p)$ is a formula, then $\forall x \varphi(x)$ and $\exists x \varphi(x)$ are formulas, for any free variable p and bound variable x.

A QPC sentence (or just sentence) is a QPC formula with no occurrence of a free variable.

EXAMPLE VII.3.1. The following is a QPC formula:

$$\forall x \exists y \big((\neg y \vee (\neg x \wedge p)) \wedge (y \vee x \vee \neg p)\big). \tag{128}$$

A truth assignment is an assignment of truth values True, False to the free variables. The truth value of a QPC formula is defined inductively, much as in the case of the Propositional Calculus. Here in the induction step, for the case of the quantifiers we use the equivalences

$$\forall x \varphi(x) \leftrightarrow (\varphi(\bot) \wedge \varphi(\top)) \quad \text{and} \quad \exists x \varphi(x) \leftrightarrow (\varphi(\bot) \vee \varphi(\top)).$$

A QPC formula is *valid* if it is true under all assignments. The notions of *satisfiability* and *logical consequence* (Definition II.1.1) generalize to QPC in the obvious way. So, for example, the formula (128) is valid (choose $y \leftrightarrow (\neg x \wedge p)$).

It is a standard result in complexity theory that the problem of determining validity of a formula of QPC is **PSPACE** complete (see Appendix A.1). Furthermore, it is natural to define a language $L \subseteq \{0, 1\}^*$ to be in nonuniform **PSPACE** if there is a polynomial size family $\langle \varphi_n(\vec{p}) \rangle$ of QPC formulas such that $\varphi_n(p_1, \ldots, p_n)$ defines the strings of length n in L. (Actually this defines the class **PSPACE**/*poly*, which is PSPACE with polynomial advice.) For this and other reasons, G (defined below) is a natural choice for a QPC proof system corresponding to the complexity class **PSPACE**. However if the number of quantifier alternations in a QPC formula is limited by some constant k, then the validity problem for such formulas is in the polynomial hierarchy.

DEFINITION VII.3.2 (Σ_i^q and Π_i^q). $\Sigma_0^q = \Pi_0^q$ is the class of quantifier-free formulas of QPC. For $i \geq 0$, Σ_{i+1}^q and Π_{i+1}^q are the smallest classes of QPC formulas satisfying

1) $\Sigma_i^q \cup \Pi_i^q \subseteq \Sigma_{i+1}^q \cap \Pi_{i+1}^q$;
2) Σ_{i+1}^q is closed under $\vee$ and $\wedge$ and existential quantification;
3) Π_{i+1}^q is closed under $\vee$ and $\wedge$ and universal quantification;
4) if $A \in \Sigma_{i+1}^q$ then $\neg A \in \Pi_{i+1}^q$;
5) if $A \in \Pi_{i+1}^q$ then $\neg A \in \Sigma_{i+1}^q$.

Thus

$$\Sigma_0^q = \Pi_0^q \subset \cdots \subset \Sigma_i^q \cap \Pi_i^q \subset \Sigma_i^q \cup \Pi_i^q \subset \Sigma_{i+1}^q \cap \Pi_{i+1}^q \subset \cdots .$$

For $i \geq 0$ every formula in Σ_{i+1}^q has a prenex form with at most i alternations of quantifiers, with the outermost quantifier being $\exists$. Similarly for Π_{i+1}^q with the outermost quantifier being $\forall$. Checking the validity of a Σ_i^q (resp. Π_i^q) sentence is Σ_i^p-complete (resp. Π_i^p-complete), for $i \geq 1$. For $i = 0$, this problem is NC^1-complete.

VII.3.1. QPC Proof Systems. We generalize Definition VII.1.2 in the obvious way to define the notion of *QPC proof system* where now F maps $\{0, 1\}^*$ onto the set of valid QPC formulas. Since the validity problem for QPC formulas is complete for *PSPACE*, the following result is proved in the same way as Theorem VII.1.4.

THEOREM VII.3.3. *There exists a polynomially bounded QPC proof system iff NP = PSPACE.*

The assertion $NP = PSPACE$ is considerably more implausible than $NP = co$-NP, but still the existence of a polynomially bounded QPC proof system is open.

The notions *p-simulate* and *p-equivalent* from Definition VII.1.5 apply in the obvious way to QPC proof systems.

VII.3.2. The System G. The QPC proof system G is a sequent system which includes the axioms and rules for *PK*, where now formulas are interpreted to be QPC formulas. It also has the following four quantifier introduction rules:

$\forall$ introduction rules:

$$\forall\text{-left: } \frac{A(B), \Gamma \longrightarrow \Delta}{\forall x A(x), \Gamma \longrightarrow \Delta} \qquad \forall\text{-right: } \frac{\Gamma \longrightarrow \Delta, A(p)}{\Gamma \longrightarrow \Delta, \forall x A(x)}$$

$\exists$ introduction rules:

$$\exists\text{-left: } \frac{A(p), \Gamma \longrightarrow \Delta}{\exists x A(x), \Gamma \longrightarrow \Delta} \qquad \exists\text{-right: } \frac{\Gamma \longrightarrow \Delta, A(B)}{\Gamma \longrightarrow \Delta, \exists x A(x)}$$

Restriction. In the rules $\forall$-right and $\exists$-left, p is a free variable called an *eigenvariable* that must not occur in the bottom sequent. For the rules $\forall$-left and $\exists$-right, $A(B)$ is the result of substituting B for all free occurrences of x in $A(x)$. The formula B is called the *target* formula and may be any quantifier-free formula (with no bound variables).

The new formulas $\exists x A(x)$ and $\forall x A(x)$ are called *principal formulas*, and the corresponding formulas in the top sequents ($A(B)$ or $A(p)$) are called *auxiliary formulas*.

Proofs in G are dags of sequents, which generalizes the treelike structure of *LK* proofs (see Subsection VII.1.1). We denote by G^* the system G restricted to treelike proofs. We will show that G and G^* are p-equivalent (Theorem VII.4.3).

The notion of free variable normal form (Definition II.2.20) readily extends to G proofs. In fact every treelike G proof can be easily transformed to one in free variable normal form by renaming variables and substituting the constant $\perp$ for some variables.

THEOREM VII.3.4 (Soundness and Completeness of G). *A sequent of G is valid iff it has a G proof. In fact, valid sequents have cut-free G proofs.*

PROOF. Soundness is easy: Provable sequents of G are valid because the axioms of G are valid, and the rules preserve validity.

For completeness, we first point out that a valid quantifier-free sequent of QPC has a cut-free G proof, by the *PK* Completeness Theorem II.1.8. In general, we prove the result by induction on the maximum quantifier depth of the formulas in the sequent (and then induction on the number of formulas in the sequent of maximum quantifier depth). We have just proved the base case, where the sequent is quantifier-free. For the induction step, the interesting cases are where the sequent is of the form

$$\forall x A(x), \Gamma \longrightarrow \Delta \qquad \text{or} \qquad \Gamma \longrightarrow \Delta, \exists x A(x).$$

These two cases are dual. So consider the sequent

$$\forall x A(x), \Gamma \longrightarrow \Delta. \tag{129}$$

We can reduce the quantifier depth in $\forall x A(x)$ by showing that (129) is valid iff the sequent

$$A(\top), A(\perp), \Gamma \longrightarrow \Delta \tag{130}$$

is valid.   □

EXERCISE VII.3.5. Carry out the details in the induction step in the above proof of the completeness of G.

The proof above shows that actually G remains complete when the target formulas B in $\forall$-left and $\exists$-right are restricted to be in the set $\{\top, \perp\}$. In fact, the restricted system is p-equivalent to G. This can be shown with the help of the following exercise.

EXERCISE VII.3.6. Show that the following sequents has cut-free G proofs of size $\mathcal{O}(|A(B)|^2)$, where A and B are any QPC formulas.
  (a) $B, A(B) \longrightarrow A(\top)$.
  (b) $A(B) \longrightarrow A(\perp), B$.
  (c) $B, A(\top) \longrightarrow A(B)$.
  (d) $A(\perp) \longrightarrow A(B), B$.
(Hint: Prove by structural induction on A for (a) and (c) simultaneously. Similarly for (b) and (d).)

EXERCISE VII.3.7 (Morioka [80]). Let *KPG* be the modification of G resulting from relaxing the condition that the target formula B in the rules $\forall$-left and $\exists$-right must be quantifier-free (so B is allowed to be any QPC formula). Show that G p-simulates *KPG*. Show that the same holds even if G is restricted so that the target formulas B in the rules $\forall$-left and $\exists$-right are restricted to be in the set $\{\top, \bot\}$. Use Exercise VII.3.6.

The original system G defined in [73] is actually KPG as defined in the above exercise. Thus the original G and our G are p-equivalent.

The proof of completeness in Theorem VII.3.4 could yield proofs of doubly exponential size. For example if the formula $\forall x A(x)$ in (129) begins with k universal quantifiers, then eliminating them all using (130) would yield 2^k copies of A, and the resulting valid sequent could require a proof exponential in its length. We now prove a singly-exponential upper bound for G proofs which allow cuts on atomic formulas.

We say that an occurrence of a symbol in a formula is *positive* (resp. *negative*) if it is in the scope of an even (resp. odd) number of $\neg$'s.

DEFINITION VII.3.8 (Sequent Length). An occurrence of a connective c in a sequent $\Gamma \longrightarrow \Delta$ is *general* if c is $\wedge$ or $\forall$ and occurs positively in Δ or negatively in Γ, or if c is $\vee$ or $\exists$ and c occurs negatively in Δ or positively in Γ. A *restricted* occurrence is defined similarly, except Δ and Γ are interchanged. For a sequent S, $|S|_g$ (resp. $|S|_r$) denotes the number of occurrences in S of general connectives (resp. $\neg$'s and restricted connectives). Also $|S|$ denotes the total number of occurrences of symbols in S, counting variables $p, q, r, \ldots, x, y, z, \ldots$ as one symbol each.

THEOREM VII.3.9. *If S is a valid sequent in the language of G with n distinct free variables, then S has a treelike G proof with $O\big(|S|_r 2^{|S|_g+n}\big)$ sequents (not counting weakenings and exchanges) in which all cut formulas are atomic and each sequent in the proof has length $O(|S|)$. If S is quantifier-free, or if all quantifier occurrences in S are general, then the proof is cut-free and the bound is improved to $O\big(|S|_r 2^{|S|_g}\big)$.*

PROOF. NOTATION. We say that a free variable p is *determined* in a sequent $A_1, \ldots, A_k \longrightarrow B_1, \ldots B_\ell$ if one of the formulas A_i or B_j is the atomic formula p. A sequent is *determined* if all of its free variables are determined.

Note that if all free variables of a sequent are determined, then there is at most one truth assignment to these free variables which fails to satisfy the sequent.

LEMMA VII.3.10. *If S is a valid sequent with all of its free variables determined, then S has a treelike G proof with $O(|S|_r 2^{|S|_g})$ sequents (not counting weakenings and exchanges) in which all cut formulas are atomic and each sequent in the proof has length $O(|S|)$. If S is quantifier-free or if all quantifier occurrences in S are general, then the same bound applies even if not all free variables in S are determined, and further the proof is treelike and cut-free.*

The second sentence of Theorem VII.3.9 follows immediately from the lemma. We now prove the first sentence of the theorem from the lemma.

Let F be the set of free variables in S. For each of the 2^n subsets K of F let S_K be the sequent resulting from S by appending a list of the variables in K to the antecedent and the variables in $F - K$ to the consequent. For example if $S = \Gamma \longrightarrow \Delta$ and $F = \{p_1, p_2, p_3\}$ and $K = \{p_2\}$, then S_K is

$$p_2, \Gamma \longrightarrow \Delta, p_1, p_3.$$

Each S_K is valid and determined, and hence by the lemma has a proof with $O(|S|_r 2^{|S|_g})$ sequents. Then S can be derived by combining these 2^n proofs with 2^{n-1} atomic cuts.   □

Proof of Lemma VII.3.10. We use induction on the total number of connectives $\wedge, \vee, \neg, \forall, \exists$ in S. The base case is immediate, since any valid sequent with no such connectives is a subsequent of an axiom.

For the induction step, we have a case for each of the connectives $\wedge, \vee, \neg, \forall, \exists$. We consider a formula A occurring in the consequent: The argument for the antecedent is dual. If A is of the form $\neg B$ then S has the form $\Gamma \longrightarrow \Delta, \neg B$. Let S' be the sequent $B, \Gamma \longrightarrow \Delta$. Then S' is valid (and determined if S is) and $|S'|_r = |S|_r - 1$, so the induction hypothesis applies and S can be derived from S' by the rule $\neg$-right. The case in which A has the form $B \vee C$ is similar, using the rule $\vee$-right.

If S has the form $\Gamma \longrightarrow \Delta, (B \wedge C)$, then $\Gamma \longrightarrow \Delta, B$ and $\Gamma \longrightarrow \Delta, C$ are each valid (and determined if S is) and have reduced $|S|_g$, and S can be derived by $\wedge$-right from these two sequents.

Suppose that S is $\Gamma \longrightarrow \Delta, \forall x A(x)$. Then $S' = \Gamma \longrightarrow \Delta, A(p)$ is valid, where p is a new free variable. Further $|S'|_g = |S|_g - 1$ and S follows from S' using $\forall$-right. This takes care of the second sentence in the lemma, but for the first sentence there is the problem that S' may not be determined, even if S is. But each of the sequents $p, \Gamma \longrightarrow \Delta, A(p)$ and $\Gamma \longrightarrow \Delta, A(p)$, p is valid and determined if S is, and by the induction hypothesis can be proved with $O(|S|_r 2^{|S|_g - 1})$ sequents. Further S can be derived from these two sequents with a cut on p and $\forall$-right, making a total of $O(|S|_r 2^{|S|_g} + 2) = O(|S|_r 2^{|S|_g})$ sequents.

Finally consider the case in which S is $\Gamma \longrightarrow \Delta, \exists x A(x)$. Since the occurrence of $\exists$ is restricted, the second sentence of the lemma does not apply, so we may assume that S is determined and valid. We claim that one of the two sequents $\Gamma \longrightarrow \Delta, A(\top)$ and $\Gamma \longrightarrow \Delta, A(\bot)$ is valid (they are both determined). To see this, note that since S is determined there is at most one truth assignment τ to the free variables of S that could falsify $\Gamma \longrightarrow \Delta$. If no such τ exists, we are done. Otherwise τ satisfies $\exists x A(x)$, and hence τ satisfies either $A(\top)$ or $A(\bot)$. Hence we may apply the induction hypothesis to one of these sequents, and obtain S using $\exists$-right.   □

VII.4. The Systems G_i and $G_i^\star$

DEFINITION VII.4.1 (G_i and $G_i^\star$). For each $i \geq 0$, G_i is the subsystem of G in which cut formulas are restricted to $\Sigma_i^q \cup \Pi_i^q$. The system $G_i^\star$ is treelike G_i.

The following result is immediate from Theorem VII.3.9.

COROLLARY VII.4.2. *Every valid QPC sequent S has a $G_0^\star$ proof of size $2^{O(|S|)}$.*

THEOREM VII.4.3. *For $i \geq 0$, $G_{i+1}^\star$ p-simulates G_i, when the systems are restricted to proving $\Sigma_i^q \cup \Pi_i^q$ formulas. $G^\star$ p-simulates G.*

PROOF. The argument is similar to the proof of Theorem VII.1.8, except for the quantifier rules $\forall$-right and $\exists$-left we can no longer argue that the conclusion is a logical consequence of the hypotheses. However for each rule deriving a sequent S from a sequent S_1 we know that $\forall A_S$ is a logical consequence of $\forall A_{S_1}$, where $\forall B$ is the universal closure of B. Thus we replace the Claim in the earlier proof by arguing that if $\pi = S_1, \ldots, S_n$ is a daglike G proof then

$$\longrightarrow \forall A_{S_1}; \quad \longrightarrow (\forall A_{S_1} \wedge \forall A_{S_2}); \ldots; \quad \longrightarrow (\forall A_{S_1} \wedge \cdots \wedge \forall A_{S_n}); \quad \longrightarrow A_{S_n} \tag{131}$$

can be augmented to a treelike G proof whose size is bounded by a polynomial in the length of π, and in which cut formulas are restricted to subformulas of formulas in the sequence. The theorem then follows from the fact if the all formulas in the sequent S are in $\Sigma_i^q \cup \Pi_i^q$ then the formula $\forall A_S$ is in Π_{i+1}^q.

Our new claim follows from Exercise VII.1.9 (b), the fact that for every axiom S of G, $\longrightarrow \forall A_S$ has an easy $G_0^\star$ proof, and the exercise below. $\quad \square$

EXERCISE VII.4.4. (a) Suppose that if S is derived from S_1 (and S_2) by an inference rule of G. Show that the following sequents have polynomial size cut-free G proofs for any formula A. (For the **PK** rules it is helpful to use Exercise VII.1.9 (b).)

(i) $A \wedge \forall A_{S_1} \longrightarrow A \wedge \forall A_S$.

(ii) $A \wedge \forall A_{S_1} \wedge \forall A_{S_2} \longrightarrow A \wedge \forall A_S$.

(b) Show that for every sequent $S = \Gamma \longrightarrow \Delta$, the sequent

$$\forall A_S, \Gamma \longrightarrow \Delta$$

has a polynomial size cut-free treelike G proof.

The next result strengthens Theorem VII.4.3 for the case $i = 0$.

Theorem VII.4.5 (Morioka [80]). $G_0^\star$ *p-simulates* G_0 *restricted to proving prenex* Σ_1^q *formulas.*

Proof sketch. Note that the proof of Theorem VII.1.8 (treelike *PK* p-simulates daglike *PK*) does not adapt to this case, because that argument requires cuts on conjunctions of earlier lines in the proof, which now would involve quantifiers.

Instead, following [80], we argue that a form of Gentzen's Midsequent Theorem can be made to work in polynomial time. Let π be a G_0 proof of a sequent

$$\longrightarrow \exists x_1 \ldots \exists x_m C(\vec{p}, x_1, \ldots, x_m) \tag{132}$$

where $C(\vec{p}, x_1, \ldots, x_m)$ is quantifier-free. Since all cut formulas in π are quantifier-free, it follows that every quantified formula in π is an ancestor of the conclusion, and must occur on the RHS and must have the form

$$\exists x_k \ldots \exists x_m C(\vec{p}, B_1 \ldots B_{k-1}, x_k, \ldots, x_m) \tag{133}$$

for some quantifier-free formulas $B_1, \ldots, B_{k-1}$ and some k, $1 \le k \le m$. Let us call a formula a π-*prototype* if it is quantifier-free and is the auxiliary formula in an $\exists$-right rule (so it is the quantifier-free parent of a formula of the form (133), with $k = m + 1$). Thus a π-prototype has the form $C(\vec{p}, B_1 \ldots B_m)$.

The *Herbrand* π *disjunction* S_π is the sequent

$$\longrightarrow A_1, \ldots, A_h$$

where $A_1, \ldots, A_h$ is a list of all the π-prototypes. It turns out that S_π is a valid sequent, and in fact π can be transformed into a *PK* proof π' of S_π in polynomial time. To form π' from π, delete each quantified formula (i.e. each formula of the form (133)) from π and add formulas from the list $A_1, \ldots, A_h$ to the RHS of each sequent so that each π-prototype is in the succedent of every sequent. The result can be turned into a *PK* proof of S_π by deleting applications of the rule $\exists$-right, and adding weakenings, exchanges, and contractions.

We may assume that the *PK* proof π' of S_π is treelike, by Theorem VII.1.8. Now π' is easily augmented to a treelike proof of (132) using the rules $\exists$-right, exchange and contraction. $\square$

We now show that for $G_i^\star$ we may as well assume that all cut formulas are prenex Σ_i^q. We start by proving an easy lemma which applies to both G_i and $G_i^\star$.

Lemma VII.4.6. *If* G_i *(resp.* $G_i^\star$*) is modified so that cuts are restricted to* Σ_i^q*-formulas, then the resulting system p-simulates* G_i *(resp.* $G_i^\star$*).*

Proof. If A is a Π_i^q formula, then any application of the cut rule to A can be replaced by first moving A to the opposite side of each parent sequent using $\neg$ introduction, and then cutting $\neg A$. $\square$

THEOREM VII.4.7 (Morioka [80]). *Let $\hat{G}_i^*$ be G_i^* with cut formulas restricted to prenex Σ_i^q formulas. Then $\hat{G}_i^*$ p-simulates G_i^*.*

PROOF. Fix $i \geq 1$. Let π be a G_i^* proof. We may assume that π is in free variable normal form.

Consider an application of the cut rule in π, with cut formula A.

$$\frac{\Gamma \longrightarrow \Delta, A \qquad A, \Gamma \longrightarrow \Delta}{\Gamma \longrightarrow \Delta}$$

We may assume that A is Σ_i^q, since if A is Π_i^q we can simply insert $\neg$-introduction steps just before the cut so that the cut formula becomes $\neg A$. Our task is to show that this cut on A can be replaced with a cut on A', where A' is some prenex form of A. To do this we will replace the tree derivation of $\Gamma \longrightarrow \Delta, A$ with a similar derivation of $\Gamma \longrightarrow \Delta, A'$, and similarly replace the derivation of $A, \Gamma \longrightarrow \Delta$ by one of $A', \Gamma \longrightarrow \Delta$.

The proof of the Prenex Form Theorem II.5.12 lists ten equivalences as follows:

$$(\forall x B \wedge C) \Longleftrightarrow \forall x (B \wedge C) \qquad\qquad (\forall x B \vee C) \Longleftrightarrow \forall x (B \vee C)$$

$$(C \wedge \forall x B) \Longleftrightarrow \forall x (C \wedge B) \qquad\qquad (C \vee \forall x B) \Longleftrightarrow \forall x (C \vee B)$$

$$(\exists x B \wedge C) \Longleftrightarrow \exists x (B \wedge C) \qquad\qquad (\exists x B \vee C) \Longleftrightarrow \exists x (B \vee C)$$

$$(C \wedge \exists x B) \Longleftrightarrow \exists x (C \wedge B) \qquad\qquad (C \vee \exists x B) \Longleftrightarrow \exists x (C \vee B)$$

$$\neg \forall x B \Longleftrightarrow \exists x \neg B \qquad\qquad\qquad \neg \exists x B \Longleftrightarrow \forall x \neg B$$

(where x does not occur free in C).

To put a formula in prenex form (which is in the same class Σ_j^q or Π_j^q with the original formula), it suffices to successively transform a formula $A(B(\vec{x}))$ to $A(B'(\vec{x}))$, where $B \Longleftrightarrow B'$ is one of the above equivalences and $\vec{x}$ is a list of the variables in B which are bound by quantifiers in A.

Consider a derivation of $\Gamma \longrightarrow \Delta, A(B(\vec{x}))$ or $A(B(\vec{x})), \Gamma \longrightarrow \Delta$ in π. If we trace the ancestors of $A(B(\vec{x}))$ up through this derivation, each path either ends when the ancestor is formed by a weakening, or it includes an occurrence of $B(\vec{D})$, where $\vec{D}$ is the list of target formulas and eigenvariables used by the quantifier introduction rules in forming $A(B(\vec{x}))$ from $B(\vec{D})$.

Thus it suffices to show, for each of the above equivalences $B \Longleftrightarrow B'$, how to convert a derivation of $\Lambda \longrightarrow \Pi, B$ to one of $\Lambda \longrightarrow \Pi, B'$ and a derivation of $B, \Lambda \longrightarrow \Pi$ to one of $B', \Lambda \longrightarrow \Pi$. (In the application to the previous paragraph, B would be $B(\vec{D})$, and B' would be $B'(\vec{D})$.)

Consider, for example, converting a derivation of

$$\Lambda \longrightarrow \Pi, \neg \forall x C(x)$$

to one of

$$\Lambda \longrightarrow \Pi, \exists x \neg C(x).$$

The ancestral paths of $\neg\forall x C(x)$ which do not end in weakening include $\forall x C(x)$ in the antecedent and then $C(D)$ in the antecedent, for some target formula D. Thus we have arrived at a sequent

$$C(D), \Lambda' \longrightarrow \Pi'.$$

We modify the derivation after this point by using $\neg$-right and $\exists$-right to obtain

$$\Lambda' \longrightarrow \Pi', \exists x \neg C(x)$$

and continue the derivation as before, omitting the steps which formed $\neg\forall x C(x)$ from $C(D)$.

The argument is similar if $\neg\forall x C(x)$ is in the antecedent.

Now consider converting a derivation of

$$\Lambda \longrightarrow \Pi, \forall x C(x) \wedge D$$

to a derivation of

$$\Lambda \longrightarrow \Pi, \forall x (C(x) \wedge D).$$

The ancestral paths of $\forall x C(x) \wedge D$ which do not end in weakening split after an $\wedge$-right, where the left branch has a $\forall$-right step

$$\frac{\Lambda' \to \Pi', C(p)}{\Lambda' \to \Pi', \forall x C(x)}$$

We modify this by combining it with the right branch just after the split as follows:

$$\frac{\dfrac{\Lambda'' \longrightarrow \Pi'', C(p) \qquad \Lambda'' \longrightarrow \Pi'', D}{\Lambda'' \longrightarrow \Pi'', C(p) \wedge D}}{\Lambda'' \longrightarrow \Pi'', \forall x (C(x) \wedge D)}$$

Here it is important that the original derivation be in free variable normal form, both in order to insure that p does not occur in D, and to guarantee that the variable restrictions continue to hold in the modified derivation of $\Lambda \longrightarrow \Pi, \forall x (C(x) \wedge D)$.

The other cases are handled similarly. $\square$

A part of the reverse direction of Theorem VII.4.3 is shown in the next theorem.

THEOREM VII.4.8 (Perron [91]). *For $i \geq 1$, G_i p-simulates $G_{i+1}^\star$ (for all formulas).*

From this theorem and Theorem VII.4.3 we have:

COROLLARY VII.4.9. *For $i \geq 1$, G_i and $G_{i+1}^\star$ are p-equivalent for proving formulas in $\Sigma_i^q \cup \Pi_i^q$.*

PROOF OF THEOREM VII.4.8. Let π be a $G_{i+1}^\star$ proof of a formula A. We show how to get a suitable G_i proof π' of A from π. The idea is to replace cuts of formulas C not in $\Pi_i^q \cup \Sigma_i^q$ by cuts on simpler ancestors of C. By Theorem VII.4.7 we can assume that all cut formulas in π are prenex Σ_{i+1}^q

formulas. Furthermore, we can assume that π is in free variable normal form.

Assume that in π for all axioms of the form

$$B \longrightarrow B$$

the formula B is quantifier free. This is possible because from these axioms we can easily derive any axiom with quantified formulas. Similarly assume that only quantifier free formulas are used for the weakening rules.

An occurrence of a formula $\exists \vec{x} B(\vec{x})$ in π is said to be *tagged* if it occurs in the antecedent Γ of a sequent

$$\mathcal{S} = \Gamma \longrightarrow \Delta$$

and B is in $(\Pi_i^q - \Sigma_i^q)$ and some descendant in π of $\exists \vec{x} B(\vec{x})$ is cut. Let

$$B(\overrightarrow{q^1}), B(\overrightarrow{q^2}), \ldots, B(\overrightarrow{q^k}) \tag{134}$$

be all $(\Pi_i^q - \Sigma_i^q)$ ancestors of $\exists \vec{x} B(\vec{x})$, where the variables $\overrightarrow{q^i}$ are eigenvariables in π. By our assumptions above, every $(\Sigma_{i+1}^q - \Pi_i^q)$ ancestor of $\exists \vec{x} B(\vec{x})$ lies on a path from some sequent containing some $B(\overrightarrow{q^i})$ to $\mathcal{S}$.

Define

$$\mathcal{S}' = \Gamma' \longrightarrow \Delta$$

where Γ' is obtained from Γ by replacing *every* tagged formula $\exists \vec{x} B(\vec{x})$ in Γ (possibly for more than one formula $B(\vec{x})$) by its corresponding list (134). By free variable normal form, the eigenvariables $\overrightarrow{q^i}$ associated with distinct tagged formulas in Γ are distinct. Notice that $\mathcal{S}'$ has size bounded by the size of π.

We will describe a polynomial time algorithm which successively transforms, for each sequent $\mathcal{S}$ in π, the (treelike) derivation $\pi_{\mathcal{S}}$ of $\mathcal{S}$ to a daglike G_i derivation $\pi'_{\mathcal{S}}$ of $\mathcal{S}'$. Note that if $\mathcal{S}$ is the final sequent in π then $\mathcal{S}' = \mathcal{S}$, and the theorem is proved.

The algorithm starts with the leaves of the proof tree π and works its way down to the endsequent. The leaf sequents are axioms, which by our assumptions have no tagged formulas, so there is nothing to do. For the general step we need to consider the rule used to derive $\mathcal{S}$. If the principle formula in the rule is not tagged, then $\pi'_{\mathcal{S}}$ is constructed using the same rule applied to the transformed proof(s) of the parent(s). If the principle formula is tagged, then the rule cannot be weakening by our assumptions, so it must be one of $\exists$-left, contraction-left, or cut. For $\exists$-left or contraction-left there is nothing to do: just use the transformed proof of the parent sequent.

Hence the only non-trivial case is where $\mathcal{S}$ is derived by cutting a tagged formula. So suppose that $\mathcal{S}_3$ is a sequent in π and is derived from $\mathcal{S}_1$ and

S_2 as below:

$$\frac{S_1 \qquad S_2}{S_3} = \frac{\Gamma \longrightarrow \Delta, \exists \vec{x} B(\vec{x}) \qquad \exists \vec{x} B(\vec{x}), \Gamma \longrightarrow \Delta}{\Gamma \longrightarrow \Delta}$$

Here $B(\vec{x})$ is a formula in $(\Pi_i^q - \Sigma_i^q)$. Suppose that

$$S_3' = \Gamma' \longrightarrow \Delta.$$

Then note that

$$S_1' = \Gamma' \longrightarrow \Delta, \exists \vec{x} B(\vec{x})$$

and S_2' has the form

$$S_2' = B(\overrightarrow{q^1}), B(\overrightarrow{q^2}), \ldots, B(\overrightarrow{q^k}), \Gamma' \longrightarrow \Delta$$

where no eigenvariable in any $\overrightarrow{q_i}$ occurs in Γ' or Δ. We have previously found short G_i derivations π_{S_1}', π_{S_2}' of the sequents S_1', S_2'. The idea is to convert π_{S_1}' into a G_i derivation of $\Gamma' \longrightarrow \Delta$ by cutting 'topmost' ancestors of $\exists \vec{x} B(\vec{x})$ using substitution instances of S_2'.

First we add Γ' to the antecedent and Δ to the succedent of every sequent in π_{S_1}' (and add necessary weakenings to have a legitimate proof). Call the result π_{S_1}''.

Now consider a sequent

$$S_{11} = \Lambda \longrightarrow \Pi, B(\vec{C}) \tag{135}$$

in π where $B(\vec{C})$ is an ancestor of $\exists \vec{x} B(\vec{x})$ in S_1. (Here $\vec{C}$ consists of Σ_0^q formulas.) We say that $B(\vec{C})$ is a *topmost* ancestor if it has no further ancestor $B(\vec{C})$ in π; i.e. $B(\vec{C})$ is the principle formula in the $\forall$-right rule used to derive the sequent (135). In π_{S_1}'' S_{11}' has become

$$\Gamma', \Lambda' \longrightarrow \Delta, \Pi, B(\vec{C}).$$

Apply the Substitution Lemma VII.4.10 below and using contractions left we create for each topmost ancestor $B(\vec{C})$ of $\exists \vec{x} B(\vec{x})$ a $G_i^{\star}$ derivation of the form

$$\frac{S_2'}{B(\vec{C}), \Gamma' \longrightarrow \Delta} \tag{136}$$

(Since there may be more than one topmost ancestor with different formulas $\vec{C}$, the sequent S_2' may have to be used more than once, which is why our transformed proof may not be treelike). For each topmost ancestor $B(\vec{C})$ in turn, working from the top of π down, insert the following derivation in π_{S_1}'':

$$\frac{\Gamma', \Lambda' \longrightarrow \Delta, \Pi, B(\vec{C}) \qquad B(\vec{C}), \Gamma' \longrightarrow \Delta}{\Gamma', \Lambda' \longrightarrow \Pi, \Delta} \tag{137}$$

(where the upper right sequent is derived by (136)) and remove all descendants of $B(\vec{C})$ in the so-far transformed π_{S_1}'' as far as possible. If a descendant is the principle formula in a contraction then simply delete that contraction rule. If a descendant is a side formula in a two-parent rule, then progress must wait until the matching side formula in the other parent is removed. When this is done for each topmost ancestor, all descendants of the form $\exists \vec{x} B(\vec{x})$ will be removed, and we obtain a proof of the sequent

$$\Gamma', \Gamma' \longrightarrow \Delta, \Delta.$$

With additional applications of the contraction rules we obtain a legitimate derivation π'_{S_3} of S'_3.

Finally we verify that the final G_i proof π' has size polynomial in the size of π. Notice that all new sequents have size polynomial in the size of π. (The bottom sequent in (136) is the only sequent that might have size larger than π.) So it remains to show that the number of sequents in π' is bounded by a polynomial in the size $|\pi|$ of π.

For a sequent S in π let $n_{S'}$ denote the number of sequents used in the derivation of S' in π'. Consider the interesting case of the cut rule in the algorithm above. It suffices to show that for some polynomial p we have

$$n_{S'_3} \leq n_{S'_1} + n_{S'_2} + p(|\pi|).$$

This follows from the fact that for each sequent S (135) in π the total number of sequents in the derivations (136) and (137), as well as the number of applications of weakening and contraction rules described above are bounded above by some polynomial in $|\pi|$ independent of S_3.

□

LEMMA VII.4.10 (Substitution). *There is a polynomial size $G_i^\star$ derivation*

$$\frac{\Gamma(p), \Gamma' \longrightarrow \Delta(p), \Delta'}{\Gamma(B), \Gamma' \longrightarrow \Delta(B), \Delta'} \tag{138}$$

where B is a quantifier-free formula, all formulas in Γ and Δ are in $\Sigma_i^q \cup \Pi_i^q$, and p does not occur in the bottom sequent.

To prove the above lemma we need:

LEMMA VII.4.11 ($G_0^\star$-Replacement). *Let $A(p)$ be a quantified propositional formula, and let $A(B)$ be the result of substituting the formula B for p in $A(p)$. Then for all formulas B_1, B_2, the sequent*

$$B_1 \leftrightarrow B_2 \longrightarrow A(B_1) \leftrightarrow A(B_2)$$

has a $G_0^\star$ proof of size bounded by a polynomial in the size of its endsequent.

EXERCISE VII.4.12. Prove the Lemma. (See Exercise VII.1.11.)

PROOF OF THE SUBSTITUTION LEMMA. From the $G_0^\star$-Replacement Lemma above, we have a $G_0^\star$ proof of

$$p \leftrightarrow B, A(p) \longrightarrow A(B)$$

for each formula $A(p)$ in $\Delta(p)$. From these and

$$\Gamma(p), \Gamma' \longrightarrow \Delta(p), \Delta'$$

we obtain (by the cut rule on the formulas $A(p)$ in $\Delta(p)$)

$$p \leftrightarrow B, \Gamma(p), \Gamma' \longrightarrow \Delta(B), \Delta'. \tag{139}$$

Again, by the $G_0^\star$-Replacement Lemma, we have $G_0^\star$ derivations of

$$p \leftrightarrow B, A(B) \longrightarrow A(p)$$

for all formulas $A(p)$ in $\Gamma(p)$. From these and (139) we obtain

$$p \leftrightarrow B, \Gamma(B), \Gamma' \longrightarrow \Delta(B), \Delta'.$$

Now by the $\exists$-left rule we get

$$\exists x (x \leftrightarrow B), \Gamma(B), \Gamma' \longrightarrow \Delta(B), \Delta'.$$

Finally, it is easy to see that the sequent

$$\longrightarrow \exists x (x \leftrightarrow B)$$

can be derived in $G_0^\star$. Consequently, by the cut rule on the Σ_1^q formula $\exists x (x \leftrightarrow B)$ we obtain the bottom sequent of (138). It is clear that all the derivations above have size polynomial in the length of the endsequents. □

Unlike the situation for PK and G_0, it seems unlikely that $G_1^\star$ p-simulates G_1. To explain why, we need the notion of witnessing for QPC proof systems.

VII.4.1. Extended Frege Systems and Witnessing in $G_1^\star$. In previous chapters we proved witnessing theorems which concern the complexity of witnessing the leading existential quantifiers in a bounded $\mathcal{L}_A^2$ formula, given values for the free variables. The analogous witnessing problem for a QPC formula is trivial, because there are only finitely many possible values for the free variables. However the problem becomes interesting if we consider a family of formulas, and include a proof of the formula as part of the input.

THEOREM VII.4.13 (The Witnessing Theorem for $G_1^\star$). *There is a polynomial time function $F(\pi, \tau)$ which, given a $G_1^\star$ proof π of a formula of the form $\exists \vec{x} A(\vec{x}, \vec{p})$ (where $A(\vec{x}, \vec{p})$ is quantifier-free) and an assignment τ to $\vec{p}$, returns an extension τ' of τ such that τ' satisfies $A(\vec{x}, \vec{p})$.*

We show in Theorem X.2.33 that if π is a G_1 proof (as opposed to a $G_1^\star$ proof), then the witnessing problem becomes complete for the search class PLS (Polynomial Local Search). Since it seems unlikely that PLS problems can all be solved in polynomial time, it seems unlikely that $G_1^\star$ p-simulates G_1.

In general the problem of computing such τ' from τ without π is complete for P^{NP}, if we are required to say "no" if there is no witness. Hence it is clear that the proof π provides helpful information.

We will prove the Witnessing Theorem for $G_1^\star$ by analyzing a closely-related system ePK, a member of the class of *extended* *Frege* proof systems. In general, a line in an extended *Frege* proof has the expressive power of a Boolean circuit, and a problem in nonuniform P is presented by a polynomial size family of Boolean circuits. The connection between the extended *Frege* proof systems and P is thus analogous to that of the bounded depth *Frege* proof systems (e.g., bPK) and AC^0 that we have seen (Section VII.2), or that of the *Frege* systems and NC^1, as we discussed in the Preface.

DEFINITION VII.4.14 (Extension Cedent). The sequence of formulas

$$\Lambda = \ e_1 \leftrightarrow B_1, e_2 \leftrightarrow B_2, \ldots, e_n \leftrightarrow B_n \qquad (140)$$

is an *extension cedent* provided that for $i = 1, \ldots, n$, the atom e_i does not occur in any of the formulas $B_1, \ldots, B_i$. The atoms $e_1, \ldots, e_n$ are called *extension variables*.

Intuitively, we think of $e_1, \ldots, e_n$ as gates in a Boolean circuit, where the value of e_i is determined by B_i together with the values of the earlier gates $e_1, \ldots, e_{i-1}$. In an ePK proof of an existential statement, some of these extension variables are used to witness the existential quantifiers.

DEFINITION VII.4.15 (ePK Proof). Let $\exists \vec{x} A(\vec{x}, \vec{p})$ be a QPC formula with free variables $\vec{p}$ such that $A(\vec{x}, \vec{p})$ is quantifier-free. An ePK proof of $\exists \vec{x} A(\vec{x}, \vec{p})$ is a PK proof of any sequent of the form

$$\Lambda \longrightarrow A(\vec{e}_1, \vec{p})$$

where Λ is an extension cedent (140) in which the extension variables $\vec{e}$ are disjoint from $\vec{p}$, $\vec{e}_1$ is a subset of $\vec{e}$, and each B_i contains only variables among $\vec{e}, \vec{p}$.

This definition is interesting even in the case that the final formula is quantifier-free. Then the extension variables are not used to witness quantifiers, but they still may be useful in defining polynomial time concepts needed in the proof. As far as we know, PK does not p-simulate ePK even when the latter is restricted to proving quantifier-free formulas.

THEOREM VII.4.16 (Krajíček [72]). $G_1^\star$, *restricted to proving prenex* Σ_1^q *formulas, is* p-*equivalent to* ePK.

Before giving the proof, we show how the Witnessing Theorem for $G_1^\star$ follows from this.

PROOF OF THEOREM VII.4.13. Let π be a $G_1^\star$ proof of $\exists \vec{x} A(\vec{x}, \vec{p})$, and let τ be an assignment to $\vec{p}$, as in the statement of the Witnessing Theorem. By the preceding theorem, we can transform π to an ePK proof of $\exists \vec{x} A(\vec{x}, \vec{p})$; that is, a PK proof of a sequent

$$e_1 \leftrightarrow B_1, e_2 \leftrightarrow B_2, \ldots, e_n \leftrightarrow B_n \longrightarrow A(\vec{e}_1, \vec{p}). \qquad (141)$$

Now given the the assignment τ to $\vec{p}$, values for $e_1, e_2, \ldots, e_n$ can be computed successively by evaluating $B_1, \ldots, B_n$, and these values define the desired extension τ' of τ which satisfies $A(\vec{x}, \vec{p})$.   □

PROOF OF THEOREM VII.4.16. First we show that $G_1^\star$ p-simulates ePK. Let π be a (treelike) ePK proof of $\exists \vec{x} A(\vec{x}, \vec{p})$. Then π is a PK proof of a sequent of the form (141). We show how to extend this PK proof to make a $G_1^\star$ proof of $\exists \vec{x} A(\vec{x}, \vec{p})$. We start by repeated application of $\exists$-right to obtain a proof of

$$e_1 \leftrightarrow B_1, e_2 \leftrightarrow B_2, \ldots, e_n \leftrightarrow B_n \longrightarrow \exists \vec{x} A(\vec{x}, \vec{p}). \tag{142}$$

Now for each formula B there is a short PK proof of $\longrightarrow (B \leftrightarrow B)$, and with one application of $\exists$-right we obtain a short $G_1^\star$ proof of

$$\longrightarrow \exists x (x \leftrightarrow B). \tag{143}$$

Now apply $\exists$-left to (142) to change the formula $(e_n \leftrightarrow B_n)$ to $\exists x (x \leftrightarrow B_n)$. (Note that e_n does not occur elsewhere in (142), so the variable restriction for this rule is satisfied.) Now apply the cut rule to this and (143) to obtain

$$e_1 \leftrightarrow B_1, e_2 \leftrightarrow B_2, \ldots, e_{n-1} \leftrightarrow B_{n-1} \longrightarrow \exists \vec{x} A(\vec{x}, \vec{p}).$$

Applying this process a total of n times we may eliminate each formula $e_i \leftrightarrow B_i$ in (142) to obtain the desired $G_1^\star$ proof of size polynomial in the size of π.

Now we prove the converse. Let π be a $G_1^\star$ proof of $\longrightarrow \exists \vec{x} A(\vec{x}, \vec{p})$. We may assume that π is in free variable normal form, and by Theorem VII.4.7 we may assume that all cut formulas in π are prenex Σ_1^q, so each sequent of π has the form

$$S = \ldots, \exists \vec{x^i} \alpha_i (\vec{x^i}, \vec{r}), \ldots, \Gamma \longrightarrow \Delta, \ldots, \exists \vec{y^j} \beta_j (\vec{y^j}, \vec{r}), \ldots \tag{144}$$

where all α_i and β_j as well as all formulas in Γ and Δ are quantifier-free, and $\vec{r}$ is precisely the list of the free variables occurring in S. Notice that $\vec{r}$ may have variables not in $\vec{p}$, which are used as eigenvariables for $\exists$-left.

We transform the proof π to an ePK proof π' by transforming each such sequent S to a corresponding quantifier-free sequent S', and supplying a suitable proof of S'. To describe S', we first replace each vector $\vec{x^i}$ of bound variables by a distinct vector $\vec{q^i} = q_1^i, \ldots, q_{\ell_i}^i$ of new free variables, and similarly we replace $\vec{y^i}$ by a new vector $\vec{e^i}$. None of these new variables should occur in π. Then

$$S' = \Lambda, \ldots, \alpha_i (\vec{q^i}, \vec{r}), \ldots, \Gamma \longrightarrow \Delta, \ldots, \beta_j (\vec{e^j}, \vec{r}), \ldots, \tag{145}$$

where Λ is an extension cedent defining the extension variables $\ldots, \vec{e^j}, \ldots$.

If S is the endsequent $\longrightarrow \exists \vec{x} A(\vec{x}, \vec{p})$, then S' has the form $\Lambda \longrightarrow A(\vec{e_1}, \vec{p})$, so π' is the desired ePK proof of $\exists \vec{x} A(\vec{x}, \vec{p})$.

We define Λ and show that S' has an *ePK* proof polynomial in the size of the G_1^* proof of S, by induction on the depth of S in π.

For the base case, S is an axiom

$$\exists \vec{x} \alpha(\vec{x}, \vec{r}) \longrightarrow \exists \vec{x} \alpha(\vec{x}, \vec{r})$$

and S' is easy to obtain.

For the induction step there is one case for each rule of G_1^*.

Case I. Weakening and exchange are trivial, and contraction follows from cut. The single parent rules $\neg$ and $\wedge$-left and $\vee$-right are easy, since the principle formulas are quantifier-free, and the same rule can be applied to form S'.

Case II. For the two parent rules $\wedge$-right and $\vee$-left, the principle formulas are quantifier-free, but we face the difficulty that the extension cedents Λ for the two parents may give inconsistent definitions of the extension variables. This is similar to the difficulty for *Case VII* in the proof of Lemma V.5.5 for the V^0 witnessing theorem. There the witnessing functions for a formula in Π for the two parents might be different. We solve the problem in a similar way, by defining the extension variables to values that make them true when possible.

Specifically, consider the case of $\wedge$-right, where for simplicity we assume there is exactly one formula in the succedent beginning with existential quantifiers (that formula cannot be C or D):

$$\frac{S_1 \qquad S_2}{S} = \frac{\Gamma \longrightarrow \Delta, \exists \vec{y} \beta(\vec{y}, \overrightarrow{r^1}), C \qquad \Gamma \longrightarrow \Delta, \exists \vec{y} \beta(\vec{y}, \overrightarrow{r^2}), D}{\Gamma \longrightarrow \Delta, \exists \vec{y} \beta(\vec{y}, \vec{r}), (C \wedge D)}$$

where $\vec{r}$ is the union of the lists $\overrightarrow{r^1}, \overrightarrow{r^2}$. By the induction hypothesis, we have *ePK* proofs of the two sequents

$$S_1' = \Lambda_1, \Gamma' \longrightarrow \Delta, \beta(\vec{e}, \overrightarrow{r^1}), C$$

and

$$S_2' = \Lambda_2, \Gamma' \longrightarrow \Delta, \beta(\vec{s}, \overrightarrow{r^2}), D$$

where in the the second case we have changed the extension variables from $\vec{e}$ to $\vec{s}$. Since π is treelike, we can assume that the *ePK* derivations of S_1' and S_2' are disjoint, and hence we can change variable names in one proof without affecting the other proof. Thus we may assume that the extension variables defined in Λ_1 and Λ_2 are disjoint, and in particular $\vec{e}$ and $\vec{s}$ have no variable in common. Thus the extension cedents Λ_1 and Λ_2 are consistent. Further we may assume that the variables $\overrightarrow{q^i}$ are the same in S_1' and S_2'.

From S_1' and S_2' with $\wedge$-right we obtain

$$\Lambda_1, \Lambda_2, \Gamma' \to \Delta, \beta(\vec{e}, \vec{r}), \beta(\vec{s}, \vec{r}), (C \wedge D). \qquad (146)$$

Now we introduce new extension variables $\vec{t}$, and introduce the extension formulas

$$E_i =_{\text{def}} \left((\beta(\vec{e}, \vec{r}) \wedge e_i) \vee (\neg\beta(\vec{s}, \vec{r}) \wedge s_i) \right)$$

and define the extension cedent

$$\Lambda_3 = t_1 \leftrightarrow E_1, t_2 \leftrightarrow E_2, \ldots$$

Then define

$$S' = \Lambda_1, \Lambda_2, \Lambda_3, \Gamma' \longrightarrow \Delta, \beta(\vec{t}, \vec{r}), (C \wedge D).$$

One can show with the help of Lemma VII.1.10 that each of the sequents

$$\Lambda_3, \beta(\vec{e}, \vec{r}) \longrightarrow \beta(\vec{t}, \vec{r}), \tag{147}$$

$$\Lambda_3, \beta(\vec{s}, \vec{r}) \longrightarrow \beta(\vec{t}, \vec{r}) \tag{148}$$

has a short **PK** proof. Using these and (146) and two cuts we obtain a short **PK** derivation of S' from S_1 and S_2.

Case III. $\exists$-left is easy, since it just means changing the role of a free eigenvariable r in S'_1 to the variable q in S' corresponding to $\exists x$.

Case IV. Suppose S comes from S_1 using $\exists$-right.

$$\frac{S_1}{S} = \frac{\Gamma \longrightarrow \Delta, \exists\vec{y}\beta(B, \vec{y}, \vec{r})}{\Gamma \longrightarrow \Delta, \exists z \exists\vec{y}\beta(z, \vec{y}, \vec{r})}.$$

Here the target formula B is quantifier-free, by definition of G. Since π is in free variable normal form, no free variable can be eliminated by this rule, and so the list $\vec{r}$ of free variables in S is the same as for S_1. By the induction hypothesis, we have an **ePK** derivation of

$$S'_1 = \Lambda, \Gamma' \longrightarrow \Delta', \beta(B, \vec{e}, \vec{r}).$$

Let s be a new extension variable, and let

$$S' = \Lambda, s \leftrightarrow B, \Gamma' \longrightarrow \Delta', \beta(s, \vec{e}, \vec{r}).$$

It follows from the **PK**-Replacement Lemma VII.1.10 that S' has a short **PK** derivation from S'_1.

Case V. Suppose S comes from S_1, S_2 by cut:

$$\frac{S_1 \quad S_2}{S_3} = \frac{\Gamma \longrightarrow \Delta, C \quad C, \Gamma \longrightarrow \Delta}{\Gamma \longrightarrow \Delta}$$

Since π is in free variable normal form, every free variable in C also occurs in the conclusion S_3. Suppose first that the cut formula C is quantifier-free. Then the only difficulty is that the extension cedents Λ for the two parents may give inconsistent definitions of the extension variables witnessing quantifiers in Δ. We handle this difficulty in the same way as for *Case II* above.

The case in which C has existential quantifiers is more complicated, since the definitions of the new extension variables witnessing quantifiers in Δ now depend on witnesses for the quantifiers in C supplied by S_1'. These new definitions are similar to the new witnessing functions defined for the case of cut (*Case VI*) in the proof of Lemma V.5.5 used to prove the V^0 Witnessing Theorem. $\qquad\square$

EXERCISE VII.4.17. Carry out the details of *Case V* in the above proof.

VII.5. Propositional Translations for V^i

In this section we show that for $i \geq 1$, $G_i^\star$ is closely related to the theory V^i. In fact Theorem VII.5.2 together with results in Chapter X suggest that $G_i^\star$ restricted to Σ_i^q formulas is a nonuniform version of the Σ_i^B-fragment of V^i. We have already shown by Theorem VII.4.13 a connection between $G_1^\star$ and V^1: Σ_1^q-theorems of $G_1^\star$ can be uniformly witnessed in polynomial time, just as each Σ_1^B-theorem of V^1 can be witnessed in polynomial time.

It is straightforward to extend the propositional translation of $\Sigma_0^B(\mathcal{L}_A^2)$ formulas (Section VII.2) to a translation of any bounded $\mathcal{L}_A^2$ formula. Here every $g\Sigma_i^B$ (resp. $g\Pi_i^B$) formula $\varphi(\vec{x}, \vec{X})$, with all free variables indicated, translates into a family of Σ_i^q (resp. Π_i^q) formulas:

$$\|\varphi(\vec{x}, \vec{X})\| = \{\varphi(\vec{x}, \vec{X})[\vec{m}; \vec{n}] : \vec{m}, \vec{n} \in \mathbb{N}\}$$

so that $\varphi(\vec{x}, \vec{X})[\vec{m}; \vec{n}]$ is valid iff

$$\underline{\mathbb{N}}_2 \models \forall \vec{X}((\bigwedge |\vec{X}| = \vec{n}) \supset \varphi(\underline{\vec{m}}, \vec{X})).$$

The formula $\varphi(\vec{x}, \vec{X})[\vec{m}; \vec{n}]$ has size bounded by a polynomial $p(\vec{m}, \vec{n})$ which depends only on φ. The free propositional variables in $\varphi(\vec{x}, \vec{X})[\vec{m}; \vec{n}]$ consist of $p_j^{X_i}$, for $0 \leq j < n_i - 1$ for each $n_i \geq 2$.

We define the translation of a bounded $\mathcal{L}_A^2$ formula φ inductively, starting with the Σ_0^B formulas, which is described in Section VII.2. For the induction step, consider the case where

$$\varphi(\vec{x}, \vec{X}, Y) \equiv \exists Y \leq t \psi(\vec{x}, \vec{X}, Y)$$

(here t is a number term of the form $t(\vec{x}, |\vec{X}|)$). By the induction hypothesis, $\psi(\vec{x}, \vec{X}, Y)[\vec{m}; \vec{n}, k]$ contains the free propositional variables $p_0^Y, p_1^Y, \ldots$ for Y, in addition to $p_j^{X_i}$ (when $k < 2$, the list $p_0^Y, \ldots, p_{k-2}^Y$ is empty). Define

$$\varphi(\vec{x}, \vec{X})[\vec{m}; \vec{n}] =_{\text{def}} \exists p_0^Y \ldots \exists p_{r-2}^Y \bigvee_{k=0}^{r} \psi(\vec{x}, \vec{X}, Y)[\vec{m}; \vec{n}, k] \qquad (149)$$

where r is the numerical value of t: $r = val(t(\underline{\vec{m}}, \vec{n}))$ (recall that $\underline{i}$ is the i-th numeral). Here the free variables p_j^Y become bound, and if $r \leq 1$ then the list $p_0^Y, \ldots, p_{r-2}^Y$ is empty. Also, if any of the formulas $\psi_k(p_0^Y, \ldots, p_{k-2}^Y)$

is a logical constant $\perp$ or $\top$, then we simplify $\varphi(\vec{x}, \vec{X})[\vec{m}; \vec{n}]$ in the obvious way.

The case where $\varphi(\vec{x}, \vec{X}) \equiv \forall Y \leq t\psi(\vec{x}, \vec{X}, Y)$ is similar:

$$\varphi(\vec{x}, \vec{X})[\vec{m}; \vec{n}] =_{\text{def}} \forall p_0^Y \ldots \forall p_{r-2}^Y \bigwedge_{k=0}^{r} \psi(\vec{x}, \vec{X}, Y)[\vec{m}; \vec{n}, k]. \qquad (150)$$

(The conjunction is also simplified if any conjunct is a Boolean constant.)

The cases of the Boolean connectives $\wedge$, $\vee$, $\neg$ or the number quantifiers are the same as for Σ_0^B formulas.

PROPOSITION VII.5.1. *For each $i \geq 0$, if φ is a $g\Sigma_i^B$ (resp. $g\Pi_i^B$) formula, then the formulas in $\|\varphi\|$ are Σ_i^q (resp. Π_i^q). There is a polynomial $p(\vec{m}, \vec{n})$ which depends only on φ so that $\varphi[\vec{m}; \vec{n}]$ has size $\leq p(\vec{m}, \vec{n})$ for all $\vec{m}, \vec{n} \in \mathbb{N}$.*

The connection between the theory V^i and the proof system $G_i^\star$ is as follows.

THEOREM VII.5.2 (V^i Translation). *Let $i \geq 1$. For any bounded theorem $\varphi(\vec{x}, \vec{X})$ of V^i, there is a polytime function $F(\vec{m}, \vec{n})$ such that $F(\vec{m}, \vec{n})$ is a $G_i^\star$ proof of $\varphi(\vec{x}, \vec{X})[\vec{m}; \vec{n}]$, for all $\vec{m}, \vec{n} \in \mathbb{N}$.*

PROOF. The proof is similar to that of the Translation Theorem for V^0 VII.2.3. We consider the case where $i = 1$; the cases where $i > 1$ are handled in the same way. By Corollary VI.4.16, for every bounded theorem $\varphi(\vec{a}, \vec{\alpha})$ of V^1 there is a (treelike) anchored LK^2-$\widetilde{V}^1$ proof π of $\longrightarrow \varphi(\vec{a}, \vec{\alpha})$. If we translate each sequent of π into the corresponding QPC sequent, the result is close to a $G_1^\star$ proof. In particular, since any cut formula in LK^2-$\widetilde{V}^1$ is Σ_1^B, its translation is a Σ_1^q formula, and can be cut in $G_1^\star$.

Formally, we will prove by induction on the depth of a sequent $\mathcal{S}(\vec{b}, \vec{\beta})$ in π that there is a polytime function $F(\vec{m}, \vec{n})$ such that $F(\vec{m}, \vec{n})$ is a $G_1^\star$ proof of $\mathcal{S}[\vec{m}; \vec{n}]$. For the base case, $\mathcal{S}$ is an axiom of LK^2-$\widetilde{V}^1$. The simple axioms are sequents of Σ_0^B formulas, and these are treated as in the proof of the Translation Theorem for V^0. The remaining axioms are instances of Σ_0^B-*COMP*, so

$$\mathcal{S} = \longrightarrow \exists X \leq t \forall z < t(X(z) \leftrightarrow \eta(z))$$

and η is a Σ_0^B formula. Let $r = val(t)$. When $r \leq 1$, it is easy to see that $\mathcal{S}$ translates into a trivially valid sequent with a short G_0 proof. Otherwise, if $r \geq 2$, then $\mathcal{S}[\vec{m}; \vec{n}]$ is the sequent

$$\longrightarrow \left(\exists X \leq t \forall z < t(X(z) \leftrightarrow \eta(z)) \right)[\vec{m}; \vec{n}]$$

where (replace $[\dots]$ by $[\vec{m};\vec{n}]$):

$$(\exists X \le t \forall z < t(X(z) \leftrightarrow \eta(z)))[\dots] \equiv \exists p_0^X \dots \exists p_{r-2}^X$$

$$\bigvee_{k=0}^{r} \left(\bigwedge_{i=0}^{k-2}(p_i^X \leftrightarrow \eta(\underline{i})[\dots]) \wedge \eta(\underline{k-1})[\dots] \wedge \bigwedge_{i=k}^{r-1} \neg\eta(\underline{i})[\dots] \right)$$

where the conjunct $\eta(\underline{k-1})$ is deleted when $k = 0$ and the conjuncts

$$\bigwedge_{i=0}^{k-2}(p_i^X \leftrightarrow \eta(\underline{i})[\dots]) \quad\text{and}\quad \bigwedge_{i=k}^{r-1} \neg\eta(\underline{i})[\dots]$$

are deleted when their sets of indices are empty.

EXERCISE VII.5.3. Let $A_0, \dots, A_\ell$ be any **PK** formulas $(\ell \ge 0)$. Show that the sequent

$$\longrightarrow \bigwedge_{i=0}^{\ell} \neg A_i, A_0 \wedge \bigwedge_{i=1}^{\ell} \neg A_i, A_1 \wedge \bigwedge_{i=2}^{\ell} \neg A_i, \dots, A_{\ell-1} \wedge \neg A_\ell, A_\ell$$

has a polynomial size treelike cut-free **PK** derivation.

We get $S[\vec{m};\vec{n}]$ as follows. First we apply the above exercise for $\ell = r - 1$ and

$$A_i \equiv \eta(\underline{i})[\vec{m};\vec{n}].$$

Then note that it is straightforward to obtain polynomial size derivations for the following tautologies:

$$\bigwedge_{i=0}^{k-2}(\eta(\underline{i})[\dots] \leftrightarrow \eta(\underline{i})[\dots]).$$

Now by using the $\wedge$-right and $\vee$-right rules obtain

$$\longrightarrow \bigvee_{k=0}^{r} \left(\bigwedge_{i=0}^{k-2}(\eta(\underline{i})[\dots] \leftrightarrow \eta(\underline{i})[\dots]) \wedge \eta(\underline{k-1})[\dots] \wedge \bigwedge_{i=k}^{r-1} \neg\eta(\underline{i})[\dots] \right).$$

From this sequent, by repeatedly applying the $\exists$-right rule we obtain a polynomial size cut-free **G** proof of $S[\vec{m};\vec{n}]$.

For the induction step, we consider all rules of $LK^2\text{-}\tilde{V}^1$. In each case, assume that S is obtained from S_1 (and S_2). We will show that $S[\dots]$ has short $G_1^\star$ derivation from $S_1[\dots]$ (and $S_2[\dots]$). It is obvious that the polytime function $F(\dots)$ giving the $G_1^\star$ proof of $S[\dots]$ can be constructed from the polytime function(s) $F_1(\dots)$ for S_1 (and $F_2(\dots)$ for S_2).

All rules (including the **IND** rule) except for the string quantifier rules are treated just as in the proof of the Translation Theorem for V^0 (page 170), although now the translation will require cuts on Σ_i^q formulas in general. We consider the string $\exists$-introduction rules. The string $\forall$-introduction rules are dual, and are left as an exercise.

Case string $\exists$-right. Suppose that $\mathcal{S}$ is obtained from $\mathcal{S}_1$ by the string $\exists$-right rule. Note that in $\widetilde{V}^1$, the only string terms are string variables.

$$\frac{\mathcal{S}_1}{\mathcal{S}} = \frac{\Lambda(\gamma) \longrightarrow \Pi(\gamma), |\gamma| \leq t \wedge \psi(\gamma)}{\Lambda(\gamma) \longrightarrow \Pi(\gamma), \exists Z \leq t \; \psi(Z)}$$

We suppress all free variables except for the principle variable γ. Note that $|\gamma| \leq t[\ldots, n]$ is either $\top$ or $\bot$. Let $r = val(t)$, then

$$\mathcal{S}_1[\ldots, n] =_{\mathrm{def}} \begin{cases} \Lambda[\ldots, n] \longrightarrow \Pi[\ldots, n], \psi(\gamma)[\ldots, n] & \text{if } n \leq r, \\ \Lambda[\ldots, n] \longrightarrow \Pi[\ldots, n], \bot & \text{if } n > r. \end{cases} \quad (151)$$

By definition (see (149)),

$$\mathcal{S}[\ldots, n] =_{\mathrm{def}} \Lambda[\ldots, n] \longrightarrow \Pi[\ldots, n], \exists p_0^Z \ldots \exists p_{r-2}^Z \bigvee_{k=0}^{r} \psi(Z)[\ldots, k].$$

Consider the interesting case where $n \leq r$, First, by repeated applications of the rules weakening and $\vee$-right, we obtain from $\mathcal{S}_1[\ldots, n]$

$$\Lambda[\ldots, n] \longrightarrow \Pi[\ldots, n], \bigvee_{k=0}^{r} \psi(\gamma)[\ldots, k].$$

Then we can derive $\mathcal{S}[\ldots, n]$ using the rule $\exists$-right.

Case string $\exists$-left. Again, suppressing all other free variables:

$$\frac{\mathcal{S}_1}{\mathcal{S}} = \frac{|\gamma| \leq t \wedge \psi(\gamma), \Lambda \longrightarrow \Pi}{\exists Z \leq t \; \psi(Z), \Lambda \longrightarrow \Pi}$$

where γ does not occur in $\mathcal{S}$, and ψ is Σ_1^B. Let $r = val(t)$, then for $n \leq r$,

$$\mathcal{S}_1[\ldots, n] =_{\mathrm{def}} \psi(\gamma)[\ldots, n], \Lambda[\ldots] \longrightarrow \Pi[\ldots]. \quad (152)$$

Also,

$$\mathcal{S}[\ldots] =_{\mathrm{def}} \exists p_0^Z \ldots \exists p_{r-2}^Z \bigvee_{n=0}^{r} \psi(Z)[\ldots, n], \Lambda[\ldots] \longrightarrow \Pi[\ldots].$$

Now if $r = 0$, then we are done. Otherwise, combine the sequents $\mathcal{S}_1[\ldots, n]$ for $n = 0, \ldots, r$ by the rule $\vee$-left we obtain

$$\bigvee_{n=0}^{r} \psi(\gamma)[\ldots, n], \Lambda[\ldots] \longrightarrow \Pi[\ldots].$$

Thus we get $\mathcal{S}[\ldots]$ by $r - 1$ applications of the $\exists$-left rule. □

EXERCISE VII.5.4. Carry out the cases for the string $\forall$-introduction rules.

VII.5.1. Translating V^0 to Bounded Depth $G_0^\star$. In Section VII.2 we show that Σ_0^B theorems of V^0 translate into families of tautologies with polynomial-size bounded depth PK proofs. We generalize this and show here that the translation of every *bounded* theorem of V^0 has polynomial-size proofs in a subsystem of $G_0^\star$ that extends bPK. First we define the system.

DEFINITION VII.5.5 (Bounded Depth G_0). For each constant $d \in \mathbb{N}$, a d-G_0 proof is a G proof in which all target formulas have depth at most d and all cut formulas are quantifier-free and also have depth at most d. A *bounded depth G_0* system (or just bG_0) is any system d-G_0 for $d \in \mathbb{N}$. Treelike d-G_0 (resp. treelike bG_0) is denoted by d-$G_0^\star$ (resp. $bG_0^\star$).

Theorem VII.2.3 is generalized as follows:

THEOREM VII.5.6. *For any bounded theorem $\varphi(\vec{x}, \vec{X})$ of V^0 there is a constant d and a polytime function $F(\vec{m}, \vec{n})$ such that $F(\vec{m}, \vec{n})$ is a d-$G_0^\star$-proof of $\varphi(\vec{x}, \vec{X})[\vec{m}; \vec{n}]$, for all $\vec{m}, \vec{n} \in \mathbb{N}$.*

We prove the theorem by translating LK^2-V^0 proofs (as opposed to the LK^2-$\widetilde{V}^0$ proofs used in the proof of Theorem VII.2.3). An LK^2-V^0 proof can have cut formulas which are Σ_1^B; these are instances of the Σ_0^B-*COMP* axioms. Because in the translation we are not allowed to cut quantified formulas, these instances of Σ_0^B-*COMP* will require different translations than the translation described before Proposition VII.5.1.

The main idea is as follows. Consider an instance of Σ_0^B-*COMP*:

$$\exists X \leq t \forall i < t \big(X(i) \leftrightarrow \psi(i)\big).$$

Instead of introducing quantified Boolean variables p_i^X for the bits $X(i)$ of X we will translate $X(i)$ using the translation of $\psi(i)$. For the string eigenvariable γ that introduces X (in a string $\exists$-left rule) we also translate $\gamma(i)$ using the translation of $\psi(i)$. Now ψ may contain other eigenvariables, so they must be translated first.

Recall the notions of anchored proofs (Definition II.1.12 on page 14) and free variable normal form (Section II.2.4 on page 23 and Section IV.4.1 on page 90).

PROOF. Since φ is a theorem of V^0, there is an anchored LK^2-V^0 proof π of φ. We can assume that π is in free variable normal form and is treelike. Note that all cut formulas in π are Σ_0^B, and all non-Σ_0^B cut formulas are instances of Σ_1^B-*COMP* axioms. Here we are only interested in the instances of Σ_0^B-*COMP* that will be cut. Furthermore, we can assume that all sequents that contain an instance of Σ_0^B-*COMP* in the succedent are derived from the Σ_0^B-*COMP* axiom by weakenings:

$$\frac{\longrightarrow \exists X \forall x < t(X(x) \leftrightarrow \psi(x))}{\Gamma \longrightarrow \exists X \forall x < t(X(x) \leftrightarrow \psi(x)), \Delta} \tag{153}$$

Consider an application of the string ∃-left rule that introduces a Σ_0^B-**COMP** formula:

$$\frac{\mathcal{S}_1}{\mathcal{S}_2} = \frac{|\gamma| \leq t \wedge \forall x < t(\gamma(x) \leftrightarrow \psi(x)), \Gamma \longrightarrow \Delta}{\exists X \leq t \forall x < t(X(x) \leftrightarrow \psi(x)), \Gamma \longrightarrow \Delta} \tag{154}$$

where γ does not occur in $\mathcal{S}_2$. Since π is in free normal variable form, each variable γ is used exactly once.

NOTATION. We say that γ as above is a *comprehension variable in* π. The associated pair $\langle t, \psi \rangle$ as above is called the *defining pair* of γ.

The idea is to translate the bit $\gamma(i)$ of any comprehension variable γ in π using its defining pair (instead of using new atoms $p_0^\gamma, p_1^\gamma, \ldots$ as before). Note that two different comprehension variables may have the same defining pair (for example, comprehension variables that introduce two identical copies of a Σ_0^B-**COMP** cut formula which are merged in contraction right or in the branching rules such as $\vee$ left or $\wedge$ right). In this case they will have the same translation.

NOTATION. We say that a comprehension variable γ *depends on* a variable β (or b) if β (resp. b) occurs in the defining pair of γ.

Since π is treelike and in free variable normal form, this dependence relation forms a partial ordering of the comprehension variables. The defining pair of γ may contain other comprehension variables. For example, in (154) $\psi(x)$ might contain a comprehension variable γ', where its corresponding ψ' is in Γ. In this case γ' must be translated before γ. This motivates the following notions.

NOTATION. The *dependence degrees* of variables in π are defined as follows. All non-comprehension variables have dependence degree 0. The dependence degree of a comprehension variable γ is one plus the maximum dependence degree of all variables occurring in its defining pair.

Translation of formulas in π. The formulas in π are translated in stages as follows. In stage 0 we translate all formulas that do not involve any comprehension variables. Generally, in stage i we translate all formulas that involve some variables of dependence degree i but no variable of higher dependence degree. In each stage, the translation is by induction on the depth of the formulas. Stage 0 is the same as described at the beginning of Section VII.5.

Consider stage $(i + 1)$ (where $i \geq 0$). For the base case, all atomic formulas have been translated in the previous stage except for atomic formulas of the form $\gamma(s)$, where γ has dependence degree $(i + 1)$. For each such γ, let

$$\gamma(s)[n_\gamma, \vec{n}] =_{\text{def}} \begin{cases} \psi(s)[\vec{n}] & \text{if } j < n_\gamma - 1, \\ \top & \text{if } j = n_\gamma - 1, \\ \bot & \text{if } j \geq n_\gamma \end{cases} \quad (155)$$

where $\langle t, \psi \rangle$ is the defining pair for γ, and $j = val(s(\underline{\vec{n}}))$ (recall val from page 166).

The induction step is handled as discussed at the beginning of Section VII.5 except for the case of the Σ_0^B-**COMP** cut formulas. Intuitively these formulas are true, so they should translate into tautologies. In this case we will show that the tautologies have polynomial size d'-**PK*** proofs for some d'. Therefore we will simply delete all sequents on the right branches of the cut Σ_0^B-**COMP** rule (these are ancestors of a sequent that contains the cut Σ_0^B-**COMP** formula in its succedent). Also, we will translate all occurrences of the cut Σ_0^B-**COMP** formulas in the antecedents into the empty formula. This completes the description of our translation.

We leave as an exercise to verify that the translation formulas have polynomial sizes and constant depths as desired.

EXERCISE VII.5.7. Show that for each Σ_0^B formula $\psi(\vec{x}, \vec{X})$ in π there is a constant d_1 and a polynomial p that depend on π such that $\psi(\vec{x}, \vec{X})[\vec{m}; \vec{n}]$ have depth at most d_1 and size $p(\vec{m}, \vec{n})$, for all $\vec{m}, \vec{n}$. Show that Proposition VII.5.1 continues to hold for non-cut formulas in π.

Now we show that for all sequents S of π that are not on the right branches of a cut Σ_0^B-**COMP** rule, the families $S[\vec{m}; \vec{n}]$ have polynomial size d-**G**$_0^*$ proofs, for some constant d.

The base case, where S is a nonlogical axiom in 2-**BASIC** (Figure 2), is handled just as in Section VII.2.3, with obvious modifications when a free string variable in the axiom is a comprehension variable. The induction step is the same as in the proof of Theorem VII.5.2 except for the case where $S = S_2$ as in (154), i.e., where it is obtained by the string $\exists$-left that introduces a Σ_0^B-**COMP** cut formula.

So suppose that S_2 is obtained from S_1 as in (154). Note that for $n_\gamma \leq v$ (where $v = val(t)$)

$$S_1[\vec{m}; n_\gamma, \vec{n}] =_{\text{def}} \quad C[\vec{m}; n_\gamma, \vec{n}], \Gamma[\vec{m}; \vec{n}] \longrightarrow \Delta[\vec{m}; \vec{n}]$$

where $C[\dots]$ translates the first formula of S_1. Let $v = val(t)$. By definition:

$$C[\vec{m}; 0, \vec{n}] = \bigwedge_{i=0}^{v-1} \neg\psi(x)[\vec{m}, i; \vec{n}] \quad (i \text{ is the value of } x)$$

and for $1 \leq n_\gamma \leq v$ (here let A_i denotes $\psi(x)[\vec{m}, i; \vec{n}]$):

$$C[\vec{m}; n_\gamma, \vec{n}] = (\bigwedge_{i=0}^{n_\gamma-2} (A_i \leftrightarrow A_i)) \wedge A_{n_\gamma-1} \wedge \bigwedge_{i=n_\gamma}^{v-1} \neg A_i.$$

Here the conjuncts

$$\bigwedge_{i=0}^{n_\gamma-2} (A_i \leftrightarrow A_i) \quad \text{and} \quad \bigwedge_{i=n_\gamma}^{v-1} \neg A_i$$

are deleted when their sets of indices are empty. Also, by definition

$$\mathcal{S}_2[\vec{m}; \vec{n}] =_{\text{def}} \Gamma[\vec{m}; \vec{n}] \longrightarrow \Delta[\vec{m}; \vec{n}].$$

Using Exercise VII.5.3, we can show (by the same arguments as in the proof of Theorem VII.5.2 below Exercise VII.5.3) that there are polynomial size cut-free **PK** proofs of the tautologies:

$$\longrightarrow \bigvee_{n_\gamma=0}^{v} C[\vec{m}; n_\gamma, \vec{n}]. \tag{156}$$

Moreover, by Exercise VII.5.7 the above tautologies have depth at most d_1 for some d_1 depending only on π. Therefore the proof of Theorem VII.1.8 shows that (156) have polynomial-size d_2-**PK*** proofs, where $d_2 = d_1 + 3$. Hence, by using the $\vee$-left rule for the sequents $\mathcal{S}_1[\vec{m}; n_\gamma, \vec{n}]$ (for $0 \leq n_\gamma \leq v$) and then applying a cut for the resulting sequent with (156) we obtain $\mathcal{S}_2[\vec{m}; \vec{n}]$.

All cut formulas in our translation are either cut formulas in the d_2-$G_0^\star$ derivations mentioned above, or the translations of Σ_0^B cut formulas in π. Thus they have depth bounded by a constant depending on π. Furthermore, it can be seen that all target formulas are atomic formulas of the form

$$p_i^\alpha$$

for some noncomprehension string variable α. As a result, the translations of π are d-$G_0^\star$ proofs for some constant d depending on π. $\square$

EXERCISE VII.5.8. Reprove Theorem VII.2.3 using the translation we describe in the proof of Theorem VII.5.6.

VII.6. Notes

Definitions VII.1.2, VII.1.5 and Theorem VII.1.4 are from [46]. Also, the fact that **Frege** proof systems are p-equivalent is proved in [46]. Ajtai's superpolynomial lower bound for bounded depth **Frege** proofs of **PHP** is published in [5].

The first propositional translation of an arithmetic theory is described in [39]. The translation of Σ_0^B formulas given in Subsection VII.2.1 is from [42], and both this and the V^0 Translation Theorem VII.2.3 are based on the treatment of $I\Delta_0(R)$ by Paris and Wilkie [90].

A proof system for the Quantified Propositional Calculus was introduced by Dowd [48]. The system G and its subsystems G_i were introduced by Krajíček and Pudlák [73] (see also Section 4.6 of [72]). The original definition of G is what we refer to as KPG in Exercise VII.3.7 and the original definition of G_i is KPG restricted so that all formulas must be either Σ_i^q or Π_i^q. Our definitions are due to Morioka [80]. Theorem VII.3.9 is new. Theorem VII.4.8 is from [91].

The idea of G_i^* (treelike G_i) is from [72], and the V^1 Translation Theorem VII.5.2 is adapted from a similar theorem for S_2^1 also in [72]. Theorem VII.4.13 is from [41]. Extended Frege proof systems, which inspired the system ePK in Section VII.4.1, were introduced in [46].

Theorem VII.5.6 is new.

Chapter VIII

THEORIES FOR POLYNOMIAL TIME AND BEYOND

We present a finitely-axiomatizable "minimal" theory for polynomial time over the basic two-sorted vocabulary $\mathcal{L}_A^2$. We show that it is robust by giving three quite different axiomatizations for it under the names VP, TV^0, and V^1-$HORN$. We also present a universal conservative extension VPV for this theory which has function symbols for all polynomial time functions based on Cobham's recursion-theoretic characterization of FP. The theory V^1 from Chapter VI has the same Σ_1^B theorems as the minimal theory, but apparently has more Σ_2^B theorems. The new theories have the following inclusions:

$$VP = TV^0 = V^1\text{-}HORN \subseteq_{cons} \widehat{VP} \subseteq_{cons} VPV$$

where $T_1 \subseteq_{cons} T_2$ means that T_2 is a conservative extension of T_1.

Section VIII.3 introduces the TV^i hierarchy and concentrates on the bottom level TV^0 mentions above. Section VIII.5 is devoted to TV^1, and characterizes the Σ_1^B-definable search problems in this theory as those reducible to polynomial local search. Section VIII.6 proves a form of the Herbrand Theorem known as the KPT Witnessing Theorem, which can be used to prove (or suggest) independence results for Σ_2^B-formulas. As an application we show that V^0 does not prove the Σ_0^B-Replacement scheme, and (unless integer factoring is easy) neither does VPV.

Section VIII.7 proves a host of results on V^∞, the interleaved V^i and TV^i hierarchies. These include the finite axiomatizability of V^i and TV^i, Σ_i^B-definability results (see Table 3 page 250 for a summary), and the equivalence of the collapse of these hierarchies and the provable collapse of the polynomial hierarchy. Section VIII.8 proves 'RSUV' isomorphism theorems relating our two-sorted theories V^i and TV^i to Buss's single-sorted theories S_2^i and T_2^i.

VIII.1. The Theory VP and Aggregate Functions

The theory VP extends V^0 by adding a single axiom asserting that the gates of a given monotone Boolean circuit with specified inputs can be evaluated. We will then use the fact that the Monotone Circuit Value problem is complete for P under many-one AC^0 reductions to prove that all polynomial time functions are Σ_1^B definable in VP. We will show that V^1 extends VP, and show that the Σ_1^B theorems of V^1 and VP are the same. Later, in Section VIII.6, we give evidence that VP does not prove either the Σ_0^B-$REPL$ scheme or the Σ_1^B-$COMP$ scheme (which do not consist of Σ_1^B formulas), and hence apparently V^1 is not conservative over VP.

It seems that VP is a "minimal" theory for polynomial time reasoning because it extends our base theory V^0 by adding one axiom asserting the existence of a solution to a standard complete problem for P. We use this same method in Chapter IX to introduce minimal theories for other complexity classes.

We specify a monotone Boolean circuit (using our two-sorted vocabulary $\mathcal{L}_A^2$) by a triple (a, G, E), where the gates are numbered $0, 1, \ldots, (a - 1)$, and for $x > 1$, $G(x)$ holds iff gate x is an $\wedge$ gate (otherwise gate x is an $\vee$ gate). Gates numbered 0 and 1 are "input" gates, and always have the values 0 and 1 respectively. The edge relation E specifies the inputs to the other gates as follows:

- For $0 \leq y < x, 2 \leq x < a$, $E(y, x)$ holds iff the output of gate y is connected to an input of gate x.

The Σ_0^B formula $\delta_{MCV}(a, G, E, Y)$ asserts that $Y(x)$ holds iff the output of gate x is 1 (i.e. $\top$), and is defined as follows:

$$\delta_{MCV}(a, G, E, Y) \equiv \neg Y(0) \wedge Y(1) \wedge \forall x < a, 2 \leq x \supset$$
$$Y(x) \leftrightarrow [(G(x) \wedge \forall y < x(E(y, x) \supset Y(y))) \vee$$
$$(\neg G(x) \wedge \exists y < x(E(y, x) \wedge Y(y)))]. \quad (157)$$

DEFINITION VIII.1.1 (VP). The theory VP has vocabulary $\mathcal{L}_A^2$ and is axiomatized by the axioms of V^0 and one more axiom called MCV, where

$$MCV \equiv \exists Y \leq a + 2 \, \delta_{MCV}(a, G, E, Y).$$

The next result is immediate from the above definition and the fact that V^0 is finitely axiomatizable (Theorem V.7.1).

COROLLARY VIII.1.2. VP is finitely axiomatizable.

THEOREM VIII.1.3. V^1 is an extension of VP.

PROOF. It suffices to show that V^1 proves the axiom MCV. But MCV is a Σ_1^B-formula, and is easily proved by induction on a. $\square$

Note that MCV is a bounded formula, and hence VP is a polynomial-bounded theory (Definition V.3.3). Thus by Parikh's Theorem V.3.4 a function is provably total (i.e. Σ_1^1-definable) in VP iff it is Σ_1^B-definable in VP.

THEOREM VIII.1.4. *A function is provably total in* VP *iff it is in* FP.

One direction is proved as Theorem VIII.1.8 below, and the other direction is Corollary VIII.1.14.

We introduce a string function F_{MCV} which witnesses the existential quantifier in the axiom MCV. The defining axiom is

$$Y = F_{MCV}(a, G, E) \leftrightarrow (|Y| \leq a \wedge \delta_{MCV}(a, G, E, Y)). \qquad (158)$$

LEMMA VIII.1.5. F_{MCV} *is* Σ_0^B *definable in* VP.

PROOF. We need to show that VP proves

$$\exists! Y(|Y| \leq a \wedge \delta_{MCV}(a, G, E, Y)).$$

Existence of Y follows from the axiom MCV. Uniqueness can be proved in V^0 by induction on i using the Σ_0^B formula $\psi(i)$ asserting that the first i bits of Y are uniquely determined. □

We define the two-sorted Monotone Circuit Value problem using the relation $R_{MCV}(a, G, E)$, which holds iff the circuit specified by (a, G, E) has output 1 (the gate numbered $a \dot{-} 1$ is designated the output).

DEFINITION VIII.1.6.

$$R_{MCV}(a, G, E) \leftrightarrow \exists Y \leq a(\delta_{MCV}(a, G, E, Y) \wedge Y(a \dot{-} 1)). \qquad (159)$$

The following proposition shows that R_{MCV} is AC^0-many-one complete for P.

PROPOSITION VIII.1.7. *For any relation* $R(\vec{x}, \vec{X})$ *in* P *there are functions* a_0, G_0, E_0 *in* FAC^0 *such that*

$$R(\vec{x}, \vec{X}) \leftrightarrow R_{MCV}(a_0(\vec{x}, \vec{X}), G_0(\vec{x}, \vec{X}), E_0(\vec{x}, \vec{X})). \qquad (160)$$

PROOF SKETCH. First we point out that Circuit Value Problem CVP for Boolean circuits which have $\neg$ gates in addition to $\wedge$ and $\vee$ gates is easily reduced to the Monotone Circuit Value Problem MCV by using the method of "double-rail logic". Given a circuit C which has gates for $\wedge, \vee, \neg$ we compute (in FAC^0) a monotone circuit C' which has two inputs x and x' for each input x of C, and two gates g' and g'' for every gate g in C. This is done such that, assuming that each input x' is the negation of x, then $g' \leftrightarrow g$ and $g'' \leftrightarrow \neg g$. Given an assignment of inputs to C, suitable inputs to C' satisfying $x' \leftrightarrow \neg x$ can trivially be computed by an FAC^0 function. To design C', note that the gate g has one of the three types $\wedge, \vee, \neg$, and in each case (by De Morgan's laws) there are easy monotone circuits which compute both g' and g'' from the inputs to g and their negations.

Now to prove the Proposition it suffices to show that, given a polytime Turing machine M for computing a relation $R(\vec{x}, \vec{X})$, there is an AC^0 function F_M such that $F_M(\vec{x}, \vec{X})$ describes a circuit (allowing $\neg$ gates and with given input values) whose gate values describe the computation of M on input $\vec{x}, \vec{X}$.

One way to see how to do this is to consider equation (95) (page 138), where the variable Z describes the computation of a polytime Turing machine. Here the rows $Z^{[z]}$ of Z are computed successively using the AC^0 functions $Init_M$ and $Next_M$. All AC^0 functions are computed by uniform circuit families, which themselves are describable by AC^0 functions. □

THEOREM VIII.1.8. *Every function in* **FP** *is* Σ_1^B-*definable in* **VP**.

PROOF. It suffices to prove this for string functions, since by Proposition VI.2.1 every number function $f(\vec{x}, \vec{X})$ in **FP** has the form $|F(\vec{x}, \vec{X})|$ for some string function F in **FP**, and by Exercise V.4.5 the Σ_1^B definable functions in **VP** are closed under composition.

By Definition V.2.3, a string function $F(\vec{x}, \vec{X})$ is in **FP** iff it is p-bounded and its bit-graph is in **P**; i.e. there is an $\mathcal{L}_A^2$ term $t(\vec{x}, \vec{X})$ and a relation $B_F(i, \vec{x}, \vec{X})$ in **P** such that

$$F(\vec{x}, \vec{X})(i) \leftrightarrow (i < t(\vec{x}, \vec{X}) \wedge B_F(i, \vec{x}, \vec{X})). \tag{161}$$

Our task is to find a Σ_1^B formula $\varphi_F(\vec{x}, \vec{X}, Z)$ representing the graph of F by satisfying

$$Z = F(\vec{x}, \vec{X}) \leftrightarrow \varphi_F(\vec{x}, \vec{X}, Z)$$

and such that

$$\boldsymbol{VP} \vdash \exists ! Z \varphi_F(\vec{x}, \vec{X}, Z). \tag{162}$$

Since the bit graph B_F of F is a polytime relation, by (159), (160) there are functions a_0, G_0, E_0 in $\mathcal{L}_{FAC^0}$ such that

$$B_F(i, \vec{x}, \vec{X}) \leftrightarrow \exists Y \leq a_0(i), \, \delta_{MCV}(a_0(i), G_0(i), E_0(i), Y) \wedge Y(a_0(i) \dot{-} 1) \tag{163}$$

where we have suppressed the arguments $(\vec{x}, \vec{X})$ in a_0, G_0, E_0. We can use the function F_{MCV} defined in (158) to witness Y in the above equation, and hence the graph φ_F of F satisfies

$$\varphi_F(\vec{x}, \vec{X}, Z) \leftrightarrow \forall i < t[Z(i) \leftrightarrow F_{MCV}(a_0(i), G_0(i), E_0(i))(a_0(i) \dot{-} 1)]. \tag{164}$$

Unfortunately the formula on the right is not Σ_1^B, and although the part in brackets [...] can be made Σ_1^B, the existential string quantifier there requires the Replacement Axiom (Definition VI.3.2) to move it in front of the quantifier $\forall i < t$. In Section VIII.6 we give evidence that **VP** does not prove Σ_0^B-**REPL**.

So we use another approach. From (163) we see that for each fixed $i, 0 \leq i < t$, the parameters $(a_0(i), G_0(i), E_0(i))$ describe a circuit $C(i)$ which computes bit i of $F(\vec{x}, \vec{X})$. Our task is to describe one circuit

$C = C(\vec{x}, \vec{X})$ which combines the circuits $C(0), \ldots, C(t-1)$ to compute all of these bits together.

In order to do this it will be helpful to introduce the important notion of the aggregate function F^*_{MCV} of F_{MCV}, where the aggregate F^* of F is the string function that gathers the values of F for a polynomially long sequence of arguments. We use the notation $Z^{[x]} = Row(x, Z)$ (Definition V.4.26) and $(Z)^x = seq(x, Z)$ (Definition V.4.31).

DEFINITION VIII.1.9 (Aggregate Function). Suppose that

$$F(x_1, \ldots, x_k, X_1, \ldots, X_n)$$

is a polynomially bounded string function, i.e., for some $\mathcal{L}^2_A$ term t,

$$|F(\vec{x}, \vec{X})| \le t(\vec{x}, \vec{X}).$$

Then $F^*(b, Z_1, \ldots, Z_k, X_1, \ldots, X_n)$ is the polynomially bounded string function that satisfies

$$|F^*(b, \vec{Z}, \vec{X})| \le \langle b, t(|\vec{Z}|, \vec{X}) \rangle$$

and

$$F^*(b, \vec{Z}, \vec{X})(w) \leftrightarrow \exists i < b \exists v < t, \; w = \langle i, v \rangle \wedge$$
$$F((Z_1)^i, \ldots, (Z_k)^i, X_1^{[i]}, \ldots, X_n^{[i]})(v). \quad (165)$$

Notice that by (165)

$$\forall i < b, F^*(b, \vec{Z}, \vec{X})^{[i]} = F((Z_1)^i, \ldots, (Z_k)^i, X_1^{[i]}, \ldots, X_n^{[i]}). \quad (166)$$

The use of *seq* in (165) and (166) can be eliminated using its definition V.4.31 to obtain an equivalent $\Sigma^B_0(Row, F)$ definition of the bit graph of F^*, but in general the use of Row cannot be eliminated to get a $\Sigma^B_0(F)$ definition.

Lemma VIII.1.10 below shows that the aggregate function F^*_{MCV} of F_{MCV} is Σ^B_1-definable in *VP*. We can interpret F^*_{MCV} as assigning values to the gates of a collection $C(0), \ldots, C(b-1)$ of circuits. Thus (166) becomes

$$\forall i < b, \; F^*_{MCV}(b, Z, U, V)^{[i]} = F_{MCV}((Z)^i, U^{[i]}, V^{[i]}) \quad (167)$$

and writing

$$((Z)^i, U^{[i]}, V^{[i]}) = (a_i, G_i, E_i) \quad (168)$$

we want the the triple (a_i, G_i, E_i) to describe the circuit $C(i)$ which computes bit i of $F(\vec{x}, \vec{X})$. Thus by (164) we want

$$((Z)^i, U^{[i]}, V^{[i]}) = (a_0(i), G_0(i), E_0(i)).$$

For this we define "pseudo-aggregate" functions A_1, G_1, E_1 for the functions a_0, G_0, E_0 which, for fixed $\vec{x}, \vec{X}$, collect values for arguments $i < t(\vec{x}, \vec{X})$. Thus for $i < t$

$$(A_1(\vec{x}, \vec{X}))^i = a_0(i, \vec{x}, \vec{X}),$$
$$G_1(\vec{x}, \vec{X})^{[i]} = G_0(i, \vec{x}, \vec{X}),$$
$$E_1(\vec{x}, \vec{X})^{[i]} = E_0(i, \vec{x}, \vec{X}).$$

Since each of the functions a_0, G_0, E_0 is in $\boldsymbol{FAC}^0$, it follows easily from the $\boldsymbol{FAC}^0$ Elimination Lemma V.6.7 that the functions A_1, G_1, E_1 have Σ_0^B-definable bit graphs and hence are themselves in $\boldsymbol{FAC}^0$.

If $Y = F^*_{MCV}(t, A_1, G_1, E_1)$ (where we have suppressed the arguments $\vec{x}, \vec{X}$) then $Y^{[i]}$ gives the correct assignment to the gates of $C(i)$. Thus for each $i < t$

$$F(\vec{x}, \vec{X})(i) \leftrightarrow Y^{[i]}(a_0(i, \vec{x}, \vec{X}) \div 1).$$

So we define the $\boldsymbol{FAC}^0$ function $Extract$ by defining its bit graph as follows:

$$Extract(\vec{x}, \vec{X}, Y)(i) \leftrightarrow \left(i < t(\vec{x}, \vec{X}) \wedge Y^{[i]}(a_0(i, \vec{x}, \vec{X}) \div 1)\right).$$

Then

$$F(\vec{x}, \vec{X}) = Extract(\vec{x}, \vec{X}, F^*_{MCV}(t, A_1, G_1, E_1)) \qquad (169)$$

(again suppressing some occurrences of $\vec{x}, \vec{X}$). This, Lemma VIII.1.10 and the fact that the Σ_1^B-definable functions in a polynomial-bounded theory are closed under composition (Exercise V.4.5) show that F is Σ_1^B-definable in $\boldsymbol{VP}$. □

To complete the proof of Theorem VIII.1.8 we need the following result.

LEMMA VIII.1.10. F^*_{MCV} is Σ_1^B-definable in $\boldsymbol{VP}$, and $\boldsymbol{VP}(F_{MCV}, F^*_{MCV})$ proves (167).

PROOF. For $i < b$ let $C(i)$ be the circuit described by (a_i, G_i, E_i) as in (168). We want to embed the circuits $C(0), C(1), \ldots, C(b-1)$ into a single circuit C. Each $C(i)$ has $a_i < |Z|$ gates, and we will be generous and allot $|Z|$ gates in the embedded version of each $C(i)$, so that C has a total of $b|Z|$ gates. Thus gate j of $C(i)$ corresponds to gate $i|Z| + j$ of C.

Circuit C has the description $(\hat{a}, \hat{G}, \hat{E})$, where $\hat{a} = b|Z|$ and $\hat{G} = \hat{G}(b, Z, U)$ and $\hat{E} = \hat{E}(b, Z, V)$ are $\boldsymbol{FAC}^0$ functions. These functions are straightforward to define to satisfy the intended embedding of $C(i)$ into C, except that the gates in C corresponding to gates 0 and 1 of $C(i)$ must have constant values 0 and 1 respectively. To achieve this, these gates have no input edges and we make them OR gates and AND gates respectively.

Thus for $i, i' < b$

$$\hat{G}(b, Z, U)(i|Z| + j) \leftrightarrow (U^{[i]}(j) \wedge 2 \leq j) \vee j = 1 \text{ if } j < |Z|,$$

$$\hat{E}(b, Z, V)(i|Z| + j, i'|Z| + k) \leftrightarrow V^{[i]}(j, k) \wedge i = i' \wedge 2 \leq k \text{ if } j, k < |Z|.$$

This is easily turned into Σ_0^B-definitions of the bit graphs of $\hat{G}$ and $\hat{E}$.

Referring to (167), it remains to define an *FAC*^{0} function *Compile*(b, Z, Y) whose i-th row assigns correct values to the gates of $C(i)$, assuming that Y assigns correct values to the gates of C. Thus

$$\textit{Compile}(b, Z, Y)(i, j) \leftrightarrow (i < b \wedge j < (Z)^i \wedge Y(i|Z| + j)).$$

Finally

$$F_{MCV}^*(b, Z, U, V) = \textit{Compile}(F_{MCV}(b|Z|, \hat{G}(b, Z, U), \hat{E}(b, Z, V))) \tag{170}$$

and (167) is provable from this equation and the defining axioms for the functions involved. Also F_{MCV}^* is a composition of Σ_1^B-definable functions in *VP*, and hence is itself Σ_1^B-definable in *VP*.   □

To prove the converse to Theorem VIII.1.8 we introduce a universal conservative extension of *VP* in the next subsection.

VIII.1.1. The Theory $\widehat{VP}$. Let $\delta'_{MCV}(a, G, E, Y)$ denote a quantifier-free formula in the vocabulary $\mathcal{L}_{FAC^0}$ which $\overline{V}^0$ proves equivalent to $\delta_{MCV}(a, G, E, Y)$ (see Lemma V.6.3). The function F'_{MCV} has defining axiom

$$Y = F'_{MCV}(a, G, E) \leftrightarrow |Y| \leq a \wedge \delta'_{MCV}(a, G, E, Y). \tag{171}$$

Thus F_{MCV} (defined in (158)) and F'_{MCV} are equal as functions, although they have different defining axioms.

DEFINITION VIII.1.11 ($\widehat{VP}$). The universal theory $\widehat{VP}$ has vocabulary

$$\mathcal{L}_{\widehat{VP}} = \mathcal{L}_{FAC^0} \cup \{F'_{MCV}\}$$

and axioms those of $\overline{V}^0$ together with the defining axiom (171) for F'_{MCV}.

Since F_{MCV} is in *FP* and every function in $\mathcal{L}_{FAC^0}$ is in *FP*, it is clear that every term in the vocabulary of $\widehat{VP}$ represents a function in *FP*. The next result states the converse.

THEOREM VIII.1.12. *Every function in* **FP** *is represented by a term in the vocabulary of* $\widehat{VP}$.

PROOF. Equation (170) expresses F_{MCV}^* as a term involving F_{MCV} and functions in $\mathcal{L}_{FAC^0}$, and equation (169) (finishing the proof of Theorem VIII.1.8) expresses an arbitrary string function F in *FP* as a term involving F_{MCV}^* and functions in $\mathcal{L}_{FAC^0}$. By Proposition VI.2.1 every number function $f(\vec{x}, \vec{X})$ in *FP* has the form $|F(\vec{x}, \vec{X})|$ for some string function F in *FP*.   □

Theorem VIII.1.13. *The theory $\widehat{VP}$ is a universal conservative extension of VP.*

Proof. The formula $\delta_{MCV}(a, G, E, F'_{MCV}(a, G, E))$ is provable in $\widehat{VP}$ by (171) and implies the axiom MCV for VP, and hence $\widehat{VP}$ is an extension of VP.

$VP + \overline{V}^0$ is conservative over VP because $\overline{V}^0$ is conservative over V^0 (Theorem V.6.5). $\widehat{VP}$ can be obtained from $VP + \overline{V}^0$ by adding the defining axiom for F'_{MCV}, and F'_{MCV} is definable in $VP + \overline{V}^0$ by Lemma VIII.1.5 (note that $\overline{V}^0$ proves the equivalence of the defining axioms for F_{MCV} and F'_{MCV}). Thus by Theorem V.4.2, $\widehat{VP}$ is conservative over $VP + \overline{V}^0$ and hence over VP. $\qquad\square$

Corollary VIII.1.14. *Every function Σ_1^1-definable in VP or $\widehat{VP}$ is in FP.*

Proof. As observed above every term of $\widehat{VP}$ stands for a function in FP. Since $\widehat{VP}$ is a universal theory, it follows from the Herbrand Theorem that the existential quantifiers in any Σ_1^1 theorem of $\widehat{VP}$ can be witnessed by a combination of terms and hence by functions in FP (see Section V.6.1 for this argument applied to $\overline{V}^0$). Therefore every Σ_1^1-definable function in $\widehat{VP}$ is in FP. Since $\widehat{VP}$ is an extension of VP, the same is true of VP. $\qquad\square$

We wish to show that $\widehat{VP}$ proves the $\Sigma_0^B(\mathcal{L}_{\widehat{VP}})$-*IND* and $\Sigma_0^B(\mathcal{L}_{\widehat{VP}})$-*COMP* schemes. Note that Lemma VIII.1.10 easily follows from the fact that $\widehat{VP}$ proves $\Sigma_0^B(\mathcal{L}_{\widehat{VP}})$-*COMP*, and to prove this scheme we need a general result about aggregate functions which will also play an important role in Chapter IX.

Theorem VIII.1.15 (Aggregate Function). *Let $\mathcal{T}$ be a theory with vocabulary $\mathcal{L}$ which extends $V^0(Row)$ and proves $\Sigma_0^B(\mathcal{L})$-COMP. Suppose that F and $F^\star$ are definable in $\mathcal{T}$ (Definition V.4.1) and $\mathcal{T}(F, F^\star)$ proves (166). Then $\mathcal{T}(F)$ proves $\Sigma_0^B(\mathcal{L} \cup \{F\})$-COMP.*

Proof. Since $\mathcal{T}$ proves $\Sigma_0^B(\mathcal{L})$-*COMP*, by Lemma V.4.25 it proves the Multiple Comprehension axioms for $\Sigma_0^B(\mathcal{L})$.

Claim. For any $\mathcal{L}$-terms $\vec{s}, \vec{T}$ that contain variables $\vec{z}$, $\mathcal{T}(F)$ proves

$$\exists Y \forall z_1 < b_1 \ldots \forall z_m < b_m,\; Y^{[\vec{z}]} = F(\vec{s}, \vec{T}) \qquad (172)$$

where $Y^{[\vec{z}]}$ denotes $Y^{[\langle \vec{z} \rangle]}$.

Proof of the Claim. Since $\mathcal{T}$ proves the Multiple Comprehension axiom scheme for $\Sigma_0^B(\mathcal{L})$ formulas, it proves the existence of $\vec{X}$ such that $X_j^{[\vec{z}]} = T_j$, for $1 \le j \le n$. It also proves the existence of Z_i such that $(Z_i)^{\langle \vec{z} \rangle} = s_i$, for $1 \le i \le k$. Now the value of Y that satisfies (172) is just $F^\star(\langle \vec{b} \rangle, \vec{Z}, \vec{X})$. $\qquad\square$

Let $\mathcal{L}' = \mathcal{L} \cup \{F\}$. We show by induction on the quantifier depth of a $\Sigma_0^B(\mathcal{L}')$ formula ψ that $\mathcal{T}(F)$ proves

$$\exists Z \leq \langle b_1, \ldots, b_m\rangle \forall z_1 < b_1 \ldots \forall z_m < b_m(Z(\vec{z}) \leftrightarrow \psi(\vec{z})) \qquad (173)$$

where $\vec{z}$ are all free number variables of ψ. It follows that

$$\mathcal{T}(F) \vdash \Sigma_0^B(\mathcal{L}')\text{-}\boldsymbol{COMP}.$$

For the base case, ψ is quantifier-free. The idea is to replace every occurrence of a term $F(\vec{s}, \vec{T})$ in ψ by a new string variable W which has the intended value of $F(\vec{s}, \vec{T})$. The resulting formula is $\Sigma_0^B(\mathcal{L})$, and we can apply the hypothesis.

Formally, suppose that $F(\vec{s}_1, \vec{T}_1), \ldots, F(\vec{s}_k, \vec{T}_k)$ are all occurrences of F in ψ. Note that the terms $\vec{s}_i, \vec{T}_i$ may contain $\vec{z}$ as well as nested occurrences of F. Assume further that these F-terms are ordered by depth so that $\vec{s}_1, \vec{T}_1$ do not contain F, and for $1 < i \leq k$, any occurrence of F in $\vec{s}_i, \vec{T}_i$ must be of the form $F(\vec{s}_j, \vec{T}_j)$, for some $j < i$. We proceed to eliminate F from ψ by using its defining axiom.

Let $W_1, \ldots, W_k$ be new string variables. Let $\overrightarrow{s_1'} = \vec{s}_1$, $\overrightarrow{T_1'} = \vec{T}_1$, and for $2 \leq i \leq k$, $\overrightarrow{s_i'}$ and $\overrightarrow{T_i'}$ be obtained from $\vec{s}_i$ and $\vec{T}_i$ respectively by replacing every maximal occurrence of any $F(\vec{s}_j, \vec{T}_j)$, for $j < i$, by $W_j^{[\vec{z}]}$. Thus F does not occur in any $\overrightarrow{s_i'}$ and $\overrightarrow{T_i'}$, but for $i \geq 2$, $\overrightarrow{s_i'}$ and $\overrightarrow{T_i'}$ may contain $W_1, \ldots, W_{i-1}$.

By the Claim above, for $1 \leq i \leq k$, $\mathcal{T}(F)$ proves the existence of W_i such that

$$\forall z_1 < b_1 \ldots \forall z_m < b_m, \ W_i^{[\vec{z}]} = F(\overrightarrow{s_i'}, \overrightarrow{T_i'}). \qquad (174)$$

Let $\psi'(\vec{z}, W_1, \ldots, W_k)$ be obtained from $\psi(\vec{z})$ by replacing each maximal occurrence of $F(\vec{s}_i, \vec{T}_i)$ by $W_i^{[\vec{z}]}$, for $1 \leq i \leq k$. Then

$$\mathcal{T} \vdash \exists Z \leq \langle b_1, \ldots, b_m\rangle \forall z_1 < b_1 \ldots \forall z_m < b_m(Z(\vec{z}) \leftrightarrow$$
$$\psi'(\vec{z}, W_1, \ldots, W_k)).$$

Such Z satisfies (173) when each W_i is defined by (174).

The induction step is straightforward. Consider for example the case $\psi(\vec{z}) \equiv \forall x < t\lambda(\vec{z}, x)$. By the induction hypothesis,

$$\mathcal{T}(F) \vdash \exists Z' \forall z_1 < b_1 \ldots \forall z_m < b_m \forall x < t, \ Z'(\vec{z}, x) \leftrightarrow \lambda(\vec{z}, x).$$

Now

$$V^0 \vdash \exists Z \forall z_1 < b_1 \ldots \forall z_m < b_m, \ Z(\vec{z}) \leftrightarrow \forall x < t Z'(\vec{z}, x)$$

and hence $\mathcal{T}(F) \vdash \exists Z \forall \vec{z} < \vec{b}\,(Z(\vec{z}) \leftrightarrow \psi(\vec{z}))$. $\square$

COROLLARY VIII.1.16. $\widehat{VP}$ proves the $\Sigma_0^B(\mathcal{L}_{\widehat{VP}})$-$IND$ and $\Sigma_0^B(\mathcal{L}_{\widehat{VP}})$-$COMP$ axioms.

Proof. In Theorem VIII.1.15 take $\mathcal{T} = VP \cup \overline{V}^0$ and $F = F'_{MCV}$. Then $\mathcal{L} = \mathcal{L}_{FAC^0}$ so $\mathcal{T}$ proves $\Sigma_0^B(\mathcal{L})\text{-}COMP$ by Lemma V.6.4. Also $\mathcal{T}$ proves the defining equations (158) and (171) for F_{MCV} and F'_{MCV} are equivalent, so by Lemmas VIII.1.5 and VIII.1.10 F and F^* are Σ_1^B-definable in $\mathcal{T}$, and $\mathcal{T}(F, F^*)$ proves (166). Thus the corollary follows from the theorem, since $\widehat{VP} = \mathcal{T}(F)$ (and V^0 proves $\Sigma_0^B\text{-}IND$). □

Note that the formulas $\Sigma_0^B(\mathcal{L}_{\widehat{VP}})$ represent precisely the polynomial time relations, so Corollary VIII.1.16 together with Theorem VIII.1.4 suggest that $\widehat{VP}$ (and hence VP) "capture" polynomial time reasoning. Also VP seems to be a minimal such theory (relative to the base theory V^0), since surely polynomial time reasoning should be able to prove the basic axiom MCV, that the monotone circuit value problem is complete for P. In the next sections we will also prove that VP is a robust theory, by giving several equivalent axiomatizations for it.

VIII.2. The Theory VPV

The universal theory VPV is based on the single-sorted theory PV [39], which historically was the first theory designed to capture polynomial time reasoning. It is an extension of $\widehat{VP}$, and (unlike $\widehat{VP}$) it has a function symbol (and not just a term) for every string function in FP. We will show that VPV is a conservative extension of $\widehat{VP}$. The vocabulary of VPV extends that of $\overline{V}^0$, with additional function symbols introduced based on Cobham's characterization of FP (Theorem VI.2.12).

Following Definition VI.2.11, we can write the defining equations for a string function $F(y, \vec{x}, \vec{X})$ defined by limited recursion from $G(\vec{x}, \vec{X})$ and $H(y, \vec{x}, \vec{X}, Z)$ as

$$F(0, \vec{x}, \vec{X}) = G(\vec{x}, \vec{X}), \tag{175}$$

$$F(y + 1, \vec{x}, \vec{X}) = (H(y, \vec{x}, \vec{X}, F(y, \vec{x}, \vec{X})))^{<t(y,\vec{x},\vec{X})} \tag{176}$$

where the bounding term $t(y, \vec{x}, \vec{X})$ is in $\mathcal{L}_A^2$ and the notation $Z^{<y}$ refers to $Cut(y, Z)$ (the first y bits of Z, page 139).

For convenience we repeat the defining axiom (86) for the functions $F_{\varphi(z),t}$ introduced in Section V.6 to define $\overline{V}^0$.

$$F_{\varphi(z),t}(\vec{x}, \vec{X})(z) \leftrightarrow \left(z < t(\vec{x}, \vec{X}) \wedge \varphi(z, \vec{x}, \vec{X})\right). \tag{177}$$

Definition VIII.2.1. The vocabulary $\mathcal{L}_{FP}$ is the smallest set that satisfies

(1) $\mathcal{L}_{FAC^0} \subseteq \mathcal{L}_{FP}$.
(2) For each open formula $\varphi(z, \vec{x}, \vec{X})$ over $\mathcal{L}_{FP}$ and term $t = t(\vec{x}, \vec{X})$ of $\mathcal{L}_A^2$ there is a string function $F_{\varphi(z),t}$ in $\mathcal{L}_{FP}$.

(3) For each triple G, H, t, where $G(\vec{x}, \vec{X})$ and $H(y, \vec{x}, \vec{X}, Z)$ are functions in $\mathcal{L}_{FP}$ and $t = t(y, \vec{x}, \vec{X})$ is a term in $\mathcal{L}_A^2$, there is a function $F_{G,H,t}$ in $\mathcal{L}_{FP}$ (with defining equations (175), (176)).

To simplify this definition we have not introduced new number functions of the from $f_{\varphi(z),t}$ that were used along with $F_{\varphi(z),t}$ in the inductive definition of $\mathcal{L}_{FAC^0}$ (although by 1) everything in $\mathcal{L}_{FAC^0}$ remains in $\mathcal{L}_{FP}$). Nevertheless by Cobham's Theorem it is easy to see that semantically the string functions of $\mathcal{L}_{FP}$ comprise the polytime string functions in **FP**. In particular every string term T over $\mathcal{L}_{FP}$ is represented by a function symbol of the form $F_{\varphi(z),t}$ in $\mathcal{L}_{FP}$, where (referring to (177)) $\varphi \equiv T(z)$ and t is a suitable bounding term. Note also that every number function in **FP** has the form $|F|$ for some function F in $\mathcal{L}_{FP}$.

We now define the theory *VPV* in the style of Definition V.6.2 of $\overline{V}^0$.

DEFINITION VIII.2.2. *VPV* is the theory over $\mathcal{L}_{FP}$ whose axioms are those of $\overline{V}^0$ together with the defining axioms (177) for each function $F_{\varphi(z),t}$ in $\mathcal{L}_{FP}$ and defining axioms (175), (176) for each function $F_{G,H,t}$ in $\mathcal{L}_{FP}$.

Thus *VPV* is a universal theory which extends $\overline{V}^0$. Every function introduced in Definition VIII.2.1 is explicitly bounded by a term in $\mathcal{L}_A^2$, and hence *VPV* is a polynomial-bounded theory.

The following general result can be proved by structural induction on φ in the same way as Lemma III.3.19 and Lemma V.6.3. Our immediate intended application is to take $\mathcal{T} = VPV$.

LEMMA VIII.2.3. *Let $\mathcal{T}$ be a theory with vocabulary $\mathcal{L}$ such that $\mathcal{T}$ extends V^0 and for every open formula $\varphi(z, \vec{x}, \vec{X})$ over $\mathcal{L}$ and term $t(\vec{x}, \vec{X})$ over $\mathcal{L}_A^2$ there is a function $F_{\varphi(z),t}$ in $\mathcal{L}$ such that*

$$\mathcal{T} \vdash F_{\varphi(z),t}(\vec{x}, \vec{X})(z) \leftrightarrow (z < t \wedge \varphi(z, \vec{x}, \vec{X})).$$

Then for every $\Sigma_0^B(\mathcal{L})$ formula φ there is an open $\mathcal{L}$-formula φ^+ such that $\mathcal{T} \vdash \varphi \leftrightarrow \varphi^+$.

Next we state a general witnessing theorem for universal theories, which applies to *VPV*.

THEOREM VIII.2.4 (Witnessing). *Let $\mathcal{T}$ be a universal polynomial-bounded theory which extends V^0, with vocabulary $\mathcal{L}$, such that for every open formula $\varphi(z, \vec{x}, \vec{X})$ over $\mathcal{L}$ and term $t(\vec{x}, \vec{X})$ over $\mathcal{L}_A^2$ there is a function $F_{\varphi(z),t}$ in $\mathcal{L}$ such that*

$$\mathcal{T} \vdash F_{\varphi(z),t}(\vec{x}, \vec{X})(z) \leftrightarrow \left(z < t \wedge \varphi(z, \vec{x}, \vec{X})\right).$$

Then for every theorem of $\mathcal{T}$ of the form $\exists Z \varphi(\vec{x}, \vec{X}, Z)$, where φ is an open formula, there is a function F in $\mathcal{L}$ such that

$$\mathcal{T} \vdash \varphi(\vec{x}, \vec{X}, F(\vec{x}, \vec{X})).$$

Proof. The proof is based on the Herbrand Theorem, and is very similar to the alternative proof of the witnessing theorem for V^0 given in Section V.6.1. This proof defines the witnessing function F by cases, and in fact F has the form $F_{\varphi(z),t}$ for suitable φ, t. By our assumption that $\mathcal{T}$ is polynomial-bounded, we know that there is a bounding term t for $F_{\varphi(z),t}$ in $\mathcal{L}_A^2$ (as opposed to $\mathcal{L}$).   □

Corollary VIII.2.5 (Witnessing for VPV). Every $\Sigma_1^1(\mathcal{L}_{FP})$ theorem of VPV is witnessed in VPV by functions in $\mathcal{L}_{FP}$.

Proof. It is clear that VPV satisfies the hypotheses for the theory $\mathcal{T}$ in the theorem. Although the theorem only states that formulas of the form $\exists Z\varphi$ (where φ is quantifier-free) can be witnessed, it is easy to generalize it to witness an arbitrary $\Sigma_1^1(\mathcal{L}_{FP})$ formula $\exists \vec{z}\exists \vec{Z}\varphi$. (See Lemma V.5.5 and how it is used to prove the witnessing theorem for V^0.)   □

This witnessing result immediately implies the following.

Corollary VIII.2.6. Every function Σ_1^1-definable in VPV is in FP.

Of course this holds whether we interpret Σ_1^1-definable to mean $\Sigma_1^1(\mathcal{L}_A^2)$-definable, or more generally $\Sigma_1^1(\mathcal{L}_{FP})$-definable. The converse of the latter, that every polytime function is $\Sigma_1^1(\mathcal{L}_{FP})$-definable in VPV, is obvious, since $\mathcal{L}_{FP}$ comprises the polytime functions. However we are interested in the stronger converse, that every $\mathcal{L}_{FP}$-function has a $\Sigma_1^B(\mathcal{L}_A^2)$ definition, provably in VPV. This is not straightforward to prove, mainly because we do not have the Σ_0^B-$REPL$ axioms available in VPV (Theorem VIII.6.3). Section VI.3.1 shows how we could proceed if Σ_0^B-$REPL$ were available, and Theorem IX.2.10 shows how we could proceed using aggregate functions. But here we take a different approach: Since V^1 proves the Σ_0^B-$REPL$ axioms it is relatively easy to show that every $\mathcal{L}_{FP}$ function is $\Sigma_1^B(\mathcal{L}_A^2)$-definable in V^1. From this we use the fact that Σ_1^B theorems of V^1 are witnessed in VPV to get our desired result (Theorem VIII.2.15). The next result is proved in the same way as Lemma V.6.4.

Lemma VIII.2.7. VPV proves the $\Sigma_0^B(\mathcal{L}_{FP})$-$COMP$, $\Sigma_0^B(\mathcal{L}_{FP})$-$IND$, $\Sigma_0^B(\mathcal{L}_{FP})$-$MIN$, and $\Sigma_0^B(\mathcal{L}_{FP})$-$MAX$ axiom schemes.

Definition VIII.2.8 (Δ_i^B Formula). Let $\mathcal{T}$ be a theory over $\mathcal{L} \supseteq \mathcal{L}_A^2$. We say that a formula φ is $\Delta_i^B(\mathcal{L})$ in $\mathcal{T}$ if there is a $\Sigma_i^B(\mathcal{L})$ formula φ_1 and a $\Pi_i^B(\mathcal{L})$ formula φ_2 such that $\mathcal{T} \vdash \varphi \leftrightarrow \varphi_1$ and $\mathcal{T} \vdash \varphi \leftrightarrow \varphi_2$.

Corollary VIII.2.9. If φ is $\Delta_1^B(\mathcal{L}_{FP})$ in VPV then $VPV \vdash \varphi \leftrightarrow \varphi_0$ for some open $\mathcal{L}_{FP}$-formula φ_0.

Proof. Suppose that φ is $\Delta_1^B(\mathcal{L}_{FP})$ in VPV, and let φ_1 and φ_2 be as in the definition. Then using pairing functions we may assume that φ_1 and φ_2 each have single string quantifiers, so for some $\Sigma_0^B(\mathcal{L}_{FP})$-formulas

ψ_1, ψ_2 we have

$$\varphi_1 \equiv \exists Y \le t_1 \psi_1(\vec{x}, \vec{X}, Y),$$

$$\varphi_2 \equiv \forall Z \le t_2 \psi_2(\vec{x}, \vec{X}, Z).$$

Since $VPV \vdash \varphi_2 \supset \varphi_1$ we have

$$VPV \vdash \exists Y \exists Z, \ \psi_2(\vec{x}, \vec{X}, Z) \supset \psi_1(\vec{x}, \vec{X}, Y).$$

By Corollary VIII.2.5 there are FP-functions F and G such that

$$VPV \vdash \psi_2(\vec{x}, \vec{X}, F(\vec{x}, \vec{X})) \supset \psi_1(\vec{x}, \vec{X}, G(\vec{x}, \vec{X})).$$

Then $VPV \vdash \varphi \leftrightarrow \varphi_0$, where $\varphi_0 \equiv \psi_1(\vec{x}, \vec{X}, G(\vec{x}, \vec{X}))$. By Lemma VIII.2.3 we may assume ψ_1 is an open $\mathcal{L}_{FP}$-formula, as required.   □

VIII.2.1. Comparing VPV and V^1. Here we prove that every $\mathcal{L}_A^2$-theorem of VPV is provable in V^1. We also prove a partial converse, that every Σ_1^1 theorem of V^1 is provable in VPV. In Section VIII.6 we show evidence that not all Σ_2^B theorems of V^1 are provable in VPV.

We establish the first assertion by defining an extension $V^1(VPV)$ of both V^1 and VPV, and showing that it is conservative over V^1. We establish the partial converse by showing that every Σ_1^1 theorem of V^1 can be, provably in VPV, witnessed by functions in $\mathcal{L}_{FP}$.

DEFINITION VIII.2.10. For $i \ge 1$, the theory $V^i(VPV)$ has vocabulary $\mathcal{L}_{FP}$, and axioms the union of the axioms for V^i and for VPV.

THEOREM VIII.2.11.   (a) *Every function in $\mathcal{L}_{FP}$ is Σ_1^B-definable in V^1.*
(b) *For $i \ge 1$, every $\Sigma_i^B(\mathcal{L}_{FP})$-formula is provably equivalent in $V^1(VPV)$ to a $\Sigma_i^B(\mathcal{L}_A^2)$-formula.*
(c) *For $i \ge 1$, $V^i(VPV)$ is conservative over V^i.*

COROLLARY VIII.2.12. *For $i \ge 1$, $V^i(VPV)$ proves the $\Sigma_i^B(\mathcal{L}_{FP})$-COMP, $\Sigma_i^B(\mathcal{L}_{FP})$-IND, $\Sigma_i^B(\mathcal{L}_{FP})$-MIN, and $\Sigma_i^B(\mathcal{L}_{FP})$-MAX axiom schemes.*

PROOF. The corollary follows immediately from part (b) of the theorem, since by Corollary V.1.8 V^i proves these schemes for $\Sigma_i^B(\mathcal{L}_A^2)$-formulas.   □

PROOF OF THEOREM VIII.2.11. Part (a) of the Theorem is essentially proved in Subsection VI.2.2. Part (b) for general i follows immediately from the case $i = 1$. Now parts (b) and (c) follow from Corollary VI.3.11, where we take $\mathcal{T}_0$ to be $V^1(Row)$, or $V^i(Row)$ for part (c) (we can get rid of the function Row by Lemma V.4.27), and the extensions $\mathcal{T}_1, \mathcal{T}_2, \ldots$ are introduced by successively adding the functions in $\mathcal{L}_{FP}$ and their defining axioms. The fact that the new function introduced in $\mathcal{T}_{i+1}$ is Σ_1^1-definable in $\mathcal{T}_i$ (and even in $\mathcal{T}_0$) is proved in Section VI.2.2.   □

THEOREM VIII.2.13. *Every $\Sigma_1^1(\mathcal{L}_{FP})$ theorem of $V^1(VPV)$ is witnessed in VPV by functions in $\mathcal{L}_{FP}$.*

Proof. A slight modification of the proof of the Witnessing Theorem for V^1 given in Section VI.4.2 proves this theorem. Note that every witnessing function introduced is in FP, and, noting that VPV proves $\Sigma_0^B(\mathcal{L}_{FP})$-*IND* (by Lemma VIII.2.7), we see that VPV proves the desired sequents.   □

The following corollary is immediate from Theorem VIII.2.13.

Corollary VIII.2.14. VPV and $V^1(VPV)$ have the same $\Sigma_1^1(\mathcal{L}_{FP})$ theorems.

In particular, every Σ_1^B theorem of V^1 is provable in VPV. From this and Corollary VIII.2.6 and part (a) of Theorem VIII.2.11 we have the following:

Theorem VIII.2.15 (Σ_1^1-Definability for VPV). A function is $\Sigma_1^1(\mathcal{L}_A^2)$-definable in VPV iff it is in FP.

Finally, from Corollary VIII.2.14 and part (b) of Theorem VIII.2.11 we have

Theorem VIII.2.16. Every $\Sigma_1^B(\mathcal{L}_{FP})$-formula is provably equivalent in VPV to a $\Sigma_1^B(\mathcal{L}_A^2)$-formula.

VIII.2.2. VPV Is Conservative over VP.

Theorem VIII.2.17. VPV is a conservative extension of $\widehat{VP}$ and VP.

Proof. By definition the vocabulary and axioms of VPV include the vocabulary and axioms of $\overline{V}^0$. Also it is easy to see that F'_{MCV} can be defined from functions in $\mathcal{L}_{FAC^0}$ using limited recursion (175), (176), and its defining axiom (171) is provable in VPV from these recursion equations using induction (Lemma VIII.2.7). Therefore VPV is an extension of $\widehat{VP}$ (and VP).

We now show that VPV is conservative over $\widehat{VP}$, and hence by Theorem VIII.1.13 over VP. The functions of $\mathcal{L}_{FP}$ can be introduced successively, each one either by a Σ_0^B bit definition or by limited recursion, in terms of previously-defined functions. Thus VPV is the union of theories $\mathcal{T}_i$ satisfying

$$\mathcal{T}_0 \subset \mathcal{T}_1 \subset \mathcal{T}_2 \subset \cdots \tag{178}$$

where $\mathcal{T}_0$ is $\widehat{VP}$ and for $i > 0$ each $\mathcal{T}_i$ is obtained from $\mathcal{T}_{i-1}$ by adding the defining equation for one new function F_i. We show by induction on i that each new string function F_i is definable in $\widehat{VP}$ by a $\Sigma_0^B(\mathcal{L}_{\widehat{VP}})$-formula $\alpha_{F_i}(\vec{x}, \vec{X}, Y)$ satisfying

$$Z = F_i(\vec{x}, \vec{X}) \leftrightarrow \alpha_{F_i}(\vec{x}, \vec{X}, Z). \tag{179}$$

Also $\mathcal{T}_{i-1}$ together with (179) prove the original defining axiom for F_i in $\mathcal{T}_i$.

This shows that each $\mathcal{T}_i$ is conservative over $\mathcal{T}_{i-1}$, and hence $\bigcup \mathcal{T}_i$ is conservative over $\widehat{VP}$.

Setting $F \equiv F_i$, the formula α_F in (179) for a general string function $F(\vec{x}, \vec{X})$ is based on a family of Boolean circuits $C_F(n_1, n_2)$ which compute F, where n_1 is an upper bound on the length of each argument in $(\vec{x}, \vec{X})$ and n_2 is an upper bound on $|F(\vec{x}, \vec{X})|$. The circuit expects unary notation for the number inputs, so $n_1 \geq x_i$ for each x_i in $\vec{x}$ and $n_1 \geq |X_i|$ for each X_i in $\vec{X}$. $C_F(n_1, n_2)$ is described by a triple

$$(a_F(n_1, n_2), G_F(n_1, n_2), E_F(n_1, n_2))$$

of FAC^0 functions, using the (a, G, E) notation explained in Section VIII.1. The circuit is monotone, and is based on the "double-rail logic" described in the proof sketch of Proposition VIII.1.7, so each of the inputs in $(\vec{x}, \vec{X})$ must be presented twice; once using the expected bit string and once as the string of negations of those bits. In fact C_F expects its inputs to be the values of gate numbers $2, 3, \ldots, 2n_1 k_F + 1$, where k_F is the number of input variables in $\vec{x}, \vec{X}$ (recall that gates 0 and 1 always have the constant values 0 and 1 respectively).

Let the $\mathcal{L}_A^2$ term $t_F(n_1)$ be an upper bound on $|F(\vec{x}, \vec{X})|$, when n_1 is an upper bound on each of the input lengths $\vec{x}, \vec{X}$, and let the $\mathcal{L}_A^2$ term $g_F(n_1)$ be an upper bound on the number of "computing" gates in $C_F(n_1, n_2)$, not counting gates used for inputs and outputs. Then there are $2n_2$ output gates right after the computing gates, which store both $F(\vec{x}, \vec{X})$ and the negations of these bits.

The FAC^0 output function $Out(c, d, Y)$ extracts bits c through $d \dot{-} 1$ of Y (bits $Y(c), Y(c+1), \ldots, Y(d \dot{-} 1)$), so

$$Out(c, d, Y)(i) \leftrightarrow \big(c \leq i \wedge i + 1 \leq d \wedge Y(i)\big).$$

The FAC^0 input function $In_F(n_1, \vec{x}, \vec{X}, E) = E'$ augments the edge relation E for C_F, so E' is the same as E except edges from gates 0 and 1 to the input gates $2, 3, \ldots, 2n_1 k_F + 1$ are set so that these gates code the values $\vec{x}, \vec{X}$. Thus

$$F(\vec{x}, \vec{X}) = Out\big(c, d, F_{MCV}(a_F(n_1, n_2), G_F(n_1, n_2),$$
$$In_F(n_1, \vec{x}, \vec{X}, E_F(n_1, n_2))\big)\big) \quad (180)$$

where

$$n_1 = max\{\vec{x}, |\vec{X}|\},$$
$$n_2 = t_F(n_1),$$
$$a_F(n_1, n_2) = 1 + 2n_1 k_F + g_F(n_1) + 2n_2,$$
$$c = 2n_1 k_F + g_F(n_1) + 2,$$
$$d = c + 2n_2.$$

Notice that (180) (with the specified $\mathcal{L}_{FAC^0}$ terms for the variables other than $\vec{x}, \vec{X}$) expresses $F(\vec{x}, \vec{X})$ as a term of $\mathcal{L}_{\widehat{VP}}$.

Now the formula α_{F_i} in (179) (with $F \equiv F_i$) is given by

$$\alpha_F(\vec{x}, \vec{X}, Z) \equiv \forall j < t, Z(j) \leftrightarrow T(j) \tag{181}$$

where the term T is the RHS of equation (180), and the quantifier bound $t(\vec{x}, \vec{X})$ is $t_F(max\{\vec{x}, |\vec{X}|\})$.

It remains to show that we can define the triple a_F, G_F, E_F of $\textbf{FAC}^0$ functions specifying the circuits $C_F(n_1, n_2)$ for every function F (or f) in $\mathcal{L}_{FP}$, in such a way that $\widehat{VP}$ proves their defining axioms (in terms of earlier functions). In order to show this, we follow Definition VIII.2.1, specifying $\mathcal{L}_{FP}$. We start with $\mathcal{L}_{FAC^0}$. The initial functions in $\mathcal{L}_A^2 \cup \{pd, f_{SE}\}$ have straightforward circuits (recall that the number inputs for $+, \times, pd$ are given in unary notation). After that functions are introduced successively using parts (2) and (3) of the definitions of $\mathcal{L}_{FAC^0}$ and $\mathcal{L}_{FP}$, where part (2) introduces functions $F_{\varphi,t}$ and (in the case of $\mathcal{L}_{FAC^0}$) $f_{\varphi,t}$, where φ is a a quantifier-free formula involving previously-defined functions, and part (3) introduces the function $F_{G,H,t}$ defined from G and H by limited recursion, where G, H are previously-defined.

To illustrate how to build circuits for new functions in terms of old functions we consider a simple example of composition. Suppose

$$F(\vec{x}, \vec{X}) = H(K(\vec{x}, \vec{X})) \tag{182}$$

and suppose that we have a circuits C_K specified by (a_K, G_K, E_K) computing K, and circuits C_H specified by (a_H, G_H, E_H) computing H, where all functions are in $\textbf{FAC}^0$. Then we can combine these circuits to form C_F by placing C_K in its original position and adding $2n_1 k_K + g_K(n_1)$ to each gate number of C_H, so that the input gates of the shifted C_H coincide with the output gates of C_K. Now $\textbf{FAC}^0$ descriptor functions (a_F, G_F, E_F) for C_F are easily bit-defined by Σ_0^B formulas in terms of (a_K, G_K, E_K) and (a_H, G_H, E_H). In particular the size of C_F is given by

$$a_F(n_1, n_2) = a_K(n_1, t_K(n_1)) + a_H(t_K(n_1), n_2).$$

Note that if the composition (182) is more complicated, say $F = H(K_1, K_2)$, then to describe the circuit C_F for F the circuit C_{K_2} for K_2 needs to be shifted and the original inputs $\vec{x}, \vec{X}$ need to be copied to the input gates for the shifted C_{K_2}, and the outputs for both C_{K_1} and C_{K_2} need to be copied to the inputs for the shifted C_H. But all this is easily accomplished with $\textbf{FAC}^0$ functions, using the techniques developed in Section VIII.1.

Each new function F introduced via circuits has a definition given by (179) and (181) and hence satisfies (180). However the theory $\mathcal{T}_i$ in the sequence (178) should be able to prove the defining axioms for F as given in Definition VIII.2.2 for $\textbf{VPV}$. A simple example is when $F \equiv F_{\varphi,t}$ and

$$\varphi(z) \equiv H(K(\vec{x}, \vec{X}))(z).$$

In this case we may assume as an induction hypothesis that T_i proves (180) when F is replaced by either H or K, and since $T_i(F)$ defines F by combining the circuits for H and K as explained above we may assume that $T_i(F)$ proves (180) as it stands. We must show that $T_i(F)$ proves (182), which amounts to showing that the combined circuits for H and K compute their composition, as intended.

The main lemma needed for this and similar correctness proofs is roughly that if C and C' are two circuits, and gates $a', \ldots, b'$ of C' are the same as gates $a, \ldots, b$ of C but with their numbers shifted by a constant c, and if Y and Y' are correct assignments to C and C' respectively (i.e. (157) (page 202) holds), and if Y and Y' agree on the 'inputs' to C and C', then $Y(i) \leftrightarrow Y(i+c)$ for $a \leq i \leq b$. This kind of lemma can be proved in $\overline{V}^0$ by induction on a $\Sigma_0^B(\mathcal{L}_{FAC^0})$ formula.

In case the new function F is introduced by limited recursion, then $F \equiv F_{G,H,t}$, and F, G, H satisfy (175) and (176) (page 210). The circuit $C_F(n_1, n_2)$ for F is built by combining the circuit $C_G(n_1, n_2)$ for G with n_1 shifted copies of the circuit $C_H(max\{n_1, n_2\}, n_2)$ for H, interleaved with circuits computing the sequence of values $0, 1, \ldots, (n_1 - 1)$ for the first argument for H.

The output of C_G is $F(0)$ (i.e. $F(0, \vec{x}, \vec{X})$), and successive outputs of the shifted circuits C_H comprise the sequence $F(1), \ldots, F(n_1)$. The output gates for C_F select from this sequence of outputs the correct output $F(y)$ based on the input argument y. Thus the i-th output gate of C_F is an OR of AND-gates, where the j-th AND-gate has one input from the i-th bit of $F(j)$ and the other input from a selector gate which is on iff $y = j$. This selector gate is the AND of bit j of the input y (which is presented in unary) and bit $j + 1$ of the negated bits of y (which are also part of the input to C_F). □

COROLLARY VIII.2.18. V^1 is Σ_1^B-conservative over VP.

PROOF. This is immediate from Theorem VIII.2.17 and Corollary VIII.2.14. □

VIII.3. TV^0 and the TV^i Hierarchy

We now introduce the TV^i hierarchy, where for $i > 0$ TV^i is the two-sorted version of Buss's [20] single-sorted theory T_2^i. For $i = 0$ it turns out that $TV^0 = VP$, although the two theories have very different axioms.

For $i \geq 0$ the theory TV^i is the same as V^i, except instead of the Σ_i^B-COMP axioms we introduce the Σ_i^B "string induction" axiom scheme. Here we view a string X as the number $\sum_i X(i)2^i$, and define the string zero $\varnothing$ (empty string) and string successor function $S(X)$ as in Example V.4.17. Thus $S(X)$ has Σ_0^B-bit definition

$$S(X)(i) \leftrightarrow \varphi_S^{bit}(i, X) \tag{183}$$

where

$$\varphi_S^{bit}(i, X) \equiv i \leq |X| \wedge [(X(i) \wedge \exists j < i \neg X(j)) \vee (\neg X(i) \wedge \forall j < i X(j))].$$

DEFINITION VIII.3.1 (String Induction Axiom). If Φ is a set of formulas, then the string induction axiom scheme, denoted Φ-**SIND**, is the set of all formulas

$$[\varphi(\varnothing) \wedge \forall X(\varphi(X) \supset \varphi(S(X)))] \supset \varphi(Y) \tag{184}$$

where $\varphi(X)$ is in Φ, and may have free variables other than X.

Since we want the theories $\boldsymbol{TV}^i$ to have underlying vocabulary $\mathcal{L}_A^2$, in case Φ has vocabulary $\mathcal{L}_A^2$ we will interpret (184) as a formula over $\mathcal{L}_A^2$, using the standard method of eliminating Σ_0^B-bit-definable function symbols (Lemma V.4.15).

DEFINITION VIII.3.2. For $i \geq 0$, $\boldsymbol{TV}^i$ is the theory over $\mathcal{L}_A^2$ with axioms those of $\boldsymbol{V}^0$ together with the Σ_i^B-**SIND** scheme.

Although the induction scheme (184) has an unbounded string quantifier, it is easy to see that the theory $\boldsymbol{TV}^i$ remains the same if that quantifier $\forall X$ is replaced by the bounded quantifier $\forall X \leq |Y|$ (see Exercise III.1.16). Hence $\boldsymbol{TV}^i$ is a polynomial-bounded theory, axiomatized by Σ_{i+1}^B-formulas.

LEMMA VIII.3.3. *For $i \geq 0$, $\boldsymbol{TV}^i$ proves Σ_i^B-**IND**.*

PROOF. We are to show that $\boldsymbol{TV}^i$ proves

$$[\varphi(0) \wedge \forall x(\varphi(x) \supset \varphi(x + 1))] \supset \varphi(z)$$

where $\varphi(x)$ is Σ_i^B.
We need the following easily verified fact:

$$\boldsymbol{V}^0 \vdash (|S(X)| = |X| \vee |S(X)| = |X| + 1). \tag{185}$$

Reasoning in $\boldsymbol{TV}^i$, assume

$$[\varphi(0) \wedge \forall x(\varphi(x) \supset \varphi(x + 1))].$$

From this and (185) we conclude

$$[\psi(\varnothing) \wedge \forall X(\psi(X) \supset \psi(S(X)))]$$

where $\psi(X) \equiv \varphi(|X|)$. Hence $\psi(X_z)$ follows by Σ_i^B-**SIND**, where X_z is a string with length z. Hence $\varphi(z)$.   □

THEOREM VIII.3.4. *For $i \geq 0$, $V^i \subseteq TV^i$.*

PROOF. We generalize Definition VI.4.6 to define $\tilde{V}^i$ to be $V^0 + \Sigma^B_i$-**IND**. The proof of Theorem VI.4.8 easily generalizes to show $V^i = \tilde{V}^i$. Hence the theorem follows from Lemma VIII.3.3. □

Just as V^i proves the number minimization and maximization axioms for Σ^B_i-formulas (Corollary V.1.8), TV^i proves the stronger string minimization and maximization axioms for Σ^B_i-formulas. First, we define the ordering relation for strings.

DEFINITION VIII.3.5 (String Ordering). The string relation $X \leq Y$ has defining axiom

$$X \leq Y \leftrightarrow [X = Y \vee (|X| \leq |Y| \wedge \exists z \leq |Y| \, (Y(z) \wedge \neg X(z) \wedge$$
$$\forall u \leq |Y|, z < u \supset (X(u) \supset Y(u))))]. \quad (186)$$

Often our vocabularies do not contain extra relation symbols outside $\mathcal{L}^2_A$. Thus the syntactic formula $X \leq Y$ will be an abbreviation for the RHS of Equation (186). Also, $X < Y$ stands for $X \leq Y \wedge \neg(X = Y)$.

EXERCISE VIII.3.6. Show that the following are theorems of V^0 (where $\varnothing, S, +$ are defined in Example V.4.17):

(a) $X \leq Y \vee Y \leq X$ ($X \leq Y$ is a total order).
(b) $(X \leq Y \wedge Y \leq X) \supset X = Y$ ($X \leq Y$ is irreflexive).
(c) $\varnothing \leq X$.
(d) $X \leq Y \leftrightarrow X + Z \leq Y + Z$.
(e) $X < Y \supset S(X) \leq Y$.

For a string term T, we define $\exists X \leq T \, \varphi(X)$ as an abbreviation for $\exists X (X \leq T \wedge \varphi(X))$. Similarly, $\forall X \leq T \, \varphi(X)$ is an abbreviation for $\forall X (X \leq T \supset \varphi(X))$. Note that the bounding term T is for the *value* of X, while the bounding term t in $\exists X \leq t \ldots$ or $\forall X \leq t \ldots$ is for the length of X (Definition IV.3.1).

DEFINITION VIII.3.7 (String Minimization & Maximization Axioms). The string minimization axiom scheme for Φ, denoted Φ-**SMIN**, is

$$\varphi(Y) \supset \exists X \leq Y, \varphi(X) \wedge \neg \exists Z < X \varphi(Z)$$

where φ is a formula in Φ. Similarly the string maximization axioms scheme for Φ, denoted Φ-**SMAX**, is

$$\varphi(\varnothing) \supset \exists X \leq Y, \varphi(X) \wedge \neg \exists Z \leq Y(X < Z \wedge \varphi(Z))$$

where φ is a formula in Φ.

THEOREM VIII.3.8. *For $i \geq 0$, TV^i proves the Σ^B_i-**SMIN** and Σ^B_i-**SMAX** axioms.*

Proof. To prove Σ_i^B-*SMAX*, let $\varphi(X)$ be a Σ_i^B-formula. Let $\varphi'(X)$ be the Σ_i^B-formula obtained by taking a prenex form of

$$X \leq Y \supset \exists U \leq Y (X \leq U \wedge \varphi(U)).$$

Then the *SMAX* axiom for $\varphi(X)$ follows from the *SIND* axiom (184) applied to $\varphi'(X)$.

The proof of Σ_i^B-*SMIN* is similar, but uses the binary subtraction function $Z \dot- Y$. □

Exercise VIII.3.9. Show that the limited subtraction function for string $Z \dot- Y$ is Σ_0^B-bit-definable, where the intended meaning of $Z \dot- Y$ is $\varnothing$ if $Z \leq Y$, and $(Z \dot- Y) + Y = Z$ otherwise.

We now concentrate on TV^0.

Theorem VIII.3.10. $TV^0 = VP$.

Proof. Subsection VIII.3.1 shows that $TV^0 \subset VPV$, and by Theorem VIII.2.17 VPV is conservative over VP. Hence $TV^0 \subseteq VP$. The reverse inclusion is shown in Subsection VIII.3.2. □

By Theorem VIII.3.10 we know the properties of VP proved in Section VIII.1 also hold for TV^0. In particular TV^0 is finitely axiomatizable, the functions Σ_1^B-definable in TV^0 comprise FP, and by Corollary VIII.2.18 V^1 is Σ_1^B-conservative over TV^0.

In the following corollary, $TV^i(VPV)$ is defined analogously to $V^i(VPV)$ in Definition VIII.2.10, namely it has the vocabulary of VPV and the axioms are the union of the axioms for TV^i and VPV. (See also Theorem VIII.2.11 and Corollary VIII.2.12).

Corollary VIII.3.11. *For $i \geq 0$, $TV^i(VPV)$ is a conservative extension of TV^i.*

Proof. For $i = 0$ this follows from the fact that VPV is a conservative extension of TV^0 (Theorems VIII.2.17 and VIII.3.10). For $i \geq 1$ we know $V^1 \subseteq TV^i$, and hence TV^i Σ_1^B-defines all functions in $\mathcal{L}_{FP}$, and also TV^i proves Σ_1^B-*REPL* by Corollary VI.3.8. Therefore the corollary follows from Corollary VI.3.11. □

VIII.3.1. $TV^0 \subseteq VPV$. In this subsection we use the string addition function $X + Y$ introduced in Chapter V and use some of its simple properties stated in Exercise V.4.19. We also need the string relation $X \leq Y$ (Definition VIII.3.5) and the string function $POW2(x)$ defined below. The intended meaning of $POW2(x)$ is such that (see Notation on page 85) $bin(POW2(x)) = 2^x$.

Example VIII.3.12. The string function $POW2(x)$, also denoted by $\{x\}$, has bit defining axiom

$$POW2(x)(i) \leftrightarrow i = x.$$

EXERCISE VIII.3.13. Show that $\overline{V}^0$ proves the following:

$$X + POW2(0) = S(X),$$
$$X < POW2(|X|),$$
$$POW2(i) + POW2(i) = POW2(i + 1).$$

The following theorem suffices to prove $TV^0 \subset VPV$. That VPV proves the open string induction axioms may seem surprising, since unwinding the induction requires exponentially many steps.

THEOREM VIII.3.14. VPV proves the $\Sigma_0^B(\mathcal{L}_{FP})$-$SIND$ axioms.

PROOF. By Lemma VIII.2.3 we may assume that $\varphi(X)$ in (184) is an open $\mathcal{L}_{FP}$-formula. Let $\vec{y}, \vec{Y}$ be a list of the parameters in $\varphi(X)$. We use binary search to define in VPV an $\mathcal{L}_{FP}$ function $G(\vec{y}, \vec{Y}, X)$ such that VPV proves

$$(\varphi(\varnothing) \wedge \neg\varphi(X)) \supset (\varphi(G(\vec{y}, \vec{Y}, X)) \wedge \neg\varphi(S(G(\vec{y}, \vec{Y}, X)))) \quad (187)$$

from which (184) follows immediately.

In more detail, we use the string functions $X + Y$ and $POW2(x)$ and the string relation $X \leq Y$ defined above.

In the following we suppress mention of the parameters $\vec{y}, \vec{Y}$.

Define the formula

$$\varphi'(X, Z) \equiv \varphi(Z) \wedge Z \leq X.$$

Now we use limited recursion (175), (176) (page 210) to define in VPV the binary search function $H(i, X)$, whose value is the left end of the interval $[A, B]$ of length $POW2(|X| \doteq i)$ satisfying $\varphi'(X, A) \wedge \neg\varphi'(X, B)$. (Recall the number function $x \doteq y$ (limited subtraction), Section III.3.3).

Let $n = |X|$.

$$H(0, X) = \varnothing,$$

$$H(i + 1, X) = \begin{cases} H(i, X) & \text{if } \neg\varphi'(X, H(i, X) + POW2(n \doteq (i + 1))), \\ H(i, X) + POW2(n \doteq (i + 1)) & \text{otherwise.} \end{cases}$$

We can use $|X|$ as a bounding term to limit this recursion. Now define

$$G(X) = H(|X|, X).$$

The following two formulas can be proved in VPV by induction on i (Lemma VIII.2.7), using Exercises V.4.19 and VIII.3.13. The first formula justifies $|X|$ as a length bound for the recursion.

$$X \neq \varnothing \supset (H(i, X) + POW2(0)) \leq X,$$

$$(\varphi(\varnothing) \wedge \neg\varphi(X) \wedge i \leq n) \supset$$
$$(\varphi'(X, H(i, X)) \wedge \neg\varphi'(X, H(i, X) + POW2(n \doteq i))).$$

Then (187) follows from these two formulas and $X + POW2(0) = S(X)$ (Exercise VIII.3.13). $\qquad\qquad\qquad\qquad\qquad\qquad\qquad\qquad\qquad\qquad$ □

Recall the notion of a Δ_i^B formula in a theory (Definition VIII.2.8).

DEFINITION VIII.3.15. Let $\mathcal{T}$ be a theory with vocabulary $\mathcal{L}$. Let AX denote any of the axiom schemes $COMP$, IND, $SIND$, etc. We say that $\mathcal{T}$ proves Δ_i^B-AX if for any $\Delta_i^B(\mathcal{L})$ formula φ in $\mathcal{T}$, $\mathcal{T}$ proves the AX axiom for φ.

From Theorem VIII.3.14 and Corollary VIII.2.9 we have

COROLLARY VIII.3.16. VPV proves Δ_1^B-$SIND$.

VIII.3.2. Bit Recursion. In order to show that $VP \subseteq TV^0$ we introduce a bit-recursion scheme and show that it is provable in TV^0.

For each formula $\varphi(i, X)$ (possibly with other free variables) we define a formula $\varphi^{rec}(y, X)$ which says that each bit i of X is defined in terms of the preceding bits of X using φ. That is, using the notation $X^{<i}$ for $Cut(i, X)$ (see (97) on page 139)

$$\varphi^{rec}(y, X) \equiv \forall i < y(X(i) \leftrightarrow \varphi(i, X^{<i})).$$

In case $\varphi(i, X)$ is an $\mathcal{L}_A^2$-formula we can interpret $\varphi^{rec}(y, X)$ as an $\mathcal{L}_A^2$-formula by eliminating occurrences of $Cut(i, X)$ using the standard method of eliminating Σ_0^B-bit-definable function symbols (Lemma V.4.15).

If $\varphi(i, X)$ is in Σ_0^B it is easy to see that V^0 can use induction on y to prove that the condition $\varphi^{rec}(y, X)$ uniquely determines bits $X(0), \ldots, X(y-1)$ of X.

DEFINITION VIII.3.17. If Φ is a set of formulas, then the bit recursion axiom scheme, denoted Φ-BIT-REC, is the set of formulas

$$\exists X \varphi^{rec}(y, X) \qquad\qquad\qquad\qquad (188)$$

where $\varphi(i, X)$ is in Φ, and may have free variables other than X.

We will show that $TV^0 = V^0 + \Sigma_0^B$-BIT-REC.

THEOREM VIII.3.18. TV^0 proves the Σ_0^B-BIT-REC-scheme.

PROOF. We use Σ_0^B-$SMAX$ to prove the existence of X in (188). Informally, imagine computing the bits $X(0), \ldots, X(y-1)$ of X in that order. Suppose that false negative is allowed, but there is no false positive. That is, we consider strings Y that satisfy

$$\forall i < y, \ Y(i) \supset \varphi(i, Y^{<i}).$$

The idea is that the maximal string Y guaranteed by $SMAX$ cannot have any false negative bit, and thus must be the correct string.

To actually use the $SMAX$ principle we need a twist in the above argument. This is because we compute X in (188) from bit 0, while string comparison starts with high order bits. Thus, let the string reversal function $Rev(y, X)$ have bit-defining axiom

$$Rev(y, X)(i) \leftrightarrow i < y \wedge X(y \dot{-} i \dot{-} 1)$$

where $\dot{-}$ is limited subtraction (Section III.3.3). Then $Rev(y, X)$ is the reverse of the string $X(0) \ldots X(y - 1)$.

Let $\varphi'(y, Y)$ be the formula

$$\forall i < y, \ Rev(y, Y)(i) \supset \varphi(i, (Rev(y, Y))^{<i}). \tag{189}$$

We can tacitly assume that $\varphi'(y, Y)$ is Σ_0^B (by Lemma V.4.15). It is easy to see that $\varphi'(y, \varnothing)$. Thus, by Σ_0^B-*SMAX*, there is a maximal string $X' \leq POW2(y)$ that satisfies (189). It is also easy to show (in V^0) that X' in fact satisfies

$$\forall i < y, \ Rev(y, X')(i) \leftrightarrow \varphi(i, (Rev(y, X'))^{<i}).$$

As a result, the string $X = Rev(y, X')$ satisfies (188). $\square$

Lemma VIII.3.19. $VP \subseteq V^0 + \Sigma_0^B$-*BIT-REC*

Proof. Observe that the axiom *MCV* for VP (Definition VIII.1.1) is an instance of Σ_0^B-*BIT-REC*. $\square$

This lemma completes the proof of Theorem VIII.3.10, showing that $VP = TV^0$.

Corollary VIII.3.20. TV^0 *proves its* Δ_1^B-*SIND axioms.* V^1 *proves its* Δ_1^B-*SIND axioms.*

Proof. The first sentence follows from $VP = TV^0$ and Corollary VIII.3.16. The second sentence follows from the first, since by Corollary VIII.2.18 any Σ_1^B-formula that is Δ_1^B in V^1 is also Δ_1^B in TV^0. $\square$

VIII.4. The Theory V^1-*HORN*

This section will not be needed for any later results, but it is interesting in that gives more evidence for the robustness of VP by giving yet another axiomatization.

The theory V^1-*HORN* [43] is the same as VP and TV^0 but presented with very different axioms. The of ideal of V^1-*HORN* comes from a theorem of Grädel in descriptive complexity theory, characterizing the class P as the sets of finite models of certain second-order formulas. We will formulate Grädel's theorem as a representation theorem over $\mathcal{L}_A^2$. We start with some definitions and examples.

Definition VIII.4.1. A *Horn formula* is a propositional formula in conjunctive normal form such that each clause (i.e. conjunct) is a *Horn clause*, i.e. it contains at most one positive occurrence of a variable.

Horn formulas are important because the satisfiability problem Horn-Sat (given a Horn formula, determine whether it is satisfiable) is complete for **P**. A polytime algorithm for HornSat can be described as follows.

HornSat Algorithm. To test whether a given Horn formula A is satisfiable, initialize a truth assignment τ by assigning $\bot$ to each atom of A. Now repeat the following until satisfiability is determined: If τ satisfies all clauses of A then decide that A is satisfiable. Otherwise select a clause C of A not satisfied by τ. If C has no positive occurrence of any atom then decide that A is unsatisfiable. Otherwise C has a unique positive occurrence of some atom p, in which case flip the value of τ on p from $\bot$ to $\top$.

EXERCISE VIII.4.2. Show that the above algorithm runs in polynomial time and correctly determines whether a given Horn formula A is satisfiable.

The HornSat algorithm suggests that a Horn clause $(p \vee \neg q_1 \vee \cdots \vee \neg q_k)$ can be written as an assignment statement

$$p \leftarrow (q_1 \wedge \cdots \wedge q_k).$$

(In fact some logic-based programming languages such as Prolog use this idea.)

We now indicate why HornSat is complete for **P**. It suffices to show that a known complete problem CVP (Circuit Value Problem) can be reduced to HornSat. Given a Boolean circuit C with binary gates $\wedge, \vee$ and unary gates $\neg$, and given a value $v(x) \in \{0,1\}$ for each input x to C, we want to find a Horn formula A which is satisfiable iff C has output 1 for the given inputs $v(x)$. The formula A uses double rail logic (see the proof of Proposition VIII.1.7) to evaluate C: for each gate and each input x of C the formula has two atoms x^+ and x^- asserting that the gate or input is 1 or 0, respectively. For each such x, A has a Horn clause $(\neg x^+ \vee \neg x^-)$ to insure that not both atoms are true. For each input x, A has a unit clause x^+ if $v(x) = 1$ and unit clause x^- if $v(x) = 0$. For each gate in C, A has up to three Horn clauses which assert that the output of the gate has the appropriate value with respect to its inputs. For example, if x is the $\vee$ of inputs y, z, then the clauses are

$$(x^+ \leftarrow y^+) \wedge (x^+ \leftarrow z^+) \wedge (x^- \leftarrow (y^- \wedge z^-)). \tag{190}$$

Finally A has the unit clause x_{out}^+, where x_{out} is the output gate.

It turns out that the collection of propositional Horn formulas that correspond to a given polytime problem can be represented by single Σ_1^B formula as follows.

DEFINITION VIII.4.3. A Σ_1^B-***Horn*** formula is an $\mathcal{L}_A^2$-formula of the form

$$\varphi \equiv \exists Z_1 \ldots \exists Z_k \forall y_1 \leq t_1 \ldots \forall y_m \leq t_m \psi \tag{191}$$

where $k, m \geq 0$ and ψ is quantifier-free in conjunctive normal form and each clause contains at most one positive occurrence of a literal of the form $Z_i(t)$. No term of the form $|Z_i|$ may occur in φ, although φ may contain free string variables X (and free number variables) with no restriction on occurrences of $|X|$, and any clause of ψ may contain any number of positive (or negative) literals of the form $X(t)$.

We will show that Σ_1^B-Horn formulas represent polynomial time relations in their free variables.

EXAMPLE VIII.4.4 ($Parity_{Horn}(X)$). This is a Σ_1^B-*Horn*-formula which holds iff the string X contains an odd number of 1's. $Parity_{Horn}(X)$ encodes a dynamic-programming algorithm for computing the parity of X: $Z_{odd}(i)$ is true (and $Z_{even}(i)$ is false) iff the prefix of X of length i contains an odd number of 1's.

$$\exists Z_{even} \exists Z_{odd} \forall i < |X| \, Z_{even}(0) \wedge \neg Z_{odd}(0) \wedge Z_{odd}(|X|) \wedge$$

$$(\neg Z_{even}(i+1) \vee \neg Z_{odd}(i+1)) \wedge (\neg Z_{even}(i) \vee \neg X(i) \vee Z_{odd}(i+1)) \wedge$$

$$(\neg Z_{odd}(i) \vee \neg X(i) \vee Z_{even}(i+1)) \wedge (\neg Z_{even}(i) \vee X(i) \vee Z_{even}(i+1)) \wedge$$

$$(\neg Z_{odd}(i) \vee X(i) \vee Z_{odd}(i+1)).$$

EXERCISE VIII.4.5. Prove that $Parity_{Horn}(X)$ has the stated property.

In Section IV.3.2 we showed how the complexity classes AC^0 and the members Σ_i^P of the polynomial hierarchy can be characterized by representation theorems involving the formula classes Σ_i^B. Now we state a similar theorem characterizing P.

THEOREM VIII.4.6 (Grädel). *A relation $R(\vec{x}, \vec{X})$ is polynomial time iff it is represented by some Σ_1^B-Horn-formula.*

PROOF SKETCH. ($\Longleftarrow$) Suppose that the formula $\varphi(\vec{x}, \vec{X})$ has the form (191). We outline an algorithm that runs in time polynomial in $(\vec{x}, |\vec{X}|)$ which, given values for $\vec{x}, \vec{X}$, determines whether $\varphi(\vec{x}, \vec{X})$ holds (in the standard model). First note that once values for $\vec{x}, \vec{X}$ are given, the bounding terms $t_i = t_i(\vec{x}, \vec{X})$ can be evaluated to numbers bounded by polynomials in $(\vec{x}, |\vec{X}|)$. We expand the quantifier prefix $\forall y_1 \leq t_1 \ldots \forall y_m \leq t_m$ by giving all possible m-tuples of values $(y_1, \ldots, y_m)$ satisfying the bounding terms, and form the conjunction $\Psi(Z_1, \ldots, Z_k)$ of all instances $\psi(\vec{y})$, as $\vec{y}$ ranges over all these tuples. (Note that the number of such tuples is bounded by a polynomial in $(\vec{x}, |\vec{X}|)$.)

Then $\Psi(Z_1, \ldots, Z_k)$ can be made into a propositional conjunctive normal form formula Ψ' involving only literals of the form $Z_i(j)$ and $\neg Z_i(j)$ for specific numbers j, since all terms and all other variables in ψ have been evaluated. (Here it is important that we have disallowed occurrences of $|Z_i|$ in φ.) The arguments j in $Z_i(j)$ and $\neg Z_i(j)$ are values of terms

t, for each $Z_i(t)$ or $\neg Z_i(t)$ that is a literal in the original formula ψ. Let B be an upper bound on the possible values of j (so B is a polynomial in $(\vec{x}, \vec{X})$). Then Ψ' is a Horn formula whose propositional variables are all in the set $\{Z_i(j) : i \leq k, j \leq B\}$. Thus the problem of checking for the existence of $Z_1, \ldots, Z_k$ reduces to the polytime HornSat problem of deciding whether Ψ' is satisfiable.

($\Longrightarrow$) Let $R(\vec{x}, \vec{X})$ be a polytime relation and let M be a deterministic polytime Turing machine that recognizes R in time $t(\vec{x}, \vec{X})$. By choosing t large enough, the entire computation of M on input $\vec{x}, \vec{X}$ can be represented (using the pairing function) by an array $Z(i, j)$ with t rows and columns, where the i-th row specifies the tape configuration at time i. Thus $R(\vec{x}, \vec{X})$ is represented by the Σ_1^B-*Horn*-formula

$$\exists Z \exists \tilde{Z} \forall i \leq t \forall j \leq t \psi(i, j, \vec{x}, \vec{X}, Z, \tilde{Z}).$$

Here the variable $\tilde{Z}$ is forced to be $\neg Z$ in the same way that Z_{even} and Z_{odd} are forced to be complementary in the parity example above. The formula ψ satisfies the conditions in Definition VIII.4.3 and each clause specifies a local condition on the computation. □

DEFINITION VIII.4.7. The theory V^1-*HORN* has vocabulary $\mathcal{L}_A^2$ and axioms those of V^0 together with Σ_1^B-*Horn-COMP*.

The original definition of V^1-*HORN* in [43] was a little different. Recall that V^0 has axioms 2-*BASIC* together with Σ_0^B-*COMP* (Definition V.1.3). The original definition was essentially V^1-*HORN* = 2-*BASIC* + Σ_1^B-*Horn-COMP*. It was shown with some effort that V^1-*HORN* proves Σ_0^B-*COMP*, so the two definitions are equivalent.

The next theorem follows from results in [43].

THEOREM VIII.4.8. V^1-*HORN* = *VP*.

PROOF SKETCH. V^1-*HORN* ⊆ *VP*: It suffices to show

$$VP \vdash \Sigma_1^B\text{-}Horn\text{-}COMP.$$

Since *VPV* is a conservative extension of *VP* (Theorem VIII.2.17), it suffices to show $VPV \vdash \Sigma_1^B$-*Horn-COMP*. Since $VPV \vdash \Sigma_0^B(\mathcal{L}_{FP})$-*COMP* (Lemma VIII.2.7), it suffices to show that for every Σ_1^B-*Horn*-formula φ there is a $\Sigma_0^B(\mathcal{L}_{FP})$ formula φ' such that $VPV \vdash \varphi \leftrightarrow \varphi'$.

So let φ be a Σ_1^B-*Horn*-formula as in (191), where we write $\psi(Z_1, \ldots, Z_k)$ simply as ψ, and let $\vec{x}, \vec{X}$ be the free variables in φ. The idea is to find a "witnessing function" $F_i(\vec{x}, \vec{X})$ in $\mathcal{L}_{FP}$ for each Z_i such that *VPV* proves $\varphi \leftrightarrow \varphi'$, where

$$\varphi' \equiv \forall y_1 \leq t_1 \ldots \forall y_m \leq t_m \psi(F_1(\vec{x}, \vec{X}), \ldots, F_k(\vec{x}, \vec{X})).$$

To define F_i we refer to the direction $\Longleftarrow$ in the proof of Theorem VIII.4.6. There the algorithm to evaluate $\varphi(\vec{x}, \vec{X})$ computes a propositional Horn formula Ψ' whose propositional variables have the form $Z_i(j)$, and then applies the HornSat algorithm to determine whether Ψ' is satisfiable. This algorithm computes a truth assignment τ to the atoms $Z_i(j)$ of Ψ' such that Ψ' is satisfiable iff τ satisfies Ψ'. Thus it suffices to define the string $F_i(\vec{x}, \vec{X})$ to be the array of truth values that τ gives to Z_i. That is, the the bit definition of each F_i is

$$F_i(\vec{x}, \vec{X})(j) \leftrightarrow j \leq B \wedge \tau(Z_i(j)).$$

The algorithm outlined to compute F_i is clearly polytime and hence corresponds to some function in *FP*. The missing details in the proof are to show that *VPV* proves the correctness of the algorithm; i.e. $VPV \vdash \varphi \supset \varphi'$.

$VP \subseteq V^1$-*HORN*: By Definition VIII.1.1 it suffices to show that

$$V^1\text{-}\textbf{\textit{HORN}} \vdash MCV.$$

We indicated earlier (190) how propositional Horn clauses can be used to evaluate circuit gates. Now we show how to use a Σ_1^B-*Horn* formula to evaluate the circuit C described by parameters a, G, E as described in Section VIII.1. In essence, the new atoms x^+, x^-, etc. in (190) are encoded by the (existentially quantified) string variables Z in the Σ_1^B-*Horn* formula. Note that the algorithm outlined on page 224 is for circuits with binary gates, while here the circuit may have unbounded fan-ins.

Thus we want to define an array $Z(x)$ (and its negation $\tilde{Z}(x)$) to evaluate gate x in C. We will put in the clause

$$\neg Z(x) \vee \neg \tilde{Z}(x)$$

to make sure that not both are true. For gates 0 and 1 (with constant values 0 and 1 respectively) we put in the four clauses

$$\tilde{Z}(0), \qquad \neg Z(0), \qquad \neg \tilde{Z}(1), \qquad Z(1). \qquad (192)$$

Next, consider gate x. Suppose that this is an $\vee$-gate, i.e., $\neg G(x)$ holds. Then we need several clauses. The first is

$$\big(\neg G(x) \wedge y < x \wedge E(y, x) \wedge Z(y)\big) \supset Z(x)$$

which assures that $Z(x)$ holds if at least one of the inputs to gate x is 1. To ensure that $\tilde{Z}(x)$ holds if all inputs to gate x are 0 is more involved. In fact, we formalize a simple algorithm that runs through the inputs of gate x to check if all of them are 0. We use a string variable P, where $P(x, y)$ is intended to mean that all gates u which are input to x, where $u < y$, output 0. The formalization is as follows:

$$P(x, 0),$$
$$(P(x, y) \wedge \neg E(y, x)) \supset P(x, y + 1),$$
$$(P(x, y) \wedge \tilde{Z}(y)) \supset P(x, y + 1),$$
$$(P(x, x) \wedge E(y, x)) \supset \tilde{Z}(y),$$
$$(\neg G(x) \wedge P(x, x)) \supset \tilde{Z}(x).$$

Let $\psi_\vee$ denote the set of the six clauses described above for the case where the gate (x) is an $\vee$-gate. Also, let ψ_I be the set of clauses in (192). The set $\psi_\wedge$ of clauses for handling the case where (x) is an $\wedge$-gate is similar to $\psi_\vee$, using an extra variable Q instead of P.

EXERCISE VIII.4.9. Give the six clauses of $\psi_\wedge$.

Now we can show in V^0 that a string Y that is computed by

$$Y(i) \leftrightarrow \exists Z \exists \tilde{Z} \exists P \exists Q \forall x < a \forall y < a, (\neg Z(x) \vee \neg \tilde{Z}(x)) \wedge$$
$$\psi_I \wedge \psi_\wedge \wedge \psi_\vee \wedge Z(i) \quad (193)$$

(for $i < a$) satisfies $\delta_{MCV}(a, G, E, Y)$. The following exercise is helpful.

EXERCISE VIII.4.10. Let the string variables $Z, \tilde{Z}, P, Q$ satisfy the RHS of (193), and Y' satisfy $\delta_{MCV}(a, G, E, Y')$. Show by induction on i that for $i < a$,

$$\neg Z(i) \supset \neg Y'(i) \quad \text{and} \quad \neg \tilde{Z}(i) \supset Y'(i).$$

EXERCISE VIII.4.11. Prove by number induction that the string Y described above satisfies the recursion in $\delta_{MCV}(a, G, E, Y)$.

Finally, the existence of Y in MCV follows from the existence of Y that satisfies (193), and the latter follows from Σ_1^B-*Horn-COMP*. This completes the proof that $VP \subseteq V^1$-*HORN*.   □

VIII.5. TV^1 and Polynomial Local Search

It follows from Theorem VIII.3.4 that $V^1 \subseteq TV^1$, and hence TV^1 can Σ_1^B-define all polynomial time functions. But there is no known nice characterization of the set of *all* functions Σ_1^B-definable in TV^1. There is however a nice characterization of the set of all *search problems* Σ_1^B-definable in TV^1.

A search problem is essentially a multivalued function, and the associated computational problem is to find one of the possible values. Here we are concerned with *total* search problems, which means that the set of possible values is always nonempty. We present a search problem by its graph. The search problem is definable in a theory if the theory proves its totality. In the two-sorted setting the set of possible values is a set of strings.

DEFINITION VIII.5.1. A *search problem* Q_R is a multivalued function with graph $R(\vec{x}, \vec{X}, Z)$, so

$$Q_R(\vec{x}, \vec{X}) = \{Z : R(\vec{x}, \vec{X}, Z)\}.$$

Here the arity of either or both of $\vec{x}, \vec{X}$ may be zero. The search problem is *total* if the set $Q_R(\vec{x}, \vec{X})$ is non-empty for all $\vec{x}, \vec{X}$. The search problem is a *function problem* if $|Q_R(\vec{x}, \vec{X})| = 1$ for all $\vec{x}, \vec{X}$. A function $F(\vec{x}, \vec{X})$ *solves* Q_R if

$$F(\vec{x}, \vec{X}) \in Q_R(\vec{x}, \vec{X})$$

for all $\vec{x}, \vec{X}$.

Here we will be concerned only with total search problems. The following notion of reduction preserves totality.

DEFINITION VIII.5.2. A search problem Q_{R_1} is many-one reducible to a search problem Q_{R_2}, written $Q_{R_1} \leq_{AC^0} Q_{R_2}$, provided there are FAC^0-functions $\vec{f}, \vec{F}, G$ such that $G(\vec{x}, \vec{X}, Z) \in Q_{R_1}(\vec{x}, \vec{X})$ for all $Z \in Q_{R_2}(\vec{f}(\vec{x}, \vec{X}), \vec{F}(\vec{x}, \vec{X}))$.

We note that the usual definition states the weaker requirement that $\vec{f}, \vec{F}, G$ are polytime functions. However experience shows that when reductions are needed they can be made to meet our stronger requirement.

EXERCISE VIII.5.3. Show that $\leq_{AC^0}$ is a transitive relation. Also show that if $Q_{R_1} \leq_{AC^0} Q_{R_2}$ and Q_{R_2} is solvable by a polytime function, then Q_{R_1} is solvable by a polytime function.

Local search is a method of finding a local maximum of a function by starting at a point in the domain of the function, finding a neighbor of the point that increases the value of the function, and continuing this process until no such neighbor exists. Polynomial Local Search (**PLS**) formalizes this as a search problem in case the function is polytime and suitable neighboring points can be found in polynomial time. Recall that $\varnothing$ denotes the empty set (Example V.4.17).

DEFINITION VIII.5.4. A **PLS** problem Q is specified by the following:

1) A polytime relation $\varphi_Q(\vec{x}, \vec{X}, Z)$ and an $\mathcal{L}_A^2$-term $t(\vec{x}, \vec{X})$ satisfying the two conditions

$$\varphi_Q(\vec{x}, \vec{X}, \varnothing),$$

$$\varphi_Q(\vec{x}, \vec{X}, Z) \supset |Z| \leq t(\vec{x}, \vec{X}).$$

$(\{Z : \varphi_Q(\vec{x}, \vec{X}, Z)\}$ is the set of *candidate solutions* for problem instance $(\vec{x}, \vec{X})$.)

2) Polytime string functions $P_Q(\vec{x}, \vec{X}, Z)$ and $N_Q(\vec{x}, \vec{X}, Z)$ satisfying the two conditions

$$\varphi_Q(\vec{x}, \vec{X}, Z) \supset \varphi_Q(\vec{x}, \vec{X}, N_Q(\vec{x}, \vec{X}, Z)),$$

$$N_Q(\vec{x}, \vec{X}, Z) \neq Z \supset P_Q(\vec{x}, \vec{X}, Z) < P_Q(\vec{x}, \vec{X}, N_Q(\vec{x}, \vec{X}, Z)).$$

(N_Q is a heuristic for finding a neighbor of Z which increases the profit P_Q. $N_Q(\vec{x}, \vec{X}, Z) = Z$ is taken to mean that Z is locally optimal. Recall that $X < Y$ stands for $X \leq Y \wedge \neg X = Y$, where $X \leq Y$ is defined in Definition VIII.3.5.)

Then

$$Q(\vec{x}, \vec{X}) = \{Z : \varphi_Q(\vec{x}, \vec{X}, Z) \wedge N_Q(\vec{x}, \vec{X}, Z) = Z\}. \tag{194}$$

The problem Q is an AC^0-**PLS** problem if φ_Q, N_Q, P_Q are AC^0-relations and functions.

It is easy to see that a **PLS** problem is a total search problem. For fixed $\vec{x}, \vec{X}$, the set of candidate solutions Z (those satisfying $\varphi_Q(\vec{x}, \vec{X}, Z)$) is nonempty and bounded. Thus given $\vec{x}, \vec{X}$, any candidate solution Z that maximizes the profit $P_Q(\vec{x}, \vec{X}, Z)$ is a member of $Q(\vec{x}, \vec{X})$.

We will concentrate on a subclass of **PLS** called **ITERATION**, which is complete for **PLS**.

Definition VIII.5.5. An **ITERATION** problem $Q = Q_F$ is specified by a polytime function $F(\vec{x}, \vec{X}, Z)$ and a bounding term $t(\vec{x}, \vec{X})$. The graph relation R is specified by a formula $\psi_F(\vec{x}, \vec{X}, Z)$ which is (suppressing the parameters $\vec{x}, \vec{X}$):

$$\psi_F(Z) \equiv (Z = \varnothing \wedge F(\varnothing) = \varnothing) \vee$$

$$|Z| \leq t \wedge Z < F(Z) \wedge (t < |F(Z)| \vee F(F(Z)) \leq F(Z)). \tag{195}$$

Then

$$Q_F(\vec{x}, \vec{X}) = \{Z : \psi_F(\vec{x}, \vec{X}, Z)\}. \tag{196}$$

The problem Q_F is an AC^0-**ITERATION** problem if F is an AC^0-function.

To see that Q_F is a total search problem, note that the largest $Z \leq t$ such that $(Z = \varnothing \vee Z < F(Z))$ is always a solution. The next exercise shows that every polytime function can be interpreted as an AC^0-**ITERATION** problem with exactly one solution.

Exercise VIII.5.6. Show that for each polytime function $G(\vec{x}, \vec{X})$ there is an AC^0 function $F(\vec{x}, \vec{X}, Z)$ and an $\mathcal{L}_A^2$ term $t(\vec{x}, \vec{X})$ so that provably in **VPV**, the only solution to Q_F is $G(\vec{x}, \vec{X})$. (Hint: Consider the computation of a Turing machine that computes G.)

LEMMA VIII.5.7. *Every **ITERATION** problem is a **PLS** problem.*

PROOF. Let Q_F be an **ITERATION** problem as above. Then Q_F can be specified as a **PLS** problem using the following definitions:

$$\varphi_Q(Z) \equiv |Z| \leq t \wedge (Z = \varnothing \vee Z < F(Z)),$$

$$P_Q(Z) = Z,$$

$$N_Q(Z) = \begin{cases} F(Z) \text{ if } |F(Z)| \leq t \text{ and } Z < F(Z) < F(F(Z)), \\ Z \text{ otherwise.} \end{cases}$$

Then (196) follows from (194). Notice that if Q_F is an AC^0-**ITERATION** problem then the corresponding problem is an AC^0-**PLS** problem. □

THEOREM VIII.5.8. *Every **PLS** problem is many-one reducible to some **ITERATION** problem. Every AC^0-**PLS** problems is many-one reducible to some AC^0-**ITERATION** problem.*

PROOF. Let Q be a **PLS** problem and let t, φ_Q, P_Q, N_Q be as in Definition VIII.5.4.

We give the following Σ_0^B-definition of the concatenation function $X *_z Y$, which is the first z bits of X followed by Y:

$$(X *_z Y)(i) \leftrightarrow i < z + |Y| \wedge [(i < z \wedge X(i)) \vee (z \leq i \wedge Y(i \dot{-} z))].$$

We wish to define an **ITERATION** problem Q_F with bounding term t' whose solutions yield solutions of Q. The idea is to let the domain of F consist of concatenations $U *_t V$ where U is a candidate solution for Q and V is its profit. Note that if $V_1 < V_2$ then $U_1 *_t V_1 < U_2 *_t V_2$ for all U_1, U_2.

In the following we suppress the parameters $\vec{x}, \vec{X}$.

Let $u = u(\vec{x}, \vec{X})$ be an $\mathcal{L}_A^2$-term large enough so that $|P_Q(N_Q(Z))| \leq u$ for $|Z| \leq t$. Then define

$$t' = t + u$$

and

$$F(U *_t V) = \begin{cases} N_Q(U) *_t P_Q(N_Q(U)) \text{ if } V = P_Q(U) \text{ and } \varphi_Q(U), \\ U *_t V \text{ otherwise.} \end{cases}$$

The term t' is chosen so that if U satisfies $\varphi_Q(U)$ then $|F(U *_t P_Q(U))| \leq t'$.

Here we redefine P_Q so that $P_Q(\varnothing) = \varnothing$. Note that the result is a **PLS** problem with the same solutions as the original problem.

Now suppose Z is a solution to the **ITERATION** problem Q_F. We show how to obtain a solution $G(Z) (= G(\vec{x}, \vec{X}, Z))$ to the original **PLS** problem Q. We write $Z = U *_t V$ where U, V are uniquely determined by Z (for $|U| \leq t$ and $|V| \leq u$). Then from (194), (196) and our definitions we see that $G(U *_t V) = N_Q(U)$ is a solution to Q.

Hence by Definition VIII.5.2 we conclude $Q \leq_{AC^0} Q_F$, where f, F take $\vec{x}, \vec{X}$ to itself and $G(\vec{x}, \vec{X}, Z) = N_Q(\vec{x}, \vec{X}, Z^{<t(\vec{x}, \vec{X})})$.   □

DEFINITION VIII.5.9. If S is a set of search problems, then $CC(S)$ is the set of search problems many-one reducible to S.

THEOREM VIII.5.10.

$$CC(\textbf{ITERATION}) = CC(\textbf{PLS}) =$$
$$CC(AC^0\text{-}\textbf{ITERATION}) = CC(AC^0\text{-}\textbf{PLS}).$$

PROOF. The first and last equalities follow from the preceding definition and theorem. The middle equality follows from these and Theorem VIII.5.12 below.   □

DEFINITION VIII.5.11. Let $Q(\vec{x}, \vec{X})$ be a search problem with graph $R(\vec{x}, \vec{X}, Z)$. We say that Q is Φ-definable in a theory $\mathcal{T}$ if there is a formula $\psi_R(\vec{x}, \vec{X}, Z)$ in Φ such that

$$\psi_R(\vec{x}, \vec{X}, Z) \supset R(\vec{x}, \vec{X}, Z)$$

and

$$\mathcal{T} \vdash \exists Z \psi_R(\vec{x}, \vec{X}, Z).$$

THEOREM VIII.5.12. *The following are equivalent for a search problem Q:*
(a) Q *is Σ_1^B-definable in* $\textbf{TV}^1$.
(b) Q *is in* $CC(\textbf{PLS})$.
(c) Q *is in* $CC(AC^0\text{-}\textbf{PLS})$.

PROOF. (a) $\Longrightarrow$ (c) follows from Theorem VIII.5.13 below (Witnessing for $\textbf{TV}^1$) and Lemma VIII.5.7. (c) $\Longrightarrow$ (b) is obvious. Hence it suffices to show (b) $\Longrightarrow$ (a).

By Theorems VIII.5.8 and VIII.2.16 and Corollary VIII.3.11 it suffices to show that every problem in $CC(\textbf{ITERATION})$ is $\Sigma_1^B(\mathcal{L}_{FP})$-definable in $\textbf{TV}^1(\textbf{VPV})$. We start by showing this for every $\textbf{ITERATION}$ problem Q_F. Let $\psi_F(\vec{x}, \vec{X}, Z)$ be the formula (195) defining Q_F. We may assume that F is an $\mathcal{L}_{FP}$-function, and hence ψ_F is a $\Sigma_1^B(\mathcal{L}_{FP})$-formula. Let

$$\eta(\vec{x}, \vec{X}, Z) \equiv (Z = \varnothing \vee Z < F(\vec{x}, \vec{X}, Z)).$$

Then $\textbf{VPV}$ proves η is equivalent to a Σ_1^B-formula (Theorem VIII.2.16), and hence by Σ_1^B-$\textbf{SMAX}$ (Theorem VIII.3.8), $\textbf{TV}^1(\textbf{VPV})$ proves the existence of a largest $Z \leq t$ satisfying $\eta(Z)$. Thus $\textbf{TV}^1(\textbf{VPV})$ proves that this Z satisfies $\psi_F(Z)$.

This shows that every $\textbf{ITERATION}$ problem is $\Sigma_1^B(\mathcal{L}_{FP})$-definable in $\textbf{TV}^1(\textbf{VPV})$. Now suppose the search Q_{R_1} is many-one reducible to some $\textbf{ITERATION}$ problem Q_{R_2}. Define the formula $\psi_{R_1}(\vec{x}, \vec{X}, Z)$ by (suppressing $\vec{x}, \vec{X}$)

$$\psi_{R_1}(Z) \equiv \exists W \leq t(Z = G(W) \wedge \psi_{R_2}(\vec{f}, \vec{F}, W))$$

where t is the bounding term for Q_{R_2} and ψ_{R_2} is a $\Sigma_1^B(\mathcal{L}_{FP})$-formula which defines Q_{R_2} in $TV^1(VPV)$, and $\vec{f}, \vec{F}, G$ show $Q_{R_1} \leq_{AC^0} Q_{R_2}$ according to Definition VIII.5.2. Then ψ_{R_1} is equivalent to a $\Sigma_1^B(\mathcal{L}_{FP})$-formula, and by Definition VIII.5.2

$$\psi_{R_1}(\vec{x}, \vec{X}, Z) \supset R_1(\vec{x}, \vec{X}, Z).$$

Since by assumption $TV^1(VPV)$ proves $\exists W \leq u \, \psi_{R_2}(W)$ (where u is a bounding term from Parikh's Theorem) it follows that $TV^1(VPV)$ proves $\exists Z \psi_{R_1}(Z)$, as required. $\square$

THEOREM VIII.5.13 (Witnessing for TV^1). *Suppose that $\varphi(\vec{x}, \vec{X}, Z)$ is a Σ_1^1-formula such that*

$$TV^1 \vdash \exists Z \varphi(\vec{x}, \vec{X}, Z).$$

Then there is an AC^0-ITERATION problem Q_F with graph $\psi_F(\vec{x}, \vec{X}, Z)$ from (195) *and an FAC^0-function G such that*

$$\overline{V}^0 \vdash \psi_F(\vec{x}, \vec{X}, Z) \supset \varphi(\vec{x}, \vec{X}, G(\vec{x}, \vec{X}, Z)).$$

PROOF. By using pairing functions we may assume that φ is Σ_0^B. The proof is similar to the proof of the Witnessing Theorem for V^1 (Section VI.4). Thus we define a sequent system $LK^2\text{-}TV^1$, which is the same as $LK^2\text{-}\widetilde{V}^1$ except that we replace the *IND* Rule by the *single-Σ_1^B-SIND* Rule, defined below. Recall (Example V.4.17) the AC^0 functions $\varnothing$ (empty set) and $S(X)$ (successor of X). For the next definition, when Φ is $\Sigma_i^B(\mathcal{L}_A^2)$ (for $i \geq 0$) the formulas $A(S(\delta))$ and $A(\varnothing)$ are understood to be the equivalent $\Sigma_i^B(\mathcal{L}_A^2)$ formulas as stated by the *FAC^0* Elimination Lemma V.6.7.

DEFINITION VIII.5.14 (The *SIND* Rule). For a set Φ of formulas, the Φ-*SIND* rule consists of the inferences of the form

$$\frac{\Gamma, A(\delta) \longrightarrow A(S(\delta)), \Delta}{\Gamma, A(\varnothing) \longrightarrow A(T), \Delta} \tag{197}$$

where A is a formula in Φ and T is a string term.
Restriction. The variable δ is called an *eigenvariable* and does not occur in the bottom sequent.

The proof that $LK^2\text{-}TV^1$ is a complete system for TV^1 is the same as the proof that $LK^2\text{-}\widetilde{V}^1$ is a complete system for $\widetilde{V}^1$, with obvious modifications. Further the proof of Theorem VI.4.15, Anchored Completeness for $LK^2\text{+}IND$, works for $LK^2\text{-}TV^1$, so every theorem of TV^1 has an anchored $LK^2\text{-}TV^1$ proof.

Now we proceed as in the proof of the Witnessing Theorem for V^1 (Section VI.4.2) and for V^0 (Section V.5.2), with appropriate changes.

Suppose that $\exists Z \varphi(\vec{x}, \vec{X}, Z)$ is a Σ_1^1-theorem of TV^1, where φ is a Σ_0^B-formula. Then there is an anchored LK^2-TV^1 proof π of

$$\longrightarrow \exists Z \varphi(\vec{a}, \vec{\alpha}, Z).$$

We may assume that π is in free variable normal form. By the Subformula Property the formulas in π are Σ_1^1 formulas, and in fact they are Σ_0^B formulas or *single*-Σ_1^1 formulas. As a result, every sequent in π has the form

$$\mathcal{S} = \underbrace{\exists X_i \theta_i(X_i)}_{i=1,\dots,m}, \Gamma \longrightarrow \Delta, \underbrace{\exists Y_j \eta_j(Y_j)}_{j=1,\dots,n} \tag{198}$$

for $m, n \geq 0$, where θ_i and η_j and all formulas in Γ and Δ are Σ_0^B.

We will prove by induction on the depth in π of the sequent $\mathcal{S}$ that there is an AC^0-*ITERATION* problem Q_F with graph ψ_F and for $1 \leq i \leq n$ there are $\mathcal{L}_{FAC^0}$-functions G_i such that $\overline{V}^0$ proves (the semantic equivalent of) the sequent

$$\mathcal{S}' = \underbrace{\theta_i(\beta_i)}_{i=1,\dots,m}, \Gamma, \psi_F(\vec{a}, \vec{\alpha}, \vec{\beta}, \gamma) \longrightarrow \Delta, \underbrace{\eta_j(G_j(\vec{a}, \vec{\alpha}, \vec{\beta}, \gamma))}_{j=1,\dots,n} \tag{199}$$

where $\vec{a}, \vec{\alpha}$ is a list of exactly those variables with free occurrences in $\mathcal{S}$. (This list may be different for different sequents.) Also $\beta_1, \dots, \beta_m$ are distinct new free variables corresponding to the bound variables $X_1, \dots, X_m$, although the latter variables may not be distinct. When $\mathcal{S}$ is the final sequent of π, note that Γ and Δ are empty, $i = 0$, $j = 1$, and $\vec{\beta}$ is empty, so the theorem follows.

Note that this induction hypothesis is the same as in the proof for V^1 and V^0, except now each witnessing function G_j is allowed to take the argument γ, which is a solution to the *ITERATION* problem Q_F. As before, the induction step has a case for Σ_0^B-*COMP* and for each rule. The argument for Σ_0^B-*COMP* is the same as for V^0 (since the witnessing function G_j can ignore its argument γ). The argument for each rule except Σ_1^B-*SIND* is similar to that for V^0 (Section V.5.2). In the case of a rule with two parents, such as $\wedge$-right or cut, we need the following lemma to combine the two *ITERATION* problems for the two parents into a single problem for the conclusion. This lemma is stated more generally than is needed for these rules (namely insertion of U as an argument of ψ_{F_2}) in order to accommodate the Σ_1^B-*SIND* rule.

LEMMA VIII.5.15 (Combining *ITERATION* Problems). *Suppose that Q_{F_1} and Q_{F_2} are ITERATION problems with graphs $\psi_{F_1}(\vec{x}, \vec{X}, U)$ and $\psi_{F_2}(\vec{x}, \vec{X}, U, V)$. Then there is an ITERATION problem Q_F with graph $\psi_F(\vec{x}, \vec{X}, Z)$ such that F is Σ_0^B-bit-definable from F_1, F_2, and there are*

FAC^0*-functions* $G_1(\vec{x}, \vec{X})$ *and* $G_2(\vec{x}, \vec{X})$ *such that (suppressing* $\vec{x}, \vec{X})$

$$\overline{V}^0(F_1, F_2, F) \vdash \psi_F(Z) \supset \psi_{F_1}(G_1(Z)) \wedge \psi_{F_2}(G_1(Z), G_2(Z)).$$

PROOF. Assume the hypotheses of the Lemma, and let t be the bounding term for Q_{F_1} and let u be the bounding term for Q_{F_2}. Using the notation $U *_t V$ in the proof of Theorem VIII.5.8, we express the argument Z in $F(\vec{x}, \vec{X}, Z)$ in the form

$$Z = (U *_t V) *_{t+u} \delta$$

where δ is a binary string equal to 0,1,or 2. We abbreviate Z by

$$Z = U * V * \delta.$$

Then we define F by (suppressing $\vec{x}, \vec{X}$)

$$F(U * V * \delta) = \begin{cases} U * V * 2 & \text{if } \psi_{F_1}(U) \wedge \psi_{F_2}(U, V) \wedge \delta \le 1, \\ U * F_2(U, V) * 1 & \text{if } \psi_{F_1}(U) \wedge |V| \le u \wedge \\ & \quad V < F_2(U, V) \wedge \delta \le 1, \\ F_1(U) * \varnothing * \varnothing & \text{if } V = \delta = \varnothing \wedge \\ & \quad |U| \le t \wedge U < F_1(U), \\ U * V * \delta & \text{otherwise.} \end{cases}$$

Let the *ITERATION* problem Q_F have bounding term $t + u + 2$.
We claim that

$$\overline{V}^0(F_1, F_2, F) \vdash \psi_F(U * V * \delta) \supset \delta = 2 \wedge \psi_{F_1}(U) \wedge \psi_{F_2}(U, V). \quad (200)$$

To see this, note that by line 3 in the definition of F, $F(\varnothing) \ne \varnothing$, since if $F_1(\varnothing) = \varnothing$ then $\psi_{F_1}(\varnothing)$, and hence one of the first two lines applies. Hence assuming $\psi_F(U * V * \delta)$ we have by (195)

$$U * V * \delta < F(U * V * \delta) = F(F(U * V * \delta)).$$

From the definitions of ψ_{F_1} and ψ_{F_2} we see that this can only happen if line 1 applies in evaluating $F(U * V * \delta)$.

This establishes (200). To prove the lemma, we define

$$G_1(U * V * \delta) = U, \qquad G_2(U * V * \delta) = V.$$

We can make these definitions explicit by defining

$$G_1(\vec{x}, \vec{X}, Z) = Z^{<t}$$

and $G_2(\vec{x}, \vec{X}, Z)$ to be the substring $Z(0), Z(1), \dots, Z(u \dotminus 1)$:

$$G_2(\vec{x}, \vec{X}, Z) = Y \leftrightarrow (|Y| \le u \wedge \forall i < u(Y(i) \leftrightarrow Z(t + i))). \qquad \square$$

It remains to handle the case in which S is obtained by an application of the Σ_1^B-*SIND* rule. Then S is the bottom sequent of

$$\frac{S_1}{S} = \frac{\Lambda, \exists X \le r(\delta)\theta(\delta, X) \longrightarrow \exists X \le r(S(\delta))\theta(S(\delta), X), \Pi}{\Lambda, \exists X \le r(\varnothing)\theta(\varnothing, X) \longrightarrow \exists X \le r(T)\theta(T, X), \Pi}$$

where δ does not occur in $\mathcal{S}$ and θ is Σ_0^B.

By the induction hypothesis for the top sequent $\mathcal{S}_1$ it follows that $\overline{V}^0$ proves a sequent $\mathcal{S}_1'$ of the form

$$\mathcal{S}_1' = \Lambda', \eta_1, \psi_F(\delta, \beta, \gamma) \longrightarrow \eta_2, \Pi' \qquad (201)$$

where

$$\eta_1 \equiv |\beta| \leq r(\delta) \wedge \theta(\delta, \beta), \qquad (202)$$

$$\eta_2 \equiv |G(\delta, \beta, \gamma)| \leq r(S(\delta)) \wedge \theta(S(\delta), G(\delta, \beta, \gamma)) \qquad (203)$$

and ψ_F defines the graph of an $\mathbf{AC^0\text{-}ITERATION}$ problem Q_F and G is an $\mathcal{L}_{FAC^0}$-function. Here δ, β, γ do not occur in Λ', but they may occur in Π' as arguments to the witnessing functions G_j.

Our task is to use Q_F and G to find $Q_{F'}$ and G' to compute a witness for $\exists X \leq r(T)\theta(T, X)$, given a witness β_0 for $\exists X \leq r(\varnothing)\theta(\varnothing, X)$. We want $\overline{V}^0$ to prove the following sequent $\mathcal{S}'$:

$$\mathcal{S}' = \Lambda', \rho_1, \psi_{F'}(\beta_0, \gamma') \longrightarrow \rho_2, \Pi'' \qquad (204)$$

where

$$\rho_1 \equiv |\beta_0| \leq r(\varnothing) \wedge \theta(\varnothing, \beta_0), \qquad (205)$$

$$\rho_2 \equiv |G'(\beta_0, \gamma')| \leq r(T) \wedge \theta(T, G'(\beta_0, \gamma')) \qquad (206)$$

and Π'' will be given later.

We will use the technique in the proof of Lemma VIII.5.15 and assume that the search variable γ' for $Q_{F'}$ has the form

$$\gamma' = (\beta *_{r(T)} \gamma) *_{r(T)+t} \delta$$

where β, γ, δ are as in (201), and t an upper bound for γ based on the bounding term for Q_F. In the following we drop the subscripts to $*$ and write

$$\gamma' = \beta * \gamma * \delta.$$

The idea is that $Q_{F'}$ uses F and G to find witnesses β for successive string values of $\delta = 1, 2, \ldots, T$ knowing that β_0 is a witness in case $\delta = \varnothing$. $Q_{F'}$ should succeed under the assumption that (201) holds for all $\delta < T$ and all β, assuming that the formulas in Λ' are true and those in Π' are false.

We define $F'(\beta_0, \beta * \gamma * \delta)$ by cases in such a way that if η_1 holds, then it continues to hold when F' is applied repeatedly, and progress is made toward finding β' such that $\theta(T, \beta')$.

$$F'(\beta_0, \beta * \gamma * \delta) = \begin{cases} G(\delta, \beta, \gamma) * \varnothing * S(\delta) \text{ if } \eta_1 \wedge \delta < T \wedge \psi_F(\delta, \beta, \gamma), \\ \text{else } \beta * F(\beta, \delta, \gamma) * \delta \text{ if } \eta_1 \wedge \delta < T \wedge \gamma < F(\beta, \delta, \gamma), \\ \text{else } \beta_0 * \varnothing * \varnothing, \text{ if } \beta = \gamma = \delta = \varnothing, \\ \text{else } \beta * \gamma * \delta. \end{cases}$$

We define the witness-extracting function $G'(\beta_0, \gamma')$ as follows:

$$G'(\beta_0, \beta * \gamma * \delta)) = \begin{cases} \beta_0 & \text{if } T = \varnothing, \\ G(\delta, \beta, \gamma) & \text{if } T \neq \varnothing. \end{cases}$$

The following Claim asserts that a witness for $\exists X \theta(T, X)$ can be obtained from a solution $\beta * \gamma * \delta$ to $Q_{F'}$, provided (201) holds with Λ' true and Π' false.

CLAIM. $\overline{V}^0$ proves

$$T \neq \varnothing, \rho_1, \psi_{F'}(\beta_0, \beta * \gamma * \delta) \longrightarrow \eta_1 \wedge \psi_F(\delta, \beta, \gamma) \wedge (\neg \eta_2 \vee \rho_2).$$

PROOF OF THE CLAIM. We argue in $\overline{V}^0$. Assume $T \neq \varnothing, \rho_1, \psi_{F'}(\beta_0, \beta * \gamma * \delta)$. By $\psi_{F'}(\beta_0, \beta * \gamma * \delta)$ and (195) there are two possibilities. The first is that $F'(\varnothing) = \varnothing$. But this is impossible, because if $\beta = \gamma = \delta = \varnothing$ then either $\beta_0 \neq \varnothing$ and line 3 in the definition of F' applies, or $\beta_0 = \varnothing$ and one of the first two lines applies (by ρ_1 and the definition of ψ_F).

Therefore the second possibility in the definition of $\psi_{F'}(\beta_0, \beta * \gamma * \delta)$ applies, and we have

$$\beta * \gamma * \delta < F'(\beta * \gamma * \delta) = F'(F'(\beta * \gamma * \delta)). \tag{207}$$

Analyzing the definition of F' and our assumptions $(T \neq \varnothing, \rho_1)$ shows that the only way that (207) can hold is if line 1 in the definition of F' applies when evaluating $F'(\beta * \gamma * \delta)$. Thus $\eta_1 \wedge \psi_F(\delta, \beta, \gamma)$. Also since line 1 applies, if $S(\delta) < T$ then $\neg \eta_2$, for otherwise line 1 or line 2 would apply when evaluating $F'(F'(\beta * \gamma * \delta))$, contradicting the second part of (207). This proves the Claim in case $S(\delta) < T$. Finally if $S(\delta) = T$ then $\eta_2 \supset \rho_2$, and the Claim follows.

To establish that $\overline{V}^0$ proves (204) we need to specify Π'' by giving values (in terms of γ') for the variables δ, β, γ which occur as arguments to the functions G_j in Π'. Motivated by the Claim and (201) we define, for $\gamma' = \beta * \gamma * \delta$,

$$B(\gamma') = \beta, \qquad GA(\gamma') = \gamma, \qquad D(\gamma') = \delta$$

and define Π'' to be the result of replacing β, γ, δ in Π' by $B(\gamma')$, $GA(\gamma')$, $D(\gamma')$ respectively.

The fact that $\overline{V}^0$ proves (204) now follows from the Claim and by (201) with β, γ, δ replaced by $B(\gamma'), GA(\gamma'), D(\gamma')$. (The case $T = \varnothing$ follows from $(T = \varnothing \wedge \rho_1) \supset \rho_2$, which holds by definition of G'.) □

VIII.6. KPT Witnessing and Replacement

Here we present a generalization of the Herbrand Theorem from Chapter II and show how it can be used to prove the independence of the

Replacement Axiom Scheme (Section VI.3) in some cases. In Section VIII.7.3 we use it to show how the collapse of the polynomial hierarchy follows from the collapse of the bounded arithmetic hierarchy V^i.

Form 2 of the Herbrand Theorem (Corollary II.5.5) applies to a $\forall\exists$ consequence of a universal theory. The next result is a generalization which applies to $\forall\exists\forall$ consequences. We call it the KPT Witnessing Theorem, after the authors of [75], who used it to prove the first part of Theorem VIII.7.20.

THEOREM VIII.6.1 (KPT Witnessing). *Let T be a universal two-sorted theory with vocabulary $\mathcal{L}$. Let φ be an open formula and suppose*

$$T \vdash \forall X \exists Y \forall Z \varphi(X, Y, Z).$$

Then there exists a finite sequence $T_1, \ldots, T_k$ of string terms over $\mathcal{L}$ such that

$$T \vdash \varphi(X, T_1(X), Z_1) \vee \varphi(X, T_2(X, Z_1), Z_2) \vee \cdots \vee$$
$$\varphi(X, T_k(X, Z_1, \ldots, Z_{k-1}), Z_k)$$

where the notation $T_i(X, Z_1, \ldots, Z_{i-1})$ means that only the displayed variables may occur in T_i.

In our applications of this theorem each term T_i is a function

$$F_i(X, Z_1, \ldots, Z_{i-1})$$

in some complexity class such as $\mathbf{FAC}^0$ or $\mathbf{FP}$. The "student-teacher" interpretation of the theorem [74] is a useful way to think of it. The student is given X and wants to find Y satisfying $\forall Z \varphi(X, Y, Z)$, but has computing power limited to the relevant complexity class. The student starts by trying $Y = F_1(X)$. The teacher either approves, or comes up with a counter-example Z_1 such that $\neg\varphi(X, F_1(X), Z_1)$. The student next tries $Y = F_2(X, Z_1)$, and the teacher either agrees or supplies a counter-example Z_2. This process continues for at most k steps after which the student finds a value of Y that works for all Z.

PROOF OF THEOREM VIII.6.1. Let $B, C_1, C_2, \ldots$ be a list of new string constants, and let $U_1, U_2, \ldots$ be an enumeration of all terms built from the functions of $\mathcal{L}$ together with $B, C_1, C_2, \ldots$, where the only new constants in U_k are among $\{B, C_1, \ldots, C_{k-1}\}$. We will show that

$$T \cup \{\neg\varphi(B, U_1, C_1), \neg\varphi(B, U_2, C_2), \ldots, \neg\varphi(B, U_k, C_k)\}$$

is unsatisfiable for some k, from which the theorem follows (let T_i be U_i with B replaced by X and each C_j replaced by Z_j).

Suppose otherwise. Then by compactness

$$T \cup \{\neg\varphi(B, U_1, C_1), \neg\varphi(B, U_2, C_2), \ldots\} \qquad (208)$$

has a model $\mathcal{M}$. Since $\mathcal{T}$ is universal, the substructure $\mathcal{M}'$ consisting of the denotations of the terms $U_1, U_2, \ldots$ is also a model for (208). It is easy to see that

$$\mathcal{M}' \models \mathcal{T} + \forall Y \exists Z \neg \varphi(B, Y, Z)$$

and hence $\mathcal{T} \not\vdash \forall X \exists Y \forall Z \varphi(X, Y, Z)$. □

VIII.6.1. Applying KPT Witnessing. Following [47] we now outline the method for using the KPT Witnessing Theorem to show that a universal theory $\mathcal{T}$ which extends $\overline{V}^0$ and has a vocabulary $\mathcal{L}$ associated with certain complexity classes cannot prove the $\Sigma_0^B(\mathcal{L})$-***REPL*** axioms (sometimes subject to complexity assumptions). Our main examples are $\mathcal{T} = \overline{V}^0$ and $\mathcal{T} = \boldsymbol{VPV}$. That $\boldsymbol{VPV}$ is unlikely to prove Σ_0^B-Replacement may seem surprising, since V^1 proves it (Corollary VI.3.8), and V^1 and $\boldsymbol{VPV}$ have the same Σ_1^B-theorems.

Choose a function F which is in the relevant complexity class but whose inverse probably is not. Suppose $\mathcal{T}$ proves the following instance of replacement (which has W as a parameter, and $t = t(W)$ and $u = u(W)$ as terms):

$$(\forall i < t \, \exists Z < u \, F(Z) = W^{[i]}) \supset \exists Y \forall j < t \, F(Y^{[j]}) = W^{[j]}. \quad (209)$$

We can rewrite this as

$$\exists i < t \, \exists Y \forall Z < u \big(F(Z) = W^{[i]} \supset \forall j < t \, F(Y^{[j]}) = W^{[j]} \big).$$

Applying the KPT Witnessing Theorem we get a positive integer k and functions $g_1, \ldots, g_k, H_1, \ldots, H_k$ such that $\mathcal{T}$ proves

$$(F(Z_1) = W^{[g_1(W)]} \supset \forall j < t \, F(H_1(W)^{[j]}) = W^{[j]}) \vee$$
$$(F(Z_2) = W^{[g_2(W,Z_1)]} \supset \forall j < t \, F(H_2(W,Z_1)^{[j]}) = W^{[j]}) \vee \cdots \vee$$
$$(F(Z_k) = W^{[g_k(W,Z_1,\ldots,Z_{k-1})]} \supset \forall j < t \, F(H_k(W,Z_1,\ldots,Z_{k-1})^{[j]}) = W^{[j]}).$$

This allows the "student", given an input W (considered as a sequence $W^{[0]}, \ldots, W^{[t-1]}$), to compute Y coding a sequence of pre-images of F of all t elements of W, by asking the "teacher" for pre-images of at most k elements of W.

The student proceeds as follows. Let $Y = H_1(W)$. If $\forall j < t \, F(Y^{[j]}) = W^{[j]}$ then output Y and halt. Otherwise compute $g_1(W)$ and ask the teacher for a pre-image Z_1 of $W^{[g_1(W)]}$. Let $Y = H_2(W,Z_1)$. If $\forall j < t \, F(Y^{[j]}) = W^{[j]}$ then output Y and halt. Otherwise compute $g_2(W,Z_1)$ and ask the teacher for a pre-image Z_2 of $W^{[g_2(W,Z_1)]}$, and so on. By our assumption the algorithm will run for at most k steps of this form before it outputs a suitable Y.

THEOREM VIII.6.2 ([47]). V^0 and $\overline{V}^0$ do not prove Σ_0^B-**REPL**.

PROOF. Since $\overline{V}^0$ extends V^0 and every $\Sigma_i^B(\mathcal{L}_{FAC^0})$-formula is provably equivalent to a Σ_i^B-formula (Lemma V.6.7), it suffices to prove the theorem for the case $\overline{V}^0$.

Recall that $PARITY(X)$ holds iff the string X has an odd number of ones. We have pointed out that $PARITY$ is not an AC^0 relation, but in fact is is known [52, 3] that $PARITY$ is not even in nonuniform AC^0; i.e. it cannot be computed by any polynomial size bounded depth family of Boolean circuits. We will show using the student-teacher method outlined above that if $\overline{V}^0$ proves Σ_0^B-**REPL** then there is a randomized AC^0 algorithm which on each input X, with probability at least one-half, correctly outputs $PARITY(X)$, and if it does not output $PARITY(X)$ it outputs an 'abort' message, meaning the computation failed. From this it follows using a standard argument that $PARITY$ is in nonuniform AC^0. For each input length n, the circuit for computing $PARITY(X)$ for $|X| = n$ is obtained by repeating the randomized computation $n + 1$ times with independent random bits, to obtain a randomized AC^0 algorithm that computes $PARITY(X)$ with abort probability at most 2^{-n-1}. Hence there must be some fixed setting of the random bits which aborts on at most a fraction 2^{-n-1} inputs X of length n; which means this setting of random bits allows the circuit to correctly compute $PARITY(X)$ on all inputs X of length n.

Let PAR be the function that maps a binary string of length m to its parity vector. That is, $PAR(m, X) = Y$ if $|Y| \leq m$ and, for each $i < m$, $Y(i)$ is the parity of the string $X(0) \ldots X(i)$. In what follows we take m to be a parameter, assume X is a string of length at most m, and suppress the argument m from $PAR(m, X)$. Note that for fixed m, PAR is a bijection from the set of strings of length at most m to itself.

Although $PAR(X)$ cannot be computed in AC^0, its inverse, which we will call $UNPAR$, is in (uniform) FAC^0: the ith bit of $UNPAR(Y)$ is given by the Σ_0^B-formula

$$(i = 0 \wedge Y(i)) \vee (i > 0 \wedge Y(i-1) \oplus Y(i)).$$

Here $UNPAR$ has an argument m, which we suppress. Then

$$UNPAR(PAR(X)) = X$$

and

$$PAR(UNPAR(Y)) = Y.$$

Notice also that for all m-bit strings A, B, C, writing $\oplus$ for bitwise XOR, if $A = B \oplus C$ then $PAR(A) = PAR(B) \oplus PAR(C)$.

Assuming that $\overline{V}^0$ proves Σ_0^B-**REPL** we can apply the argument of Section VIII.6.1 and assume that $\overline{V}^0$ proves (209) for the case in which

F is $UNPAR$. We can assume that the parameter m is coded by the parameter W; specifically $m = |W^{[0]}|$, where $W^{[0]}$ is a string of 0's except bit $m - 1$ is 1. (Note that $PAR(W^{[0]}) = W^{[0]}$.) Also we define the terms $t = m + 1$, and $u = m$. Then for some fixed k there is a uniform AC^0 algorithm which, for any sequence $W^{[1]}, \ldots, W^{[m]}$ of binary strings of length at most m makes k queries of the form "what is $PAR(W^{[i]})$?" and outputs the sequence of parity vectors of W.

Suppose $m \geq 2k$. We will show how to use this algorithm to compute the parity of a single string I, $|I| \leq m$, in uniform randomized AC^0.

Choose m strings $U_1, \ldots, U_m$ of m bits each uniformly at random, and for each i compute $V_i = UNPAR(U_i)$. Choose a number r, $1 \leq r \leq m$, uniformly at random. For $1 \leq i \leq m$ define $W^{[i]}$ by the condition

$$W^{[i]} = \begin{cases} V_i & \text{if } i \neq r, \\ I \oplus V_r & \text{if } i = r. \end{cases}$$

Since for each m the function $UNPAR$ defines a bijection from the set $\{0, 1\}^m$ to itself, and since for each I with $|I| < m$ the map $X \mapsto I \oplus X$ also defines a bijection from that set to itself, it follows that the string W defined above, interpreted as an $m \times m$ bit matrix, is uniformly distributed over all such matrices.

Now run our student-teacher AC^0 algorithm on W. If the student asks "what is $PAR(W^{[i]})$?" for $i \neq r$, reply with U_i (or $W^{[0]}$ if $i = 0$) (which is the correct answer). If the algorithm queries "what is $PAR(Y^{[r]})$?", then abort the computation.

Since $PAR(W^{[i]})$ is queried for at most k different values of i and since for each input I each pair (W, r) is equally likely to have been chosen, it follows that the computation will be aborted with probability at most $k/m \leq 1/2$.

Hence with probability at least $1/2$ the algorithm is not aborted, we are able to answer all the queries correctly, and we obtain Y such that $Y^{[r]} = PAR(W^{[r]}) = PAR(I \oplus V_r)$. But $I = V_r \oplus (I \oplus V_r)$ and hence

$$PAR(I) = PAR(V_r) \oplus PAR(I \oplus V_r)$$
$$= U_r \oplus Y^{[r]}.$$

We use this to compute $PAR(I)$ and use bit $m - 1$ of $PAR(I)$ to determine $PARITY(I)$.

For each input I the algorithm succeeds with probability at least $1/2$, where the probability is taken over its random input bits. If the algorithm aborts, this is reflected in the output. As explained earlier, this implies the existence of a nonuniform AC^0 algorithm for $PARITY(I)$. Since no such algorithm exists, it follows that V^0 does not prove the Σ_0^B-Replacement scheme.   □

We now show that VPV seems unlikely to prove Σ_0^B-$REPL$ because a consequence would be that integer factoring is easy. This contrasts with V^1, which proves the stronger $g\Sigma_1^B$-$REPL$ scheme (Corollary VI.3.8).

We adapt the proof [94] that cracking Rabin's cryptosystem based on squaring modulo N is as hard as factoring. Let N be the product of distinct odd primes P and Q. Suppose $0 < X_1 < N$ and $\gcd(X_1, N) = 1$. Let $C = X_1^2$. Then C has precisely four square roots X_1, X_2, X_3, X_4 modulo N. This can be seen as follows: let $X_P = (X_1 \bmod P)$ and $X_Q = (X_1 \bmod Q)$. By the Chinese remainder theorem there are uniquely determined numbers X_1, X_2, X_3, X_4 with $0 < X_i < N$ such that

$$
\begin{aligned}
X_1 &\equiv X_P \pmod{P} & X_1 &\equiv X_Q \pmod{Q} \\
X_2 &\equiv X_P \pmod{P} & X_2 &\equiv -X_Q \pmod{Q} \\
X_3 &\equiv \ X_P \pmod{P} & X_3 &\equiv X_Q \pmod{Q} \\
X_4 &\equiv -X_P \pmod{P} & X_4 &\equiv -X_Q \pmod{Q}.
\end{aligned}
$$

Now $X_1 - X_2 \equiv 0 \pmod{P}$ and $X_1 - X_2 \equiv 2X_Q \not\equiv 0 \pmod{Q}$, so $\gcd(X_1 - X_2, N) = P$. So from X_1 and X_2 we can recover P, and similarly from X_1 and X_3 we can recover Q.

Hence if we have one square root of C, and are then given a square root at random, we can factor N with probability $1/2$.

THEOREM VIII.6.3 ([47]). *If VPV proves Σ_0^B-$REPL$ then factoring (of products of two odd primes) is possible in probabilistic polynomial time.*

PROOF. We will argue as in the proof of the previous theorem, this time taking squaring modulo N as our function F (so F has N as a parameter).

If VPV proves Σ_0^B-$REPL$ then there is polynomial time algorithm which, for some fixed k, given any sequence $W^{[0]}, \ldots, W^{[m-1]}$ of squares (modulo N) (where $m = |N|$), makes at most k queries of the form "what is the square root of $W^{[i]}$?" and, if these are answered correctly, outputs square roots of all the $W^{[i]}$s.

Now suppose N is large enough that $m = |N| > k$. Choose numbers $X_0, \ldots, X_{m-1}$ uniformly at random with $0 < X_i < N$. We may assume that $\gcd(X_i, N) = 1$ for all i, since otherwise we can immediately find a factor of N.

Choose W so that for each i, $W^{[i]} = (X_i^2 \bmod N)$. Notice that each X_i is distributed uniformly among the four square roots of $W^{[i]}$.

Run our algorithm, and to each query "what is the square root of $W^{[i]}$?", answer with X_i. We will get as output Y coding a sequence $Y^{[0]}, \ldots, Y^{[m-1]}$ of square roots of $W^{[0]}, \ldots, W^{[m-1]}$.

If we think of N as fixed, the value of Y depends only on the inputs given to the algorithm, namely W and the k many numbers X_i that we gave as replies. Let i be some index for which X_i was not used. Then X_i is distributed at random among the square roots of $W^{[i]}$, and $Y^{[i]}$ is a square root of $W^{[i]}$ that was chosen without using any information about which square root X_i is. Hence $\gcd(X_i - Y^{[i]}, N)$ is a factor of N with probability $1/2$.   □

VIII.7. More on V^i and TV^i

VIII.7.1. Finite Axiomatizability. V^0 is finitely axiomatizable by Theorem V.7.1. By the discussion following Theorem VIII.3.10 we know that TV^0 is finitely axiomatizable, as are the $\forall \Sigma_1^B$-consequences of V^1. Here we show that V^i and TV^i are finitely-axiomatizable for all $i \geq 0$. (In Chapter X we will give other proofs using the Reflection Principles for G_i and G_i^*.) We start by proving the existence of a universal polynomial time function.

THEOREM VIII.7.1 (Universal Function). *There is an $\mathcal{L}_{FP}$ function*

$$Univ(X, W)$$

such that for every $\mathcal{L}_{FP}$-function $F(X)$ there is an $\mathcal{L}_{FAC^0}$-function $H_F(n)$ such that

$$VPV \vdash |X| < n \supset F(X) = Univ(X, H_F(n)).$$

In particular VPV proves $F(X) = Univ(X, H_F(|X|))$.

PROOF. We use the machinery of (180) (page 215). The value of $Univ(X, W)$ is the output of the circuit C described by W, where C expects an input string of length at most n (specified by W), and (assuming $|X| < n$) $Univ(X, W)$ supplies X to the input gates of C. Then $H_F(n)$ describes a circuit which computes $F(X)$ for $|X| < n$.   □

To help prove the next result, we introduce a string pairing function.

DEFINITION VIII.7.2. $\langle X, Y \rangle$ is the $\mathcal{L}_{FAC^0}$-function defined by

$$\langle X, Y \rangle(i) \leftrightarrow \exists j \leq i, (i = \langle 0, j \rangle \wedge X(j)) \vee (i = \langle 1, j \rangle \wedge Y(j)).$$

More generally $\langle X_1, \ldots, X_n \rangle$ is defined inductively by

$$\langle X_1, \ldots, X_{n+1} \rangle = \langle \langle X_1, \ldots, X_n \rangle, X_{n+1} \rangle.$$

Finally we define

$$\langle x_1, \ldots, x_k, \vec{X} \rangle = \langle POW2(x_1), \ldots, POW2(x_k), \vec{X} \rangle.$$

Note that $\overline{V}^0$ proves

$$\langle X, Y \rangle = Z \supset (X = Z^{[0]} \wedge Y = Z^{[1]}).$$

THEOREM VIII.7.3. V^i and TV^i are finitely axiomatizable for all $i \geq 0$.

PROOF. We have already proved this for $i = 0$. For the general case we start with the finitely axiomatizable theory VP and add one Σ_i^B-**COMP**-axiom to get V^i and one Σ_i^B-**SIND**-axiom to get TV^i. The axioms in question involve universal formulas. For notational simplicity we treat the case $i = 1$; the general case will be clear.

For V^1 we define the $\Sigma_1^B(\mathcal{L}_{FP})$ formula

$$UV(i, a, X, W) \equiv \exists Y \leq a\, Univ(\langle i, X, Y\rangle, W)(0)$$

and let $UV'(i, a, X, W)$ be the equivalent Σ_1^B formula according to Theorem VIII.2.16 (so $\textbf{VPV} \vdash UV \leftrightarrow UV'$).

Let $\mathcal{T}$ be the finitely axiomatizable theory extending $\textbf{VP}$ by the comprehension axiom for formula $UV'(i, a, X, W)$ (where a, X, W are parameters). Obviously $\mathcal{T} \subseteq V^1$. To prove the reverse inclusion, since $\textbf{VPV}$ is conservative over $\textbf{VP}$, it suffices to show $V^1 \subseteq \mathcal{T} + \textbf{VPV}$.

Let $\varphi(i, \vec{x}, \vec{X})$ be a Σ_1^B-formula. Then there is an $\mathcal{L}_{FP}$-function F such that (using Theorem VIII.7.1) $\textbf{VPV}$ proves

$$\varphi(i, \vec{x}, \vec{X}) \leftrightarrow \exists Y \leq t\, F(\langle i, \langle \vec{x}, \vec{X}\rangle, Y\rangle)(0)$$

$$\leftrightarrow \exists Y \leq t\, Univ(\langle i, \langle \vec{x}, \vec{X}\rangle, Y\rangle, \Pi_F(|\langle i, \langle \vec{x}, \vec{X}\rangle, Y\rangle|))(0)$$

$$\leftrightarrow UV'(i, t, \langle \vec{x}, \vec{X}\rangle, \Pi_F(|\langle i, \langle \vec{x}, \vec{X}\rangle, Y\rangle|)).$$

Hence $\textbf{VPV}$ proves the comprehension for φ from the comprehension axiom for UV'. It follows that $V^1 = \mathcal{T}$.

The argument is similar for $\textbf{TV}^1$. This time we define

$$UT(a, X, Z, W) \equiv \exists Y \leq a\, Univ(\langle X, Z, Y\rangle, W)(0)$$

and axiomatize $\textbf{TV}^1$ by the string induction axiom for $UT'(X)$, where a, Z, W are parameters and UT' is a Σ_1^B-formula equivalent to UT.   □

Since $\textbf{VP} \subseteq V^1$ (Theorem VIII.1.3) and $\textbf{TV}^0 = \textbf{VP}$ (Theorem VIII.3.10) it follows that $\textbf{TV}^0 \subseteq V^1$. This is generalized in the following result.

THEOREM VIII.7.4. *For* $i \geq 0$

$$V^i \subseteq \textbf{TV}^i \subseteq V^{i+1}.$$

PROOF. The first inclusion is Theorem VIII.3.4. For the second inclusion, by definition VIII.3.2 it suffices to show that V^{i+1} proves the Σ_i^B-$\textbf{SIND}$ induction scheme

$$[\varphi(\varnothing) \wedge \forall X(\varphi(X) \supset \varphi(S(X)))] \supset \varphi(Y)$$

where $\varphi(X)$ is a Σ_i^B-formula. Reasoning in $V^{i+1}(\textbf{VPV})$ (which by Theorem VIII.2.11 is conservative over V^{i+1}) assume

$$\varphi(\varnothing) \wedge \forall X(\varphi(X) \supset \varphi(S(X))). \qquad (210)$$

Define the Π_{i+1}^B-formula

$$\psi(i) \equiv \forall Z \leq i\, \forall W \leq |Y|((|W + Z| \leq |Y| \wedge \varphi(W)) \supset \varphi(W + Z)).$$

By Corollary VI.1.4 applied to V^{i+1} we are justified in using number induction on $\psi(i)$. The base case $\psi(0)$ is easy since $\varphi(W) \supset \varphi(S(W))$ by assumption (210). The induction step $\psi(i) \supset \psi(i + 1)$ is proved by using the hypothesis $\psi(i)$ twice, first with (W, Z) set to (W, Z') with

$Z' = \lfloor \frac{1}{2} Z \rfloor$ and then with (W, Z) set to $(W + Z', Z')$ to infer $\varphi(W + 2Z')$, from which $\psi(i + 1)$ follows (using the assumption (210) if $2Z' \neq Z$). Finally $\varphi(Y)$ follows from $\psi(|Y| + 1)$ and $\varphi(\varnothing)$. □

From the above theorem we have the hierarchy

$$V^0 \subset TV^0 \subseteq V^1 \subseteq TV^1 \subseteq V^2 \subseteq TV^2 \subseteq V^3 \subseteq \cdots$$

so the unions of V^i and TV^i are the same. We use the notation

$$V^\infty = \bigcup_i V^i = \bigcup_i TV^i. \tag{211}$$

The next result follows from Theorem VIII.7.3 and compactness.

COROLLARY VIII.7.5. V^∞ is finitely axiomatizable iff V^∞ collapses to V^i or TV^i for some i.

See Section VIII.7.3 for consequences of the collapse of V^∞.

It is not known whether V^∞ (or equivalently S_2) and $I\Delta_0$ are finitely axiomatizable, although it is known that their relativized versions $S_2(R)$ and $I\Delta_0(R)$ (or equivalently $\widetilde{V}^0$, Section VII.2.2) are not [75]. (For $I\Delta_0(R)$ and $\widetilde{V}^0$ this follows also from the results of [71, 78].)

VIII.7.2. Definability in the V^∞ Hierarchy. See Table 3 page 250 for a partial summary of the results in this section.

Recall that for $i \geq 1$, Σ_i^P is the set of (two-sorted) relations in level i of the polynomial hierarchy, and that these are precisely the relations represented by Σ_i^B-formulas (Theorem IV.3.7). Also $FP^{\Sigma_i^P}$ is the set of functions computable by a polynomial time Turing machine with a Σ_i^P oracle. (See also Appendix A.3.) For $i = 0$, we will take $FP^{\Sigma_0^P}$ to be simply FP (this is consistent with taking Σ_0^P to be either P or AC^0). We will show that for $i \geq 0$, $FP^{\Sigma_i^P}$ is the class of functions Σ_{i+1}^B-definable in TV^i, and also in V^{i+1}. (We have already shown this for $i = 0$.)

We start by generalizing the universal theory VPV to VPV^i, for $i \geq 1$. Here $VPV^1 = VPV$, and for $i \geq 0$ VPV^{i+1} has function symbols for all functions in $FP^{\Sigma_i^P}$. We use $\mathcal{L}_{FP^i}$ to denote the vocabulary of VPV^i.

Since $\mathcal{L}_{FP^i}$ includes the vocabulary $\mathcal{L}_{FAC^0}$ of $\overline{V}^0$, it includes symbols for the string functions $\varnothing, S, +$ defined using Σ_0^B-formulas in Example V.4.17. Since we want the theories VPV^i to be universal we take the defining axioms for these functions to be the quantifier-free axioms for these functions in $\overline{V}^0$. Also for the present purposes string ordering $X \leq Y$ as given in Definition VIII.3.5 is replaced by its equivalent quantifier-free definition in $\overline{V}^0$. See Example VIII.3.12 for the definition of $POW2(x)$.

We need functions to witness bounded existential string quantifiers, just as $f_{\varphi(z),t}$ as defined in (87) (page 125) is used to witness bounded existential number quantifiers. Thus let $G_{\varphi,t}(\vec{x}, \vec{X})$ be the least Y with

$|Y| \leq t(\vec{x}, \vec{X})$ such that $\varphi(\vec{x}, \vec{X}, Y)$ holds, or $POW2(t)$ if there is no such Y. Then $G_{\varphi,t}$ has defining axiom (suppressing $\vec{x}, \vec{X}$)

$$G_{\varphi,t} \leq POW2(t) \wedge \left(G_{\varphi,t} < POW2(t) \supset \varphi(G_{\varphi,t}) \right) \wedge$$
$$\left(Y < G_{\varphi,t} \supset \neg\varphi(Y) \right). \quad (212)$$

The definition of vocabularies $\mathcal{L}_{FP^i}$ is similar to Definition VIII.2.1 for $\mathcal{L}_{FP}$.

DEFINITION VIII.7.6. The vocabularies $\mathcal{L}_{FP^1} \subset \mathcal{L}_{FP^2} \subset \cdots$ are defined as follows.

(i) $\mathcal{L}_{FP^1} = \mathcal{L}_{FP}$.
(ii) For $i \geq 1$ $\mathcal{L}_{FP^{i+1}}$ is the smallest set that satisfies
    (1) $\mathcal{L}_{FP^{i+1}} \supseteq \mathcal{L}_{FP^i}$.
    (2) For each open formula $\varphi(\vec{x}, \vec{X}, Y)$ over $\mathcal{L}_{FP^i}$ and term $t(\vec{x}, \vec{X})$ of $\mathcal{L}_A^2$ there is a string function $G_{\varphi,t}$ in $\mathcal{L}_{FP^{i+1}}$.
    (3) For each open formula $\varphi(z, \vec{x}, \vec{X})$ over $\mathcal{L}_{FP^{i+1}}$ and term $t = t(\vec{x}, \vec{X})$ of $\mathcal{L}_A^2$ there is a string function $F_{\varphi(z),t}$ in $\mathcal{L}_{FP^{i+1}}$.
    (4) For each triple G, H, t, where $G(\vec{x}, \vec{X})$ and $H(y, \vec{x}, \vec{X}, Z)$ are functions in $\mathcal{L}_{FP^{i+1}}$ and $t = t(y, \vec{x}, \vec{X})$ is a term in $\mathcal{L}_A^2$, there is a function $F_{G,H,t}$ in $\mathcal{L}_{FP^{i+1}}$.

DEFINITION VIII.7.7. For $i \geq 1$ the universal theory VPV^i has vocabulary $\mathcal{L}_{FP^i}$ and (i) $VPV^1 = VPV$ and (ii) for $i \geq 1$ VPV^{i+1} contains VPV^i and has as (sometimes) additional defining axioms (212) for each function $G_{\varphi,t}$ in $\mathcal{L}_{FP^{i+1}}$ and (177) for each function $F_{\varphi(z),t}$ in $\mathcal{L}_{FP^{i+1}}$ and (175), (176) (page 210) for each function $F_{G,H,t}$ in $\mathcal{L}_{FP^{i+1}}$.

LEMMA VIII.7.8. *For all $i \geq 0$ for every Σ_i^B-formula ψ there is an open formula φ over $\mathcal{L}_{FP^{i+1}}$ such that VPV^{i+1} proves $\psi \leftrightarrow \varphi$.*

PROOF. We use induction on i. For $i = 0$ this is clear. Now suppose $i > 0$ and ψ is a Σ_i^B-formula. Then we may assume that $\psi \equiv \exists Y < t \, \eta(Y)$, where η is Π_{i-1}^B. By the induction hypothesis there is an open formula φ over $\mathcal{L}_{FP^i}$ such that VPV^i proves $\eta \leftrightarrow \varphi$. Then VPV^{i+1} proves

$$\varphi(G_{\varphi,t}) \leftrightarrow \exists Y < t\varphi(Y)$$

so $VPV^i \vdash \psi \leftrightarrow \varphi(G_{\varphi,t})$ and hence $\varphi(G_{\varphi,t})$ satisfies the Lemma. $\square$

THEOREM VIII.7.9. *For $i \geq 0$ the string function symbols F of $\mathcal{L}_{FP^{i+1}}$ represent precisely the string functions in $FP^{\Sigma_i^P}$ and terms of the form $|F(\vec{x}, \vec{X})|$ represent precisely the number functions of $FP^{\Sigma_i^P}$.*

PROOF. The part about number functions follows from the part about string functions, so we prove the latter. We use induction in i. For $i = 0$ this was observed when introducing $\mathcal{L}_{FP}$. For the induction step the proof is similar to the proof of Cobham's Theorem. To see that every string function in $\mathcal{L}_{FP^{i+1}}$ is in $FP^{\Sigma_i^p}$ it suffices to show that this is true for each of the cases in part (ii) of Definition VIII.7.6. For 3) and 4) this is true because the functions computable by a polynomial time Turing machine with a Σ_i^p oracle are closed under composition and limited recursion, and such a machine can evaluate an open formula whose functions are so computable. For 2), observe that such a machine can query its Σ_i^p oracle to find out for $W \leq POW2(t)$ whether there is $Y \leq W$ satisfying $\varphi(\vec{x}, \vec{X}, Y)$, and hence use binary search to find the least such Y (if any).

Conversely, to see that every string function in $FP^{\Sigma_i^p}$ is represented by a function symbol in $\mathcal{L}_{FP^{i+1}}$, use limited recursion to define a function like $Conf_M$ in the proof of Cobham's Theorem VI.2.12 to compute the configurations of the oracle Turing machine, where now the Σ_i^p oracle queries are answered with the help of open formulas in Lemma VIII.7.8. Now the value of F can be extracted using the output function Out_M as in (100) (page 140), so $F(\vec{x}, \vec{X}) = T(\vec{x}, \vec{X})$ for some term T of $\mathcal{L}_{FP^{i+1}}$. Then $F \equiv F_{\varphi(z),t}$ where $\varphi(z) \equiv T(z)$ and t is a bounding term for F. □

The next result generalizes Theorem VIII.2.7.

THEOREM VIII.7.10. *For* $i \geq 1$ VPV^i *proves the* $\Sigma_0^B(\mathcal{L}_{FP^i})$-*COMP*, $\Sigma_0^B(\mathcal{L}_{FP^i})$-*IND*, $\Sigma_0^B(\mathcal{L}_{FP^i})$-*MIN*, *and* $\Sigma_0^B(\mathcal{L}_{FP^i})$-*MAX* *axiom schemes.*

PROOF. By Corollary V.1.8 it suffices to prove this for the case of *COMP*. For every $\Sigma_0^B(\mathcal{L}_{FP^i})$-formula φ there is an open $\mathcal{L}_{FP^i}$-formula φ^+ such that VPV^i proves $\varphi \leftrightarrow \varphi^+$ (see the proof of Lemma V.6.3). The function $F_{\varphi,y}$ is easily used to prove the comprehension axiom for an open formula φ. □

THEOREM VIII.7.11. *For* $i \geq 0$, *every function in* $\mathcal{L}_{FP^{i+1}}$ *is* Σ_{i+1}^B-*definable in* TV^i, *and* VPV^{i+1} *is a conservative extension of* TV^i.

PROOF. For $i = 0$ this follows from Theorems VIII.2.11 a), VIII.2.17, VIII.3.10, and Corollary VIII.2.18.

In general we show VPV^{i+1} extends TV^i, by showing VPV^{i+1} proves the Σ_i^B-*SIND* axioms (184). By Lemma VIII.7.8 we may assume that φ is an open $\mathcal{L}_{FP^{i+1}}$-formula. Now proceed exactly as in the proof of Theorem VIII.3.14, replacing VPV by VPV^{i+1}.

To show that the extension is conservative, and to show that every $\mathcal{L}_{FP^{i+1}}$-function is Σ_{i+1}^B-definable in TV^i, by Theorem VIII.7.9 it suffices to show that all functions in $FP^{\Sigma_i^p}$ are Σ_{i+1}^B-definable in TV^i, and (for conservativity) that this can be done in such a way that the defining axioms for the $\mathcal{L}_{FP^{i+1}}$-functions are provable. We omit the latter (which amounts to formalizing in TV^i part of the proof of Theorem VIII.7.9), and concentrate on the former.

Let $F(\vec{x}, \vec{X})$ be a function in $FP^{\Sigma_i^p}$, where $i \geq 1$. Then some polynomial time oracle Turing machine M computes F using an oracle $\varphi(W)$, where φ is (represented by) a Σ_i^B-formula.

Let $t = t(\vec{x}, \vec{X})$ be a suitable $\mathcal{L}_A^2$ bounding term and let

$$Comp_\mathsf{M}(\vec{x}, \vec{X}, U, W, Z)$$

be a Σ_0^B-formula which asserts that U codes a computation of M on input $\vec{x}, \vec{X}$ where for all $i < t$, $W^{[i]}$ is the ith oracle query (if any) and $Z(t \dot- i)$ is the answer to this query.

Define $\psi(\vec{x}, \vec{X}, Z, Y)$ to be

$$\exists U \le t \exists W \le t\, Comp_\mathsf{M}(\vec{x}, \vec{X}, U, W, Z) \wedge Y = Ans_\mathsf{M}(U) \wedge$$
$$\forall i < t(Z(t \dot- i) \supset \varphi(W^{[i]}))$$

where $Ans_\mathsf{M}(U)$ is the output of the computation coded by U. Let $\psi'(\vec{x}, \vec{X}, Z)$ be $\exists Y < t\psi(\vec{x}, \vec{X}, Z, Y)$. Then ψ' is a $g\Sigma_i^B$-formula, which TV^i proves equivalent to a Σ_i^B-formula by the Replacement scheme (Corollary VI.3.8). Note that if $\psi'(\vec{x}, \vec{X}, Z)$ holds then the 'true' query answers coded by Z must be correct, but 'false' answers may not be correct. However the largest Z satisfying ψ' must code all correct answers, since if the ith query is the first incorrect answer then changing $Z(t \dot- i)$ from 'false' to 'true' would increase Z no matter how the subsequent answers are changed.

Since TV^i proves the Σ_i^B-$SMAX$ axioms (Theorem VIII.3.8) and VPV proves $\psi'(\vec{x}, \vec{X}, \varnothing)$ it follows that TV^i proves the existence of a largest Z, $|Z| < t$, satisfying $\psi'(\vec{x}, \vec{X}, Z)$.

Thus we may use the following definition for $F(\vec{x}, \vec{X})$.

$$Y = F(\vec{x}, \vec{X}) \leftrightarrow \exists Z < t\big(\psi(\vec{x}, \vec{X}, Z, Y) \wedge$$
$$\forall Z' < t(Z < Z' \supset \neg\psi'(\vec{x}, \vec{X}, Z'))\big).$$

The formula $\eta(\vec{x}, \vec{X}, Y)$ on the RHS is equivalent to a Σ_{i+1}^B-formula. Also TV^i proves the existence of a largest $Z < t$ satisfying $\psi'(\vec{x}, \vec{X}, Z)$ and hence the existence of Y satisfying $\eta(\vec{x}, \vec{X}, Y)$. Finally V^0 (and hence TV^i) proves the uniqueness of Y, since obviously there is at most one largest Z satisfying ψ', and by Σ_0^B-IND this Z uniquely determines U, W in $Comp_\mathsf{M}$ and hence Y. □

THEOREM VIII.7.12. *For $i \ge 0$ the following are equivalent for a string function F:*

(i) *F is in $FP^{\Sigma_i^P}$.*
(ii) *F is Σ_{i+1}^B-definable in TV^i.*
(iii) *F is Σ_{i+1}^B-definable in V^{i+1}.*
(iv) *F is Σ_{i+1}^B-definable in VPV^{i+1}.*
(v) *F is $\Sigma_1^B(\mathcal{L}_{FP^{i+1}})$-definable in VPV^{i+1}.*

Similarly for a number function f.

Proof. (i) $\implies$ (ii) by Theorems VIII.7.9 and VIII.7.11. (ii) $\implies$ (iii) by Theorem VIII.7.4. (iii) $\implies$ (ii) by Theorem VIII.7.13. (ii) $\implies$ (iv) by Theorem VIII.7.11. (iv) $\implies$ (v) by Lemma VIII.7.8. (v) $\implies$ (i) by Theorems VIII.2.4 and VIII.7.9.   □

Theorem VIII.7.13. *For* $i \geq 0$ V^{i+1} *is* Σ_{i+1}^B-*conservative over* TV^i.

Proof. By Lemma VIII.7.8 every Σ_{i+1}^B-formula φ is provably equivalent in VPV^{i+1} to a $\Sigma_1^B(\mathcal{L}_{FP^{i+1}})$-formula φ'. Thus if V^{i+1} proves φ then $V^{i+1} + VPV^{i+1}$ proves φ', and, arguing as in the proof of Theorem VIII.2.13, VPV^{i+1} proves that φ' can be witnessed by functions in $\mathcal{L}_{FP^{i+1}}$. Thus VPV^{i+1} proves φ' and φ, and so TV^i proves φ by Theorem VIII.7.11.   □

The next result generalizes Theorem VIII.5.12. We define a $PLS^{\Sigma_i^P}$ problem Q to be the same as in Definition VIII.5.4 except now the relation φ_Q and the functions N_Q and P_Q are allowed to be polynomial time with a Σ_i^P-oracle.

Theorem VIII.7.14. *For* $i \geq 1$ *a search problem* Q *is* Σ_i^B-*definable in* TV^i *iff* $Q \leq_{AC^0} Q'$ *for some* $PLS^{\Sigma_{i-1}^P}$ *problem* Q'.

Proof. The proof is very similar to the proof of Theorem VIII.5.12. Instead of an AC^0-*ITERATION* problem we need an $FP^{\Sigma_{i-1}^P}$-*ITERATION* problem, in which the function F is allowed to be in $FP^{\Sigma_{i-1}^P}$. To see that every $PLS^{\Sigma_{i-1}^P}$ problem is many-one reducible to some $FP^{\Sigma_{i-1}^P}$-*ITERATION* problem, we need to slightly alter the proof of Theorem VIII.5.8. The difficulty in that proof is that the reducing function G is defined by $G(U *_t V) = N_Q(U)$, where the neighborhood function N_Q is now allowed to be in $FP^{\Sigma_{i-1}^P}$ instead of in FAC^0. To fix this, we change the iterating function F in the proof to a function F'. The idea behind F was to let its domain be concatenations $U *_t V$ where U is a candidate solution for Q and V is its profit. The idea behind F' is that its domain consists of concatenations $(U *_t W) *_{2t} V$ where U and V are as before, and $W = N_Q(U)$. Now we can define the reducing function G by $G((U *_t W) *_{2t} V) = W$.

Exercise VIII.7.15. Work out the details in the definition of F'.

Continuing the proof of Theorem VIII.7.14, it remains to generalize the witnessing theorem VIII.5.13 so that the assumption is

$$TV^i \vdash \exists Z \varphi(\vec{x}, \vec{X}, Z)$$

where now φ is Π_{i-1}^B and Q_F is an $FP^{\Sigma_{i-1}^P}$-*ITERATION*-problem. By Theorems VIII.7.11 and VIII.7.12 it suffices to replace TV^i by $TV^i + VPV^i$ and φ by an open formula in $\mathcal{L}_{FP^i}$, and $\overline{V}^0$ by VPV^i. The proof is a straightforward modification of the proof of Theorem VIII.5.13, where now the string induction rule (197) applies to $\Sigma_1^B(\mathcal{L}_{FP^i})$-formulas A.   □

The previous results characterize the search problems Σ^B_i-definable in V^j and TV^j when $i = j$ and sometimes when i and j differ by one. In order to specify these search problems for more general i and j we need to define a generalization of oracle Turing machines. (See also Appendix A.3.)

DEFINITION VIII.7.16. A *witness query* Y to an oracle $\exists W \le t\, R(Y, W)$ returns a witness $W \le t$ satisfying $R(Y, W)$ if such exists, and otherwise returns "NO". For $i \ge 1$, $FP^{\Sigma^P_i}[wit, O(g(n))]$ is the class of search problems Q solvable by a polynomial time Turing machine that makes $O(g(n))$ witness queries to a Σ^P_i oracle $\exists W \le t\, R(Y, W)$ for $R \in \Pi^P_{i-1}$.

Notice that a witness query can be simulated by polynomial many Boolean queries, using binary search. Hence

$$FP^{\Sigma^P_i}[wit, n^{O(1)}] = FP^{\Sigma^P_i}.$$

However as far as we know, the class $FP^{\Sigma^P_i}[wit, O(g(n))]$ cannot be specified without referring to witness queries when $g(n)$ is $O(\log n)$.

	V^0	TV^0	V^1	TV^1
Σ^B_1	FAC^0	FP	FP	$CC(PLS)$
Σ^B_2	$FP^{\Sigma^P_1}[wit, \mathcal{O}(1)]$	$FP^{NP}[wit, \mathcal{O}(1)]$	$FP^{NP}[wit, \mathcal{O}(\log n)]$	FP^{NP}
Σ^B_3	$FP^{\Sigma^P_2}[wit, \mathcal{O}(1)]$	$FP^{\Sigma^P_2}[wit, \mathcal{O}(1)]$	$FP^{\Sigma^P_2}[wit, \mathcal{O}(1)]$	$FP^{\Sigma^P_2}[wit, \mathcal{O}(1)]$
Σ^B_4	$FP^{\Sigma^P_3}[wit, \mathcal{O}(1)]$	$FP^{\Sigma^P_3}[wit, \mathcal{O}(1)]$	$FP^{\Sigma^P_3}[wit, \mathcal{O}(1)]$	$FP^{\Sigma^P_3}[wit, \mathcal{O}(1)]$
Σ^B_5	$FP^{\Sigma^P_4}[wit, \mathcal{O}(1)]$	$FP^{\Sigma^P_4}[wit, \mathcal{O}(1)]$	$FP^{\Sigma^P_4}[wit, \mathcal{O}(1)]$	$FP^{\Sigma^P_4}[wit, \mathcal{O}(1)]$

	V^2	TV^2	V^3	TV^3
Σ^B_1	$CC(PLS)$			
Σ^B_2	FP^{NP}	$CC(PLS)^{NP}$	$CC(PLS)^{NP}$	
Σ^B_3	$FP^{\Sigma^P_2}[wit, \mathcal{O}(\log n)]$	$FP^{\Sigma^P_2}$	$FP^{\Sigma^P_2}$	$CC(PLS)^{\Sigma^P_2}$
Σ^B_4	$FP^{\Sigma^P_3}[wit, \mathcal{O}(1)]$	$FP^{\Sigma^P_3}[wit, \mathcal{O}(1)]$	$FP^{\Sigma^P_3}[wit, \mathcal{O}(\log n)]$	$FP^{\Sigma^P_3}$
Σ^B_5	$FP^{\Sigma^P_4}[wit, \mathcal{O}(1)]$	$FP^{\Sigma^P_4}[wit, \mathcal{O}(1)]$	$FP^{\Sigma^P_4}[wit, \mathcal{O}(1)]$	$FP^{\Sigma^P_4}[wit, \mathcal{O}(1)]$

TABLE 3. Definable search problems.

THEOREM VIII.7.17.   (i) *For $i \ge 1$, a search problem Q is Σ^B_{i+1}-definable in V^i iff $Q \in FP^{\Sigma^P_i}[wit, O(\log n)]$.*
(ii) *For $j \ge 2$ and $V^0 \subseteq \mathcal{T} \subseteq TV^{j-2}$, a search problem Q is Σ^B_j-definable in $\mathcal{T}$ iff $Q \in FP^{\Sigma^P_{j-1}}[wit, O(1)]$.*

PROOF OF (ii). By Theorem VIII.7.11 VPV^{j-1} extends TV^{j-2} and hence the 'only if' direction is an easy consequence of the KPT Witnessing Theorem (see Exercise VIII.7.19 below).

For the 'if' direction it suffices to show that for $j \geq 2$, V^0 can Σ_j^B-define every search problem Q in $FP^{\Sigma_{j-1}^p}[wit, O(1)]$. For concreteness we show this for $j = 2$; the general argument is essentially the same.

Let M be a polytime Turing machine which solves $Q(\vec{x}, \vec{X})$ by making at most a constant q number of queries to a Σ_1^p witness oracle represented by a Σ_1^B-formula

$$\psi(Y) \equiv \exists W < t \, \eta(Y, W)$$

where η is in Σ_0^B. It is straightforward to give a Σ_2^B-formula $\varphi(\vec{x}, \vec{X}, Z)$ which asserts that there is some computation C of M on input $\vec{x}, \vec{X}$ with correct answers to all oracle queries such that C outputs Z. However it is more difficult to find such a formula such that V^0 proves $\exists Z \varphi$.

To do this we first observe that we can find a machine M$'$ which is equivalent to M but such that M$'$ makes all of its queries in parallel, so that the answer to a query does not depend on the witness answer to any other query. The machine M$'$ makes a witness query for each of the 2^q binary strings of length q, asking whether there is an apparently-correct computation of M on input $\vec{x}, \vec{X}$ such that the YES-NO answers to the $\leq q$ queries correspond to the bits of the string (or an initial segment). Here 'apparently-correct' means that for each query Y to $\psi(Y)$ a YES answer must be supplied with a witness W satisfying $\eta(Y, W)$, although NO answers need not be verified. Each YES answer to a query to M$'$ must include a witness which codes an apparently-correct computation of M. From these witnesses M$'$ can find a truly correct computation C of M where no initial segment S of C ending in a NO answer coincides with an initial segment S' of another computation except S' ends in a YES answer to the same query.

The parallel queries made by M$'$ all have the form

$$\psi'(Y) \equiv \exists W < t \, \eta'(\vec{x}, \vec{X}, Y, W)$$

where η' is Σ_0^B-formula and Y is simply a bit string of length q. Now we describe a Σ_2^B-formula $\alpha_{\mathsf{M}}(\vec{x}, \vec{X}, Z)$ which asserts that Z is a possible output of a correct computation of M$'$ on input $\vec{x}, \vec{X}$, and such that V^0 proves $\exists Z \alpha_{\mathsf{M}}$. It suffices to describe $\alpha_{\mathsf{M}}(Z)$ as a disjunction of two Σ_2^B-formulas $\alpha_{\mathsf{M}}^1(Z) \vee \alpha_{\mathsf{M}}^2$, where $\alpha_{\mathsf{M}}^1(Z)$ makes the assertion as just stated and α_{M}^2 is false.

Let $ApCor(\vec{x}, \vec{X}, C)$ be a Σ_0^B-formula which asserts that C codes an apparently-correct computation of M$'$ on input $\vec{x}, \vec{X}$. Then $\alpha_{\mathsf{M}}^1(Z)$ asserts the intended meaning of $\alpha_{\mathsf{M}}(Z)$ in the obvious way:

$$\alpha_{\mathsf{M}}^1(Z) \equiv \exists C \forall W, \; ApCor(C) \wedge \neg Wit(W, C) \wedge Z = Out(C)$$

where we have omitted the bounds on the quantifiers and suppressed the arguments $\vec{x}, \vec{X}$, and $Wit(W, C)$ is a Σ_0^B-formula asserting that W is a witness for some query in C which was (incorrectly) answered NO, and $Out(C)$ is the output of the computation C.

Presumably V^0 does not prove $\exists Z \alpha_M^1(Z)$, so we need the false disjunct α_M^2, which asserts that there is no correct computation of M' on input $\vec{x}, \vec{X}$. Specifically α_M^2 is a Σ_2^B-formulas which asserts that there exists a sequence $W_1, \ldots, W_\ell$ of potential witnesses to the $\ell = 2^q$ parallel queries made by a computation of M' such that for all apparently-correct computations C, one of the NO queries in C is in fact witnessed by some W_i. It suffices to prove the following:

CLAIM. V^0 proves $\exists Z \left(\alpha_M^1(Z) \vee \alpha_M^2 \right)$.

Reasoning in V^0, if α_M^2 then (since Z does not occur in α_M^2) we can conclude $\exists Z \, \alpha_M^2$ and we are done.

So assume $\neg \alpha_M^2$. For each of the $\ell = 2^q$ queries Y_i (which are simply strings of length q), if $\exists W \eta'(Y_i, W)$ then let W_i be a witness satisfying $\eta'(Y_i, W_i)$; otherwise let $W_i = \varnothing$. Since $\neg \alpha_M^2$ there exists an apparently-correct computation C such that none of the NO queries in C is witnessed by any W_i. But by the way we chose W_i, this means that all NO queries are correct, and hence C is a correct computation. Let $Z = Out(C)$. Then $\alpha_M^1(Z)$. Hence $\exists Z \, \alpha_M(Z)$. $\qquad\square$

EXERCISE VIII.7.18. Explain what goes wrong if we try to extend the above proof to the case that M makes more than a constant number of witness oracle queries.

PROOF OF (i). The 'only if' direction can be proved using the same method as for Theorem VI.4.1 (witnessing for V^1); see [70, 72] for details. For the proof of the 'if' direction let M be a polytime Turing machine which solves $Q(\vec{x}, \vec{X})$ by making $O(\log n)$ queries to a witness oracle represented by a formula $\psi(Y) \equiv \exists W < t \, \eta(Y, W)$, where η is in Π_{i-1}^B. It is easy to see that there is a Σ_{i+1}^B-formula $\varphi(\vec{x}, \vec{X}, Z)$ which asserts that $Z \in Q(\vec{x}, \vec{X})$ by asserting that there is a computation of M on input $\vec{x}, \vec{X}$ with output Z such that if $Y^{[i]}$ codes the ith query and $W^{[i]}$ codes the ith answer, then either $\eta(Y^{[i]}, W^{[i]})$ or $\neg \psi(Y^{[i]})$ and $W^{[i]} = $ 'NO'.

In order to show that V^i proves $\exists Z \varphi(\vec{x}, \vec{X}, Z)$ we use the fact that V^i proves the Σ_i^B-MAX axiom scheme (Corollary VI.1.4) and argue as in the proof of Theorem VIII.7.11. Thus V^i proves there is a largest $n < t$ (for suitable t) satisfying the Σ_i^B-formula $\alpha(n, \vec{x}, \vec{X})$, which asserts that there exists a computation of M as above and there exists the query sequence Y and answer sequence W such that the bits of the reverse binary notation for n code the Boolean answers to the successive queries of the computation, and for all i, if the ith query answer is positive, then $\eta(Y^{[i]}, W^{[i]})$. $\qquad\square$

EXERCISE VIII.7.19. Show using the KPT Witnessing Theorem VIII.6.1 that for $i \geq 1$ if a search problem Q is Σ^B_{i+1}-definable in VPV^i then Q is in $FP^{\Sigma^P_i}[wit, O(1)]$.

VIII.7.3. Collapse of V^∞ vs Collapse of PH. It is an open question whether V^∞ (the union of the theories V^i) collapses to some particular V^i. Since each V^i is finitely axiomatizable (Theorem VIII.7.3), this question is equivalent to asking whether V^∞ is finitely axiomatizable. As far as we know it is possible that the polynomial hierarchy PH could collapse without V^∞ collapsing. (For example there might be a polynomial time algorithm for propositional satisfiability whose correctness is not provable in V^∞.) However if some V^i proves that PH collapses, then V^∞ collapses to V^i. That is, if for every Σ^B_{i+1}-formula φ there is a Σ^B_i-formula φ' such that V^i proves $\varphi \leftrightarrow \varphi'$, then V^i proves Σ^B_{i+1}-$COMP$, so $V^{i+1} = V^i$. But the same assumption shows that $V^{i+2} = V^{i+1}$, and so on, so $V^\infty = V^i$.

The following theorem is an application of KPT Witnessing, and shows that the converse also holds: if V^∞ collapses to V^i then V^i proves that PH collapses. (This would be obvious if a function is in $FP^{\Sigma^P_{i-1}}$ iff it is Σ^B_1-definable in V^i, as opposed to Σ^B_i-definable in V^i as stated in Theorem VIII.7.12.)

THEOREM VIII.7.20 ([75, 26, 112]). *For $i \geq 0$ if $TV^i = V^{i+1}$ then $TV^i = V^\infty$ and $\Sigma^P_{i+2} = \Pi^P_{i+2} = PH$ and TV^i proves $\Sigma^P_{i+3} = \Pi^P_{i+3} = PH$.*

COROLLARY VIII.7.21. *V^∞ is finitely axiomatizable iff some V^i proves that PH collapses.*

PARTIAL PROOF OF THEOREM VIII.7.20. For readability we treat the case $i = 0$; the general case is similar. (See the remark at the end of this proof.) Assuming $TV^0 = V^1$ we show that PH collapses to $P/poly$ and provably collapses to $NP/poly$, where $poly$ refers to polynomial "advice", as explained below. It follows from the methods of Karp and Lipton [66] that PH collapses to $\Sigma^P_2 = \Pi^P_2$ and provably collapses to $\Sigma^P_3 = \Pi^P_3$. The proof that $TV^0 = V^1$ implies $TV^0 = V^\infty$ can be obtained from [26].

Since VPV is a conservative extension of TV^0 (Theorem VIII.7.11) and $V^1(VPV) = V^1 + VPV$ (Section VIII.2.1) our assumption $TV^0 = V^1$ is equivalent to $VPV = V^1(VPV)$.

The assertion "Every sequence $\alpha_1, \ldots, \alpha_m$ of propositional formulas has an initial sequence of maximal length ℓ of satisfiable formulas" is expressible by a formula

$$\psi \equiv \forall X \exists Y \forall Z \varphi(X, Y, Z)$$

where φ is an open $\mathcal{L}_{FP}$-formula. Here X codes the sequence $\alpha_1, \ldots, \alpha_m$ and φ asserts that Y codes a sequence of satisfying assignments to the the first ℓ formulas of X for some $\ell \leq m$, and also that if $\ell < m$ then Z codes an assignment which falsifies $\alpha_{\ell+1}$.

$V^1(VPV)$ proves ψ by applying the $\Sigma_1^B\text{-}MAX$ axioms (Corollary VIII.2.12) to the $\Sigma_1^B(\mathcal{L}_{FP})$-formula expressing the condition that the first ℓ formulas coded by X are satisfiable. Hence by our assumption, VPV proves ψ, so by the KPT Witnessing Theorem there are polytime functions $F_1, \ldots, F_k$ such that VPV proves

$$\varphi(X, F_1(X), Z_1) \vee \varphi(X, F_2(X, Z_1), Z_2) \vee \cdots \vee$$
$$\varphi(X, F_k(X, Z_1, \ldots, Z_{k-1}), Z_k). \quad (213)$$

Note that each function F_i plays the role of Y, and hence should code a sequence of assignments satisfying some initial segment of the formulas coded by X. From these functions $F_1, \ldots, F_k$ we obtain polytime functions $G_1, \ldots, G_k$ such that VPV proves that for every sequence $\alpha_1, \ldots, \alpha_k$ of propositional formulas with satisfying assignments $Z_1', \ldots, Z_k'$ there is an i, $1 \leq i \leq k$, such that $G_i(X, Z_1', \ldots, Z_{i-1}')$ codes a satisfying assignment for α_i (we say that G_i 'wins' in this case).

The algorithm for evaluating $G_1(X)$ proceeds by computing $W_1 = F_1(X)$. If the sequence W_1 begins with an assignment satisfying α_1, then $G_1(X)$ is set to that assignment, so G_1 wins. Otherwise the algorithm for $G_2(X, Z_1')$ sets $Z_1 = Z_1'$, so the first disjunct $\varphi(F_1(X), Z_1)$ in (213) is false (since by assumption Z_1' satisfies α_1). Now the G_2 algorithm computes $W_2 = F_2(X, Z_1)$. If W_2 includes an assignment satisfying α_2, then $G_2(X, Z_1')$ is set to that assignment, and G_2 wins. Otherwise the algorithm for $G_3(X, Z_1', Z_2')$ sets Z_2 to either Z_1' or Z_2' depending on $F_2(X, Z_1)$, so that the second conjunct $\varphi(X, F_2(X, Z_1), Z_2)$ in (213) is false. Then G_3 is set to an assignment in $F_3(X, Z_1, Z_2)$ which satisfies α_3, if one exists. In general, if none of $G_1, \ldots, G_{i-1}$ wins, then the algorithm for G_i chooses $Z_1, \ldots, Z_{i-1}$ so the first $i-1$ disjuncts in (213) are false, and evaluates $F_i(X, Z_1, \ldots, Z_{i-1})$, looking for an assignment that satisfies α_i. At least one of $G_1, \ldots, G_k$ must win, since otherwise $Z_1, \ldots, Z_k$ can be chosen so as to falsify (213).

P/poly is the class of problems solvable by a polynomial size family of Boolean circuits, or equivalently the class of problems solvable by a polynomial time Turing machine which is allowed a polynomial length advice string A_n for each input length n. In order to show that $NP \subseteq P/poly$ it suffices to define a polytime relation $R(X, Y)$ such that for each n there is an advice string A_n of length bounded by a polynomial in n so that for every propositional formula α of length n, $R(\alpha, A_n)$ holds iff α is satisfiable.

We now explain how to use the functions $G_1, \ldots, G_k$ to define R and A_n. For each satisfiable propositional formula α of length n we associate a fixed assignment Z_α which satisfies α. We define a map H which takes a k-tuple $X = (\alpha_1, \ldots, \alpha_k)$ of distinct satisfiable propositional formulas of length n, where the formulas in X are ordered lexicographically, to a

$(k-1)$-tuple of such formulas, where $H(X)$ is obtained from X by deleting the first formula α_i such that the assignment $G_i(X, Z_{\alpha_1}, \ldots, Z_{\alpha_{i-1}})$ satisfies α_i. The domain of H has size (the number of ordered k-tuples X)

$$C(C-1)\ldots(C-k+1)$$

and the range of H has size

$$C(C-1)\ldots(C-k+2).$$

Each $(k-1)$-tuple has at most k preimages. Hence there is a $(k-1)$-tuple $(\beta_1, \ldots, \beta_{k-1})$ of formulas which is the image under H of at least $(C-k+1)/k$ different k-tuples. Part of the advice string A_n codes the sequence $\beta_1, \ldots, \beta_{k-1}, Z_{\beta_1}, \ldots, Z_{\beta_{k-1}}$. Each of the k-tuples mapping to $(\beta_1, \ldots, \beta_{k-1})$ consists of $(\beta_1, \ldots, \beta_{k-1})$ with a new formula α inserted somewhere. Further distinct such k-tuples have distinct inserted formulas α, since the formulas in the tuples are ordered lexicographically. Hence there are at least $(C-k+1)/k$ such formulas α, and each such α has a satisfying assignment which can be computed from the advice string A_n using $G_1, \ldots, G_k$.

Now delete this set of at least $(C-k+1)/k$ formulas α from the set of satisfiable formulas of length n, and apply the above process to the set of remaining formulas, obtaining another $(k-1)$-tuple of formulas and satisfying assignments to add to the advice string A_n. After $O(\log n)$ such iterations, an advice string A_n of length $O(n \log n)$ is obtained which, using the functions $G_1, \ldots, G_k$ suffices to compute a satisfying assignment to any satisfiable formula of length n. This yields the required polynomial time procedure with advice for solving the satisfiability problem.

The correctness proof for the above $P/poly$ procedure seems to require a counting argument which cannot (as far as we know) be formalized in V^∞. We now show how to define an advice string A'_n which can be used to put the satisfiability problem in $co\text{-}NP/poly$, provably in V^2. Again we use the functions $G_1, \ldots, G_k$ described above. The idea is to find the smallest ℓ, $1 \le \ell \le k$, such that there exists formulas $\alpha_{\ell+1}, \ldots, \alpha_k$ of length n (not necessarily satisfiable) such that for all tuples $(\alpha_1, \ldots, \alpha_\ell)$ of satisfiable formulas of length n, and all tuples $(Z_1, \ldots, Z_\ell)$ of satisfying assignments for $\vec{\alpha}$, there exists $i \le \ell$ such that $G_i(\alpha_1, \ldots, \alpha_k, Z_1, \ldots, Z_{i-1})$ codes a satisfying assignment for α_i. V^2 proves the existence of ℓ and $\alpha_{\ell+1}, \ldots, \alpha_k$ by the $\Sigma_2^B\text{-}MIN$ axioms (note that k is a candidate for ℓ). The advice A'_n is the tuple $\alpha_{\ell+1}, \ldots, \alpha_k$. Then an arbitrary formula α_ℓ of length n is satisfiable iff for all tuples $(\alpha_1, \ldots, \alpha_{\ell-1})$ and satisfying assignments $(Z_1, \ldots, Z_{\ell-1})$ there exists $i \le \ell$ such that $G_i(\alpha_1, \ldots, \alpha_k, Z_1, \ldots, Z_{i-1})$ codes a satisfying assignment for α_i. (The 'if' direction follows from the minimality of ℓ.) Hence we have expressed length n satisfiability with a Π_1^B

formula involving the advice A_n', which shows that the satisfiability problem is in *co-NP/poly*, and hence *PH* collapses to $NP/poly = co\text{-}NP/poly$, as desired.

To prove this theorem for $i > 0$, replace *VPV* by VPV^i, and replace the propositional formulas α by quantified propositional formulas with a quantifier prefix limited to i alternations beginning with $\exists$.   □

VIII.8. RSUV Isomorphism

Recall the hierarchies of single-sorted theories S_2^i and T_2^i (for $i \geq 1$) from Section III.5. In particular, S_2^1 characterizes the class single-sorted *P* in much the same way as V^1 characterizes the class (two-sorted) *P* (Theorem VI.2.2 and Corollary VI.2.4). Here we will show that each theory S_2^i is essentially a single-sorted version of V^i (for $i \geq 1$), i.e., they are "RSUV isomorphic" (the same is true for T_2^i and TV^i).

This section is organized as follows. First we formally define S_2^i and T_2^i. Then in Section VIII.8.2 we define the notion of an RSUV isomorphism as a bijection between classes of single-sorted and two-sorted models. These are associated with the syntactical translations of single-sorted and two-sorted formulas, defined in Subsections VIII.8.3 and VIII.8.4. Finally we sketch a proof of the RSUV isomorphism between S_2^1 and V^1.

VIII.8.1. The Theories S_2^i and T_2^i. For this subsection it is helpful to revisit Sections III.1, III.5, and IV.3.2. Recall that the vocabulary for S_2^1 is

$$\mathcal{L}_{S_2} = [0, S, +, \cdot, \#, |x|, \lfloor \tfrac{1}{2}x \rfloor; =, \leq]$$

where $|x|$ is the length of the binary representation of x, and the function $x\#y = 2^{|x| \cdot |y|}$ provides the polynomial growth in length for the terms of $\mathcal{L}_{S_2}$.

The *sharply bounded quantifiers* are bounded quantifiers (Definition III.1.6) which are of the form $\exists x \leq |t|$ and $\forall x \leq |t|$. The syntactic classes of bounded formulas of $\mathcal{L}_{S_2}$ are defined as follows.

DEFINITION VIII.8.1 (Bounded Formulas of $\mathcal{L}_{S_2}$). $\Delta_0^b = \Sigma_0^b = \Pi_0^b$ is the set of formulas whose quantifiers are sharply bounded. For $i \geq 0$, Σ_{i+1}^b and Π_{i+1}^b are the smallest sets of formulas that satisfy:

1) $\Pi_i^b \subseteq \Sigma_{i+1}^b$, $\Sigma_i^b \subseteq \Pi_{i+1}^b$.
2) If $\varphi, \psi \in \Sigma_{i+1}^b$ (or Π_{i+1}^b), then so are $\varphi \wedge \psi$, $\varphi \vee \psi$.
3) If $\varphi \in \Sigma_{i+1}^b$ (resp. $\varphi \in \Pi_{i+1}^b$), then $\neg\varphi \in \Pi_{i+1}^b$ (resp. $\neg\varphi \in \Sigma_{i+1}^b$).
4) If $\varphi \in \Sigma_{i+1}^b$ (resp. $\varphi \in \Pi_{i+1}^b$), then $\exists x \leq t\ \varphi$ and $\forall x \leq |t|\ \varphi$ are in Σ_{i+1}^b (resp. $\forall x \leq t\ \varphi$ and $\exists x \leq |t|\ \varphi$ are in Π_{i+1}^b).

Notice that different from Σ_i^B and Π_i^B (Definition IV.3.2), here the formulas in Σ_i^b and Π_i^b are not required to be in prenex form, and any bounded quantifier can occur in the scope of a sharply bounded quantifier. Nevertheless, it can be shown that for $i \geq 1$, a single-sorted relation is in the (single-sorted) class Σ_i^p if and only if it is represented by a Σ_i^b formula. In particular, a single-sorted relation is in *NP* if and only if it is represented by a Σ_1^b formula. (See Definition IV.3.3 and the Σ_i^B and Σ_1^1 Representation Theorem IV.3.7.)

The set *BASIC* of the defining axioms for symbols in $\mathcal{L}_{S_2}$ are given in Figure 3. There 1 and 2 are the numerals $S0$ and $SS0$, respectively. Note that *BASIC* is by no means optimal, i.e., it is possible to derive some of its axioms from others. Here we are not concerned with its optimality.

1. $x \leq y \supset Sx \leq Sy$	18. $\lvert x \rvert = \lvert u \rvert + \lvert v \rvert \supset$
2. $x \neq Sx$	$\quad x \# y = (u \# y) \cdot (v \# y)$
3. $0 \leq x$	19. $x \leq x + y$
4. $(x \leq y \wedge x \neq y) \leftrightarrow Sx \leq y$	20. $x \leq y \wedge x \neq y \supset$
5. $x \neq 0 \supset 2 \cdot x \neq 0$	$\quad S(2 \cdot x) \leq 2 \cdot y \wedge$
6. $x \leq y \vee y \leq x$	$\quad S(2 \cdot x) \neq 2 \cdot y$
7. $(x \leq y \wedge y \leq x) \supset x = y$	21. $x + y = y + x$
8. $(x \leq y \wedge y \leq z) \supset x \leq z$	22. $x + 0 = x$
9. $\lvert 0 \rvert = 0$	23. $x + Sy = S(x + y)$
10. $\lvert S0 \rvert = S0$	24. $(x + y) + z = x + (y + z)$
11. $x \neq 0 \supset (\lvert 2 \cdot x \rvert = S(\lvert x \rvert) \wedge$	25. $x + y \leq x + z \leftrightarrow y \leq z$
$\quad \lvert S(2 \cdot x) \rvert = S(\lvert x \rvert))$	26. $x \cdot 0 = 0$
12. $x \leq y \supset \lvert x \rvert \leq \lvert y \rvert$	27. $x \cdot Sy = (x \cdot y) + x$
13. $\lvert x \# y \rvert = S(\lvert x \rvert \cdot \lvert y \rvert)$	28. $x \cdot y = y \cdot x$
14. $0 \# x = S0$	29. $x \cdot (y + z) = (x \cdot y) + (x \cdot z)$
15. $x \neq 0 \supset (1 \# (2 \cdot x) = 2 \cdot (1 \# x)$	30. $1 \leq x \supset (x \cdot y \leq x \cdot z \leftrightarrow y \leq z)$
$\quad \wedge 1 \# S(2 \cdot x)) = 2 \cdot (1 \# x))$	31. $x \neq 0 \supset \lvert x \rvert = S(\lvert \lfloor \frac{1}{2} x \rfloor \rvert)$
16. $x \# y = y \# x$	32. $x = \lfloor \frac{1}{2} y \rfloor \leftrightarrow$
17. $\lvert x \rvert = \lvert y \rvert \supset x \# z = y \# z$	$\quad (2 \cdot x = y \vee S(2 \cdot x) = y)$

FIGURE 3. *BASIC*.

Recall the definition of an induction scheme Φ-*IND* (Definition III.1.4). For formulas of $\mathcal{L}_{S_2}$ there are other kinds of induction, namely *length induction* and *polynomially induction*, which are defined below.

DEFINITION VIII.8.2 (*LIND* and *PIND*). Let $\mathcal{L}$ be a vocabulary which extends $\mathcal{L}_{S_2}$, and Φ be a set of $\mathcal{L}$-formulas. Then Φ-*LIND* is the set of formulas of the form

$$\big(\varphi(0) \wedge \forall x(\varphi(x) \supset \varphi(x + 1))\big) \supset \forall z \varphi(\lvert z \rvert) \tag{214}$$

and $\Phi\text{-}\boldsymbol{PIND}$ is the set of formulas of the form

$$\bigl(\varphi(0) \wedge \forall x(\varphi(\lfloor\tfrac{1}{2}x\rfloor) \supset \varphi(x))\bigr) \supset \forall z\varphi(z) \qquad (215)$$

where φ is a formula in Φ, $\varphi(x)$ is allowed to have free variables other than x.

DEFINITION VIII.8.3 (S_2^i and T_2^i). For $i \geq 1$, S_2^i is the theory axiomatized by $\boldsymbol{BASIC}$ and $\Sigma_i^b\text{-}\boldsymbol{PIND}$; T_2^i is the theory axiomatized by $\boldsymbol{BASIC}$ and $\Sigma_i^b\text{-}\boldsymbol{IND}$.

We leave as an exercise the following interesting results (for (b) see also Theorem VIII.7.4):

EXERCISE VIII.8.4. Show that for $i \geq 1$:
(a) S_2^i can be axiomatized by $\boldsymbol{BASIC}$ together with $\Sigma_i^b\text{-}\boldsymbol{LIND}$.
(b) $S_2^i \subseteq T_2^i \subseteq S_2^{i+1}$.

S_2^1 and V^1 turn out to be essentially the same, as explained in the next subsection.

VIII.8.2. RSUV Isomorphism. Here we define the notion of RSUV isomorphism model-theoretically by defining the $^\flat$ and $^\sharp$ *mappings* between single-sorted and two-sorted models. These (semantic) mappings are associated with the syntactical translations between of single-sorted and two-sorted formulas, to be defined in later sections.

Recall that $BIT(i, x)$ is the relation which holds if and only if the i-th lower-order bit in the binary representation of x is 1. It is left as an exercise to show that this relation is Σ_1^b-definable in S_2^1. It follows that $S_2^1(BIT)$ is a conservative extension of S_2^1.

EXERCISE VIII.8.5. Show that $BIT(i, x)$ is Σ_1^b-definable in S_2^1, and that
$$S_2^1(BIT) \vdash \forall x \forall y, \; x = y \leftrightarrow (|x| = |y| \wedge \forall i \leq |x|, \; BIT(i, x) \leftrightarrow BIT(i, y)).$$

Now let $\mathcal{M}$ be a model of S_2^1 with universe U. We can construct from $\mathcal{M}$ a two-sorted $\mathcal{L}_A^2$-structure $\mathcal{N}$ as follows. First, expand $\mathcal{M}$ to include the interpretation of BIT. The universe $\langle U_1, U_2 \rangle$ of $\mathcal{N}$ is defined to be

$$U_2 = U, \qquad \text{and} \qquad U_1 = \{|u| : u \in U\}.$$

The constants 0 and 1 are interpreted as 0 and $S0$ respectively (which are in U_1, by the axioms 9 and 10 of $\boldsymbol{BASIC}$). The interpretations of the other symbols of $\mathcal{L}_A^2$ (except for $\in$) in $\mathcal{N}$ are exactly as in $\mathcal{M}$. (Note that by this definition, $|\;|$ is clearly a function from U_2 to U_1.) Finally $\in$ is interpreted as

$$i \in^{\mathcal{N}} x \Leftrightarrow BIT(i, x) \text{ holds in } \mathcal{M}, \qquad \text{for all } i \in U_1, x \in U_2.$$

DEFINITION VIII.8.6. For a model $\mathcal{M}$ of S_2^1, denote by $\mathcal{M}^\sharp$ the two-sorted structure $\mathcal{N}$ obtained as described above.

Conversely, suppose that $\mathcal{N}$ is a model of V^1 with universe $\langle U_1, U_2 \rangle$. We can construct from $\mathcal{N}$ a (single-sorted) $\mathcal{L}_{S_2}$-structure $\mathcal{M}$ with universe $U = U_2$ where each bounded set X in U_2 is interpreted as the number $bin(X)$ (see (46) on page 85):

$$bin(X) = \sum_i X(i)2^i.$$

In order to interpret the symbols of $\mathcal{L}_{S_2}$ in $\mathcal{M}$, we need the fact that the functions and predicates of $\mathcal{L}_{S_2}$ when interpreted as taking string arguments are respectively provably total and definable in V^1.

In fact, by Exercise VI.2.7 the string multiplication function $X \times Y$ is Σ_1^1-definable in V^1. Also, using the fact that $BIT(i, x)$ is definable in $I\Delta_0$ (Subsection III.3.3) and that V^0 is a conservative extension of $I\Delta_0$ (Theorem V.1.9), we have $BIT(i, x)$ is Σ_0^B-definable in V^0:

COROLLARY VIII.8.7. *The relation $BIT(i, x)$ is Σ_0^B-definable in V^0.*

Thus the string function $|X|_2$ whose bit-graph is

$$|X|_2(i) \leftrightarrow (i \leq |X| \wedge BIT(i, |X|))$$

is provably total in V^0.

The string relation $X \leq Y$ is defined in Definition VIII.3.5. The constant 0 is interpreted as the empty set $\varnothing$, which is defined in V^0 by Exercise V.4.19. The successor and addition functions on strings are also definable in V^0 (Exercise V.4.19). Finally, the functions $X \# Y$ and $\lfloor \frac{1}{2} X \rfloor$ can be defined in V^0 using Σ_0^B-*COMP* as follows:

$$(X \# Y)(z) \leftrightarrow z = |X| \cdot |Y| + 1, \qquad \lfloor \tfrac{1}{2} X \rfloor(z) \leftrightarrow z \leq |X| \wedge z + 1 \in X.$$

DEFINITION VIII.8.8. For a model $\mathcal{N}$ of V^1, let $\mathcal{N}^\flat$ denote the single-sorted $\mathcal{L}_{S_2}$-structure $\mathcal{M}$ constructed as above.

DEFINITION VIII.8.9 (RSUV Isomorphism). Let T_1 be a single-sorted theory over $\mathcal{L}_{S_2}$ and T_2 be a two-sorted theory over $\mathcal{L}_A^2$ so that $S_2^1 \subseteq T_1$ and $V^1 \subseteq T_2$. Then T_1 and T_2 are said to be RSUV isomorphic (denoted by $T_1 \overset{\text{RSUV}}{\simeq} T_2$) if (i) for every model $\mathcal{M}$ of T_1, $\mathcal{M}^\sharp \models T_2$, and (ii) for every model $\mathcal{N}$ of T_2, $\mathcal{N}^\flat \models T_1$.

Note that we can loosen the restrictions that $S_2^1 \subseteq T_1$ and $V^1 \subseteq T_2$ by, for example, imposing that BIT is definable in T_1, and $X \times Y$ is definable in T_2 (while maintaining that T_1 extends a certain subtheory of S_2^1, and T_2 extends V^0). This allows us to speak of the RSUV isomorphism between subtheories of S_2^1 and V^1.

The main result of this section is stated below.

THEOREM VIII.8.10. *For $i \geq 1$, S_2^i and V^i are RSUV isomorphic, and T_2^i and TV^i are RSUV isomorphic.*

Associated with the $\sharp$ and $\flat$ mappings defined above are respectively the $\flat$ and $\sharp$ translations of formulas that we will introduce shortly. For example, one direction of Theorem VIII.8.10 (for $i = 1$) requires showing that $\mathcal{M}^\sharp \models V^1$ for every model $\mathcal{M}$ of $S_2^1(BIT)$. Thus we will translate syntactically each $\mathcal{L}_A^2$ formula φ into an $\mathcal{L}_{S_2}(BIT)$ formula $\varphi^\flat$ (the $\flat$ translation) so that

$$\mathcal{M}^\sharp \models \forall\varphi \text{ if and only if } \mathcal{M} \models \forall\varphi^\flat.$$

(Recall that $\forall\varphi$ is the universal closure of φ. See Definition II.2.22.) Then we will prove that $S_2^1(BIT) \vdash \varphi^\flat$ for each axiom φ of V^1.

The $\sharp$ translation is essentially the inverse of the $\flat$ translation. The RSUV isomorphism between S_2^1 and V^1 is pictured below (Figure 4).

$$
\begin{array}{ccc}
S_2^1 & \overset{\text{RSUV}}{\cong} & V^1 \\
\hline
\mathcal{M} & \longrightarrow & \mathcal{M}^\sharp \\
\varphi^\flat & \longleftarrow & \varphi \\
\hline
\mathcal{N}^\flat & \longleftarrow & \mathcal{N} \\
\psi & \longrightarrow & \psi^\sharp
\end{array}
$$

FIGURE 4. The RSUV isomorphism between S_2^1 and V^1.

In the next two subsections we define the $\flat$ and $\sharp$ translations. The proof of Theorem VIII.8.10 will be given in Subsection VIII.8.5.

VIII.8.3. The $\sharp$ Translation. The sharply bounded quantifiers in a bounded $\mathcal{L}_{S_2}$-formula are translated into bounded number quantifiers, and other bounded quantifiers are translated into bounded string quantifiers. In other words, a bound variable is translated into a bound number variable if it is sharply bounded. (Note that the bounding term of a bounded string quantifier bounds the *length* of the quantified variable, while in single-sorted logic the bounding terms are for the *values* of the variables.)

It can be easily seen that simply translating bounded quantifiers as above results in bounded (two-sorted) formulas over the vocabulary that extends $\mathcal{L}_A^2$ by taking the functions (except 0) and predicates of $\mathcal{L}_{S_2}$ to be two-sorted functions and predicates whose arguments can be of either sort. For example, there are formally four $+$ functions: one with arity $\langle 2, 0 \rangle$, two with arity $\langle 1, 1 \rangle$ and one with arity $\langle 0, 2 \rangle$. Also, it is straightforward to determine the sorts to which these functions belong. Thus $x + Y$ and $X + Y$ are string functions, while $|x|$ is a number function.

NOTATION. Let $\mathcal{L}^+$ denote the extension of $\mathcal{L}_A^2$ described above.

The functions of $\mathcal{L}^+$ can be shown to be Σ_1^B-definable in V^1. In fact, the number functions and most of the string functions of $\mathcal{L}^+$ (except for the string multiplication function, or the multiplication functions of "mixed" sorts) are respectively Σ_0^B-definable (in V^0) and Σ_0^B-bit-definable. For example, the number functions $|x|$ and $x \# y$ are Σ_0^B-bit-definable due to the fact that the predicate $BIT(i, x)$ is Δ_0-definable in $I\Delta_0$ (Subsection III.3.3). For the fact that the afore-mentioned multiplication functions are Σ_1^B-definable in V^1, see Exercise VI.2.7 and the discussion in the previous subsection about the $^\flat$ mapping.

The next Corollary follows from Corollary VI.3.11 and Corollary VI.3.8.

COROLLARY VIII.8.11. $V^1(\mathcal{L}^+) \vdash g\Sigma_1^B(\mathcal{L}^+)\text{-}\textbf{IND}.$

Now we define for each bounded $\mathcal{L}_{S_2}$ formula $\psi(\vec{x}, \vec{y})$ a bounded $\mathcal{L}^+$ formula $\psi^\sharp(\vec{x}, \vec{Y})$ (i.e., the subset $\vec{y}$ of the free variables of ψ is selected to be translated into the free string variables of $\psi^\sharp$) so that for every model $\mathcal{N}$ of V^1,

$$\mathcal{N}^\flat \models \forall\vec{x}\forall\vec{y}\psi(|\vec{x}|, \vec{y}) \quad \text{if and only if} \quad \mathcal{N}[\mathcal{L}^+] \models \forall\vec{x}\forall\vec{Y}\psi^\sharp(\vec{x}, \vec{Y})$$

(where $\mathcal{N}[\mathcal{L}^+]$ denote the expansion of $\mathcal{N}$ by including the interpretations for $\mathcal{L}^+$). We will focus on the case where all bounding terms of ψ are of the form $t(\vec{x}, \vec{y})$ (i.e., they involve only the free variables of ψ). We need the following result whose proof is left as an exercise.

EXERCISE VIII.8.12. Let $t(\vec{x}, \vec{y})$ be an $\mathcal{L}_{S_2}$ term. Let $T(\vec{x}, \vec{Y})$ be the $\mathcal{L}^+$ term obtained from $t(\vec{x}, \vec{y})$ by replacing the variables $\vec{y}$ by new string variables $\vec{Y}$, and treating the functions occurring in t as the corresponding functions of $\mathcal{L}^+$. Then there is an $\mathcal{L}_A^2$ term $t'(\vec{x}, |\vec{Y}|)$ so that $V^1(\mathcal{L}^+) \vdash |T(\vec{x}, \vec{Y})| \leq t'(\vec{x}, |\vec{Y}|)$.

The formula $\psi^\sharp(\vec{x}, \vec{Y})$ is constructed inductively as follows. First if $\psi(\vec{x}, \vec{y})$ is an atomic formula, then $\psi^\sharp(\vec{x}, \vec{Y})$ is the atomic formula obtained from $\psi(\vec{x}, \vec{y})$ by translating the free variables $\vec{y}$ into free string variables $\vec{Y}$, and translating the symbols of $\mathcal{L}_{S_2}$ into the appropriate symbols of $\mathcal{L}^+$.

Next, if ψ is $\psi_1 \wedge \psi_2$ (resp. $\psi_1 \vee \psi_2$), then $\psi^\sharp$ is $\psi_1^\sharp \wedge \psi_2^\sharp$ (resp. $\psi_1^\sharp \vee \psi_2^\sharp$). If $\psi \equiv \neg\psi_1$, then $\psi^\sharp$ is obtained from of $\neg\psi_1^\sharp$ by pushing the $\neg$ to the atomic subformulas.

Now consider the case where $\psi(\vec{x}, \vec{y}) \equiv \exists z \leq t \, \psi_1(z, \vec{x}, \vec{y})$. Let $T(\vec{x}, \vec{Y})$ and $t'(\vec{x}, |\vec{Y}|)$ be as in Exercise VIII.8.12. Then

$$\psi^\sharp(\vec{x}, \vec{Y}) \equiv \exists Z \leq 1 + t'(\vec{x}, |\vec{Y}|)(Z \leq T(\vec{x}, \vec{Y}) \wedge \psi_1^\sharp(Z, \vec{x}, \vec{Y})).$$

Next, suppose that $\psi(\vec{x}, \vec{y}) \equiv \exists z \leq |t| \, \psi_1(z, \vec{x}, \vec{y})$. Then

$$\psi^\sharp(\vec{x}, \vec{Y}) \equiv \exists z \leq t'(\vec{x}, |\vec{Y}|)(z \leq |T(\vec{x}, \vec{Y})| \wedge \psi_1^\sharp(z, \vec{x}, \vec{Y})).$$

The cases where $\psi(\vec{x}, \vec{y}) \equiv \forall z \leq t \, \psi_1(z, \vec{x}, \vec{y})$ or $\psi(\vec{x}, \vec{y}) \equiv \forall z \leq |t| \, \psi_1(z, \vec{x}, \vec{y})$ are handled similarly. This completes our description of the $^\sharp$ translation. The proof of its desired properties are left as an exercise.

EXERCISE VIII.8.13. Let $\psi(\vec{x}, \vec{y})$ be an $\mathcal{L}_A^2$-formula.

(a) Show that if ψ is in Σ_i^b (resp. Π_i^b) for some $i \geq 0$, then $\psi^\sharp(\vec{Y})$ is in $g\Sigma_i^B(\mathcal{L}^+)$ (resp. $g\Pi_i^B(\mathcal{L}^+)$).

(b) Let $\mathcal{N}$ be a model of V^1. Show that

$$\mathcal{N}^b \models \forall \vec{x} \forall \vec{y} \psi(|\vec{x}|, \vec{y}) \quad \text{if and only if} \quad \mathcal{N}[\mathcal{L}^+] \models \forall \vec{x} \forall \vec{Y} \psi^\sharp(\vec{x}, \vec{Y}).$$

VIII.8.4. The ^{b} Translation. The ^{b} translation is essentially a syntactical counter-part of the $^\sharp$ mapping. In general we will translate bounded string quantifiers into bounded quantifiers, and bounded number quantifiers into sharply bounded quantifiers. Thus we need to find the translation t' for each bounding term t. This task is left as an exercise (see also Exercise VIII.8.12).

Exercise VIII.8.14. Let $t(\vec{x}, |\vec{Y}|)$ be an $\mathcal{L}_A^2$-term, and $t_1(\vec{x}, |\vec{y}|)$ be the $\mathcal{L}_{S_2}$-term obtained from t by replacing each the string variables $\vec{Y}$ by new variables $\vec{y}$, and replacing each occurrence of 1 by $S0$. Then there is an $\mathcal{L}_{S_2}$-term $t'(\vec{x}, \vec{y})$ so that $S_2^1 \vdash t_1(|\vec{x}|, |\vec{y}|) \leq |t'(\vec{x}, \vec{y})|$.

We also need the following results, which follows from the fact that BIT is Σ_1^b-definable in S_2^1.

Notation. Let $\mathcal{L}_{S_2}^+$ stand for $\mathcal{L}_{S_2} \cup \{BIT\}$.

Exercise VIII.8.15. Show that $S_2^1(BIT)$ proves both axiom schemes $\Sigma_1^b(BIT)$-**LIND** and $\Sigma_1^b(BIT)$-**IND**.

As in the $^\sharp$ translation, we will consider only those formulas whose bounding terms involve only the free variables. Thus suppose that $\varphi(\vec{x}, \vec{Y})$ is a bounded formula whose bounding terms are of the form $t(\vec{x}, |\vec{Y}|)$ (with all variables displayed). Then the $\mathcal{L}_{S_2}^+$ formula $\varphi^b(\vec{x}, \vec{y})$ has the same set of variables as that of φ (where each string variable Y is replaced by a new variable y) and satisfies

$$\mathcal{M}^\sharp \models \forall \vec{x} \forall \vec{Y} \varphi(\vec{x}, \vec{Y}) \quad \text{if and only if} \quad \mathcal{M} \models \forall \vec{x} \forall \vec{y} \varphi^b(|\vec{x}|, \vec{y})$$

for any model $\mathcal{M}$ of $S_2^1(BIT)$.

The formula $\varphi^b(\vec{x}, \vec{y})$ is defined inductively as follows. First, if $\varphi(\vec{x}, \vec{Y})$ is an atomic formula, then let $\varphi^b(\vec{x}, \vec{y})$ be obtained from $\varphi(\vec{x}, \vec{y})$ by

- replacing each occurrence of 1 by $S0$,
- replacing each occurrence of $Y(t)$ by $BIT(t, Y)$, and
- replacing each occurrence of a string variable Y by the corresponding new variable y.

For the induction step, if $\varphi \equiv (\varphi_1 \wedge \varphi_2)$ (resp. $(\varphi_1 \vee \varphi_2)$, $\neg \varphi_1$), then define $\varphi^b \equiv (\varphi_1^b \wedge \varphi_2^b)$ (resp. $(\varphi_1^b \vee \varphi_2^b)$, $\neg \varphi_1^b$).

Next consider the case where $\varphi(\vec{x}, \vec{Y}) \equiv \exists Z \leq t(\vec{x}, |\vec{Y}|) \, \varphi_1(\vec{x}, \vec{Y}, Z)$. Let $t'(\vec{x}, \vec{y})$ be as in Exercise VIII.8.14. Then

$$\varphi^b(\vec{x}, \vec{y}) \equiv \exists z \leq S0 + t'(\vec{x}, \vec{y})\big(|z| \leq t(\vec{x}, |\vec{y}|) \wedge \varphi_1^b(\vec{x}, \vec{y}, z)\big).$$

Now consider the case where $\varphi(\vec{x}, \vec{Y}) \equiv \exists u \leq t(\vec{x}, |\vec{Y}|)\, \varphi_1(u, \vec{x}, \vec{Y})$. Let $t'(\vec{x}, \vec{y})$ be as before. Then define

$$\varphi^\flat(\vec{x}, \vec{y}) \equiv \exists u \leq |t'(\vec{x}, \vec{y})|\, \big(u \leq t(\vec{x}, |\vec{y}|) \wedge \varphi_1^\flat(u, \vec{x}, \vec{y})\big).$$

The cases where $\varphi(\vec{x}, \vec{Y}) \equiv \forall Z \leq t(\vec{x}, |\vec{Y}|)\, \varphi_1(\vec{x}, \vec{Y}, Z)$ or $\varphi(\vec{x}, \vec{Y}) \equiv \forall u \leq t(\vec{x}, |\vec{Y}|)\, \varphi_1(u, \vec{x}, \vec{Y})$ are handled analogously. This completes our description of the $^\flat$ translation.

The desired properties of $\varphi^\flat$ can be proved by structural induction on φ. Details are left as an exercise.

EXERCISE VIII.8.16. Let $\varphi(\vec{x}, \vec{Y})$ be an $\mathcal{L}_A^2$-formula.

(a) Show that if φ is in Σ_i^B (resp. Π_i^B) for some $i \geq 0$, then $\varphi^\flat(|\vec{x}|, \vec{y})$ is in $\Sigma_i^b(BIT)$ (resp. $\Pi_i^b(BIT)$).
(b) Let $\mathcal{M}$ be a model of $S_2^1(BIT)$. Show that

$$\mathcal{M}^\sharp \models \forall \vec{x} \forall \vec{Y} \varphi(\vec{x}, \vec{Y}) \quad \text{if and only if} \quad \mathcal{M} \models \forall \vec{x} \forall \vec{Y} \varphi^\flat(|\vec{x}|, \vec{y}).$$

VIII.8.5. The RSUV Isomorphism between S_2^i and V^i. In this subsection we will sketch the proof of the RSUV isomorphism between S_2^1 and V^1. The proof of the RSUV isomorphism between S_2^i and V^i for $i \geq 2$ is similar, and is left as an exercise.

First, the next theorem is useful in proving RSUV isomorphism.

NOTATION. We will assume that the theories mentioned here are axiomatized by set of formulas whose bounding terms do not contain any bound variable.

THEOREM VIII.8.17. *Let T_1 be a single-sorted theory over $\mathcal{L}_{S_2}$ such that $S_2^1 \subseteq T_1$, and T_2 be a two-sorted theory over $\mathcal{L}_A^2$ such that $V^1 \subseteq T_2$. Suppose that (i) $T_1(BIT) \vdash \varphi^\flat$ for every axiom φ of T_2, and (ii) $T_2(\mathcal{L}^+) \vdash \psi^\sharp$ for every axiom ψ of T_1. Then $T_1 \overset{\text{RSUV}}{\simeq} T_2$.*

PROOF. We show that $\mathcal{M}^\sharp \models T_2$ for every model $\mathcal{M}$ of T_1. The other half (that $\mathcal{N}^\flat \models T_1$ for every model $\mathcal{N}$ of T_2) is similar.

Thus suppose that $\mathcal{M} \models T_1(BIT)$. Then by (i) we have $\mathcal{M} \models \varphi^\flat$ for every axiom φ of T_2. By Exercise VIII.8.16 (b) it follows that $\mathcal{M}^\sharp \models T_2$. $\square$

EXERCISE VIII.8.18. Show that $S_2^1(BIT) \vdash \psi \leftrightarrow (\psi^\sharp)^\flat$ and $V^1(\mathcal{L}^+) \vdash \varphi \leftrightarrow (\varphi^\flat)^\sharp$ for every bounded $\mathcal{L}_{S_2}$ formula ψ and bounded $\mathcal{L}_A^2$ formula φ.

Notice that by Theorem VIII.8.10, if $\mathcal{M}$ is a model of S_2^1, then $\mathcal{M}^\sharp$ is a model of V^1. Hence we can define $(\mathcal{M}^\sharp)^\flat$. Similarly, if $\mathcal{N}$ is a model of V^1, then $(\mathcal{N}^\flat)^\sharp$ is well-defined. The $^\sharp$ and $^\flat$ operations define a bijection between isomorphism classes of models of S_2^1 and V^1, as shown in the next corollary.

Corollary VIII.8.19. *Let T_1 be a single-sorted theory that extends S_2^1. Then $(\mathcal{M}^\sharp)^\flat$ and $\mathcal{M}$ are same for every model $\mathcal{M}$ of T_1. Similarly, suppose that T_2 is a two-sorted theory that extends V^1. Then $(\mathcal{N}^\flat)^\sharp$ is isomorphic to $\mathcal{N}$ for every model $\mathcal{N}$ of T_2.*

Proof sketch. First, let $\mathcal{M}$ be a model of T_1. Clearly $\mathcal{M}$ and $(\mathcal{M}^\sharp)^\flat$ have the same universe. Indeed, the mappings

$$U(\mathcal{M}) \longrightarrow U_2(\mathcal{M}^\sharp) \longrightarrow U((\mathcal{M}^\sharp)^\flat)$$

are all identity maps. (Here $U(\mathcal{M})$ and $U((\mathcal{M}^\sharp)^\flat)$ denote respectively the universe of $\mathcal{M}$ and $(\mathcal{M}^\sharp)^\flat$, and $U_2(\mathcal{M}^\sharp)$ denotes the second-sort universe of $\mathcal{M}^\sharp$.) So we need to show that the symbols of $\mathcal{L}_{S_2}$ have the same interpretations in $\mathcal{M}$ and $(\mathcal{M}^\sharp)^\flat$. This essentially follows from the fact that $\mathcal{M}^\sharp \models V^1$, the functions and relations of $\mathcal{L}^+$ are definable in V^1, and that the "extension axiom" is provable in S_2^1 (Exercise VIII.8.5).

The second statement is proved similarly. (Here $(\mathcal{N}^\flat)^\sharp$ and $\mathcal{N}$ might have different first-sort universes, but they are isomorphic.)   □

The next corollary is the converse of Theorem VIII.8.17 above.

Corollary VIII.8.20. *Let T_1 be a single-sorted theory over $\mathcal{L}_{S_2}$ and T_2 be a two-sorted theory over $\mathcal{L}_A^2$ such that $T_1 \overset{\mathrm{RSUV}}{\simeq} T_2$. Then*

(i) *$T_1(BIT) \vdash \varphi^\flat$ for every axiom φ of T_2, and*
(ii) *$T_2(\mathcal{L}^+) \vdash \psi^\sharp$ for every axiom ψ of T_1.*

Proof. For (i), let $\mathcal{M}$ be a model of T_1 and φ be an axiom of T_2. Then $\mathcal{M}^\sharp \models T_2$. Therefore by Exercise VIII.8.16 (b) $(\mathcal{M}^\sharp)^\flat \models \varphi^\flat$. Since $(\mathcal{M}^\sharp)^\flat$ and $\mathcal{M}$ are the same structure (Corollary VIII.8.19), it follows that $\mathcal{M} \models \varphi^\flat$. Hence $T_1 \vdash \varphi^\flat$.

(ii) is proved similarly using Exercise VIII.8.13 (b).   □

Theorem VIII.8.21. *Suppose that T_1 and T_2 are RSUV isomorphic. Then T_1 is finitely axiomatizable if and only if T_2 is.*

Proof. Suppose that T_1 is a finitely axiomatizable single-sorted theory. Note that by the Σ_1^B-Transformation Lemma VI.3.9, for each $\mathcal{L}^+$ formula φ there is an $\mathcal{L}_A^2$ formula φ' so that $V^1(\mathcal{L}^+) \vdash \varphi \leftrightarrow \varphi'$. We will use this notation in the following definition. Let T denote the union of the following sets:

$$\{(\psi^\sharp)' : \psi \text{ is an axiom of } T_1(BIT)\}$$

and the set of the sentences of the form

$$\forall \vec{x} \forall \vec{Y} \exists! z \varphi(\vec{x}, z, \vec{Y})$$

or

$$\forall \vec{x} \forall \vec{Y} \exists! Z \varphi(\vec{x}, Z, \vec{Y})$$

where φ the the formula in the defining axiom of a function symbol of $\mathcal{L}^+$.

We show that T_2 can be axiomatized by T. First, let ψ be an axiom of T_1. By Corollary VIII.8.20 (ii) above, $T_2(\mathcal{L}^+) \vdash \psi^\sharp$. Consequently (since T_2 extends V^1, and $T_2(\mathcal{L}^+)$ is conservative over T_2) $T_2 \vdash (\psi^\sharp)'$. The defining axioms for symbols of $\mathcal{L}^+$ are in T_2 because $V^1 \subseteq T_2$.

It remains to show that $T \vdash \varphi$ for each axiom φ of T_2.

CLAIM. For each model $\mathcal{N}$ of T, there is a model $\mathcal{M}$ of $T_1(BIT)$ so that $\mathcal{M}^\sharp = \mathcal{N}$.

The Claim follows from part (a) of the exercise below and the fact that $T \vdash (\psi^\sharp)' \leftrightarrow \psi^\sharp$ for every axiom ψ of T_1. The latter follows from a careful examination of the proof of part (c) of the Σ_1^B-Transformation Lemma VI.3.9. (Here we do not require that T proves the Replacement axiom scheme.)

Now let φ be an axiom of T_2. Let $\mathcal{N}$ be any model of T, and let $\mathcal{M}$ be as in the Claim. Since $\mathcal{M} \models T_1(BIT)$ and $T_1(BIT) \models \varphi^\flat$ we have $\mathcal{M} \models \varphi^\flat$. By Exercise VIII.8.16 (b) we have $\mathcal{N} \models \varphi$.   □

EXERCISE VIII.8.22. (a) Suppose that T_1 is a single-sorted theory that extends S_2^1. Show that for every two-sorted model $\mathcal{N}$ of the set $\{\psi^\sharp : \psi$ is an axiom of $T_1\}$ there is a model $\mathcal{M}$ of T_1 so that $\mathcal{M}^\sharp = \mathcal{N}$.
(b) Similarly, let T_2 be a two-sorted theory that extends V^1, and $T_2' = \{\varphi^\flat : \varphi$ is an axiom of $T_2\}$. Show that for every model $\mathcal{M}$ of T_2' there is a model $\mathcal{N}$ of T_2 so that $\mathcal{M} = \mathcal{N}^\flat$.

PROOF SKETCH OF $S_2^1 \overset{\text{RSUV}}{\simeq} V^1$. We need to show that $V^1(\mathcal{L}^+)$ proves the $\sharp$ translations of the axioms in BASIC as well as Σ_1^b-LIND (see Exercise VIII.8.4). The former is straightforward and is left as an exercise.

EXERCISE VIII.8.23. Show that $V^1(\mathcal{L}^+)$ proves the $\sharp$ translations of the BASIC axioms.

Now we consider the Σ_1^b-LIND axiom scheme. We will show that $\mathcal{N}$ satisfies the $\sharp$ translations of the following bounded length induction for Σ_1^b formulas, which logically imply Σ_1^b-LIND:

$$[\varphi(0) \wedge \forall x \leq |z|, \; \varphi(x) \supset \varphi(x+1)] \supset \forall z \varphi(|z|) \qquad (216)$$

(where φ is a Σ_1^b formula).

Using Exercise VIII.8.13 (a) it is easy to see that instances of (216) translate into $g\Sigma_1^B(\mathcal{L}^+)$-IND. Hence the conclusion follows from Corollary VIII.8.11.

Consider the next half of the RSUV isomorphism. By Theorem VI.4.8 it suffices to show that $S_2^1(BIT)$ satisfies the $\flat$ translations of the 2-BASIC axioms and Σ_1^B-IND axioms. The Σ_1^B-IND axioms translate into $\Sigma_1^b(BIT)$-LIND which is provable in $S_2^1(BIT)$ by Exercise VIII.8.15. Thus the following simple exercise completes our proof of the RSUV isomorphism between S_2^1 and V^1.   □

EXERCISE VIII.8.24. Show that $S_2^1(BIT)$ proves the $\flat$ translation of the 2-BASIC axioms.

EXERCISE VIII.8.25. Complete the proof of Theorem VIII.8.10 by showing that $S_2^i \overset{\text{RSUV}}{\simeq} V^i$ for $i \geq 2$.

VIII.9. Notes

The theory VP in Section VIII.1 is from [82], but the theory $\widehat{VP}$ is new. The theory VPV defined in Section VIII.2 is based on the single-sorted equational theory PV [39]. The results in Section VIII.2.1 were first proved in single-sorted versions in Chapter 6 of [20].

In Section VIII.3 the TV^i hierarchy for $i \geq 1$ is the two-sorted version of Buss's [20] T_2^i hierarchy. The theory TV^0 was introduced in [42] where the results of Section VIII.3 are outlined, except Theorem VIII.3.10 is from [82].

The theory V^1-$HORN$ was introduced in [43], where versions of the results of Section VIII.4 are proved.

The PLS problems were introduced in [65]. The results in Section VIII.5 are mostly two-sorted versions of results from [30]. However our Witnessing Theorem VIII.5.13 is stronger than the one in [30], in that our witnessing function G is in the small class FAC^0, and the weak theory $\overline{V}^0$, as opposed to TV^1, proves the witnessing.

The results from Section VIII.6.1 are from [47].

Results and definitions in Section VIII.7 have single-sorted precursors as follows. Theorem VIII.7.4 is from [20]. The theories VPV^i are (for $i \geq 2$) two-sorted versions of the theories PV^i introduced in [75]. Theorems VIII.7.12 and VIII.7.13 are from [20, 23, 75]. Theorem VIII.7.14 is from [30, 33]. Definition VIII.7.16 (witness oracles) is from [31]. Theorem VIII.7.17 (i) is from [70] and (ii) is from [93] and [80].

Table 3 is inspired by Table 2.1 in [80].

Theorem VIII.7.20 is recently improved in [61] (under the same assumption, TV^i proves that PH collapses to the Boolean closure of Σ_{i+1}^P).

Buss [20] introduced the hierarchies S_2, T_2, and more generally, S_k, T_k (for $k \geq 2$). (The index k indicates the presence of the function $\#_k$, where $\#_2 = \#$, and $x\#_{k+1}y = 2^{|x|\#_k|y|}$.) He also introduced the hierarchy U_2, V_2, where U_2^1 and V_2^1 capture $PSPACE$ and $EXPTIME$, respectively. (The theories V^i in this book is sometimes called V_1^i.) The equivalence between S_{k+1}^i and V_k^i was first realized in [69, 107]. The name "RSUV isomorphism" was introduced by Takeuti in [108], where he also introduced the hierarchies R_k, and proved the equivalences between R_{k+1}^i and U_k^i and between S_{k+1}^i and V_k^i. The $S - V$ equivalence was also proved in [95]. The syntactic translations $\flat$ and $\sharp$ are called *interpretations* in [107, 95] (the symbols $\flat$ and $\sharp$ were introduced in [95]).

Chapter IX

THEORIES FOR SMALL CLASSES

In this chapter we develop subtheories of VP that are associated with the following subclasses of P:

$$AC^0(m) \subseteq TC^0 \subseteq NC^1 \subseteq L \subseteq NL \subseteq NC.$$

For each class C we obtain a minimal theory VC in the style of VP (Section VIII.1). Here each theory VC is axiomatized by the axioms of V^0 and a single axiom that asserts the existence of a solution for a complete problem of C. Thus VC is finitely axiomatizable, since V^0 is finitely axiomatizable. Our theories VC are minimal in the sense that they are axiomatized by some Σ_0^B and Σ_1^B formulas which appear to be necessary to establish the basic properties of the functions in the associated class C. In contrast V^1 appears *not* to be a minimal theory for FP, since by Theorem VIII.7.20 it has Σ_2^B axioms which are not provable in VP (assuming that the polynomial hierarchy does not collapse).

In this chapter completeness is with respect to AC^0-Turing reductions, which are more general than the AC^0-many-one reductions used in Section VIII.1 (see Proposition VIII.1.7). Therefore our results apply to classes such as TC^0 that are not known to have any AC^0-many-one complete problem.

The theory VP in Section VIII.1 can be seen as a member of the family VC here. In general we consider classes C that are the AC^0-closure of a polytime function F_C. Together with VC we will obtain two universal theories $\widehat{VC}$ and $\overline{VC}$ whose classes of provably total functions both are equal to FC. Following the development in Section VIII.1, after defining VC we introduce $\widehat{VC}$ and show that it is a conservative extension of VC. The vocabulary of $\widehat{VC}$ is $\mathcal{L}_{FAC^0} \cup \{F_C\}$, and the terms of $\widehat{VC}$ represent precisely the functions in FC. Using the Herbrand Theorem it follows that both the Σ_1^B-definable functions of VC and of $\widehat{VC}$ are FC (and hence all relations in C are Δ_1^B-definable in both theories). Then the theory $\overline{VC}$ is obtained from $\widehat{VC}$ by including symbols for all string functions in FC, similar to the way in which V^0 is extended to $\overline{V}^0$ in Section V.6. The defining axioms for functions in $\overline{VC}$ are based on the AC^0-reductions to the function F_C. We will show that $\overline{VC}$ is indeed a conservative extension of $\widehat{VC}$, and conclude from this that $\overline{VC}$ characterizes FC as mentioned.

For some subclasses C of L we are able to obtain universal theories VCV using recursion schemes similar to the limited recursion scheme given in Definition VI.2.11 used to obtain VPV. Here VCV has symbols for all string functions in FC, but their defining axioms are based on a particular recursion scheme rather than on AC^0 reductions as in the case of $\overline{VC}$. We will prove that in each case VCV is conservative over VC, giving evidence of the robustness of our definition of VC. The conservativity results also justify the "minimality" of our theories for characterizing C: The axioms consist essentially of 2-*BASIC* (page 96) and axioms defining the functions in FC (either using AC^0-reductions to the complete problem of C or using the recursion scheme that characterizes FC).

The fact that a theory VCV is a universal conservative extension of VC also implies that our theory VC proves the recursion scheme for the functions in FC. We will formalize in our theories proofs of a number of other mathematical theorems, such as the Pigeonhole Principle (PHP), the discrete version of Jordan Curve Theorem (JCT), and Bondy's Theorem. Some other theorems are of the form $C_1 \subseteq C_2$; for these we need to show that the defining axioms of VC_1 are provable in VC_2. We identify the research area of formalizing mathematical results in theories (preferably the weakest possible theories) of Bounded Arithmetic as "Bounded Reverse Mathematics", and we mention some open problems in this area in Section IX.7.

Part of the interest in establishing the provability of a principle such as PHP and JCT in a theory VC is that this implies the existence of polynomial size propositional proofs, in the proof system corresponding to the complexity class C, of the propositional tautologies expressing these principles (Section X.1).

This chapter is organized as follows. We start by formally defining the notion of AC^0 reduction in Section IX.1. Then in Section IX.2 we introduce the families VC, $\widehat{VC}$ and $\overline{VC}$. In the subsequent sections we define the theories for the classes mentioned above and carry out several formalizations in these theories: theories for TC^0 are presented in Section IX.3, theories for $AC^0(m)$ are presented in Section IX.4, theories for the NC hierarchy are presented in Section IX.5, and theories for NL and L are given in Section IX.6. For each of these sections it is helpful to specialize the meta-theorems that we prove for VC, $\widehat{VC}$ and $\overline{VC}$ in Section IX.2. Finally some open problems are listed in Section IX.7.

IX.1. AC^0 Reductions

Roughly speaking a function F is AC^0-reducible to a collection $\mathcal{L}$ of functions if F can be computed by a uniform polynomial size constant depth family of circuits which have unbounded fan-in gates computing functions from $\mathcal{L}$, in addition to Boolean gates (see for example [10]). This is a Turing style reduction, and generalizes the more restrictive many-one style. The class P and all classes that we consider in this chapter are closed under AC^0 reductions. Below we will formalize the notion of AC^0-reducible and show that in standard settings the FAC^0 closure of a set of functions is the same as closure under composition and a comprehension operator.

Recall that a function F (resp. f) is Σ_0^B-definable from $\mathcal{L}$ if it is polynomially bounded, and its bit graph (resp. graph) is represented by a $\Sigma_0^B(\mathcal{L})$ formula (Definition V.4.12). The following definition generalizes the notion of Σ_0^B-definability.

DEFINITION IX.1.1 (AC^0 Reduction). We say that a string function F (resp. a number function f) is AC^0-reducible to $\mathcal{L}$ if there is a sequence of string functions $F_1, \ldots, F_n$ ($n \geq 0$) such that

$$F_i \text{ is } \Sigma_0^B\text{-definable from } \mathcal{L} \cup \{F_1, \ldots, F_{i-1}\}, \text{ for } i = 1, \ldots, n; \quad (217)$$

and F (resp. f) is Σ_0^B-definable from $\mathcal{L} \cup \{F_1, \ldots, F_n\}$. A relation R is AC^0-reducible to $\mathcal{L}$ if there is a sequence $F_1, \ldots, F_n$ as above, and R is represented by a $\Sigma_0^B(\mathcal{L} \cup \{F_1, \ldots, F_n\})$ formula.

EXERCISE IX.1.2. Show that a number function f is AC^0-reducible to $\mathcal{L}$ if and only if $f = |F|$ for some string function F which is AC^0-reducible to $\mathcal{L}$.

If in the above definition $\mathcal{L}$ consists only of functions in FAC^0, then a single iteration ($n = 1$) is enough to obtain any function in FAC^0, and by Corollary V.4.16 no more functions are obtained by further iterations. However, as we shall see in the next section, if we start with a function such as *numones*, then repeated iterations generate the complexity class TC^0. It is an open question whether there is a bound on the number of iterations needed.

DEFINITION IX.1.3 (FAC^0- and AC^0-Closure). For a vocabulary $\mathcal{L}$, the FAC^0 *closure* of $\mathcal{L}$ is the class of functions which are AC^0-reducible to $\mathcal{L}$. The AC^0 *closure* of $\mathcal{L}$ is the class of relations which are AC^0-reducible to $\mathcal{L}$.

All complexity classes of interest here are closed under AC^0 reductions, because the corresponding function classes are closed under Σ_0^B-definability. For the case of FAC^0, this follows from Corollary V.4.16.

Corollary IX.1.4. *The FAC^0 closure of FAC^0 is FAC^0. The AC^0 closure of AC^0 is AC^0.*

For a complexity class C, recall that FC is the corresponding function class (Definition V.2.3). The following lemma is straightforward consequence of the definitions involved.

Lemma IX.1.5. *Suppose that a complexity class C is the AC^0 closure of a vocabulary $\mathcal{L}$. Then FC is the FAC^0 closure of $\mathcal{L}$.*

The composition of two functions is AC^0 reducible to the functions, because a term representing the composition can be used in a $\Sigma_0^B(\mathcal{L})$-formula defining the composition. We now define another operation which preserves AC^0 reducibility and which will be used together with composition to give a characterization of AC^0 reducibility. The new operation takes a number function and collects a bounded number of its values in a set to form a string function. This notion and Theorem IX.1.7 below will be useful in Section IX.3.3.

Definition IX.1.6 (String Comprehension). For a number function $f(x)$ (which may contain other arguments), the *string comprehension of f* is the string function $F(y)$ such that

$$F(y) = \{f(x) : x \le y\}.$$

(See (49) on page 96 for this set-theoretic notation.) Note that if f is polynomially bounded, then so is F.

For example, recall that the Σ_0^B formula $\varphi_{parity}(X, Y)$ (80) on page 118 asserts that for $0 \le i < |X|$, bit $Y(i + 1)$ is 1 iff the number of 1's among bits $X(0), \ldots, X(i)$ is odd. Here Y can be expressed as a function of X by $Y = F(|X|, X)$, where F is obtained from the following function f by string comprehension:

$$f(x, X) = \begin{cases} x & \text{if } x > 0 \text{ and the number of 1 bits in} \\ & \quad X(0), \ldots, X(x - 1) \text{ is odd,} \\ |X| + 1 & \text{otherwise.} \end{cases}$$

Theorem IX.1.7. *Suppose that $\mathcal{L}$ is a class of polynomially bounded functions that includes FAC^0. Then a function is AC^0-reducible to $\mathcal{L}$ iff it can be obtained from $\mathcal{L}$ by finitely many applications of composition and string comprehension.*

Proof. For the IF direction, it suffices to prove that a function obtained from input functions by either of the operations composition or string comprehension is Σ_0^B-definable from the input functions.

For composition, suppose

$$F(\vec{x}, \vec{X}) = G(h_1(\vec{x}, \vec{X}), \ldots, h_k(\vec{x}, \vec{X}), H_1(\vec{x}, \vec{X}), \ldots, H_m(\vec{x}, \vec{X}))$$

where G and $h_1, \ldots, h_k, H_1, \ldots, H_m$ are polynomially bounded. Then F is also polynomially bounded, and its bit graph $F(\vec{x}, \vec{X})(z)$ is represented

by the open formula
$$G(h_1(\vec{x}, \vec{X}), \ldots, h_k(\vec{x}, \vec{X}), H_1(\vec{x}, \vec{X}), \ldots, H_m(\vec{x}, \vec{X}))(z).$$
(A similar argument works for a number function f.)

For string comprehension, suppose that $f(x)$ is a polynomially bounded number function. As noted before, the string comprehension $F(y)$ of f is also polynomially bounded, and it has bit graph
$$F(y)(z) \leftrightarrow z < t \wedge \exists x \leq y \; z = f(x)$$
where t is the bounding term for F. Hence F is also Σ_0^B-definable from f.

For the ONLY IF direction, it suffices to show that if $\mathcal{L} \supseteq FAC^0$ and F (or f) is Σ_0^B-definable from $\mathcal{L}$, then F (resp. f) can be obtained from $\mathcal{L}$ by composition and string comprehension.

CLAIM. If $\mathcal{L} \supseteq FAC^0$ and $\varphi(\vec{z}, \vec{X})$ is a $\Sigma_0^B(\mathcal{L})$ formula, then the characteristic function c_φ defined by
$$c_\varphi(\vec{z}, \vec{Z}) = \begin{cases} 1 & \text{if } \varphi(\vec{z}, \vec{Z}), \\ 0 & \text{otherwise} \end{cases}$$
can be obtained from $\mathcal{L}$ by composition.

The Claim is holds because $c_\psi(\vec{x}, \vec{X})$ is in FAC^0 for every $\Sigma_0^B(\mathcal{L}_A^2)$-formula ψ, and (by structural induction on φ) it is clear that for every $\Sigma_0^B(\mathcal{L})$-formula $\varphi(\vec{z}, \vec{Z})$ there is a $\Sigma_0^B(\mathcal{L}_A^2)$-formula $\psi(\vec{x}, \vec{X})$ such that
$$\varphi(\vec{z}, \vec{Z}) \leftrightarrow \psi(\vec{s}, \vec{T})$$
for some $\mathcal{L}$-terms $\vec{s}$ and $\vec{T}$. Hence
$$c_\varphi(\vec{z}, \vec{Z}) = c_\psi(\vec{s}, \vec{T}).$$
Now suppose that F is Σ_0^B-definable from $\mathcal{L}$, so
$$F(\vec{z}, \vec{X})(x) \leftrightarrow x < t \wedge \varphi(x, \vec{z}, \vec{X})$$
where $t = t(\vec{z}, \vec{X})$ is an $\mathcal{L}_A^2$ term and φ is a $\Sigma_0^B(\mathcal{L})$ formula.

Define the number function f by cases as follows:
$$f(x, \vec{z}, \vec{X}) = \begin{cases} x & \text{if } \varphi(x, \vec{z}, \vec{X}), \\ t & \text{if } \neg\varphi(x, \vec{z}, \vec{X}). \end{cases}$$
Then by the Claim, f can be obtained from $\mathcal{L}$ by composition as follows. Define the FAC^0 function g by
$$g(x, y, z, w) = x \cdot y + z \cdot w.$$
Thus
$$f(x, \vec{z}, \vec{X}) = g(x, c_\varphi, t, c_{\neg\varphi}).$$
Now
$$F(\vec{z}, \vec{X}) = Cut(t, G(t, \vec{z}, \vec{X}))$$

where $G(y, \vec{z}, \vec{X})$ is the string comprehension of $f(x, \vec{z}, \vec{X})$, and Cut (see (97) on page 139) is the $\mathbf{FAC^0}$ function defined by

$$Cut(x, X)(z) \leftrightarrow z < x \wedge X(z).$$

It remains to show that if a number function f is Σ_0^B-definable from $\mathcal{L}$ then f can be obtained from $\mathcal{L}$ by composition and string comprehension. Suppose f satisfies

$$y = f(\vec{z}, \vec{X}) \leftrightarrow y < t \wedge \varphi(y, \vec{z}, \vec{X})$$

where $t = t(\vec{z}, \vec{X})$ is a $\mathcal{L}_A^2$ term and φ is a $\Sigma_0^B(\mathcal{L})$ formula. Use the Claim to define $c_\varphi(y, \vec{z}, \vec{X})$ by composition from $\mathcal{L}$, and define g by

$$g(x, \vec{z}, \vec{X}) = x \cdot c_\varphi(x, \vec{z}, \vec{X}).$$

Then

$$f(\vec{z}, \vec{X}) = |G(t, \vec{z}, \vec{X})| \dot{-} 1$$

where $G(y, \vec{z}, \vec{X})$ is the string comprehension of $g(x, \vec{z}, \vec{X})$. $\qquad\square$

IX.2. Theories for Subclasses of P

In this section, we show how to develop finitely axiomatizable theories for a number of uniform subclasses of P in the style of VP (Section VIII.1). Recall that VP is obtained from the base theory V^0 by adding the axiom MCV which states the existence of a value for F_{MCV}, a function which is AC^0-many-one complete for P. Here we obtain a theory VC for any class C which is the AC^0 closure of a polytime function F_C. The provably total functions of VC are precisely the functions in FC and the Δ_1^B-definable relations in VC are precisely the relations in C. Thus the function F_{MCV} plays the role of F_C when $C = P$.

In Section IX.2.1 we define VC and state the definability theorems for VC. In Section IX.2.2 we follow the discussion in Section VIII.1 and introduce the universal theory $\widehat{VC}$ in the same style as $\widehat{VP}$. The vocabulary $\mathcal{L}_{\widehat{VC}}$ of $\widehat{VC}$ is $\mathcal{L}_{FAC^0}$ together with the new function F_C. We show that $\widehat{VC}$ is a conservative extension of VC. We also prove that the terms in $\mathcal{L}_{\widehat{VC}}$ represent precisely functions in FC and hence the relations in C are represented by open formula of $\mathcal{L}_{\widehat{VC}}$. Consequently we derive our definability theorems for both $\widehat{VC}$ and VC.

In Section IX.2.3 we introduce a universal theory $\overline{VC}$. The vocabulary $\mathcal{L}_{FC}$ of $\overline{VC}$ contains all string functions of FC. (Note that by Exercise IX.1.2 the number functions in FC are represented by $\mathcal{L}_{FC}$-terms of the form $|G|$, for string functions G in $\mathcal{L}_{FC}$.) We show that $\overline{VC}$ is a conservative extension of both $\widehat{VC}$ and VC, and therefore it also characterizes C.

In Section IX.2.4 we discuss a general way of applying our results above to the subclasses of P mentioned at the beginning of this chapter.

IX.2.1. The Theories VC. In the following discussion the intended function F_C will be simply denoted by F. So suppose that F is a polytime function with a Σ_0^B graph:

$$Y = F(X) \leftrightarrow (|Y| \leq t \wedge \delta_F(X, Y)) \tag{218}$$

for some $\mathcal{L}_A^2$ term t and $\Sigma_0^B(\mathcal{L}_A^2)$ formula δ_F. Suppose further that V^0 proves the uniqueness of the value of F:

$$V^0 \vdash \forall Y_1 \forall Y_2 \big((|Y_1| \leq t \wedge |Y_2| \leq t \wedge \delta_F(X, Y_1) \wedge \delta_F(X, Y_2)) \supset Y_1 = Y_2\big).$$

Let C be the class of two-sorted relations which are AC^0-reducible to F. By Lemma IX.1.5, the class FC (Definition V.2.3) can be equivalently defined as the FAC^0 closure of F (Definition IX.1.3).

Our functions F_C introduced later in this chapter often have more than one argument, but they can be easily defined using a one argument-function as above. For example, we can easily encode the arguments (a, G, E) of F_{MCV} into a single string argument X and let F be the resulting function:

$$F(X) = F_{MCV}(a, G, E) \quad \text{whenever } X \text{ encodes } (a, G, E).$$

Then we have $C = P$.

DEFINITION IX.2.1 (VC). The theory VC has vocabulary $\mathcal{L}_A^2$ and is axiomatized by the axioms of V^0 and the following axiom:

$$\exists Y \leq b \forall i < b \, \delta_F(X^{[i]}, Y^{[i]}). \tag{219}$$

The notation $X^{[i]}$ technically involves the function Row (Definition V.4.26). But according to the Row Elimination Lemma V.4.27, $\delta_F(X^{[i]}, Y^{[i]})$ is easily equivalent to a Σ_0^B-formula $\delta_F'(i, X, Y)$, so we will interpret this axiom to be $\exists Y \leq b \forall i < b \, \delta_F'(i, X, Y)$, which is a formula over $\mathcal{L}_A^2$.

Note that VC is a polynomial-bounded finitely axiomatizable theory, because V^0 is, and the new axiom is bounded.

Recall the notion of aggregate function (Definition VIII.1.9). Notice that (219) states the existence of the value for the aggregate function $F^\star$ of F. Even though δ_{MCV} (157) (page 202) is only the graph of F_{MCV} (as opposed to $F_{MCV}^\star$), the fact that $F_{MCV}^\star$ is Σ_1^B-definable in VP (Lemma VIII.1.10) shows that VP is equivalent to a theory VC defined as above. In Section IX.2.4 we explain how to design theories for the other classes mentioned at the beginning of this chapter. In each case, we will be able to use the (simpler) defining axiom for F instead of the axioms of the form (219). This is because we can prove that for each of our theories $F^\star$ (for the F associated with the theory) is definable (although the proofs are different for each theory).

The next lemma is straightforward:

Lemma IX.2.2. *The functions F and $F^\star$ are Σ_0^B-definable in VC, and*

$$VC(F, F^\star) \vdash \forall b \forall X \forall i < b \; F^\star(b, X)^{[i]} = F(X^{[i]}).$$

Our first goal is to prove the following theorem (recall from Corollary V.4.4 that a function is provably total in *VC* iff it is Σ_1^B-definable in *VC*):

Theorem IX.2.3. *A function is provably total in VC iff it is in FC.*

Corollary IX.2.4. *A relation is in C iff it is Δ_1^B-definable in VC iff it is Δ_1^1-definable in VC.*

Proof. From Theorems IX.2.3 and V.4.35. □

We prove Theorem IX.2.3 by introducing the universal conservative extension $\widehat{VC}$ of *VC*, an analog of $\widehat{VP}$. The proof is given on page 278.

IX.2.2. The Theory $\widehat{VC}$. Here we define the universal theory $\widehat{VC}$ and show that it is a conservative extension of *VC*. We start by obtaining a quantifier-free defining axiom for F. For this we need a quantifier-free formula that is equivalent to $\delta_F(X, Y)$. So let $\delta_F'(X, Y)$ be the quantifier-free formula over $\mathcal{L}_{FAC^0}$ which $\overline{V}^0$ proves equivalent to $\delta_F(X, Y)$ (by Lemma V.6.3). Formally, we will not change the defining axiom for F. Therefore let F' be the function with the same value as F but has the following quantifier-free defining axiom:

$$Y = F'(X) \leftrightarrow (|Y| \le t \wedge \delta_F'(X, Y)). \tag{220}$$

Definition IX.2.5 ($\widehat{VC}$). $\widehat{VC}$ is the universal theory over the vocabulary $\mathcal{L}_{\widehat{VC}} = \mathcal{L}_{FAC^0} \cup \{F'\}$, and is axiomatized by the axioms of $\overline{V}^0$ and (220) for F'.

The next theorem is proved in the same way as Theorem VIII.1.13 using Lemma IX.2.2 above.

Theorem IX.2.6. *The theory $\widehat{VC}$ is a universal conservative extension of VC.*

The next corollary follows from Theorem VIII.1.15 in the same way as Corollary VIII.1.16:

Corollary IX.2.7. *The theory $\widehat{VC}$ proves the axiom schemes:*
$$\Sigma_0^B(\mathcal{L}_{\widehat{VC}})\text{-}COMP, \; \Sigma_0^B(\mathcal{L}_{\widehat{VC}})\text{-}IND, \text{ and } \Sigma_0^B(\mathcal{L}_{\widehat{VC}})\text{-}MIN.$$

The following theorem generalizes Theorem VIII.1.12.

Theorem IX.2.8. (a) *A function is in FC if and only if it is represented by a term in $\mathcal{L}_{\widehat{VC}}$.*
(b) *A relation is in C if and only if it is represented by an open formula of $\mathcal{L}_{\widehat{VC}}$ iff it is represented by a $\Sigma_0^B(\mathcal{L}_{\widehat{VC}})$ formula.*

Proof. It is straightforward to prove (b) from (a). So below we will only prove (a). Here *FC* is the FAC^0-closure of F. First we prove by induction based on Definition IX.1.1 that the functions in *FC* are represented by $\mathcal{L}_{\widehat{VC}}$ terms. The base case is obvious: F is represented by the term $F(\vec{x}, \vec{X})$. For the induction step, by Exercise IX.1.2 it suffices to consider the case of a string function. Suppose that $G(\vec{x}, \vec{X})$ is Σ_0^B-definable from $\mathcal{L} = \{F_1 = F, F_2, \ldots, F_n\}$, and that each F_i is represented by a term T_i in $\mathcal{L}_{\widehat{VC}}$. By definition there is a $\Sigma_0^B(\mathcal{L})$ formula $\varphi(z, \vec{x}, \vec{X})$ that represents the bit graph of G, i.e.,

$$G(\vec{x}, \vec{X})(z) \leftrightarrow z \leq t \wedge \varphi(z, \vec{x}, \vec{X})$$

for some $\mathcal{L}_A^2$-term t.

Let $\varphi'(z, \vec{x}, \vec{X})$ be the $\mathcal{L}_{\widehat{VC}}$-formula obtained from $\varphi(z, \vec{x}, \vec{X})$ by substituting $T_i(\vec{s}, \vec{S})$ for all occurrences of $F_i(\vec{s}, \vec{S})$. Let $F(\vec{s}_1, \vec{S}_1), \ldots, F(\vec{s}_m, \vec{S}_m)$ be all maximal occurrences of F in φ'. Thus

$$\varphi'(\vec{x}, \vec{X}) \equiv \varphi''(\vec{x}, \vec{X}, F(\vec{s}_1, \vec{S}_1), \ldots, F(\vec{s}_m, \vec{S}_m))$$

where $\varphi''(\vec{x}, \vec{X}, Y_1, \ldots, Y_m)$ is a $\Sigma_0^B(\mathcal{L}_{FAC^0})$-formula. Then G is represented by the $\mathcal{L}_{\widehat{VC}}$-term $H(\vec{x}, \vec{X}, F(\vec{s}_1, \vec{S}_1), \ldots, F(\vec{s}_m, \vec{S}_m))$, where H is the AC^0 function with bit graph

$$H(\vec{x}, \vec{X}, Y_1, \ldots, Y_m)(z) \leftrightarrow z \leq t \wedge \varphi''(z, \vec{x}, \vec{X}, Y_1, \ldots, Y_m).$$

For the other direction, we prove by induction on the nesting depth of an $\mathcal{L}_{\widehat{VC}}$-term that it represents a function in FC. The base case (the nesting depth is 0) is obvious, so consider the induction step. Let $T(\vec{x}, \vec{X})$ be an $\mathcal{L}_{\widehat{VC}}$ string term of nesting depth $d \geq 1$. (The case of a number term is similar.) Then

$$T(\vec{x}, \vec{X}) = H(s_1(\vec{x}, \vec{X}), \ldots, s_n(\vec{x}, \vec{X}), T_1(\vec{x}, \vec{X}), \ldots, T_m(\vec{x}, \vec{X}))$$

for $\mathcal{L}_{\widehat{VC}}$-terms s_i, T_j of nesting depth at most $d - 1$, and $H = F$ or H is an AC^0 function. By the induction hypothesis, s_i and T_j represent C functions f_i and G_j, respectively (for $1 \leq i \leq n, 1 \leq j \leq m$). For $1 \leq i \leq n$ let the string functions F_i in FC be such that $f_i = |F_i|$ (see Exercise IX.1.2). Then T represents the function K which is Σ_0^B-definable from $H, F_1, \ldots, F_n, G_1, \ldots, G_m$ as follows:

$$K(\vec{x}, \vec{X})(z) \leftrightarrow$$
$$z \leq t \wedge H(|F_1(\vec{x}, \vec{X})|, \ldots, |F_n(\vec{x}, \vec{X})|, G_1(\vec{x}, \vec{X}), \ldots, G_m(\vec{x}, \vec{X}))(z)$$

for some appropriate $\mathcal{L}_A^2$-term t. This shows that T represents a function in FC. $\square$

COROLLARY IX.2.9.   (a) *A function is* $\Sigma_1^B(\mathcal{L}_{\widehat{VC}})$-*definable in* $\widehat{VC}$ *iff it is in* FC.

    (b) *A relation is* C *iff it is* $\Delta_1^B(\mathcal{L}_{\widehat{VC}})$-*definable in* $\widehat{VC}$.

Proof. (a) This follows from Theorem IX.2.8 (a) and the Herbrand Theorem (see the proof of Corollary VIII.1.14).

(b) Follows from (a) and Theorem V.4.35. □

The next result is important for replacing the $\Sigma_1^B(\mathcal{L}_{\widehat{FC}})$ (and $\Pi_1^B(\mathcal{L}_{\widehat{FC}})$) formulas from Corollary IX.2.9 above by just Σ_1^B (i.e., $\Sigma_1^B(\mathcal{L}_A^2)$) (and Π_1^B) formulas. (In Corollary IX.3.20 we will prove similar theorem for number functions f, $f^\star$.)

Theorem IX.2.10 (First Elimination Theorem). *Let T be a theory with vocabulary $\mathcal{L}$ which extends $V^0(Row)$ and proves $\Sigma_0^B(\mathcal{L})$-COMP. Suppose that F and $F^\star$ are Σ_1^B-definable in T (Definition V.4.1) and $T(F, F^\star)$ proves* (166):

$$\forall i < b, \ F^\star(b, \vec{Z}, \vec{X})^{[i]} = F((Z_1)^i, \ldots, (Z_k)^i, X_1^{[i]}, \ldots, X_n^{[i]}).$$

Suppose also that every $\Sigma_0^B(\mathcal{L})$ formula is equivalent in T to a $\Sigma_1^B(\mathcal{L}_A^2)$ formula. Then every $\Sigma_1^B(\mathcal{L} \cup \{F\})$ formula is equivalent in $T(F)$ to a $\Sigma_1^B(\mathcal{L}_A^2)$ formula.

Proof. It suffices to prove the the last statement for Σ_0^B-formulas. Let

$$\varphi^+ \equiv Q_1 z_1 \leq r_1 \ldots Q_n z_n \leq r_n \psi(\vec{z})$$

be a $\Sigma_0^B(\mathcal{L}, F)$ formula, where $Q_1, \ldots, Q_n \in \{\exists, \forall\}$ and ψ is a quantifier-free formula. We show that there is a $\Sigma_1^B(\mathcal{L}_A^2)$ formula φ so that

$$T(F) \vdash \varphi^+ \leftrightarrow \varphi.$$

As in the base case in the proof of Theorem VIII.1.15, the idea here is to replace every occurrence of a term $F(\vec{s}, \vec{T})$ in ψ by a new string variable W which has the intended value of $F(\vec{s}, \vec{T})$. We need to state the existence of such strings, and this contributes to the string quantifiers in the resulting Σ_1^B formula.

So suppose that $F(\vec{s}_1, \vec{T}_1), \ldots, F(\vec{s}_k, \vec{T}_k)$ are all occurrences of F in ψ. Note that the terms $\vec{s}_i$, $\vec{T}_i$ may contain $\vec{z}$ as well as nested occurrences of F. Assume further that these F-terms are ordered by depth so that $\vec{s}_1$, $\vec{T}_1$ do not contain F, and for $1 < i \leq k$, any occurrence of F in $\vec{s}_i$, $\vec{T}_i$ must be of the form $F(\vec{s}_j, \vec{T}_j)$, for some $j < i$.

Let $W_1, \ldots, W_k$ be new string variables. Let $\vec{s'_1} = \vec{s}_1$, $\vec{T'_1} = \vec{T}_1$, and for $2 \leq i \leq k$, $\vec{s'_i}$ and $\vec{T'_i}$ be obtained from $\vec{s}_i$ and $\vec{T}_i$ respectively by replacing every maximal occurrence of any $F(\vec{s}_j, \vec{T}_j)$, for $j < i$, by $W_j^{[z]}$. Thus F does not occur in any $\vec{s'_i}$ and $\vec{T'_i}$, but for $i \geq 2$, $\vec{s'_i}$ and $\vec{T'_i}$ may contain $W_1, \ldots, W_{i-1}$.

Let $\psi'(\vec{z}, W_1, \ldots, W_k)$ be obtained from $\psi(\vec{z})$ by replacing each maximal occurrence of $F(\vec{s}_i, \vec{T}_i)$ by $W_i^{[\vec{z}]}$, for $1 \leq i \leq k$. Obviously,

$$\mathcal{T}(F) \vdash \left[\exists W_1 \ldots \exists W_k \big((\forall z_1 \leq r_1 \ldots \forall z_n \leq r_n \bigwedge W_i^{[\vec{z}]} = F(\vec{s_i'}, \vec{T_i'})) \wedge \right.$$

$$\left. Q_1 z_1 \leq r_1 \ldots Q_n z_n \leq r_n \psi'(\vec{z}, \vec{W}))\right] \supset Q_1 z_1 \leq r_1 \ldots Q_n z_n \leq r_n \psi(\vec{z}).$$

Notice that $W_i = F^\star(\langle \vec{r} \rangle, \vec{U}_i, \vec{V}_i)$ satisfy

$$\forall z_1 \leq r_1 \ldots \forall z_n \leq r_n \bigwedge W_i^{[\vec{z}]} = F(\vec{s_i'}, \vec{T_i'})$$

where $U_{i,j}$ and $V_{i,\ell}$ are unique strings such that

$$|U_{i,j}| \leq t \wedge \forall z_1 \leq r_1 \ldots \forall z_k \leq r_k (U_{i,j})^{\vec{z}} = s_{i,j}', \tag{221}$$

$$|V_{i,\ell}| \leq t \wedge \forall z_1 \leq r_1 \ldots \forall z_k \leq r_k V_{i,\ell}^{[\vec{z}]} = T_{i,\ell}' \tag{222}$$

and t is an $\mathcal{L}_A^2$-term such that $t \geq \langle \vec{r}, max\{|\vec{T_i'}|, \vec{s_i'}\}\rangle$ for all $1 \leq i \leq k$.

Denote the conjunction of (221) and (222) for all i, j, ℓ by $\theta(\vec{U}, \vec{V})$. Then

$$\mathcal{T}(F, F^\star) \vdash \left[\exists \vec{U} \exists \vec{V} \exists \vec{W} \big(\theta(\vec{U}, \vec{V}) \wedge \bigwedge W_i = F^\star(\langle \vec{r} \rangle, \vec{U}_i, \vec{V}_i) \wedge \right.$$

$$\left. Q_1 z_1 \leq r_1 \ldots Q_n z_n \leq r_n \psi'(\vec{z}, \vec{W}))\right] \supset Q_1 z_1 \leq r_1 \ldots Q_n z_n \leq r_n \psi(\vec{z}).$$

On the other hand, since $F^\star$ is definable in $\mathcal{T}$ and since $\mathcal{T} \vdash \Sigma_0^B(\mathcal{L})$-$\textbf{COMP}$, we have

$$\mathcal{T}(F, F^\star) \vdash \exists \vec{U} \exists \vec{V} \exists \vec{W} \big(\theta(\vec{U}, \vec{V}) \wedge \bigwedge W_i = F^\star(\langle \vec{r} \rangle, \vec{U}_i, \vec{V}_i)\big). \tag{223}$$

Therefore

$$\mathcal{T}(F, F^\star) \vdash Q_1 z_1 \leq r_1 \ldots Q_n z_n \leq r_n \psi(\vec{z}) \supset \left[\exists \vec{U} \exists \vec{V} \exists \vec{W} \big(\theta(\vec{U}, \vec{V}) \wedge \right.$$

$$\left. \bigwedge W_i = F^\star(\langle \vec{r} \rangle, \vec{U}_i, \vec{V}_i) \wedge Q_1 z_1 \leq r_1 \ldots Q_n z_n \leq r_n \psi'(\vec{z}, \vec{W}))\right].$$

As a result,

$$\mathcal{T}(F, F^\star) \vdash Q_1 z_1 \leq r_1 \ldots Q_n z_n \leq r_n \psi(\vec{z}) \leftrightarrow \left[\exists \vec{U} \exists \vec{V} \exists \vec{W} \big(\theta(\vec{U}, \vec{V}) \wedge \right.$$

$$\left. \bigwedge W_i = F^\star(\langle \vec{r} \rangle, \vec{U}_i, \vec{V}_i) \wedge Q_1 z_1 \leq r_1 \ldots Q_n z_n \leq r_n \psi'(\vec{z}, \vec{W}))\right].$$

Finally the strings in (223) are bounded by some $\mathcal{L}_A^2$-terms and are provably unique in $\mathcal{T}(F^\star)$. Therefore (223) is equivalent in $\mathcal{T}(F^\star)$ to a $\Sigma_1^B(\mathcal{L}_A^2)$ formula. Also, by the hypothesis,

$$Q_1 z_1 \leq r_1 \ldots Q_n z_n \leq r_n \psi'(\vec{z}, \vec{W})$$

is equivalent in $\mathcal{T}$ to a $\Sigma_1^B(\mathcal{L}_A^2)$ formula. As a result,

$$\exists \vec{U} \exists \vec{V} \exists \vec{W} \big(\theta(\vec{U}, \vec{V}) \wedge \bigwedge W_i = F^\star(\langle \vec{r} \rangle, \vec{U}_i, \vec{V}_i) \wedge$$

$$Q_1 z_1 \leq r_1 \ldots Q_n z_n \leq r_n \psi'(\vec{z}, \vec{W}))$$

is equivalent in $\mathcal{T}(F, F^\star)$ to a Σ_1^B formula. The conclusion follows from the fact that $\mathcal{T}(F, F^\star)$ is conservative over $\mathcal{T}(F)$. □

COROLLARY IX.2.11 ($\Sigma_1^B(\mathcal{L}_{\widehat{VC}})$ Elimination). *For each $\Sigma_1^B(\mathcal{L}_{\widehat{VC}})$ formula φ^+ there is a $\Sigma_1^B(\mathcal{L}_A^2)$ formula φ such that $\widehat{VC} \vdash \varphi^+ \leftrightarrow \varphi$.*

PROOF. We apply Theorem IX.2.10 for $\mathcal{T} = VC + \overline{V}^0$ and $\mathcal{L} = \mathcal{L}_{FAC^0}$. The hypothesis that every $\Sigma_0^B(\mathcal{L}_{FAC^0})$ formula is equivalent in $\overline{V}^0$ to a Σ_1^B formula is from Lemma V.6.7, the facts that F and $F^\star$ are both Σ_1^B-definable in VC, and that $VC + \overline{V}^0$ proves (166) are established in Lemma IX.2.2. □

The next corollary follows from Corollaries IX.2.9 and IX.2.11.

COROLLARY IX.2.12. (a) *A function is in FC iff it is $\Sigma_1^B(\mathcal{L}_A^2)$-definable in $\widehat{VC}$.*

(b) *A relation is in C iff it is $\Delta_1^B(\mathcal{L}_A^2)$-definable in $\widehat{VC}$.*

Now we are able to prove Theorem IX.2.3.

PROOF OF THEOREM IX.2.3. The proof is straightforward using Theorem IX.2.6 and Corollary IX.2.12. □

IX.2.3. The Theory $\overline{VC}$. Here we introduce $\overline{VC}$, another universal conservative extension of VC and $\widehat{VC}$. Its vocabulary $\mathcal{L}_{FC}$ contains symbols for all string functions in FC. The defining axioms for functions in $\mathcal{L}_{FC}$ are based on their AC^0-reductions to the function F_C. Recall the function F' and its quantifier-free defining axiom (220).

DEFINITION IX.2.13 ($\overline{VC}$). $\mathcal{L}_{FC}$ is the smallest set containing $\mathcal{L}_{FAC^0} \cup \{F'\}$ and satisfying the following condition: for each open formula $\varphi(z, \vec{x}, \vec{X})$ over $\mathcal{L}_{FC}$ and $\mathcal{L}_A^2$-term $t = t(\vec{x}, \vec{X})$, there is a string function $F_{\varphi(z),t}$ in $\mathcal{L}_{FC}$ with defining axiom (86)

$$F_{\varphi(z),t}(\vec{x}, \vec{X})(z) \leftrightarrow z < t(\vec{x}, \vec{X}) \wedge \varphi(z, \vec{x}, \vec{X}). \tag{224}$$

$\overline{VC}$ is the universal theory over $\mathcal{L}_{FC}$ whose axioms consist of the axioms of $\overline{V}^0$, (220) for F', and the above defining axioms for the functions $F_{\varphi(z),t}$.

Note that Lemma VIII.2.3 and Theorem VIII.2.4 (Witnessing) apply to $\overline{VC}$.

The proof of the next theorem uses the property of aggregates stated in Theorem VIII.1.15.

THEOREM IX.2.14. *$\overline{VC}$ is a conservative extension of $\widehat{VC}$ and VC.*

PROOF. It suffices to show that $\overline{VC}$ is a conservative extension of $\widehat{VC}$, because $\widehat{VC}$ is a conservative extension of VC.

First, $\overline{VC}$ is an extension of $\widehat{VC}$ because all axioms of $\widehat{VC}$ are axioms of $\overline{VC}$. Thus $\overline{VC}$ is the union

$$\overline{VC} = \bigcup_{i \geq 0} \mathcal{T}_i$$

where $\mathcal{T}_0 = \widehat{VC}$ and for $i \geq 0$ each $\mathcal{T}_{i+1}$ is obtained from $\mathcal{T}_i$ by adding a new function F_{i+1} of the form $F_{\varphi(z),t}$, with defining axiom (224), where φ is a quantifier-free formula in the vocabulary of $\mathcal{T}_i$. We will show that F_{i+1} is definable in $\mathcal{T}_i$, which implies that $\mathcal{T}_{i+1}$ is a conservative extension of $\mathcal{T}_i$, and hence $\overline{VC}$ is a conservative extension of $\widehat{VC}$. We need the following lemma, whose proof is straightforward.

LEMMA IX.2.15. *Let $\mathcal{T}$ be an extension of $V^0(Row)$ with vocabulary $\mathcal{L}$ such that $\mathcal{T}$ proves $\Sigma_0^B(\mathcal{L})$-COMP. Let $F_{\varphi(z),t}$ be the function with defining axiom (224) where φ is any $\Sigma_0^B(\mathcal{L})$ formula. Then both $F_{\varphi(z),t}$ and $F_{\varphi,t}^{\star}$ are $\Sigma_0^B(\mathcal{L})$-definable in $\mathcal{T}$, and $\mathcal{T}(F_{\varphi(z),t}, F_{\varphi,t}^{\star})$ proves (166) for $F_{\varphi(z),t}$ and $F_{\varphi,t}^{\star}$:*

$$\forall i < b, \; F_{\varphi,t}^{\star}(b, \vec{Z}, \vec{X})^{[i]} = F_{\varphi(z),t}((Z_1)^i, \ldots, (Z_k)^i, X_1^{[i]}, \ldots, X_n^{[i]}).$$

Let $\mathcal{L}_i$ denote the vocabulary of $\mathcal{T}_i$. It follows by induction on $i \geq 0$, using the above lemma and the Aggregate Function Theorem VIII.1.15, that $\mathcal{T}_i$ proves $\Sigma_0^B(\mathcal{L}_i)$-*COMP*. Then the fact that F_{i+1} is definable (in fact, $\Sigma_0^B(\mathcal{L}_i)$-definable) in $\mathcal{T}_i$ follows from Lemma IX.2.15.   □

LEMMA IX.2.16. *The theory $\overline{VC}$ proves the axiom schemes*

$$\Sigma_0^B(\mathcal{L}_{FC})\text{-}COMP, \; \Sigma_0^B(\mathcal{L}_{FC})\text{-}IND \; and \; \Sigma_0^B(\mathcal{L}_{FC})\text{-}MIN.$$

PROOF. By Corollary V.1.8 it suffices to show that $\overline{VC}$ proves the $\Sigma_0^B(\mathcal{L}_{FC})$-*COMP* axioms. This follows from the proof of Theorem IX.2.14 above, but it also has simple proof as follows.

Let $\varphi(z, \vec{x}, \vec{X})$ be a $\Sigma_0^B(\mathcal{L}_{FC})$ formula. By Lemma VIII.2.3 there is a quantifier-free $\mathcal{L}_{FC}$-formula $\varphi^+(z, \vec{x}, \vec{X})$ so that

$$\overline{VC} \vdash \varphi^+(z, \vec{x}, \vec{X}) \leftrightarrow \varphi(z, \vec{x}, \vec{X}).$$

Let $Y = F_{\varphi^+,y}(\vec{x}, \vec{X})$, where $F_{\varphi^+,y}$ is the string function of $\mathcal{L}_{FC}$ with defining axiom

$$F_{\varphi^+,y}(\vec{x}, \vec{X})(z) \leftrightarrow z < y \wedge \varphi^+(z, \vec{x}, \vec{X}).$$

Then

$$\overline{VC} \vdash |Y| \leq y \wedge \forall z < y\big(Y(z) \leftrightarrow \varphi(z, \vec{x}, \vec{X})\big).$$

Hence $\overline{VC}$ proves the comprehension axiom for φ.   □

THEOREM IX.2.17. (a) *A string function is in **FC** if and only if it is represented by a string function symbol in $\mathcal{L}_{FC}$.*

(b) *A relation is in **C** iff it is represented by an open formula of $\mathcal{L}_{FC}$, iff it is represented by a $\Sigma_0^B(\mathcal{L}_{FC})$ formula.*

PROOF. Part (b) follows from (a), so below we will prove (a). First, we prove by induction using Definition IX.1.1 that every string function in **FC** is represented by a string function in $\mathcal{L}_{FC}$. The base case is simple because F and every function in $\mathcal{L}_{FAC^0}$ are in $\mathcal{L}_{FC}$. For the induction step, suppose that $G(\vec{x}, \vec{X})$ is Σ_0^B-definable from $\mathcal{L} = \{F_1 = F', F_2, \ldots, F_n\}$, and that each $F_i \in \mathcal{L}_{FC}$, for $1 \leq i \leq n$. By definition, there is a $\Sigma_0^B(\mathcal{L})$ formula φ and an $\mathcal{L}_A^2$-term t such that

$$G(\vec{x}, \vec{X})(z) \leftrightarrow z \leq t \wedge \varphi(z, \vec{x}, \vec{X}).$$

By Lemma VIII.2.3 there is a quantifier-free $\mathcal{L}_{FC}$-formula $\varphi^+(z, \vec{x}, \vec{X})$ that is equivalent (in $\boldsymbol{VC}$) to $\varphi(z, \vec{x}, \vec{X})$. Hence

$$G(\vec{x}, \vec{X})(z) \leftrightarrow z \leq t \wedge \varphi^+(z, \vec{x}, \vec{X}).$$

So G is equal to the function $F_{\varphi^+, t}$ in $\mathcal{L}_{FC}$.

For the other direction, we prove by induction (using Definition IX.2.13) that every string function in $\mathcal{L}_{FC}$ represents a string function in $\boldsymbol{FC}$. For the base case, the functions in $\mathcal{L}_{FAC^0} \cup \{F'\}$ obviously represent functions in $\boldsymbol{FC}$. For the induction step, let $F_{\varphi(z), t}$ be a function in $\mathcal{L}_{FC}$, where all functions $F_1, F_2, \ldots, F_n$ in φ represent functions in $\boldsymbol{FC}$. Then $F_{\varphi(z), t}$ is AC^0-reducible to $F_1, F_2, \ldots, F_n$, hence $F_{\varphi(z), t}$ also represents a function in $\boldsymbol{FC}$. □

The next result is a corollary of Theorem IX.2.10.

COROLLARY IX.2.18 ($\Sigma_1^B(\mathcal{L}_{FC})$ Elimination). *Every* $\Sigma_1^B(\mathcal{L}_{FC})$ *formula* φ^+ *is equivalent in* $\overline{VC}$ *to a* $\Sigma_1^B(\mathcal{L}_A^2)$ *formula* φ.

PROOF. Let the theories $\mathcal{T}_i$ and their vocabularies $\mathcal{L}_i$ be as in the proof of Theorem IX.2.14. It suffices to prove by induction on i that for each $\Sigma_0^B(\mathcal{L}_i)$ formula φ^+ there is a $\Sigma_1^B(\mathcal{L}_A^2)$ formula φ such that $\mathcal{T}_i$ proves equivalent to φ^+.

The base case is Corollary IX.2.11. For the induction step, suppose that the statement is true for some $i \geq 0$. The statement for $i + 1$ is proved by applying Theorem IX.2.10 for $\mathcal{T} = \mathcal{T}_i$ and $\mathcal{L} = \mathcal{L}_i$. The hypothesis of Theorem IX.2.10 is satisfied by Lemma IX.2.15 and the fact (from the proof of Theorem IX.2.14) that $\mathcal{T}_i$ proves $\Sigma_0^B(\mathcal{L}_i)$-$\boldsymbol{COMP}$. □

The characterization of C by $\overline{VC}$ follows from Theorem IX.2.17, Corollary IX.2.18, the Herbrand Theorem, and Theorem V.4.35.

COROLLARY IX.2.19.   (a) *A function is in* $\boldsymbol{FC}$ *iff it is* $\Sigma_1^B(\mathcal{L}_{FC})$-*definable in* $\overline{VC}$ *iff it is* $\Sigma_1^B(\mathcal{L}_A^2)$-*definable in* $\overline{VC}$.

(b) *A relation is in* C *iff it is* $\Delta_1^B(\mathcal{L}_{FC})$-*definable in* $\overline{VC}$ *iff it is* $\Delta_1^B(\mathcal{L}_A^2)$-*definable in* $\overline{VC}$.

Note that Theorem IX.2.3 also follows from Theorem IX.2.14 (that $\overline{VC}$ is a universal conservative extension of VC), Theorem IX.2.17 and Corollary IX.2.18.

IX.2.4. Obtaining Theories for the Classes of Interest. The results so far in this chapter show how to obtain a theory VC for each class C mentioned in the introduction to this chapter. In fact, for each class C of interest, there is a polytime Turing machine M such that the function

$$F(X) = \text{"the computation of M on input } X\text{"}$$

is AC^0 complete for C. For example, for $C = P$ we can take the machine that computes $F_{MCV}(a, G, E)$ by computing inductively the bits of Y that satisfies (157) on page 202.

The $\Sigma_0^B(\mathcal{L}_A^2)$ defining axiom (218) for F can be obtained using the following AC^0 functions (which can be eliminated by Lemma V.6.7):

- $Init_M(X)$ is the initial configuration of M given input X,
- $Next_M(U)$ is the next configuration of the configuration U, and
- $Cut(t, Z)$ is the set of all elements of Z that are less than t with defining axiom (97) (page 139):

$$Cut(t, Z) = \{z : z \in Z \wedge z < t\}.$$

Let t be an $\mathcal{L}_A^2$ term that bounds the running time of M. We have

$$F(X) = Y \leftrightarrow \big(|Y| \leq \langle t, t \rangle \wedge Y^{[0]} = Cut(t, Init_M(X)) \wedge$$
$$\forall x < t,\ Y^{[x+1]} = Cut(t, Next_M(Y^{[x]}))\big).$$

By eliminating $Init_M$, $Next_M$ and Cut, the above formula has the required form (218) and it is easy to prove in V^0 the uniqueness for Y (by proving by induction on $x \leq t$ that the rows $Y^{[x]}$ are unique).

Although the axiom (219) states the existence of the value for the function $F^\star$, for each class C that we consider we are able to simplify (219) so that it only states the existence of the value for F. Thus we will need to prove the analogue of Lemma VIII.1.10, i.e., that $F^\star$ is definable using the simplified axiom and V^0. It turns out that the proof is different for each class that we consider.

In the remaining of this chapter we will develop instances of VC as discussed here without referring to any specific machine M; they are implicit in the additional defining axioms of the instances that we introduce.

IX.3. Theories for TC^0

The class TC^0 (see definition in Section IX.3.1 below) is the smallest class with nice closure properties that contains problems such as sorting, integer multiplication and division (when the input integer arguments are presented in binary). Here we define VTC^0, $\widehat{VTC^0}$ and $\overline{VTC}^0$ (Section IX.3.2) in the style of the theories VC, $\widehat{VC}$ and $\overline{VC}$ in Section IX.2. In Section IX.3.3 we define the bounded number recursion (BNR). Then in Section IX.3.4 we use number summation, a special case of BNR, to characterize TC^0 and develop VTC^0V in the style of VPV (see Section VIII.2). This is another universal conservative extension of VTC^0. We formalize a proof of the Pigeonhole Principle in VTC^0 in Section IX.3.5. We define the string multiplication function $X \times Y$ and prove its properties in VTC^0 in Section IX.3.6. Finally in Section IX.3.7 we show that VTC^0 proves the finite case of Szpilrajn's Theorem (every finite partial order can be extended to a total order).

In Chapter X we will prove the Propositional Translation Theorem for VTC^0.

IX.3.1. The Class TC^0. The class nonuniform TC^0 (or $TC^0/poly$) consists of languages that are accepted by a family of polynomial-size constant-depth circuits whose gates can be Boolean gates or the *majority* gates. A majority gate has unbounded fan-in and which outputs 1 if and only if the number of 1 inputs is more than the number of 0 inputs. We are interested in FO-uniform TC^0 (or just TC^0) where the family can be described by an FO-formula (Section IV.1). A formal definition is given in Appendix A.5.

Instead of the majority gates, TC^0 can be equivalently defined using *counting* gates or *threshold* gates. A counting gate C_k (for $k \in \mathbb{N}$) has unbounded fan-in, and $C_k(x_1, x_2, \ldots, x_n)$ is true if and only if there are exactly k inputs x_i that are true. Similarly, for $k \in \mathbb{N}$, the threshold gate Th_k has unbounded fan-in, and $Th_k(x_1, x_2, \ldots, x_n)$ is true if and only if there are at least k inputs that are true.

There are several equivalent characterizations of TC^0 in descriptive complexity theory [10] (see also Section IV.1 for the descriptive characterization of AC^0). They are obtained by augmenting the first-order logic FO with quantifiers that correspond to the majority, counting or threshold gates described above. For example, let $\mathcal{L}_{FO(M)}$ denote the set of formulas over the vocabulary $\mathcal{L}_{FO}$ (41):

$$[0, max; X, BIT, \leq, =]$$

where a new quantifier M is allowed. The meaning of this quantifier is as follows: for a finite structure $\mathcal{M}$ and a $\mathcal{L}_{FO(M)}$ formula $Mx\varphi(x)$,

$$\mathcal{M} \models Mx\varphi(x)$$

iff $\mathcal{M} \models \varphi(a)$ for at least half of the elements a in the universe of $\mathcal{M}$. Let

$$FO(M) = \{L : L = L(\varphi) \text{ for some } \mathcal{L}_{FO(M)}\text{-sentence } \varphi\}$$

and define $FO(COUNT)$, $FO(THRESHOLD)$ similarly. Then it can be shown that

$$TC^0 = FO(M) = FO(COUNT) = FO(THRESHOLD).$$

TC^0 can also be defined using other computation models, such as the so-called Threshold Turing machines, but we will not go into detail here. The proposition below uses the notion of AC^0-reducibility defined in Section IX.1 and is based on the fact that $TC^0 = FO(COUNT)$, or in other words, *numones* is AC^0-complete for TC^0. (Recall the function $numones(y, X)$ defined on page 149: $numones(y, X)$ is the number of elements of X that are $< y$.)

Proposition IX.3.1 ([10]). TC^0 is the AC^0 closure of numones. FTC^0 is the FAC^0 closure of numones.

Below we will introduce the theories VTC^0, $\widehat{VTC^0}$ and $\overline{VTC}^0$. The above proposition will be used to justify the association between these theories and TC^0.

IX.3.2. The Theories VTC^0, $\widehat{VTC^0}$, and $\overline{VTC}^0$. Here we specialize the general treatment of VC given in Section IX.2 to the case $C = TC^0$. The theory VTC^0 is similar to VP (Definition VIII.1.1) in the sense that it is axiomatized by V^0 and a defining axiom for the function numones (which is AC^0 complete for TC^0). The following defining axiom for numones is given in (108) on page 149:

$$numones(y, X) = z \leftrightarrow z \leq y \land \exists Z \leq 1 + \langle y, y \rangle, (Z)^0 = 0 \land$$
$$(Z)^y = z \land \forall u < y, (X(u) \supset (Z)^{u+1} = (Z)^u + 1) \land$$
$$(\neg X(u) \supset (Z)^{u+1} = (Z)^u). \quad (225)$$

(Recall that $(Z)^u$ denotes $seq(u, Z)$, the u-th element of the bounded sequence of numbers coded by Z, see Definition V.4.31.) We want to define the theory VTC^0 by introducing an axiom which defines a string function $Numones(y, X) = Z$, where Z is the string asserted to exist in the above formula (225) defining numones.

However there is a technical difficulty here because not all of the bits of Z are uniquely determined when we simply specify the values $(Z)^u$. To solve this problem, we introduce a formula $SEQ(y, Z)$ which asserts that Z is the lexicographically first string which codes a given sequence of numbers.

$$SEQ(y, Z) \equiv$$
$$\forall w < |Z|(Z(w) \leftrightarrow \exists i \leq y \exists j < |Z|(w = \langle i, j \rangle \land j = (Z)^i)). \quad (226)$$

Let $\delta_{NUM}(y, X, Z)$ be the $\Sigma_0^B(\mathcal{L}_A^2)$ formula obtained from (227) below by eliminating seq as described in Lemmas V.4.15 and V.6.7:

$$SEQ(y, Z) \land (Z)^0 = 0 \land \forall u < y((X(u) \supset (Z)^{u+1} = (Z)^u + 1) \land$$
$$(\neg X(u) \supset (Z)^{u+1} = (Z)^u)). \quad (227)$$

Informally, we can think of Z as a "counting sequence" for X:

$$(Z)^u = z \leftrightarrow numones(u, X) = z \qquad \text{for } u \leq y.$$

Definition IX.3.2 (VTC^0). Let $NUMONES$ denote

$$\exists Z \leq 1 + \langle y, y \rangle \delta_{NUM}(y, X, Z).$$

The theory VTC^0 has vocabulary $\mathcal{L}_A^2$ and is axiomatized by the axioms of V^0 and $NUMONES$.

Note that V^0 (like VC in general) is a polynomial-bounded finitely axiomatizable theory.

To develop $\widehat{VTC^0}$, we will use the "string version" of *numones*, denoted by *Numones*, that has the defining axiom:

$$Numones(y, X) = Z \leftrightarrow |Z| \leq 1 + \langle y, y \rangle \wedge \delta_{NUM}(y, X, Z).$$

By the above discussion for *SEQ*, *Numones* is uniquely specified by its defining equation, and hence (using the axiom *NUMONES*) it is Σ_1^B-definable in VTC^0.

Since *numones* and *Numones* are AC^0-definable from each other, Proposition IX.3.1 remains true if we replace *numones* by *Numones*.

Recall the notion of aggregate functions in Definition VIII.1.9. The next lemma shows that VTC^0 is indeed an instance of the family VC (because it shows that the existence of the value of $Numones^\star$ is provable in VTC^0).

LEMMA IX.3.3. *The functions numones, Numones, and Numones* are* Σ_1^B-*definable (and hence also* Σ_1^1-*definable) in* VTC^0, *and the theory* $V^0(Row, Numones, Numones^\star)$ *proves*

$$\forall i < b, Numones^\star(b, Y, X)^{[i]} = Numones((Y)^i, X^{[i]}). \tag{228}$$

PROOF. The fact that *numones* and *Numones* are provably total in VTC^0 is obvious. We will show that *Numones** is Σ_1^B-definable in VTC^0. The fact that $V^0(Row, Numones, Numones^\star)$ (which extends VTC^0) proves (228) will be clear from the proof below.

For convenience, we use the functions *Row* and *seq* in the defining axiom for *Numones** described below; it is straightforward to eliminate *Row* and *seq* from the axiom (Lemmas V.4.27 and V.6.7).

Intuitively we need to show that $VTC^0(Row, seq)$ proves the existence of Z such that for all $i < b$, $Z^{[i]}$ is the "counting sequence" for $X^{[i]}$:

$$VTC^0(Row, seq) \vdash \exists Z \forall i < b \, \delta_{NUM}((Y)^i, X^{[i]}, Z^{[i]}).$$

The idea is to (i) concatenate the first $(Y)^i$ bits of the rows $X^{[i]}$, for $i < b$, to form a "big" string X', (ii) obtain the counting sequence Z' for X', and (iii) extract the desired array of counting sequences $Z^{[i]}$ from Z'.

We will use $|Y|$ as an upper bound for $(Y)^i$, for $i < b$. Thus, let X' be defined by

$$X'(i|Y| + x) \leftrightarrow x < (Y)^i \wedge X^{[i]}(x), \quad \text{for } i < b.$$

In other words, for $i < b$, the bit string

$$X'(i|Y|)\, X'(i|Y| + 1) \ldots X'(i|Y| + (Y)^i - 1)$$

is a copy of

$$X^{[i]}(0)\, X^{[i]}(1) \ldots X^{[i]}((Y)^i - 1)$$

and $X'(i|Y| + (Y)^i), \ldots, X'((i + 1)|Y| - 1)$ are all 0. Therefore, for $u \le (Y)^i$,

$$numones(u, X^{[i]}) = numones(i|Y| + u, X') - numones(i|Y|, X').$$

Let Z' be such that $\delta_{NUM}(b|Y|, X', Z')$ holds, i.e., Z' is the "counting sequence" for X'. Then

$$numones(u, X^{[i]}) = z \leftrightarrow (Z')^{i|Y|+u} \dot{-} (Z')^{i|Y|} = z.$$

Thus,

$$Numones^\star(b, Y, X) = Z \supset$$
$$\forall i < b \forall u \le (Y)^i \left((Z^{[i]})^u = (Z')^{i|Y|+u} \dot{-} (Z')^{i|Y|} \right).$$

Although the RHS does not uniquely specify all the bits in Z, we can add clauses which set all the undefined bits to 0 using the method in (226) used to define SEQ. It follows easily that $Numones^\star(b, Y, X)$ is provably total in VTC^0. $\qquad\qquad\square$

EXERCISE IX.3.4. Similar to the aggregate of a string function, we can define the aggregate of a number function as follows. Suppose that $f(x_1, \ldots, x_k, X_1, \ldots, X_n)$ is a polynomially bounded number function, i.e., for some $\mathcal{L}_A^2$ term t,

$$f(\vec{x}, \vec{X}) \le t(\vec{x}, |\vec{X}|).$$

Then $f^\star(b, \vec{Z}, \vec{X})$ is the polynomially bounded *string* function that satisfies

$$|f^\star(b, \vec{Z}, \vec{X})| \le \langle b, 1 + t \rangle$$

and

$$f^\star(b, \vec{Z}, \vec{X})(w) \leftrightarrow \exists u < b, \, w = \langle u, f((Z_1)^u, \ldots, (Z_k)^u, X_1^{[u]}, \ldots, X_n^{[u]}) \rangle.$$
$$(229)$$

Show that $numones^\star$ is provably total in VTC^0.

Following Section IX.2.2, to define $\widehat{VTC^0}$ we need a quantifier-free defining axiom for $Numones$. (Formally we will not change the defining axiom for $Numones$ but will introduce a function $Numones'$ that has the same value as $Numones$ and has a quantifier-free defining axiom.) So let $\delta'_{NUM}(y, X, Z)$ be a quantifier-free $\mathcal{L}_{FAC^0}$-formula (from Lemma V.6.3) which $\overline{V}^0$ proves equivalent to $\delta_{NUM}(y, X, Z)$.

Let $Numones'(y, X)$ be defined by

$$Numones'(y, X) = Z \leftrightarrow |Z| \le 1 + \langle y, y \rangle \wedge \delta'_{NUM}(y, X, Z). \quad (230)$$

Thus $Numones'(y, X) = Numones(y, X)$, but they have different defining axioms.

DEFINITION IX.3.5 ($\widehat{VTC^0}$). $\widehat{VTC^0}$ is the universal theory over the vocabulary $\mathcal{L}_{\widehat{VTC^0}} = \mathcal{L}_{FAC^0} \cup \{Numones'\}$ and is axiomatized by the axioms of $\overline{V}^0$ and (230).

We define $\overline{VTC}^0$ using the number function $numones'$ instead of the string function $Numones'$. Here $numones'$ has the same value as $numones$ but it has the following quantifier-free defining axioms:

$$numones'(0, X) = 0, \tag{231}$$

$$X(z) \supset numones'(z + 1, X) = numones'(z, X) + 1, \tag{232}$$

$$\neg X(z) \supset numones'(z + 1, X) = numones'(z, X). \tag{233}$$

DEFINITION IX.3.6. $\mathcal{L}_{FTC^0}$ is the smallest set that contains $\mathcal{L}_{FAC^0} \cup \{numones'\}$ such that for every quantifier-free $\mathcal{L}_{FTC^0}$-formula $\varphi(z, \vec{x}, \vec{X})$ and every $\mathcal{L}_A^2$-term $t = t(\vec{x}, \vec{X})$, there is a string function $F_{\varphi(z),t}$ in $\mathcal{L}_{FTC^0}$ with defining axiom (86):

$$F_{\varphi(z),t}(\vec{x}, \vec{X})(z) \leftrightarrow z < t(\vec{x}, \vec{X}) \wedge \varphi(z, \vec{x}, \vec{X}). \tag{234}$$

$\overline{VTC}^0$ is the theory over $\mathcal{L}_{FTC^0}$ and is axiomatized by the axioms of $\overline{V}^0$ together with (231), (232) and (233) for $numones'$, and (234) for each function $F_{\varphi(z),t}$.

It is easy to see that $Numones = F_{\varphi(z),t}$ for some $F_{\varphi(z),t} \in \mathcal{L}_{FTC^0}$. On the other hand, it is also easy to see that $numones = |T|$ for some term $T \in \mathcal{L}_{\widehat{VTC^0}}$. Therefore the results in Section IX.2 apply for $\widehat{VTC^0}$ and $\overline{VTC}^0$. We summarize the Definability Theorems for these theories as follows:

THEOREM IX.3.7. *Assume either $\mathcal{L}$ is $\mathcal{L}_{\widehat{VTC^0}}$ and T is $\widehat{VTC^0}$, or $\mathcal{L}$ is $\mathcal{L}_{FTC^0}$ and T is $\overline{VTC}^0$. Then*

(a) *A function is in FTC^0 iff it is represented by a term in $\mathcal{L}_{\widehat{VTC^0}}$. A string function is in FTC^0 iff it is in $\mathcal{L}_{FTC^0}$. A relation is in TC^0 iff it is represented by an open (or a Σ_0^B) formula in $\mathcal{L}$.*

(b) *For every $\Sigma_1^B(\mathcal{L})$ formula φ^+ there is a Σ_1^B-formula φ such that $T \vdash \varphi^+ \leftrightarrow \varphi$.*

(c) *T proves $\Sigma_0^B(\mathcal{L})$-COMP, $\Sigma_0^B(\mathcal{L})$-IND, and $\Sigma_0^B(\mathcal{L})$-MIN.*

(d) *$\overline{VTC}^0$ is a universal conservative extension of $\widehat{VTC^0}$, which is in turn a universal conservative extension of VTC^0.*

(e) *A function is in FTC^0 iff it is Σ_1^B-definable in VTC^0 iff it is Σ_1^B-definable in T.*

(f) *A relation is in TC^0 iff it is Δ_1^B-definable in VTC^0 iff it is Δ_1^B-definable in T.*

COROLLARY IX.3.8. VTC^0 *is a proper extension of* V^0. *In fact,* VTC^0 *is not* Σ_0^B-*conservative over* V^0.

PROOF. The first sentence follows from the second, which is true because VTC^0 proves the Pigeonhole Principle (Section IX.3.5 below), while this principle is not provable in V^0 (Corollary VII.2.4).

Another way of proving the the first sentence is to use Theorem IX.3.7 (e) above. Recall that the number function $parity(X)$, which is the parity of the total number of elements in X (Section V.5.1), is not in FAC^0. Hence V^0 does not prove the defining axiom for $parity$. On the other hand, $parity$ is in FTC^0, since it can be easily computed using $numones$:

$$parity(X) = numones(|X|, X) \quad \text{mod } 2.$$

So VTC^0 proves the defining axiom for $parity$. □

The problem of sorting a given sequence of natural numbers into non-decreasing order is complete for TC^0 [32]. Let (y, Z) encode a sequence of $(y + 1)$ numbers as in (226). For $x \le y$ let $rank(x, y, Z)$ be the number that appears at the x-th position when X is sorted in nondecreasing order (positions start from 0), and let $rank(x, y, Z) = 0$ if $x > y$. The next exercise shows that $rank$ is provably total in VTC^0.

EXERCISE IX.3.9. Give a Σ_1^B defining axiom $\varphi(x, y, z, Z)$ for $rank(x, y, Z)$, i.e., for y, Z such that $SEQ(y, Z)$ holds:

$$\forall x \left(rank(x, y, Z) = z \Leftrightarrow \varphi(x, y, z, Z) \right)$$

and show that

$$VTC^0 \vdash \forall x \forall y \forall Z \exists! z \le |Z| \varphi(x, y, z, Z).$$

Hint: For each v define

$$V^{[v]} = \{u \; : \; u \text{ is a value in } Z \text{ and } u < v\}.$$

Now let $c(v)$ be the cardinality of $V^{[v]}$. Show that for $x \le y$, $rank(x, y, Z)$ is precisely the smallest value v in the sequence Z such that $x \le c(v)$.

IX.3.3. Number Recursion and Number Summation. The number recursion operation produces a new number function from existing number functions. This operation is similar to *limited recursion* (Definition VI.2.11) but the latter defines new *string functions* from existing string functions. It is useful in characterizing FL and a number of its subclasses (later we will use this to develop the theory VTC^0V, an analogue of VPV (Section VIII.2). See Sections IX.4.4, IX.4.8, IX.5.5, IX.6.4 and IX.3.4 below). The number summation operation is a special instance of number recursion and is useful in characterizing FTC^0.

DEFINITION IX.3.10 (Number Recursion). A number function $f(y, \vec{x}, \vec{X})$ is obtained by *number recursion* from $g(\vec{x}, \vec{X})$ and $h(y, z, \vec{x}, \vec{X})$ if

$$f(0, \vec{x}, \vec{X}) = g(\vec{x}, \vec{X}), \tag{235}$$

$$f(y + 1, \vec{x}, \vec{X}) = h(y, f(y, \vec{x}, \vec{X}), \vec{x}, \vec{X}). \tag{236}$$

If further $f(y, \vec{x}, \vec{X}) < t(y, \vec{x}, \vec{X})$, then we also say that f is obtained by t-bounded number recursion (t-BNR) from g and h. In particular, if f is polynomially bounded then we say that f is obtained from g and h by *polynomial-bounded number recursion* (pBNR).

DEFINITION IX.3.11 (Number Summation). For a number function $f(y, \vec{x}, \vec{X})$, define the number function $\mathrm{sum}_f(y, \vec{x}, \vec{X})$ by

$$\mathrm{sum}_f(y, \vec{x}, \vec{X}) = \sum_{z=0}^{y} f(z, \vec{x}, \vec{X}).$$

The function sum_f is said to be defined from f by *number summation*, or just *summation*.

THEOREM IX.3.12. *A function is in* **FTC⁰** *iff it is obtained from* **FAC⁰** *functions by finitely many application of composition, string comprehension, and number summation iff it is obtained from* **FAC⁰** *by* **AC⁰** *reduction and number summation.*

PROOF. By Theorem IX.1.7 it suffices to prove that a function is in **FTC⁰** iff it is obtained from **FAC⁰** by **AC⁰** reduction and number summation.

For the ONLY IF direction, by Proposition IX.3.1 we need only show that *numones* can be obtained by number summation from **AC⁰** functions. This fact is straightforward:

$$numones(y, X) = \sum_{z=0}^{y} f_X(z, X)$$

where f_X (the "characteristic function of X") is the **AC⁰** function defined by

$$f_X(z, X) = w \leftrightarrow \big((X(z) \supset w = 1) \wedge (\neg X(z) \supset w = 0)\big). \tag{237}$$

We prove the other direction by induction on the number of applications of the number summation operation. The base case (number summation is not used) is obvious. For the induction step, it suffices to show that sum_f is **AC⁰** reducible to f and *numones*. Thus we **AC⁰**-define a string function $W_f(y)$ from f that contains the right number of bits; namely if $W = W_f(y)$ then

$$W(xa + v) \leftrightarrow x \leq y \wedge v < f(x)$$

where $a = max(\{f(x) : x < y\})$. Then it is easy to verify that $\mathrm{sum}_f(y) = numones((y + 1)a, W)$. □

IX.3.4. The Theory VTC^0V. We define here the theory VTC^0V, another universal conservative extension of VTC^0. The vocabulary of VTC^0V contains a symbol for each functions in FTC^0, but here, except for the FAC^0 functions, they are defined using the number summation scheme (based on Theorem IX.3.12).

DEFINITION IX.3.13 ($\mathcal{L}_{VTC^0V}$). The vocabulary $\mathcal{L}_{VTC^0V}$ is the smallest set that contains $\mathcal{L}_{FAC^0}$ such that:

1) For every number function $f(y, \vec{x}, \vec{X})$ in $\mathcal{L}_{VTC^0V}$ the function $\text{sum}_f(y, \vec{x}, \vec{X})$ is also in $\mathcal{L}_{VTC^0V}$ with defining axioms

$$\text{sum}_f(0, \vec{x}, \vec{X}) = 0 \wedge$$
$$\text{sum}_f(y + 1, \vec{x}, \vec{X}) = \text{sum}_f(y, \vec{x}, \vec{X}) + f(y, \vec{x}, \vec{X}). \quad (238)$$

2) For every $\mathcal{L}_A^2$-term t and quantifier-free $\mathcal{L}_{VTC^0V}$-formula φ the function $F_{\varphi(z),t}$ is in $\mathcal{L}_{VTC^0V}$ with defining axiom (86):

$$F_{\varphi(z),t}(\vec{x}, \vec{X})(z) \leftrightarrow z < t(\vec{x}, \vec{X}) \wedge \varphi(z, \vec{x}, \vec{X}). \quad (239)$$

3) For every string function $F(\vec{x}, \vec{X})$ in $\mathcal{L}_{VTC^0V}$ the number function $f_F(\vec{x}, \vec{X})$ is also in $\mathcal{L}_{VTC^0V}$ with defining axiom

$$f_F(\vec{x}, \vec{X}) = |F(\vec{x}, \vec{X})|. \quad (240)$$

COROLLARY IX.3.14.   (a) *A function is in FTC^0 iff it is represented by a function symbol in $\mathcal{L}_{VTC^0V}$.*

(b) *A relation is in TC^0 iff it is represented by an open (or a Σ_0^B) formula in $\mathcal{L}_{VTC^0V}$.*

PROOF. Part (b) follows from part (a), and part (a) follows from Theorem IX.3.12. Note that closure of $\mathcal{L}_{VTC^0V}$ under string comprehension (Definition IX.1.6) follows from item 2) in Definition IX.3.13 and closure under composition follows from items 2) and 3).   □

DEFINITION IX.3.15. The theory VTC^0V has vocabulary $\mathcal{L}_{VTC^0V}$ and axioms those of $\overline{V}^0$ and (238), (239), and (240) for the functions sum_f, $F_{\varphi(z),t}$, and f_F, respectively.

The following Lemma is proved in the same way as Lemma V.6.4.

LEMMA IX.3.16. *VTC^0V proves*

$$\Sigma_0^B(\mathcal{L}_{VTC^0V})\text{-}COMP, \Sigma_0^B(\mathcal{L}_{VTC^0V})\text{-}IND \text{ and } \Sigma_0^B(\mathcal{L}_{VTC^0V})\text{-}MIN.$$

The next result is proved in the same way as Theorem IX.2.14.

THEOREM IX.3.17. *VTC^0V is a universal conservative extension of VTC^0.*

PROOF. First, by definition, VTC^0V extends $\overline{V}^0$. As noted in the proof of Theorem IX.3.12 (the ONLY IF direction), $numones' = \text{sum}_{f_X}$ where f_X is defined in (237):

$$f_X(z, X) = w \leftrightarrow \big((X(z) \supset w = 1) \wedge (\neg X(z) \supset w = 0)\big).$$

It is easy to see that $VTC^0 V$ proves the defining axioms (231), (232), (233) for *numones'*. It follows that $VTC^0 V$ extends VTC^0.

Now we show that $VTC^0 V$ is conservative over VTC^0. Since $VTC^0 V$ extends VTC^0, we have

$$VTC^0 V = \bigcup_{i \geq 0} \mathcal{T}_i$$

where $\mathcal{T}_0 = VTC^0$ and each $\mathcal{T}_{i+1}$ is obtained from $\mathcal{T}_i$ by adding the defining axiom for a new function sum_f, $F_{\varphi(z),t}$, or f_F. We show that $\mathcal{T}_{i+1}$ is conservative over $\mathcal{T}_i$ by showing that the new function of $\mathcal{T}_{i+1}$ is definable in $\mathcal{T}_i$.

Let $\mathcal{L}_i$ denote the vocabulary of $\mathcal{T}_i$. Consider the case where the new function in $\mathcal{T}_{i+1}$ has the form $F_{\varphi(z),t}$ for some quantifier-free $\mathcal{T}_i$-formula φ and $\mathcal{L}_A^2$-term t. It is easy to see that $F_{\varphi(z),t}$ is definable in $\mathcal{T}_i$ if

$$\mathcal{T}_i \vdash \Sigma_0^B(\mathcal{L}_i)\text{-}COMP. \tag{241}$$

Similarly suppose that the new function in $\mathcal{T}_{i+1}$ has the form sum_f for some number function $f \in \mathcal{L}_i$. Following the IF direction of the proof of Theorem IX.3.12, the fact that sum_f is definable in $\mathcal{T}_i$ also follows from (241). In fact using (241) above it can be shown that $\text{sum}_f^\star$ is Σ_1^B-definable in $\mathcal{T}_i$. This is left as an exercise. Recall the notion of aggregate function for a number function in Exercise IX.3.4.

EXERCISE IX.3.18. Suppose that (241) holds. Show that both sum_f and $\text{sum}_f^\star$ are definable in $\mathcal{T}_i$.

It remains to prove (241). The proof is by induction on i. The base case is Theorem IX.3.7 (c). The induction step follows from Theorem VIII.1.15 (using Lemma IX.2.15) and Corollary IX.3.19 below (using Exercise IX.3.18).   □

The next result refers to number aggregates (Exercise IX.3.4) and is a corollary of the Aggregate Function Theorem VIII.1.15.

COROLLARY IX.3.19 (Aggregate Number Function Theorem). *Let $\mathcal{T}$ be a theory with vocabulary $\mathcal{L}$ which extends $V^0(Row)$ and proves $\Sigma_0^B(\mathcal{L})$-COMP. Suppose that f and $f^\star$ are definable in $\mathcal{T}$ (Definition V.4.1) and $\mathcal{T}(f, f^\star)$ proves (229). Then $\mathcal{T}(f)$ proves $\Sigma_0^B(\mathcal{L} \cup \{f\})$-COMP.*

PROOF. Let F be the string function that contains at most one element and $|F| = f$:

$$(f = 0 \supset |F| = 0) \wedge (f > 0 \supset (|F| = f \wedge \forall z < f(F(z) \leftrightarrow z + 1 = f))).$$

Then both F and $F^\star$ are definable in $\mathcal{T}$. So the corollary follows easily from Theorem VIII.1.15.   □

The above proof method can be used to show that the next corollary follows from the First Elimination Theorem IX.2.10.

COROLLARY IX.3.20 (Second Elimination Theorem). *Let T be a theory with vocabulary $\mathcal{L}$ which extends $V^0(Row)$ and proves $\Sigma_0^B(\mathcal{L})$-COMP. Suppose that f and $f^\star$ are Σ_1^B-definable in T and $T(f, f^\star)$ proves (229). Suppose also that every $\Sigma_0^B(\mathcal{L})$ formula is equivalent in T to a $\Sigma_1^B(\mathcal{L}_A^2)$ formula. Then every $\Sigma_0^B(\mathcal{L} \cup \{f\})$ formula is equivalent in $T(f)$ to a $\Sigma_1^B(\mathcal{L}_A^2)$ formula.*

COROLLARY IX.3.21 ($\Sigma_1^B(\mathcal{L}_{VTC^0 V})$ Elimination). *For each $\Sigma_1^B(\mathcal{L}_{VTC^0 V})$ formula φ^+ there is a Σ_1^B formula φ so that $VTC^0 V \vdash \varphi^+ \leftrightarrow \varphi$.*

PROOF. The argument is similar to the proof of Corollary IX.2.18 from the First Elimination Theorem IX.2.10 (for string functions) but now we also need to use the Second Elimination Theorem IX.3.20 (for number functions). Let T_i and $\mathcal{L}_i$ be as in the proof of Theorem IX.3.17. It suffices to prove the statement of the present corollary with $\mathcal{L}_{VTC^0 V}$ and $VTC^0 V$ replaced by $\mathcal{L}_i$ and T_i. The proof is by induction on i as in the proof of Corollary IX.2.18, but now we have the extra case of sum_f to consider, using Exercise IX.3.18 and Corollaries IX.3.19 and IX.3.20. □

The definability theorems for $VTC^0 V$ are as follows.

COROLLARY IX.3.22. (a) *A function is in FTC^0 iff it is Σ_1^B-definable in $VTC^0 V$.*
(b) *A relation is in TC^0 iff it is Δ_1^B-definable in $VTC^0 V$.*

PROOF. The corollary follows either from Theorems IX.3.7(e) and IX.3.17, or directly from Theorem IX.3.12 and Lemma IX.3.16 using the Herbrand Theorem. □

IX.3.5. Proving the Pigeonhole Principle in VTC^0. We present a proof of the Pigeonhole Principle (Section VII.1.2) in VTC^0. As mentioned in the proof of Corollary VII.2.4 this implies that VTC^0 is a proper extension of V^0. In Chapter X we show that each Σ_0^B theorem of VTC^0 translates into a family of tautologies having polysize bounded depth PTK proofs (Corollary X.4.19). It follows that the family PHP (Definition VII.1.12) has polysize bounded depth PTK proofs. This separates bounded depth PK from bounded depth PTK. Furthermore we show in Section IX.5.4 that VNC^1 extends VTC^0. Therefore PHP is also provable in VNC^1. The Propositional Translation Theorem for VNC^1 (Theorem X.3.1 and Corollary X.3.4) thus allow us to derive a theorem of Buss [21] that PHP has polysize $Frege$ proofs.

The formula $PHP(a, X)$ is defined in Example VII.2.1 as follows:

$$\forall x \leq a \exists y < aX(x, y) \supset$$
$$\exists x \leq a \exists z \leq a \exists y < a(x \neq z \wedge X(x, y) \wedge X(z, y)). \quad (242)$$

THEOREM IX.3.23. $VTC^0 \vdash PHP(a, X)$.

Proof. Since $VTC^0(numones)$ is conservative over VTC^0, it suffices to show that
$$VTC^0(numones) \vdash PHP(a, X).$$
We prove by contradiction, so assume that
$$\forall x \leq a \exists y < aX(x, y) \tag{243}$$
and
$$\forall x \leq a \forall z \leq a \forall y < a((x \neq z \wedge X(x, y)) \supset \neg X(z, y)). \tag{244}$$

Let P be the set of pigeons:
$$P = \{0, 1, 2, \ldots, a\}.$$
Let $\varphi(x, y)$ be the following formula which asserts that y is the first hole that pigeon x occupies:
$$\varphi(x, y) \equiv x \leq a \wedge y < a \wedge X(x, y) \wedge \forall v < y \neg X(x, v).$$
Then by (243) and (244) φ defines an injective function from P into the set of holes $\{0, 1, 2, \ldots, a - 1\}$, i.e., VTC^0 proves
$$\forall x \leq a \exists! y < a\varphi(x, y) \wedge \forall x \leq a \forall z \leq a \forall y < a((x \neq z \wedge \varphi(x, y)) \supset$$
$$\neg\varphi(z, y)).$$

Let H be the image of P (defined using $\Sigma_0^B\text{-}COMP$):
$$|H| \leq a \wedge \forall y < a(H(y) \leftrightarrow \exists x \leq a\varphi(x, y)).$$
Then it is easy to see that φ defines a bijection between P and H (i.e., φ satisfies the premise of (245) below for $b = a + 1$). Lemma IX.3.24 below shows that P and H have the same cardinality:
$$VTC^0(numones) \vdash numones(a + 1, P) = numones(a + 1, H).$$
However, it is easy to show in VTC^0 that $numones(a + 1, P) = a + 1$ and $numones(a + 1, H) \leq a$, a contradiction. $\square$

For the following lemma, informally we show that if there is a bijection between two sets P and H that is described by a Σ_0^B formula $\varphi(x, y)$, then provably in $VTC^0(numones)$ the sets have the same cardinality.

Lemma IX.3.24. *For any $\Sigma_0^B(numones)$ formula $\varphi(x, y)$, the following is a theorem of $VTC^0(numones)$*:
$$(\forall x < b(P(x) \supset \exists! y < b(\varphi(x, y) \wedge H(y)) \wedge \forall y < b(H(y) \supset$$
$$\exists! x < a(\varphi(x, y) \wedge P(x))) \supset numones(b, P) = numones(b, H). \tag{245}$$

Proof. Let Z be the array whose rows $Z^{[i]}$ are the images of the initial segments $Cut(i, P)$ of P under the bijection, i.e.,
$$\forall i < b \forall y < b\big(Z^{[i]}(y) \leftrightarrow \exists x < i(\varphi(x, y) \wedge P(x))\big).$$

Since φ is Σ_0^B (*numones*) and VTC^0(*numones*) $\vdash \Sigma_0^B$ (*numones*)-**COMP** (by Theorem IX.3.7 (c)), VTC^0(*numones*) proves the existence of such Z.
Now we prove by induction on $i < b$ that

$$numones(i, P) = numones(b, Z^{[i]}). \tag{246}$$

It will follow that $numones(b, P) = numones(b, Z^{[b]})$, and since $Z^{[b]} = Cut(b, H)$ we have $numones(b, P) = numones(b, Cut(b, H))$, so

$$numones(b, P) = numones(b, H).$$

The base case ($i = 0$) is obvious. For the induction step, assume that (246) is true for some $i \geq 0$. We show that it is also true for $i + 1$. There are two cases: either $i \in P$ or $i \notin P$.
First, suppose that $i \in P$, then $numones(i + 1, P) = numones(i, P) + 1$. Let $j \in Z^{[i+1]}$ be such that $\varphi(i, j)$ holds. Then $j \notin Z^{[i]}$, and it can be shown by induction on y that

$$numones(y, Z^{[i+1]}) = \begin{cases} numones(y, Z^{[i]}) & \text{if } y \leq j, \\ numones(y, Z^{[i]}) + 1 & \text{if } y \geq j + 1. \end{cases}$$

Hence $numones(b, Z^{[i+1]}) = numones(b, Z^{[i]}) + 1$, and we are done.
The other case is similar. $\qquad\qquad\square$

IX.3.6. Defining String Multiplication in VTC^0. Recall that $bin(X)$ is the integer value associated with a string X (46) (page 85):

$$bin(X) = \sum_{i \in X} X(i)2^i.$$

The string multiplication function, $X \times_2 Y$ (or simply $X \times Y$) is defined so that

$$bin(X \times Y) = bin(X) \times bin(Y).$$

Exercise VI.2.7 shows that this function is Σ_1^B-definable in V^1. Here we will show that it is actually Σ_1^B-definable in VTC^0 by formalizing in VTC^0 a TC^0 algorithm that computes $X \times Y$. Furthermore, VTC^0 proves the usual properties of this function, such as commutativity, distributivity over $X + Y$, etc.

Notice that the "school" algorithm described in Exercise VI.2.7 is a polytime algorithm. The main component of this algorithm is the polytime process that computes the sum of all rows of the table $X \otimes Y$. The TC^0 algorithm for $X \times Y$ is obtained by replacing this polytime process by a uniform family of TC^0 circuits. First, we outline this TC^0 algorithm and formalize it in VTC^0 by showing that the function *Sum* defined below is Σ_1^B-definable in VTC^0.

For the formalizations, recall the string functions $\varnothing, S(X), X + Y$ given in Example V.4.17, $Cut(x, X)$ on page 139, and the number function $\lceil \log(x + 1) \rceil$ in Exercise III.3.30.

IX.3.6.1. *Adding n Strings.* Suppose that we are to add n integers written as n binary strings, each of length m. The idea is to write these binary strings as rows in a table of n rows and m columns, then divide the columns of the table into blocks of ℓ columns each (for some parameter ℓ to be determined later) so that the sum of the rows in each block can be easily computed in TC^0, and the desired result can be computed from these sums by a TC^0 circuit.

More precisely let $\ell = \lceil \log_2(n+1) \rceil$. Then in each block B_i, each row can be seen as a number with value $\leq 2^\ell - 1$. Therefore the sum of the rows in B_i is at most

$$n(2^\ell - 1) < 2^{2\ell}$$

and hence has a binary representation of length at most 2ℓ. It is important that this sum can be defined as the number of 1-bits in a long string easily obtained from B_i.

Now let b_i be the sum of the rows in the block B_i. Then the required sum is

$$\sum_i 2^{i\ell} b_i. \tag{247}$$

Write each b_i as a binary string of length exactly 2ℓ (add preceding 0's if necessary). Then

$$b_0 + 2^{2\ell} b_2 + 2^{4\ell} b_4 + \dots \tag{248}$$

is simply the concatenation of the strings $b_0, b_2, b_4, \dots$, and similarly for

$$2^\ell b_1 + 2^{3\ell} b_3 + 2^{5\ell} b_5 + \dots. \tag{249}$$

As a result, (247) can be computed in AC^0 by adding the above two sums.

IX.3.6.2. *Formalization.* For the formalization, we will use the function *numones* and some AC^0 functions such as the length function for numbers $|x| = \lceil \log_2(x+1) \rceil$ (see Exercise III.3.30, Section III.3.3). It will be clear that the functions defined here belong to $\mathcal{L}_{FTC^0}$.

Suppose that the n input strings are given as the rows $Z^{[0]}, \dots, Z^{[n-1]}$ in an array Z. We will define in VTC^0 the function $Sum(n, m, Z)$ that satisfies

$$Sum(0, m, Z) = \varnothing, \tag{250}$$

$$Sum(n+1, m, Z) = Sum(n, m, Z) + Cut(m, Z^{[n]}) \tag{251}$$

where $Cut(x, X)$ is the first x bits of X (97):

$$Cut(x, X)(z) \leftrightarrow z < x \wedge X(z).$$

We define the columns of Z as strings using the function *Transpose* defined as follows:

$$Transpose(n, m, X) = Y \leftrightarrow (|Y| \leq \langle m, n \rangle \wedge$$
$$\forall z < \langle m, n \rangle \big(Y(z) \leftrightarrow \exists i < m \exists j < n (z = \langle i, j \rangle \wedge X(j, i)) \big)). \quad (252)$$

Let $V = Transpose(n, m, Z)$. Then the sum of the bits in column i of Z is

$$c_i = numones(n, V^{[i]}).$$

Let

$$\ell = |n|, \qquad k = \lceil m/2\ell \rceil.$$

Note that ℓ is an AC^0 function of n (Exercise III.3.30). We want the sequence B: $(B)^0 = b_0, (B)^1 = b_1, \ldots, (B)^{2k} = b_{2k}$ (see (247)) so that

$$(B)^i = \sum_{j=0}^{\ell-1} 2^j c_{i\ell+j}.$$

We show below how to define each $(B)^i$ by a Σ_1^B formula. It follows from Exercise IX.3.4 that B is also Σ_1^B-definable in VTC^0.

To define $(B)^i$, it suffices to define a string U that contains exactly $(B)^i$ 1-bits. Then

$$(B)^i = numones(|U|, U).$$

Notice that $c_i \leq n$ for $0 \leq i < m$. The string U consists of

$$1 + 2^1 + 2^2 + \cdots + 2^{\ell-1} = 2^\ell - 1$$

substrings and each has n bits, so that for $j < \ell$, 2^j substrings contain exactly $c_{i\ell+j}$ 1-bits. Thus U can be defined as follows:

$$|U| \leq 2^\ell n \wedge \forall j < \ell \forall u < 2^j \forall v < n \big(U((2^j - 1)n + un + v) \leftrightarrow v < c_{i\ell+j} \big).$$

Now the sum (248) is formally defined as a string L with bit definition:

$$|L| \leq 2k\ell \wedge \forall x < 2k\ell \big(L(x) \leftrightarrow \exists i < k \exists y < 2\ell (x = 2i\ell + y \wedge$$
$$BIT(y, (B)^{2i})) \big)$$

(where BIT is the Δ_0 formula defined in Section III.3.3). Similarly, the sum (249), denoted by H, is defined as follows:

$$|H| \leq 2k\ell \wedge \forall x < 2k\ell \big(H(x) \leftrightarrow \exists i < k \exists y < 2\ell (x = (2i+1)\ell + y \wedge$$
$$BIT(y, (B)^{2i+1})) \big).$$

Finally

$$Sum(n, m, Z) = L + H.$$

Lemma IX.3.25. *The theory* $\overline{VTC}^0$ *proves* (250) *and* (251).

Proof. We reason in $\overline{VTC}^0$. If $n = 0$ then $\ell = 0$, so it is easy to see that $Sum(0, m, Z) = \varnothing$. This establishes (250).

We prove (251) by induction on m. The base case ($m = 0$) is obvious. For the induction step, we need the exercise below. Here $Shift(x, y)$ is the string obtained by shifting all bits of the binary representation of y by x positions to the left:

$Shift(x, y) = U \leftrightarrow$

$$\left(|U| \leq x + |y| \wedge \forall z < x + |y|(U(z) \leftrightarrow \exists i < |y|BIT(i, y))\right).$$

Example IX.3.26 (Provable in $\overline{V}^0$).

$$Shift(x, y + z) = Shift(x, y) + Shift(x, z).$$

Exercise IX.3.27. Show that it is provable in $\overline{VTC}^0$ that

$$Sum(n, m + 1, Z) = Sum(n, m, Z) + Shift(m, c_m)$$

where c_m is the sum of the first n bits in column m of Z:

$$c_m = numones(n, V^{[m]}) \qquad \text{where } V = Transpose(n, m + 1, Z).$$

Reasoning in $\overline{VTC}^0$, the induction step follows from Exercise IX.3.27 as follows. Suppose that we need to prove (251) for $m + 1$. Let c'_m be the sum of the first $(n + 1)$ bits in column m of Z:

$$c'_m = numones(n + 1, Transpose(n + 1, m, Z)^{[m]}).$$

Then

$$c'_m = c_m + Z^{[n]}(m). \tag{253}$$

By Exercise IX.3.27 we need to prove

$Sum(n + 1, m, Z) + Shift(m, c'_m) =$

$$Sum(n, m, Z) + Shift(m, c_m) + Cut(m + 1, Z^{[n]}).$$

By the induction hypothesis,

$$Sum(n + 1, m, Z) = Sum(n, m, Z) + Cut(m, Z^{[n]}).$$

Also,

$$Cut(m + 1, Z^{[n]}) = Cut(m, Z^{[n]}) + Shift(m, Z^{[n]}).$$

So we need to show that

$$Shift(m, c'_m) = Shift(m, c_m) + Shift(m, Z^{[n]}(m)).$$

This follows from Example IX.3.26 and (253). ☐

IX.3.6.3. *Defining* $X \times Y$. To define $X \times Y$ we use the table $X \otimes Y$ given in Exercise VI.2.7 and can be equivalently defined as follows:

$$X \otimes Y = Z \leftrightarrow \big(|Z| \leq \langle |Y|, |X| + |Y| \rangle \wedge$$
$$\forall i < |Y|((\neg Y(i) \supset Z^{[i]} = \varnothing) \wedge (Y(i) \supset Z^{[i]} = Shift(i, X))))\big)$$

where $Shift(x, Y)$ is the string obtained from Y by shifting all bits x positions to the left

$$Shift(x, Y) = Z \leftrightarrow$$
$$|Z| \leq x + |Y| \wedge \forall z < x + |Y|(Z(z) \leftrightarrow \exists u < |Y|, \; Y(u) \wedge z = x + u).$$

(Notice that $Shift(x, y)$ and $Shift(x, Y)$ have different arity, so even though they have the same name, their meaning will be clear from context.)

We define

$$X \times Y = Sum(|Y|, |X| + |Y|, X \otimes Y).$$

LEMMA IX.3.28. $\overline{VTC^0} \vdash X \times Y = Y \times X$.

PROOF. By definition, we need to show that

$$Sum(|Y|, |X| + |Y|, X \otimes Y) = Sum(|X|, |X| + |Y|, Y \otimes X).$$

Therefore it suffices to show that the columns of $X \otimes Y$ and $Y \otimes X$ have the same number of 1-bits:

$$numones(|Y|, V^{[i]}) = numones(|X|, W^{[i]}) \tag{254}$$

for $i < |X| + |Y|$ and

$$V = Transpose(|X|, |X| + |Y|, X \otimes Y),$$
$$W = Transpose(|Y|, |X| + |Y|, Y \otimes X).$$

Notice that there is a bijection between $V^{[i]}$ and $W^{[i]}$ defined by

$$V^{[i]}(z) \leftrightarrow W^{[i]}(i \dot- z) \qquad \text{for } z \leq i$$

because

$$V^{[i]}(z) \leftrightarrow Y(z) \wedge X(i \dot- z) \quad \text{and} \quad W^{[i]}(z) \leftrightarrow X(z) \wedge Y(i \dot- z).$$

So the conclusion follows from Lemma IX.3.24. $\qquad \square$

LEMMA IX.3.29. $\overline{VTC^0} \vdash X \times (Y + Z) = X \times Y + X \times Z$.

PROOF. We will prove by induction on $i \leq |X|$ that

$$Cut(i, X) \times (Y + Z) = Cut(i, X) \times Y + Cut(i, X) \times Z. \tag{255}$$

The lemma follows by letting $i = |X|$.

For the base case, $i = 0$, we have $Cut(0, X) = \varnothing$. So this case follows from Exercise IX.3.31 (a) below and Lemma IX.3.28.

For the induction step, suppose that (255) holds for some $i \geq 0$. We prove it for $i + 1$. There are two cases: either $i \in X$ or $i \notin X$. In the second case $Cut(i+1, X) = Cut(i, X)$ so the conclusion if obvious. Thus we consider the case where $i \in X$. We have

$$Cut(i+1, X) = Cut(i, X) + \{i\}.$$

We need the following results:

EXERCISE IX.3.30. Show that the following are theorems of $\overline{VTC}^0$:

(a) $|X| \le i \supset (X + \{i\}) \times Y = X \times Y + \{i\} \times Y$.
(b) $\{i\} \times X = \{x + i \ : \ x \in X\}$.
(c) $\{i\} \times (Y + Z) = \{i\} \times Y + \{i\} \times Z$.

Now $|Cut(i, X)| \le i$. Using Exercises IX.3.30 and V.4.19 and the induction hypothesis we have

$$Cut(i+1, X) \times (Y + Z)$$
$$= (Cut(i, X) + \{i\}) \times (Y + Z)$$
$$= Cut(i, X) \times (Y + Z) + \{i\} \times (Y + Z)$$
$$= (Cut(i, X) \times Y + Cut(i, X) \times Z) + (\{i\} \times Y + \{i\} \times Z)$$
$$= (Cut(i, X) \times Y + \{i\} \times Y) + (Cut(i, X) \times Z + \{i\} \times Z)$$
$$= (Cut(i, X) + \{i\}) \times Y + (Cut(i, X) + \{i\}) \times Z$$
$$= Cut(i+1, X) \times Y + Cut(i+1, X) \times Z.$$

So (255) holds for $i + 1$.   □

EXERCISE IX.3.31. Show that the following are theorems of $\overline{VTC}^0$:

(a) $X \times \varnothing = \varnothing$.
(b) $X \times S(Y) = (X \times Y) + X$.

EXERCISE IX.3.32. Show that

$$\overline{VTC}^0 \vdash (X \times Y) \times Z = X \times (Y \times Z).$$

Hint: First prove the equation for Z of the form $\{i\}$. Then prove by induction on i that

$$(X \times Y) \times Cut(i, Z) = X \times (Y \times Cut(i, Z)).$$

IX.3.7. Proving Finite Szpilrajn's Theorem in VTC^0. One version of Szpilrajn's Theorem states that every partial order can be extended to a total order. Here we show that VTC^0 proves this for finite partial orders.

A *partial order* is a binary relation on a set S which is reflexive, anti-symmetric, and transitive. A partial order is *total* if every two elements are comparable. We represent a partial order $\preceq$ on $\{0, 1, \ldots, n-1\}$ by an array X, where $X(i, j)$ holds iff $i \preceq j$. The above properties are expressed by the following four Σ_0^B-formulas:

$$Reflex(n, X) \equiv \forall x < n X(x, x),$$

$$Anti(n, X) \equiv \forall x, y < n\big((X(x, y) \wedge X(y, x)) \supset x = y\big),$$

$$Trans(n, X) \equiv \forall x, y, z < n\big((X(x, y) \wedge X(y, z)) \supset X(x, z)\big),$$

$$Total(n, X) \equiv \forall x, y < n(X(x, y) \vee X(y, x)).$$

Thus X represents a partial order on $\{0, 1, \ldots, n - 1\}$ if it satisfies the formula $Partial(n, X)$, where

$$Partial(n, X) \equiv Reflex(n, X) \wedge Anti(n, X) \wedge Trans(n, X).$$

Finite Szpilrajn's Theorem can be expressed by the formula

$$Partial(n, X) \supset \exists Z\big(\forall x, y < n(X(x, y) \supset Z(x, y)) \wedge$$
$$Partial(n, Z) \wedge Total(n, Z)\big). \quad (256)$$

THEOREM IX.3.33. VTC^0 proves (256).

PROOF. Assume $Partial(n, X)$, and define an array Y which stores the sums of the columns of X (interpreted as a 0-1 matrix). That is, $(Y)^x = numones(n, \hat{X}^{[x]})$, where $\hat{X} = Transpose(n, n, X)$. Now we define $Z(x, y)$ to hold iff

$$x < n \wedge y < n \wedge \big((Y)^x < (Y)^y \vee ((Y)^x = (Y)^y \wedge X(x, y))\big).$$

The following exercise completes the proof. □

EXERCISE IX.3.34. Show that $\overline{VTC}^0$ (and hence VTC^0) proves (256) when Z is defined as above.

IX.3.8. Proving Bondy's Theorem. Consider a 2×2 0-1 matrix whose rows are distinct, e.g.:

0	1
0	0

It is easy to see that there is always a column whose removal from the matrix results in a column of two distinct bits. On the other hand if we start with a 3×2 0-1 matrix of distinct rows, then after removing any column there is always a pair of rows that contain the same bit.

Bondy's Theorem [16] states more generally that for any $n \times n$ 0-1 matrix whose rows are distinct we can always delete a column so that the remaining $n \times (n - 1)$ matrix still has n distinct rows. (It is easy to construct a $(n + 1) \times n$ matrix with $(n + 1)$ distinct rows such that deleting any column results in a pair of identical row.)

Frankl's Theorem [51] generalizes further by specifying a maximal value for m such that any $m \times n$ matrix with distinct rows contains a column whose deletion results in a $m \times (n-1)$ matrix that contains at least $m - 2^{t-1} + 1$ distinct rows. Here we will formalize Bondy's Theorem (i.e., the case $t = 1$) and show that V^0 proves its equivalence to **PHP**. It can be shown that the case $t = 2$ is also equivalent to **PHP** over V^0. Thus these cases are provable in **VTC**^{0}. However, it is not known whether the same is true for other cases.

To formulate Bondy's Theorem, an $m \times n$ 0-1 matrix will be encoded by a string X: $X(i, j)$ holds iff the entry with indices (i, j) is 1, for $0 \le i < m$, $0 \le j < n$. We will also write $X(i, j) = 1$ for $X(i, j)$ and $X(i, j) = 0$ for $\neg X(i, j)$. The following Σ_0^B formula states that the rows of the $m \times n$ matrix X are distinct:

$$DISTINCT(m, n, X) \equiv \forall i_1 < m \forall i_2 < i_1 \exists j < n(X(i_1, j) \leftrightarrow \neg X(i_2, j)).$$
(257)

Let $BONDY(n, X)$ denote

$$DISTINCT(n, n, X) \supset$$
$$\exists j < n \forall i_1 < n \forall i_2 < i_1 \exists k < n(k \neq j \wedge (X(i_1, k) \leftrightarrow \neg X(i_2, k))).$$

Let **BONDY** (resp. **PHP**) denote the universal closure of $BONDY(n, X)$ (resp. $PHP(y, X)$).

THEOREM IX.3.35 ([17]). $V^0 \vdash$ **BONDY** $\leftrightarrow$ **PHP**.

To prove the theorem we use the following principle, called the *surjective* (or also *dual*) Pigeonhole Principle, which says that there is no surjective (single-valued) mapping from n "holes" to $(n + 1)$ "pigeons" (S stands for surjective):

$$SPHP(n, X) \equiv \forall i < n \exists! j \le n X(i, j) \supset \exists j \le n \forall i < n \neg X(i, j).$$

By considering inverse maps the following fact is easy to prove.

EXERCISE IX.3.36. Show that $V^0 \vdash$ **SPHP** $\leftrightarrow$ **PHP** (where **SPHP** is the universal closure of $SPHP(n, X)$).

(In Section IX.4.3 we will define **OPHP**, a weaker version of the Pigeonhole Principle and show that it is provable in the theory $V^0(2)$ (Section IX.4). On the other hand, it is not known whether **PHP** is provable in $V^0(2)$.)

PROOF OF THEOREM IX.3.35. By Exercise IX.3.36 it suffices to show that

$$V^0 \vdash SPHP \leftrightarrow BONDY.$$

First we show that

$$V^0 \vdash SPHP \supset BONDY.$$

Given an $n \times n$ matrix X with distinct rows we need to show that there exists a column j that can be removed without creating identical rows. A key observation is as follows. Suppose that we order the rows (regarded as binary strings) lexicographically in increasing order (comparing the left-most bits first). Associate each row i of the first $(n-1)$ rows (i.e., $i < n-1$) with the left-most column j_i that distinguishes row i and row $(i+1)$. Then the columns $\{j_0, j_1, \ldots, j_{n-2}\}$ suffice to distinguish all n rows. By **SPHP** there is at least one column not listed in the set, and hence it can be removed without creating any identical rows.

This observation can be proved as follows. By the lexicographical ordering of the rows we must have

$$X(i, j_i) = 0 \wedge X(i+1, j_i) = 1.$$

It suffices to show that for any $i \neq i'$ $(i, i' < n-1)$ there is i'' so that rows i and i' are different on column $j_{i''}$. Suppose without loss of generality that $i < i'$, and let j be the left-most position where rows i and i' differ. (See Figure 5.) Then we must have

$$X(i, j) = 0 \wedge X(i', j) = 1.$$

Let i'' be the largest number such that row i'' agrees with row i up to (including) column j. Then it can be seen that $j_{i''} = j$, so rows i and i' differ on column $j_{i''}$.

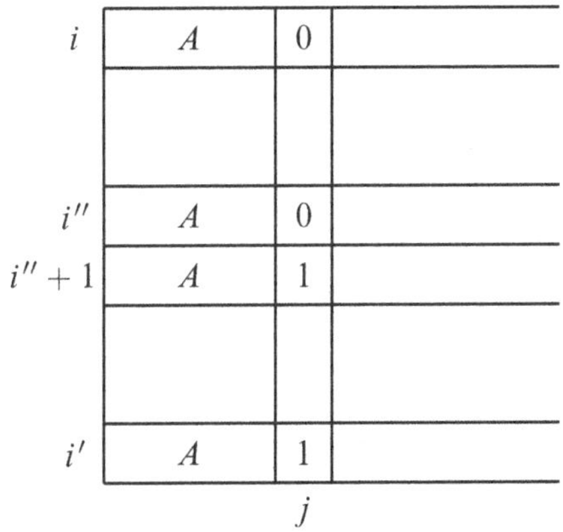

FIGURE 5. An i'' such that rows i and i' differ on column $j_{i''}$. From the top down, the rows are ordered lexicographically.

In other words, the set $\{j_0, j_1, \ldots, j_{n-2}\}$ already distinguishes all rows of the given matrix. By SPHP, the map

$$\{0, 1, \ldots, n-2\} \rightarrow \{0, 1, \ldots, n-1\} : i \mapsto j_i$$

is not surjective. Therefore there is $k \leq n-1$ such that $k \notin \{j_0, j_1, \ldots, j_{n-2}\}$. Then we can remove column k without creating identical rows.

For the formalization, the main task is to define the map $i \mapsto j_i$ and to show the existence of $j_{i''}$ as above. The rows of X can be compared by the following Σ_0^B formula (notice we do not need the actual position of the rows in this ordering):

$$i \prec_{m,n,X} k \qquad (\text{or just } i \prec k) \tag{258}$$

which is true iff row i is lexicographically less than row k:

$$i \prec k \equiv \exists j < n \forall \ell < j \big(\neg X(i,j) \wedge X(k,j) \wedge (X(i,\ell) \leftrightarrow X(k,\ell)) \big).$$

Then we can define a Σ_0^B formula $PRED(i,k,m,n,X)$ which is true iff k is the next row of i in the lexicographical ordering:

$$PRED(i,k,m,n,X) \equiv i \prec k \wedge \forall k'(i \prec k' \supset k \preceq k').$$

(Here $k \preceq k'$ stands for $k = k' \vee k \prec k'$.)

We also need the fact that there is a $\prec$-maximal index i_0 (so that the map $i \mapsto j_i$ is not defined for i_0 and we can apply the SPHP). For this, observe that $BONDY(n, X)$ is equivalent to $BONDY(n, X')$ where X' is obtained from X by simultaneously flipping all bits in some columns. Therefore we can assume that row $(n-1)$ of X contains all 1's, and hence $n-1$ is $\prec$-maximum.

Now using Σ_0^B-MIN it can be shown in V^0 that every row i, where $i < n - 1$, has a unique "next" row:

$$\forall i < n - 1 \exists! k \, PRED(i,k,m,n,X).$$

EXERCISE IX.3.37. Formally define using Σ_0^B-$COMP$ the mapping j_i as a string Y: for $i < n - 1$

$$Y(i,j) \equiv j = j_i.$$

Then use $SPHP(n-1, Y)$ to derive $BONDY(n, X)$.

Now we show

$$V^0 \vdash \neg SPHP \supset \neg BONDY.$$

Consider the following $(n+1) \times n$ matrix (which can be defined using Σ_0^B-$COMP$):

$$A = \begin{pmatrix} 0 & 0 & \ldots & 0 \\ 1 & 0 & \ldots & 0 \\ 0 & 1 & \ldots & 0 \\ \vdots & \vdots & \ddots & \vdots \\ 0 & 0 & \ldots & 1 \end{pmatrix}.$$

It is easy to see that the rows of A are distinct, and removing any column j from A will result in two rows 0 and $j+1$ being identical.

Suppose that $\neg SPHP(n, Y)$ holds for some (n, Y), i.e., Y specifies a surjective mapping

$$Y : \{0, 1, \ldots, n-1\} \to \{0, 1, \ldots, n\}.$$

We define a $n \times n$ matrix X as follows. For $i < n$, row $X^{[i]}$ is the same as row $A^{[k]}$, where k is the image of i under Y, i.e., k is the unique value, $k \leq n$, such that $Y(i, k)$ holds. Since the rows of A are distinct, the rows of X are also distinct. Moreover, removing a column j from X will make rows i_0 and i_1 identical, where i_0 and i_1 are such that $Y(i_0, 0)$ and $Y(i_1, j + 1)$ hold. Hence $\neg BONDY(n, X)$. $\qquad\square$

IX.4. Theories for $AC^0(m)$ and ACC

In this section we develop the theories associated with the classes $AC^0(m), m \geq 2$ and their union ACC. These classes lie between AC^0 and TC^0. In Section IX.4.1 we define the classes. In Section IX.4.2 we define the theories $V^0(2)$, $\widehat{V^0(2)}$ and $\overline{V^0(2)}$ for $AC^0(2)$. Functions in $FAC^0(2)$ can be characterized by a bounded number recursion (BNR) scheme (see Section IX.3.3) and in Section IX.4.4 we use this to develop $VAC^0(2)V$, another universal conservative extension of $V^0(2)$. A discrete version of the Jordan Curve Theorem can be proved in $V^0(2)$ and we present the formalization in Section IX.4.5. Then in Section IX.4.6 we define theories for other classes $AC^0(m)$. Finally, the class $FAC^0(6)$ also has a recursion characterization using the BNR scheme, and we use this to develop $VAC^0(6)V$ in Section IX.4.8.

IX.4.1. The Classes $AC^0(m)$ and ACC. For each $m \in \mathbb{N}$, $m \geq 2$, the classes nonuniform/uniform $AC^0(m)$ are defined just as nonuniform/uniform TC^0 but using *modulo m* gates instead of majority gates. A modulo m gate has unbounded fan-in and outputs 1 if and only if the total number of 1 inputs is exactly 1 modulo m. Also,

$$ACC = \bigcup_{i \geq 2} AC^0(m).$$

See Appendix A.5 for a formal definition.

Obviously $AC^0 \subseteq AC^0(m)$. Furthermore, the relation $PARITY$ (Sections IV.1 and V.5.1):

$PARITY(X)$ iff X contains an odd number of elements

is in $AC^0(2)$. Since $PARITY \notin AC^0$, it follows that $AC^0 \subsetneq AC^0(2)$. It is easy to show that for $2 \leq m_1 < m_2 \in \mathbb{N}$, if $m_1 : m_2$ then

$$AC^0(m_1) \subseteq AC^0(m_2).$$

On the other hand, let $MODULO_p(X)$ be the relation

$MODULO_p(X)$ iff the number of elements of X is $= 1 \, mod(p)$.

Then

$$MODULO_p(X) \in AC^0(p).$$

A major result in complexity theory due to Razborov and Smolensky (see [19]) states that for any prime p and any $m \geq 2$ which is not a power of p,

$$MODULO_m \notin AC^0(p).$$

As a result,

$$AC^0(m) \nsubseteq AC^0(p).$$

Also modulo m gates can be easily simulated by threshold gates. Thus

$$AC^0 \subsetneq AC^0(p) \subsetneq ACC \subseteq TC^0$$

(the last inclusion is because counting gates can simulate a *modulo m* gate, for any m). On the other hand, it is an open problem whether $AC^0(m) \subsetneq ACC$ for composite $m \in \mathbb{N}$. In fact, it is not known whether $AC^0(6) \subsetneq NP$.

In descriptive complexity, uniform $AC^0(m)$ (or just $AC^0(m)$) can be characterized using the $mod(m)$ quantifier [10]. Here we use the fact that the following function is (Turing) AC^0 complete for $AC^0(m)$:

$$mod_m(x, Y) = numones(x, Y) \mod m. \tag{259}$$

The "string version" of this function, called $Mod_m(x, Y)$, is the sequence of the values of $mod_m(z, Y)$ for $z \leq x$:

$$Mod_m(x, Y) = Z \leftrightarrow SEQ(x, Z) \wedge$$

$$\left(|Z| \leq 1 + \langle x, m \rangle \wedge \forall z \leq x((Z)^z = mod_m(z, Y))\right) \tag{260}$$

where SEQ is defined in (226).

We use the following result (based on [10]) to define our theories for $AC^0(m)$.

PROPOSITION IX.4.1. *A relation is in $AC^0(m)$ iff it is AC^0-reducible to mod_m iff it is AC^0-reducible to Mod_m. A function is in $FAC^0(m)$ iff it is AC^0-reducible to mod_m iff it is AC^0-reducible to Mod_m.*

The next section treats the case $m = 2$, and Section IX.4.6 treats theories for $AC^0(m)$ for $m \geq 3$.

IX.4.2. The Theories $V^0(2)$, $\widehat{V^0(2)}$, and $\overline{V^0(2)}$. For $m = 2$ the function Mod_m as defined in (260) has a simpler version which we call $Parity(x, Y)$. The graph of this function is defined by the Σ_0^B-formula $\delta_{parity}(x, Y, Z)$ which asserts that for $1 \leq z \leq x$, $Z(z)$ holds iff there is an odd number of ones in $Y(0)Y(1)\ldots Y(z \dot{-} 1)$:

$$\delta_{parity}(x, Y, Z) \equiv \neg Z(0) \wedge \forall z < x(Z(z + 1) \leftrightarrow (Z(z) \oplus Y(z))).$$

Thus *Parity* is defined by

$$Parity(x, Y) = Z \leftrightarrow \left(|Z| \leq x + 1 \wedge \delta_{parity}(x, Y, Z)\right). \tag{261}$$

Definition IX.4.2 ($V^0(2)$). The theory $V^0(2)$ has vocabulary $\mathcal{L}_A^2$ and axioms those of V^0 and

$$\exists Z \leq x + 1 \, \delta_{parity}(x, Y, Z).$$

Since $\delta_{parity}(x, Y, Z)$ uniquely determines Z as a function of x, Y it follows from (261) that $Parity(x, Y)$ is Σ_1^B-definable in $V^0(2)$.

The next lemma shows that the aggregate $Parity^\star$ of $Parity$ (see Definition VIII.1.9) is definable in $V^0(2)$, and hence $V^0(2)$ is indeed an instance of the family VC.

Lemma IX.4.3. *The function $Parity^\star$ is Σ_1^B-definable in $V^0(2)$, and the theory $V^0(Row, Parity, Parity^\star)$ proves*

$$\forall i < b, Parity^\star(b, Y, X)^{[i]} = Parity((Y)^i, X^{[i]}).$$

Proof. The argument is very similar to the proof of Lemma IX.3.3 (showing that $Numones^\star$ is Σ_1^B-definable in VTC^0). □

Exercise IX.4.4. Carry out the above proof.

Following Section IX.2.2, to define $\widehat{V^0(2)}$ we need a quantifier-free defining axiom for $Parity$. Instead of appealing to Lemma V.6.3 to find a quantifier-free version of $\delta_{parity}(x, Y, Z)$ we will give an explicit defining axiom for $Parity'$ by introducing a new free variable u in the definition.

$$Parity'(x, Y) = Z \supset |Z| \leq x + 1 \wedge \neg Z(0) \wedge$$
$$\big(u < x \supset (Z(u+1) \leftrightarrow (Z(u) \oplus Y(u)))\big). \quad (262)$$

Note that $Parity$ satisfies this defining axiom for $Parity'$, and $V^0(Parity, Parity')$ proves $Parity(y, X) = Parity'(y, X)$.

The universal theory $\widehat{V^0(2)}$ is defined to be $\overline{V}^0(Parity')$ with the defining axiom (262) for $Parity'$. Its vocabulary $\mathcal{L}_{FAC^0} \cup \{Parity'\}$ is denoted by $\mathcal{L}_{\widehat{V^0(2)}}$. The theory $\overline{V^0(2)}$ is defined as follows. Its vocabulary, $\mathcal{L}_{FAC^0(2)}$ is the smallest set that contains $\mathcal{L}_{\widehat{V^0(2)}}$ such that for every $\mathcal{L}_A^2$-term t and $\mathcal{L}_{FAC^0(2)}$-formula φ, there is a string function $F_{\varphi(z),t}$ with defining axiom (86). Then $\overline{V^0(2)}$ is axiomatized by the axioms of $\overline{V}^0$ together with (262) for $Parity'$ and (86) for each function $F_{\varphi(z),t}$.

The results of Section IX.2 apply to the theories just defined, so the following statements are corollaries of that section.

Theorem IX.4.5. *Assume either $\mathcal{L}$ is $\mathcal{L}_{\widehat{V^0(2)}}$ and T is $\widehat{V^0(2)}$, or $\mathcal{L}$ is $\mathcal{L}_{FAC^0(2)}$ and T is $\overline{V^0(2)}$.*

(a) *A function is in $FAC^0(2)$ iff it is represented by a term in $\mathcal{L}_{\widehat{V^0(2)}}$ (and for string functions) iff it is represented by a symbol in $\mathcal{L}_{FAC^0(2)}$. A relation is in $AC^0(2)$ iff it is represented by an open (or Σ_0^B) formula in $\mathcal{L}$.*

(b) Every $\Sigma_1^B(\mathcal{L})$ formula is equivalent in $\mathcal{T}$ to a $\Sigma_1^B(\mathcal{L}_A^2)$ formula.

(c) $\mathcal{T}$ proves $\Sigma_0^B(\mathcal{L})$-**COMP**, $\Sigma_0^B(\mathcal{L})$-**IND**, and $\Sigma_0^B(\mathcal{L})$-**MIN**.

(d) $\overline{\widehat{V^0(2)}}$ is a universal conservative extension of $\widehat{V^0(2)}$, which is in turn a universal conservative extension of $V^0(2)$.

(e) A function is in $\mathbf{FAC}^0(2)$ iff it is Σ_1^B-definable in $V^0(2)$ iff it is Σ_1^B-definable in $\mathcal{T}$.

(f) A relation is in $\mathbf{AC}^0(2)$ iff it is Δ_1^B-definable in $V^0(2)$ iff it is Δ_1^B-definable in $\mathcal{T}$.

IX.4.3. The "onto" PHP and Parity Principle. The onto pigeonhole principle, **OPHP**, states that there is no bijection between $(a+1)$ "pigeons" and a "holes". Formally, **OPHP** is the Σ_0^B formula

$$OPHP(a, X) \equiv (\forall x \le a \exists! z < a \, X(x, z)) \supset$$
$$(\neg INJ(a, X) \vee \neg SUR(a, X))$$

where

$$INJ(a, X) \equiv \forall x \le a \forall y \le a \forall z < a((X(x,z) \wedge X(y,z)) \supset x = y),$$
$$SUR(a, X) \equiv \forall z < a \exists x \le a X(x,z).$$

We will also use **OPHP** to denote the family of propositional tautologies translated from **OPHP**(a, X) (as described in Section VII.2). We mentioned in Section VII.1.2 that this family does not have polynomial size $\mathbf{bPK}$ proofs (as shown by Ajtai [5]). We state this result formally below.

THEOREM IX.4.6 (Ajtai). *The family* **OPHP** *does not have polynomial size* $\mathbf{bPK}$ *proofs.*

COROLLARY IX.4.7. *The theory* V^0 *does not prove the true* $\forall \Sigma_0^B$ *sentence* $\forall a \forall X \mathbf{OPHP}(a, X)$.

On the other hand, it is relatively easy to show that **OPHP** is provable in $V^0(2)$. In fact we will show that **OPHP** is implied by the Parity Principle (also called the Modulo 2 Counting Principle) which asserts that a set of odd cardinality cannot be partitioned into subsets of two elements each. Formally, let

$$Count_2(a, X) \equiv \neg \forall x \le 2a \exists! y \le 2a \big((x < y \wedge X(x, y)) \vee$$
$$(y < x \wedge X(y, x))\big).$$

Here $X(x, y)$ holds iff $\{x, y\}$ is a partition and $x < y$, and thus $Count_2(a, X)$ states that X is not a partitioning of the set $\{0, 1, \ldots, 2a\}$.

This principle is stronger than **OPHP** in the following sense:

EXERCISE IX.4.8. Show that V^0 proves

$$\forall a \forall X \, Count_2(a, X) \supset \forall a \forall X \, \mathbf{OPHP}(a, X). \tag{263}$$

COROLLARY IX.4.9. *The Parity Principle does not have polynomial size* **bPK** *proofs.*

PROOF SKETCH. A prenex form of (263) is

$$\forall a \forall X \exists b \exists Y (\boldsymbol{Count}_2(b, Y) \supset \boldsymbol{OPHP}(a, X)).$$

By the above exercise and Theorem V.5.1 (witnessing for V^0) there are **FAC**^{0} functions f, F such that

$$V^0(f, F) \vdash \boldsymbol{Count}_2(f(a, X), F(a, X)) \supset \boldsymbol{OPHP}(a, X).$$

By the **FAC**^{0} Elimination Lemma V.6.7 there is a Σ_0^B formula $\boldsymbol{Count}_2'(a, X)$ such that

$$V^0(f, F) \vdash \boldsymbol{Count}_2(f(a, X), F(a, X)) \leftrightarrow \boldsymbol{Count}_2'(a, X).$$

Since $V^0(f, F)$ is a conservative extension of V^0 we conclude

$$V^0 \vdash \boldsymbol{Count}_2'(a, X) \supset \boldsymbol{OPHP}(a, X).$$

The idea now is to apply the proof of the V^0 Translation Theorem VII.2.3 to the above theorem of V^0 and argue that there are polynomial size **bPK** proofs of the translations of $\boldsymbol{Count}_2'(a, X)$ from substitution instances of translations of $\boldsymbol{Count}_2(b, Y)$. Thus if the Parity Principle has polynomial size **bPK** proofs then so does **OPHP**, contradicting Ajtai's Theorem IX.4.6. □

*EXERCISE IX.4.10. Formulate and prove a generalization of the V^0 Translation Theorem VII.2.3 so that it applies to $\Sigma_0^B(\mathcal{L}_{FAC^0})$ theorems of $\overline{V}^0$ and justifies the above proof of Corollary IX.4.9.

Another theorem of Ajtai [4] states a stronger result. Here an instance of **OPHP** is a formula obtained from a member of the family **OPHP** by substituting polynomial size constant depth formulas for the variables.

THEOREM IX.4.11 (Ajtai). *The family* **Count**$_2$ *does not have polynomial size* **bPK** *proofs even when instances of* **OPHP** *are allowed as axioms.*

As a result, V^0 does not prove the reverse direction of the implication in Exercise IX.4.8.

The next corollary follows from the above exercise and the previous corollary.

COROLLARY IX.4.12. *The theory V^0 does not prove the true* $\forall \Sigma_0^B$ *sentence* $\forall a \forall X \boldsymbol{Count}_2(a, X)$.

By Exercise IX.4.8, to show that $V^0(2)$ proves **OPHP** it suffices to show that $V^0(2)$ proves **Count**$_2$. This is left as an exercise.

EXERCISE IX.4.13. Show that

$$V^0(2) \vdash \forall a \forall X \boldsymbol{Count}_2(a, X).$$

(Hint: Define the sets $Y^{[0]}, Y[1], \ldots, Y^{[2a]}$ such that $Y^{[i]}$ contains all and only elements x such that either $x \leq i$ or x is coupled with some $y \leq i$.)

COROLLARY IX.4.14. *The theory $V^0(2)$ is a proper extension of V^0. In fact, $V^0(2)$ is not Σ_0^B-conservative over V^0.*

PROOF. This is similar to Corollary IX.3.8. The first sentence follows from the second, which is true because (by Corollary IX.4.7) V^0 does not prove $\forall a \forall X\, OPHP(a, X)$, but (by Exercises IX.4.8 and IX.4.13) $V^0(2)$ does. The first sentence also follows directly from the fact that the function *Parity* is definable in $V^0(2)$ but not in V^0 (because it is not in FAC^0). □

IX.4.4. The Theory $VAC^0(2)V$. The universal theory $VAC^0(2)V$ is defined in the same way as $VTC^0 V$ and VPV. Its vocabulary has symbols for every function in $FAC^0(2)$. Their defining axioms are based on the recursion theoretic characterization of $FAC^0(2)$ using the bounded number recursion scheme 2-BNR scheme as shown in Theorem IX.4.15 below.

Here 2-BNR refers to the number recursion scheme in Definition IX.3.10, where $t = 2$ (t is the bound on the function f defined by the recursion).

THEOREM IX.4.15. *$FAC^0(2)$ is equal to the closure of FAC^0 under AC^0 reductions and 2-BNR and also equal to the closure of FAC^0 under composition, string comprehension and 2-BNR.*

PROOF. By Theorem IX.1.7 it suffices to show that a function is in $FAC^0(2)$ iff it can be obtained from FAC^0 functions by finitely many applications of AC^0 reduction and 2-BNR.

The ONLY IF direction follows easily from the fact that the function $mod_2(x, Y)$ (259) is obtained from the characteristic function f_X ((237) on page 288) by 2-BNR.

For the IF direction, we prove by induction on the number of the applications of 2-BNR. The base case (no application of 2-BNR) is obvious. For the induction step, suppose that $f(y, \vec{x}, \vec{X})$ is obtained from $FAC^0(2)$ functions g, h by 2-BNR as in Definition IX.3.10:

$$f(0, \vec{x}, \vec{X}) = g(\vec{x}, \vec{X}),$$

$$f(y + 1, \vec{x}, \vec{X}) = h(y, f(y, \vec{x}, \vec{X}), \vec{x}, \vec{X})$$

and for all $y, \vec{x}, \vec{X}$, $f(y, \vec{x}, \vec{X}) < 2$.

For $y \geq 1$, let (we drop mention of $\vec{x}, \vec{X}$)

$$z = max(\{0\} \cup \{u < y \,:\, h(u, 0) = h(u, 1)\}),$$

$$n = mod_2(y, \{u \,:\, z < u < y \wedge h(u, 0) \neq 0\}),$$

$$v = \begin{cases} g & \text{if } z = 0, \\ h(z, 0) & \text{otherwise.} \end{cases}$$

Then $f(y) = 0$ iff either (i) $v = 0$ and $n = 0$, or (ii) $v = 1$ and $n = 1$. In other words, f can be obtained from g, h and mod_2 by AC^0 reduction. □

The next definition is analogous to Definition IX.3.13.

Definition IX.4.16. The vocabulary $\mathcal{L}_{VAC^0(2)V}$ is the smallest set that includes $\mathcal{L}_{FAC^0}$ such that for every $\mathcal{L}_A^2$-term t and open $\mathcal{L}_{VAC^0(2)V}$-formula φ the function $F_{\varphi(z),t}$ with defining axiom (86) is in $\mathcal{L}_{VAC^0(2)V}$, and for every string function F in $\mathcal{L}_{VAC^0(2)V}$ there is a number function f_F in $\mathcal{L}_{VAC^0(2)V}$ with defining axiom $f_F(\vec{x}, \vec{X}) = |F(\vec{x}, \vec{X})|$, and for every two number functions $g(\vec{x}, \vec{X}), h(y, z, \vec{x}, \vec{X})$ in $\mathcal{L}_{VAC^0(2)V}$ there is a number function $f_{g,h}$ in $\mathcal{L}_{VAC^0(2)V}$ with defining axioms (omitting $\vec{x}, \vec{X}$):

$$(g < 2 \supset f_{g,h}(0) = g) \wedge (g \geq 2 \supset f_{g,h}(0) = 0) \wedge$$
$$(h(y, f_{g,h}(y)) < 2 \supset f_{g,h}(y+1) = h(y, f_{g,h}(y))) \wedge$$
$$(h(y, f_{g,h}(y)) \geq 2 \supset f_{g,h}(y+1) = 0). \quad (264)$$

It follows from Theorem IX.4.15 that semantically the functions in $\mathcal{L}_{VAC^0(2)V}$ represent precisely the functions in $FAC^0(2)$. We have:

Corollary IX.4.17. (a) *A function is in $FAC^0(2)$ iff it is represented by a function in $\mathcal{L}_{VAC^0(2)V}$.*

(b) *A relation is in $AC^0(2)$ iff it is represented by an open formula in $\mathcal{L}_{VAC^0(2)V}$ iff it is represented by a $\Sigma_0^B(\mathcal{L}_{VAC^0(2)V})$ formula.*

Definition IX.4.18. The theory $VAC^0(2)V$ has vocabulary $\mathcal{L}_{VAC^0(2)V}$ and is axiomatized by the axioms of $\overline{V}^0$ together with (86) for the functions $F_{\varphi(z),t}$ and (240) for the functions f_F and (264) for the functions $f_{g,h}$.

Theorem IX.4.19. (a) *The theory $VAC^0(2)V$ is a universal conservative extension of $V^0(2)$.*

(b) *For every $\Sigma_0^B(\mathcal{L}_{VAC^0(2)V})$ formula φ^+ there is a $\Sigma_1^B(\mathcal{L}_A^2)$ formula φ that is equivalent in $VAC^0(2)V$ to φ^+.*

Proof sketch. Part (a) of the following theorem can be proved by formalizing the proof of Theorem IX.4.15 (see also Theorem IX.3.17). Part (b) follows from Theorem IX.2.10 and Corollary IX.3.20 (see also Corollary IX.3.21). □

The characterization of $AC^0(2)$ by $VAC^0(2)V$ can be proved as in Section IX.3.4 (for the class TC^0 and the theory VTC^0V).

Corollary IX.4.20. (a) *A function is in $FAC^0(2)$ iff it is Σ_1^B-definable in $VAC^0(2)V$.*

(b) *A relation is in $AC^0(2)$ iff it is Δ_1^B-definable in $VAC^0(2)V$.*

IX.4.5. The Jordan Curve Theorem and Related Principles. The Jordan Curve Theorem (JCT) asserts that any simple, closed curve divides the plane into exactly two connected components. Here we consider the setting where the curve lies on a grid graph and consists of only horizontal or vertical edges. The notions of grid vertex and edge can be defined using the pairing function. To state the theorem, one way is to represent the curve as a *sequence* of edges that form a simple cycle. To show that there are exactly two connected components we can show that (i) any path (represented by a sequence of edges) that connect two points on different sides of the curve must intersect the curve, and (ii) any two points on the same side of the curve can be connected by a path without intersecting the curve.

Suppose that instead of representing the curve as a sequence of edges we have a *set* of edges such that every grid vertex has degree either 0 or 2. So there may be multiple simple closed curves, and we can only show that there are *at least* two connected components. We will refer to this as the *set* setting of JCT, as opposed to the above *sequence* setting.

Surprisingly the sequence setting is a theorem of V^0 [84], but the proof of this is difficult and will not be presented here. In this section we will show that the set JCT is a theorem of $V^0(2)$.

We start by defining the notions of grid vertex (or just vertex, or point) and edge, and certain sets of edges which include closed curves, or connect grid points. All of these notions are definable by Σ_0^B-formulas, and their basic properties can be proved in V^0.

We assume a parameter n which bounds the x and y coordinates of points on the curve in question. Thus a point p is a pair (x, y) which is encoded by the pairing function $\langle x, y \rangle$ (see (69) on page 113), where $0 \le x, y \le n$. The x and y coordinates of a point p are denoted by $x(p)$ and $y(p)$ respectively. Thus if $p = \langle i, j \rangle$ then $x(p) = i$ and $y(p) = j$. An (undirected) *edge* is a pair (p_1, p_2) (represented by $\langle p_1, p_2 \rangle$) of adjacent points; i.e. either $|x(p_2) - x(p_1)| = 1$ and $y(p_2) = y(p_1)$, or $x(p_2) = x(p_1)$ and $|y(p_2) - y(p_2)| = 1$. For a horizontal edge e, we also write $y(e)$ for the (common) y-coordinate of its endpoints.

Let E be a set of edges (represented by a set of numbers representing those edges). The E-*degree* of a point p is the number of edges in E that are incident to p.

DEFINITION IX.4.21. A *curve* is a nonempty set E of edges such that the E-degree of every grid point is either 0 or 2. A set E of edges is said to *connect* two points p_1 and p_2 if the E-degrees of p_1 and p_2 are both 1 and the E-degrees of all other grid points are either 0 or 2. Two sets E_1 and E_2 of edges are said to *intersect* if there is a grid point whose E_i-degree is ≥ 1 for $i = 1, 2$.

As noted before, a curve in the above sense is actually a collection of one or more disjoint closed curves. Also if E connects p_1 and p_2 then E consists of a path connecting p_1 and p_2 together with zero or more disjoint closed curves.

We also need to define the notion of two points being on different sides of a curve. We are able to consider only points which are "close" to the curve. It suffices to consider the case in which one point is above and one point is below an horizontal edge in E. (Note that the case in which one point is to the left and one point is to the right of a vertical edge in E can be reduced to this case by rotating the $(n + 1) \times (n + 1)$ array of all grid points by 90 degrees.)

DEFINITION IX.4.22. Two points p_1, p_2 are said to be *on different sides* of E if

(i) $x(p_1) = x(p_2) \wedge |y(p_1) - y(p_2)| = 2$,
(ii) the E-degree of $p_i = 0$ for $i = 1, 2$, and
(iii) the E-degree of p is 2, where p is the point with $x(p) = x(p_1)$ and $y(p) = \frac{1}{2}(y(p_1) + y(p_2))$. (See Figure 6.)

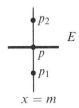

FIGURE 6. p_1, p_2 are on different sides of E.

Now we show that any set of edges that forms at least one simple curve must divide the plane into at least two connected components.

THEOREM IX.4.23. *The theory $V^0(2)$ proves the following: Suppose that B is a set of edges forming a curve, p_1 and p_2 are two points on different sides of B, and that R is a set of edges that connects p_1 and p_2. Then B and R intersect.*

PROOF. Since $\widehat{V^0(2)}$ is conservative over $V^0(2)$, it suffices to give a $\widehat{V^0(2)}$ proof of the theorem. By Theorem IX.4.5 we can use $\Sigma_0^B(Parity)$-*COMP* and hence also $\Sigma_0^B(Parity)$-*IND* and $\Sigma_0^B(Parity)$-*MIN* (Theorem V.1.8).

In the following discussion we also refer to the edges in B as "undashed" edges, and the edges in R as "dashed" edges.

We argue in $\widehat{V^0(2)}$, and prove the theorem by contradiction. Suppose to the contrary that B and R satisfy the hypotheses of the theorem, but do not intersect.

NOTATION. A horizontal edge is said to be *on column k* (for $k \leq n - 1$) if its endpoints have x-coordinates k and $k + 1$.

Let $m = x(p_1) = x(p_2)$. W.l.o.g., assume that $2 \leq m \leq n - 2$. Also, we may assume that the dashed path comes to both p_1 and p_2 from the left, i.e., the two dashed edges that are incident to p_1 and p_2 are both horizontal and on column $m - 1$ (see Figure 7). (Note that if the dashed path does not come to both points from the left, we could fix this by effectively doubling the density of the points by doubling n to $2n$, replacing each edge in B or

R by a double edge, and then extending each end of the new path by three (small) edges forming a "C" shape to end at points a distance 1 from the un-dashed curve, approaching from the left.)

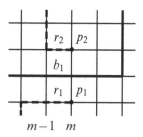

Figure 7. The dashed path must cross the un-dashed curve.

We say that edge e_1 *lies below* edge e_2 if e_1 and e_2 are horizontal and in the same column and $y(e_1) < y(e_2)$. For each horizontal dashed edge r we consider the parity of the number of horizontal un-dashed edges b that lie below r. We say that an edge is "odd" if it is dashed and there are an odd number of un-dashed edges below it. Recall that $PARITY(X)$ holds iff X contains an odd number of elements:

$$PARITY(X) \leftrightarrow Parity(|X|, X)(|X| \overset{.}{-} 1).$$

Formally we have:

Notation. For each edge r let Z_r denote the set of all horizontal un-dashed edges that lie below r. An edge r is said to be an *odd edge* if it is dashed and horizontal and $PARITY(Z_r)$.

For example, it is easy to show in $V^0(2)$ that exactly one of r_1, r_2 in Figure 7 is an odd edge.

For each $k \leq n - 1$, define using $\Sigma_0^B(Parity)$-**COMP** the set

$$X_k = \{r : r \text{ is an odd edge in column } k\}.$$

Lemma IX.4.24. *It is provable in* $V^0(2)$ *that*

(a) $PARITY(X_{m-1}) \leftrightarrow \neg PARITY(X_m)$.
(b) *For* $0 \leq k \leq n - 2, k \neq m$, $PARITY(X_k) \leftrightarrow PARITY(X_{k+1})$.

Using this lemma the proof of the theorem is completed as follows. We may assume that there are no edges in either B or R in columns 0 and $(n - 1)$, so $\neg PARITY(X_0) \wedge \neg PARITY(X_{n-1})$. On the other hand, it follows by $\Sigma_0^B(\mathcal{L}_{\widehat{V^0(2)}})$-**IND** using Lemma IX.4.24 (b) that $PARITY(X_0) \leftrightarrow PARITY(X_{m-1})$ and $PARITY(X_m) \leftrightarrow PARITY(X_{n-1})$, which contradicts (a).   □

It remains to prove Lemma IX.4.24.

Proof of Lemma IX.4.24. First we prove (b). For $k \leq n-1$ and $0 \leq j \leq n$, let $e_{k,j}$ be the horizontal edge on column k with y-coordinate j. Fix $k \leq n-2$. Define the ordered lists (see Figure 8)

$$L_0 = e_{k,0}, e_{k,1}, \ldots, e_{k,n}; \qquad L_{n+1} = e_{k+1,0}, e_{k+1,1}, \ldots, e_{k+1,n}$$

and for $1 \leq j \leq n$:

$$L_j = e_{k+1,0}, \ldots, e_{k+1,j-1}, \langle (k+1, j-1), (k+1, j) \rangle, e_{k,j}, \ldots, e_{k,n}.$$

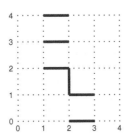

Figure 8. L_2 (for $n=4, k=1$).

A dashed edge $r \in L_j$ is said to be *odd in* L_j if there are an odd number of un-dashed edges in L_j preceding r. Formally, for each dashed edge $r \in L_j$ let W be the set of un-dashed edges in L_j that precede r. Then r is odd in L_j just in case $PARITY(W)$ is true. In particular, X_k and X_{k+1} consist of odd edges in L_0 and L_{n+1}, respectively.

For $0 \leq j \leq n+1$, let

$$Y_j = \{r : r \text{ is an odd edge in } L_j\}.$$

Thus $Y_0 = X_k$ and $Y_{n+1} = X_{k+1}$.

Claim. If $k \neq m-1$ then $PARITY(Y_j) \leftrightarrow PARITY(Y_{j+1})$ for $j \leq n$.

This is because the symmetric difference of Y_j and Y_{j+1} has either no dashed edges, or two dashed edges with the same parity.

Thus by $\Sigma_0^B(\mathcal{L}_{\widetilde{V^0(2)}})$-*IND* on j we have $PARITY(Y_0) \leftrightarrow PARITY(Y_{n+1})$, and hence $PARITY(X_k) = PARITY(X_{k+1})$.

The proof of (a) is similar. The only change here is that $PARITY(L_j)$ and $PARITY(L_{j+1})$ must differ for exactly one value of j: either $j = y(p_1)$ or $j = y(p_2)$ (because either r_1 is odd in $L_{y(p_1)}$ or r_2 is odd in $L_{y(p_2)}$, but not both).   □

IX.4.6. The Theories for $AC^0(m)$ and ACC. Now we present theories associated with $AC^0(m)$, for $m \geq 3$. They are defined in the same way as the theories $V^0(2)$, $\widehat{V^0(2)}$ and $\overline{V^0(2)}$. Let $\delta_{MOD_m}(x, Y, Z)$ be the $\Sigma_0^B(\mathcal{L}_A^2)$

equivalent (by Lemma V.6.7) of the following formula:

$$SEQ(x, Z) \wedge Z(0,0) \wedge \forall z < x, (Y(z) \supset$$
$$(Z)^{z+1} = ((Z)^z + 1) \mod m)) \wedge (\neg Y(z) \supset (Z)^{z+1} = (Z)^z).$$

Thus $\delta_{MOD_m}(x, Y, Z)$ states that $Z = Mod_m(x, Y)$, the "counting modulo m sequence" for Y (see (260) on page 304). Indeed, we take the following as the defining axiom for Mod_m:

$$Mod_m(x, Y) = Z \leftrightarrow (|Z| \leq 1 + \langle x, m \rangle \wedge \delta_{MOD_m}(x, Y, Z)).$$

Let

$$MOD_m \equiv \exists Z \leq 1 + \langle x, m \rangle \delta_{MOD_m}(x, Y, Z).$$

DEFINITION IX.4.25. For each $m \geq 3$, the theory $V^0(m)$ has vocabulary $\mathcal{L}_A^2$ and is axiomatized by V^0 and the axiom MOD_m.

The next exercise can be proved in the same way as Lemma IX.3.3.

EXERCISE IX.4.26. For each $m \geq 3$ the function $Mod_m^\star$ is Σ_1^B-definable in $V^0(m)$, and $V^0(m)(Row, Mod_m, Mod_m^\star)$ proves

$$\forall i < b, \ Mod_m^\star(b, X, Y)^{[i]} = Mod_m((X)^i, Y^{[i]}).$$

Now we define $\overline{V^0(m)}$ and $\widehat{V^0(m)}$. To define $\widehat{V^0(m)}$ we use the string function $Mod'_m(x, Y)$ defined by

$$Mod'_m(x, Y) = Z \leftrightarrow (|Z| \leq 1 + \langle x, m \rangle \wedge \delta'_{MOD_m}(x, Y, Z)) \qquad (265)$$

where $\delta'_{MOD_m}(x, Y, Z)$ is the quantifier-free $\mathcal{L}_{FAC^0}$-formula that is equivalent to $\delta_{MOD_m}(x, Y, Z)$ over $\overline{V}^0$ (see Lemma V.6.3).

DEFINITION IX.4.27 ($\widehat{V^0(m)}$). For each $m \geq 3$, the theory $\widehat{V^0(m)}$ has vocabulary $\mathcal{L}_{\widehat{V^0(m)}} = \mathcal{L}_{FAC^0} \cup \{Mod'_m\}$ and axioms that of $\overline{V}^0$ and (265).

For $\overline{V^0(m)}$ we start with the function mod'_m which is equal to mod_m (see (259) on page 304) but has the following quantifier-free defining axioms (we identify the natural number m with the corresponding numeral $\underline{m}$):

$$mod'_m(0, Y) = 0, \qquad (266)$$
$$(Y(x) \wedge mod'_m(x, Y) + 1 < m) \supset mod'_m(x + 1, Y) = mod'_m(x, Y) + 1, \qquad (267)$$
$$(Y(x) \wedge mod'_m(x, Y) + 1 = m) \supset mod'_m(x + 1, Y) = 0, \qquad (268)$$
$$\neg Y(x) \supset mod'_m(x + 1, Y) = mod'_m(x, X). \qquad (269)$$

DEFINITION IX.4.28. For each $m \geq 2$, $\mathcal{L}_{FAC^0(m)}$ is the smallest set that contains $\mathcal{L}_{FAC^0} \cup \{mod'_m\}$ such that for each quantifier-free formula

$\varphi(z, \vec{x}, \vec{X})$ of $\mathcal{L}_{FAC^0(m)}$ and term $t(\vec{x}, \vec{X})$ of $\mathcal{L}_A^2$, there is a string function $F_{\varphi(z),t}$ with defining axiom (86):

$$F_{\varphi(z),t}(\vec{x}, \vec{X})(z) \leftrightarrow z < t(\vec{x}, \vec{X}) \wedge \varphi(z, \vec{x}, \vec{X}).$$

The theory $\overline{V^0(m)}$ has vocabulary $\mathcal{L}_{FAC^0(m)}$ and is axiomatized by the axioms of $\overline{V}^0$, (266), (267), (268) and (269) for mod'_m, and (86) for each function $F_{\varphi(z),t}$.

The following Definability Theorem follows from the results in Section IX.2:

COROLLARY IX.4.29. *Here either $\mathcal{L}$ is $\mathcal{L}_{\widehat{V^0(m)}}$ and T is $\widehat{V^0(m)}$, or $\mathcal{L}$ is $\mathcal{L}_{FAC^0(m)}$ and T is $\overline{V^0(m)}$.*

(a) *A function is in $FAC^0(m)$ iff it is represented by a term in $\mathcal{L}_{\widehat{V^0(m)}}$ (and for string function) iff it is represented by a symbol in $\mathcal{L}_{FAC^0(m)}$. A relation is in $AC^0(m)$ iff it is represented by an open (or a Σ_0^B) formula of $\mathcal{L}$.*

(b) *Every $\Sigma_1^B(\mathcal{L})$ formula is equivalent in T to a $\Sigma_1^B(\mathcal{L}_A^2)$ formula.*

(c) *T proves $\Sigma_0^B(\mathcal{L})$-COMP, $\Sigma_0^B(\mathcal{L})$-IND, and $\Sigma_0^B(\mathcal{L})$-MIN.*

(d) *$\overline{V^0(m)}$ is a universal conservative extension of $\widehat{V^0(m)}$, which is in turn a universal conservative extension of $V^0(m)$.*

(e) *A function is in $FAC^0(m)$ iff it is Σ_1^B-definable in $V^0(m)$ iff it is Σ_1^B-definable in T.*

(f) *A relation is in $FAC^0(m)$ iff it is Δ_1^B-definable in $V^0(m)$ iff it is Δ_1^B-definable in T.*

COROLLARY IX.4.30. *Let p, q be two distinct prime numbers. Then*

$$V^0 \subsetneq V^0(p) \nsubseteq V^0(q).$$

PROOF. Both facts follow from Corollary IX.4.29 (f) and the results of Razborov and Smolensky mentioned in Section IX.4.1 (that $MODULO_m \notin AC^0(p)$ if p is a prime number not divisible by m). □

Theories for ACC are as follows:

DEFINITION IX.4.31. $VACC = \bigcup_{m \geq 2} V^0(m)$,

$$\mathcal{L}_{FACC} = \bigcup_{m \geq 2} \mathcal{L}_{FAC^0(2)}, \qquad \overline{VACC} = \bigcup_{m \geq 2} \overline{V^0(m)},$$

$$\mathcal{L}_{\widehat{VACC}} = \bigcup_{m \geq 2} \mathcal{L}_{\widehat{V^0(m)}}, \qquad \widehat{VACC} = \bigcup_{m \geq 2} \widehat{V^0(m)}.$$

The next Definability Theorems for the theories associated with ACC follow from Corollary IX.4.29.

COROLLARY IX.4.32. *Here either $\mathcal{L}$ is $\mathcal{L}_{\widehat{VACC}}$ and T is $\widehat{VACC}$, or $\mathcal{L}$ is $\mathcal{L}_{FACC}$ and T is $\overline{VACC}$.*

(a) *A function is in **FACC** iff it is represented by a term in $\mathcal{L}_{\overline{VACC}}$ (and for string function) iff it is represented by a symbol in $\mathcal{L}_{FACC}$. A relation is in **ACC** iff it is represented by an open (or a Σ_0^B) formula of $\mathcal{L}$.*

(b) *Every $\Sigma_1^B(\mathcal{L})$ formula is equivalent in $\mathcal{T}$ to a $\Sigma_1^B(\mathcal{L}_A^2)$ formula.*

(c) *$\mathcal{T}$ proves $\Sigma_0^B(\mathcal{L})$-**COMP**, $\Sigma_0^B(\mathcal{L})$-**IND**, and $\Sigma_0^B(\mathcal{L})$-**MIN**.*

(d) *$\overline{VACC}$ is a universal conservative extension of $\widehat{VACC}$, which in turn is a universal conservative extension of **VACC**.*

(e) *A function is in **FACC** iff it is Σ_1^B-definable in **VACC** iff it is Σ_1^B-definable in $\mathcal{T}$.*

(f) *A relation is in **FACC** iff it is Δ_1^B-definable in **VACC** iff it is Δ_1^B-definable in $\mathcal{T}$.*

EXERCISE IX.4.33. Show that for all integers $m, n \geq 2$, if $m|n$ then $AC^0(m) \subseteq AC^0(n)$ and $V^0(m) \subseteq V^0(n)$.

The next result is analogous to Corollary VIII.7.21 (relating the finite axiomatizability of V^∞ to the provable collapse of **PH**), but the proof is much easier.

THEOREM IX.4.34. *VACC is finitely axiomatizable iff for some $m \geq 2$, $V^0(m)$ proves $AC^0(m) = ACC$.*

PROOF. Since each theory $AC^0(m)$ is finitely axiomatizable, it follows from Exercise IX.4.33 and compactness that *VACC* is finitely axiomatizable iff *VACC* $= V^0(m)$ for some $m \geq 2$. If *VACC* $= V^0(m)$ then *VACC* and $V^0(m)$ have the same Σ_1^B-theorems, so by Corollaries IX.4.29 and IX.4.32 every relation in *ACC* is Δ_1^B-definable in *VACC* and hence also in $V^0(m)$ by the same formulas. Thus every relation in *ACC* is provably in $AC^0(m)$.

Conversely if $V^0(m)$ proves $AC^0(m) = ACC$ then $V^0(m)$ proves the axiom MOD_n for every $n \geq 2$, so $V^0(n) \subseteq V^0(m)$, so *VACC* $= V^0(m)$. To justify the conclusion $V^0(m)$ proves MOD_n we could argue as follows. If $V^0(m)$ proves $AC^0(m) = ACC$, then by Corollary IX.4.29(f) it follows that $AC^0(m) \Delta_1^B$-defines $MODULO_n$, so by the witnessing theorem, $V^0(m) \Sigma_1^B$-defines the characteristic function of $MODULO_n$, which in turn can be used to prove MOD_n by Corollary IX.4.29. □

EXERCISE IX.4.35. *VACC* $\subseteq$ *VTC^0*, and $V^0(p) \subsetneq$ *VACC* for any prime p.

IX.4.7. The Modulo m Counting Principles. Recall the Parity Principle (or also Modulo 2 Counting Principle) from Section IX.4.3. Generally, for $m \in \mathbb{N}$, $m \geq 2$, the Modulo m Counting Principle, denoted by *Count$_m$*, states that a set of cardinality which is $(1 \mod m)$ cannot be partitioned into disjoint subsets of exactly m elements each. The formula *Count$_m$*(a, X) below expresses the fact that the m-dimensional array X does not encode a partition of the set $\{0, 1, \ldots, ma\}$ into m-element subsets. Here the encoding of the partition is such that $X(x_1, x_2, \ldots, x_m)$

holds iff $\{x_1, x_2, \ldots, x_m\}$ is a subset in the partition and $x_1 < x_2 < \cdots < x_m$.

$$\textbf{\textit{Count}}_m(a, X) \equiv \neg \forall x \leq ma \exists! y_{m-1} \leq ma \exists! y_{m-2} < y_{m-1} \ldots \exists! y_1 < y_2$$

$$\bigvee_{t=1}^{m-2} (y_t < x \wedge x < y_{t+1} \wedge X(y_1, \ldots, y_t, x, y_{t+1}, \ldots, y_{m-1})) \vee$$

$$\left(x < y_1 \wedge X(x, y_1, \ldots, y_{m-1})\right) \vee \left(y_{m-1} < x \wedge X(y_1, \ldots, y_{m-1}, x)\right).$$

We will also write $\textbf{\textit{Count}}_m$ for the family of tautologies

$$\{\textbf{\textit{Count}}_m(a, X)[n, \langle n, n, \ldots, n\rangle + 1] \; : \; n \geq 1\}.$$

Recall the onto Pigeonhole Principle $\textbf{\textit{OPHP}}$ from Section IX.4.3. The following exercise generalizes Exercise IX.4.8.

EXERCISE IX.4.36. Let $m \in \mathbb{N}$, $m \geq 2$. Show that V^0 proves

$$\forall a \forall X \, \textbf{\textit{Count}}_m(a, X) \supset \forall a \forall X \, \textbf{\textit{OPHP}}(a, X).$$

Ajtai's Theorem IX.4.11 holds for $\textbf{\textit{Count}}_m$ in general and shows that V^0 does not prove the reverse implication. (See the summary in Theorem IX.4.39 and Corollary IX.4.40 below.)

Another theorem of Ajtai [6] is that the family $\textbf{\textit{Count}}_p$ does not have polynomial size $\textbf{\textit{bPK}}$ proofs even when instances of $\textbf{\textit{Count}}_{q_1}, \textbf{\textit{Count}}_{q_2}, \ldots,$ $\textbf{\textit{Count}}_{q_k}$ are used as axioms, for distinct prime numbers $p, q_1, q_2 \ldots, q_k$. It follows that the $\forall \Sigma_0^B$ sentence $\forall a \forall X \, \textbf{\textit{Count}}_p(a, X)$ is not provable from V^0 and the sentences

$$\{\forall a \forall X \, \textbf{\textit{Count}}_{q_t}(a, X) \; : \; 1 \leq t \leq k\}.$$

On the other hand, it can be shown that $\textbf{\textit{Count}}_m$ is provable in $V^0(m)$. (This generalizes Exercise IX.4.13.) The proofs of this and some other facts are left as an exercise.

EXERCISE IX.4.37. Let $m \in \mathbb{N}$, $m \geq 2$. Show that

$$V^0(m) \vdash \forall a \forall X \, \textbf{\textit{Count}}_m(a, X).$$

EXERCISE IX.4.38.   (a) Show that if $m, m' \in \mathbb{N}$, $m \geq 2$ and $m | m'$, then V^0 proves

$$\forall a \forall X \, \textbf{\textit{Count}}_m(a, X) \supset \forall a \forall X \, \textbf{\textit{Count}}_{m'}(a, X).$$

(b) Let $m, m' \in \mathbb{N}$ and m, m' have a common divisor $p > 1$. Show that

$$V^0(m) \vdash \textbf{\textit{Count}}_{m'}(a, X).$$

Now we summarize some of Ajtai's Theorems and their corollaries.

THEOREM IX.4.39 (Ajtai [5, 4, 6]).   (a) *For $m \geq 2$, the family $\textbf{\textit{Count}}_m$ does not have polynomial size $\textbf{\textit{bPK}}$ proof even when instances of $\textbf{\textit{OPHP}}$ are used as axioms.*

(b) *For distinct prime numbers $p, q_1, q_2, \ldots, q_k$, the family $\mathbf{Count}_p$ does not have polynomial size $\mathbf{bPK}$ proof even when instances of $\{\mathbf{Count}_{q_t} : 1 \leq t \leq k\}$ are used as axioms.*

As a result, we have the following independence results:

COROLLARY IX.4.40.   (a) *For $m \geq 2$, the theory V^0 does not prove*

$$\forall a \forall X \, \mathbf{OPHP}(a, X) \supset \forall a \forall X \, \mathbf{Count}_m(a, X).$$

(b) *Let $p, q_1, q_2, \ldots, q_k$ be distinct prime numbers. The theory V^0 does not prove the following implication*:

$$\left(\bigwedge_{t=1}^{k} \forall a \forall X \, \mathbf{Count}_{q_t}(a, X) \right) \supset \forall a \forall X \, \mathbf{Count}_p(a, X).$$

The next corollary is proved in the same way as Corollary IX.4.14.

COROLLARY IX.4.41. *For $m \geq 2$, the theory $V^0(m)$ is a proper extension of V^0. In fact, $V^0(m)$ is not Σ_0^B-conservative over V^0.*

See Section IX.7.4 for an open problem regarding $V^0(m)$ and the counting principles.

IX.4.8. The Theory $\mathbf{VAC^0(6)V}$. Here we develop $\mathbf{VAC^0(6)V}$ in the same way as $\mathbf{VAC^0(2)V}$, using Theorem IX.4.42 below. Recall the bounded number recursion (BNR) from Section IX.3.3. The following result if from [82].

THEOREM IX.4.42. *A function is in $\mathbf{FAC^0(6)}$ iff it is obtained from $\mathbf{FAC^0}$ by finitely many applications of $\mathbf{AC^0}$ reduction and 3-BNR iff it is obtained from $\mathbf{FAC^0}$ by finitely many applications of $\mathbf{AC^0}$ reduction and 4-BNR.*

Thus, the functions in $\mathcal{L}_{VAC^0(6)V}$ defined below represent precisely the functions in $\mathbf{FAC^0(6)}$.

DEFINITION IX.4.43. The vocabulary $\mathcal{L}_{VAC^0(6)V}$ is the smallest set that includes $\mathcal{L}_{FAC^0}$ such that for every $\mathcal{L}_A^2$-term t and open $\mathcal{L}_{VAC^0(6)V}$-formula φ the function $F_{\varphi(z),t}$ with defining axiom (86) (page 125) is in $\mathcal{L}_{VAC^0(6)V}$, and for every string function F in $\mathcal{L}_{VAC^0(6)V}$ there is a number function f_F in $\mathcal{L}_{VAC^0(6)V}$ with defining axiom $f_F(\vec{x}, \vec{X}) = |F(\vec{x}, \vec{X})|$, and for every two number functions $g(\vec{x}, \vec{X}), h(y, z, \vec{x}, \vec{X}) \in \mathcal{L}_{VAC^0(6)V}$ there is a number function $f_{g,h}$ in $\mathcal{L}_{VAC^0(6)V}$ with defining axiom (omitting $\vec{x}, \vec{X}$):

$$(g < 3 \supset f_{g,h}(0) = g) \wedge (g \geq 3 \supset f_{g,h}(0) = 0) \wedge$$
$$(h(y, f_{g,h}(y)) < 3 \supset f_{g,h}(y+1) = h(y, f_{g,h}(y))) \wedge$$
$$(h(y, f_{g,h}(y)) \geq 3 \supset f_{g,h}(y+1) = 0). \quad (270)$$

The next corollary follows from Theorem IX.4.42. It states that semantically the functions in $\mathcal{L}_{VAC^0(6)V}$ represent precisely the functions in $\mathbf{FAC^0(6)}$.

COROLLARY IX.4.44.   (a) *A function is in $FAC^0(6)$ iff it is represented by an $\mathcal{L}_{VAC^0(6)V}$-term.*

(b) *A relation is in $AC^0(6)$ iff it is represented by an open formula in $\mathcal{L}_{VAC^0(6)V}$ iff it is represented by a $\Sigma_0^B(\mathcal{L}_{VAC^0(6)V})$ formula.*

DEFINITION IX.4.45. The theory $VAC^0(6)V$ has vocabulary $\mathcal{L}_{VAC^0(6)V}$ and is axiomatized by the axioms of $\overline{V}^0$ together with (86) for the functions $F_{\varphi(z),t}$ and (240) for the functions f_F and (270) for the functions $f_{g,h}$.

THEOREM IX.4.46.   (a) *The theory $VAC^0(6)V$ is a universal conservative extension of $V^0(6)$.*

(b) *Every $\Sigma_0^B(\mathcal{L}_{VAC^0(6)V})$ formula φ^+ is equivalent in $VAC^0(6)V$ to a Σ_1^B formula φ.*

PROOF SKETCH. Part (a) is proved by formalizing both directions of the proof of Theorem IX.4.42 in appropriate theories ($V^0(6)$ or $VAC^0(6)V$). Part (b) is proved in the same way as Corollary IX.3.21.   □

The characterization of $AC^0(6)$ by $VAC^0(6)V$ can be proved as in Section IX.3.4 (for the class TC^0 and the theory VTC^0V).

COROLLARY IX.4.47.   (a) *A function is in $FAC^0(6)$ iff it is Σ_1^B-definable in $VAC^0(6)V$.*

(b) *A relation is in $AC^0(6)$ iff it is Δ_1^B-definable in $VAC^0(6)V$.*

IX.5. Theories for NC^1 and the NC Hierarchy

The classes NC^k and AC^k form a hierarchy inside P as follows:

$$AC^0 \subsetneq NC^1 \subseteq AC^1 \subseteq NC^2 \subseteq \cdots.$$

We have already developed the theory V^0 for AC^0. Here we develop theories for the other classes in the hierarchy, with a focus on NC^1.

The class NC^1 plays an important role in propositional logic because the NC^1 relations can be characterized as those that can be expressed by a uniform polynomial size family of propositional formulas. The Σ_0^B theorems of the theory VNC^1 translate into families of tautologies with polynomial size PK proofs (Section X.3).

In Section IX.5.1 we define the classes and characterize them in terms of alternating Turing machines. In Section IX.5.2 we define the Boolean Sentence Value Problem, which is complete for NC^1. In Section IX.5.3 we develop the theories VNC^1, $\widehat{VNC^1}$ and $\overline{VNC}^1$ that characterize NC^1 following the method presented in Section IX.2. In Section IX.5.4 we show that VNC^1 extends VTC^0. In Section IX.5.5 we use the bounded number recursion (BNR) operation (see Section IX.3.3) to develop VNC^1V, and use the fact that this theory can formalize the proof of Barrington's

Theorem to prove that it is a universal conservative extension of VNC^1. Finally, the theories for other classes in the NC hierarchy are defined in Section IX.5.6.

In Section X.3.1 we will prove the Propositional Translation Theorem for VNC^1.

IX.5.1. Definitions of the Classes. (See also Appendix A.5.) Recall the definition of AC^0 using uniform families of circuits in Section IV.1. In general for $k \geq 0$, uniform AC^k (or just AC^k) is the class of problems decidable using a uniform family $\langle C_n \rangle$ of polynomial size Boolean circuits, where each circuit C_n has n input bits and $\mathcal{O}((\log n)^k)$ depth, and the gates in C_n have unbounded fan-in. The class uniform NC^k (or simply NC^k) is defined in the same way, except the gates have fan-in two.

It is easy to see that for $k \geq 0$:

$$AC^k \subseteq NC^{k+1} \subseteq AC^{k+1}.$$

Furthermore $AC^0 \subsetneq NC^1$ because $PARITY(X)$ is in NC^1 but not in AC^0. These classes form the NC hierarchy:

$$NC = \bigcup_{k>0} NC^k = \bigcup_{k \geq 0} AC^k \subseteq P.$$

Ruzzo [100] discusses at length alternative notions of uniformity that can be used to define these classes. For $k \geq 1$ the classes AC^k and NC^{k+1} remain the same under a wide choice of uniformity conditions, from AC^0 to log space. However NC^1 appears to be more sensitive to the notion of uniformity used. The standard definition is quite strong, and requires that the *extended connection language* (ECL) be recognizable in time $\mathcal{O}(\log n)$ (where n is the number of input bits for the circuit). Here ECL specifies for each gate g and each string $p \in \{L, R\}^*$ of length $\log n$ the gate that is reached by following the path specified by p starting from g and proceeding to the left input for L and the right input for R.

Fortunately under the standard notions of uniformity these classes are robust in the sense that they can be characterized using a different model of computation, namely alternating Turing machines (ATMs) (see Appendix A.4). Let *ASpace-Alt*(s, r) denote the class of languages accepted by ATMs in space s with r alternations. Then for $k \geq 1$ [40, 110]

$$AC^k = ASpace\text{-}Alt(\mathcal{O}(\log n), \mathcal{O}((\log n)^k)). \tag{271}$$

(Recall Theorem IV.1.2 states that AC^0 consists of languages accepted by ATMs working in time $\mathcal{O}(\log n)$ and constant alternations.)

Similarly, let *ASpace-Time*(s, t) denote the class of languages accepted by ATMs in simultaneous space s and time t. Then for $k \geq 1$ [100]

$$NC^k = ASpace\text{-}Time(\mathcal{O}(\log n), \mathcal{O}((\log n)^k)) \tag{272}$$

and in particular

$$NC^1 = ATime(\mathcal{O}(\log n)). \qquad (273)$$

For this reason NC^1 is also called *ALogTime* in the literature.

Here we take the above three equations as definitions of the classes AC^k and NC^k.

IX.5.2. BSVP and NC^1. The Boolean Sentence Value Problem (BSVP) is to decide the truth value of a Boolean sentence given its infix representation. In Section X.3.2 we will show that that BSVP is AC^0-many-one complete for NC^1. In fact, the problem remains AC^0-many-one complete for NC^1 even for monotone formulas that have a "balanced" structure when viewed as a binary tree. We use this fact here to define the function *Fval* (*Fval* stands for "formula value") whose AC^0 closure is NC^1.

Consider the following encoding of a balanced monotone Boolean sentence using the heap data structure. We view the sentence as a balanced binary tree with $(2a - 1)$ nodes, including a leaves numbered

$$a, (a + 1), \dots, (2a - 1)$$

and $(a - 1)$ inner nodes numbered

$$1, 2, \dots, (a - 1).$$

Each inner node (or gate) is either an $\wedge$-gate or an $\vee$-gate, and each leaf is labeled with a Boolean value. The two children (inputs) of an inner node x are $2x$ and $(2x + 1)$ (as in the heap data structure). Therefore the sentence can be encoded by (a, G, I), where $G(x)$ specifies the label of node x: for $1 \leq x < a$,

if $G(x)$ holds then node x is an $\wedge$-gate, otherwise x is an $\vee$-gate,

and I specifies the values at the leaves: for $x < a$,

$I(x)$ is the value labeling leaf $(a + x)$.

We will also refer to the binary tree (a, G) as a tree-like circuit and refer to I as its inputs.

The function $Fval(a, G, I)$ gives the value of the sentence encoded by (a, G, I) as well as the values of all nodes in the associated tree. The function is computed by a polytime procedure which evaluates these nodes inductively, starting with the leaves. In the formula $\delta_{MFV}(a, G, I, Y)$ below $Y(0)$ always holds, and $Y(x)$ is the value of gate x for $0 < x < 2a$. (MFV stands for "monotone formula value".)

$$\delta_{MFV}(a, G, I, Y) \equiv \forall x < a \big[(Y(x + a) \leftrightarrow I(x)) \wedge Y(0) \wedge$$
$$0 < x \supset \big(Y(x) \leftrightarrow ((G(x) \wedge Y(2x) \wedge Y(2x + 1)) \vee$$
$$(\neg G(x) \wedge (Y(2x) \vee Y(2x + 1)))))\big]. \quad (274)$$

Figure 9 depicts a computation of (the bits of) Y for $a = 6$. Here $Y(1), \dots, Y(5)$ are the values of gates $G(1), \dots, G(5)$.

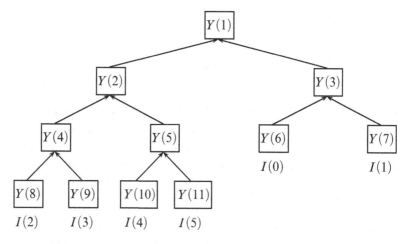

FIGURE 9. Computing Y which satisfies $\delta_{MFV}(a, G, I, Y)$ for $a = 6$.

DEFINITION IX.5.1.

$$R_{MFV}(a, G, I) \leftrightarrow \exists Y \leq 2a(\delta_{MFV}(a, G, I, Y) \wedge Y(1)).$$

The following proposition shows that R_{MFV} is AC^0-many-one complete for NC^1. For a proof see [22] and [10, Lemma 6.2, page 287].

PROPOSITION IX.5.2. *A Relation* $R(\vec{x}, \vec{X})$ *is in* NC^1 *iff there are* AC^0 *functions* a_0, G_0, I_0 *such that*

$$R(\vec{x}, \vec{X}) \leftrightarrow R_{MFV}(a_0(\vec{x}, \vec{X}), G_0(\vec{x}, \vec{X}), I_0(\vec{x}, \vec{X})).$$

THEOREM IX.5.3. NC^1 *is the* AC^0 *closure of* R_{MFV}.

PROOF. It follows from Proposition IX.5.2 that NC^1 is included in the AC^0 closure of R_{MFV}. Thus it suffices to show that NC^1 is closed under AC^0 reductions, and for this it suffices to show that FNC^1 is closed under AC^0 reductions. By (273) and Theorem IX.1.7 our task is to show that the class of functions computed by alternating Turing machines (ATMs) in time $\mathcal{O}(\log n)$ is closed under composition and string comprehension.

We use the restricted form of log time ATMs given in Section 6 of [10]. In particular only one input bit is queried on any one computation path, and that query occurs at the end of that path. Further, although inputs of number sort are presented in unary, an ATM can easily write on its work tape the binary notation of a unary input x in time $\mathcal{O}(\log x)$ and constant alternations, simply by guessing that notation and verifying it with a couple of queries using its index tape.

For composition, for notational simplicity consider the case

$$F(\vec{x}, \vec{X}) = G(\vec{x}, \vec{X}, H(\vec{x}, \vec{X}))$$

where G and H are computed by ATMs in time $\mathcal{O}(\log n)$. The machine M computing F actually computes the bit graph $F(\vec{x}, \vec{X})(i)$ of F. On input $(\vec{x}, \vec{X}, i)$ M simulates the machine computing the ith bit of G. Whenever a computation path of that machine ends with an input query for a bit j of $H(\vec{x}, \vec{X})$, M simply simulates the machine computing H on input $(\vec{x}, \vec{X}, j)$.

For string comprehension, suppose

$$F(y) = \{f(x) : x \leq y\}$$

(where F and f may contain other arguments) and suppose f is computed by an ATM in time $\mathcal{O}(\log n)$. Then to compute $F(y)(i)$ (where as above we may assume binary notation for y and i) a machine M guesses x, verifies $x \leq y$, then for each bit number j (using universal states) M computes bit j of $f(x)$ and verifies that it is the same as bit j of i.   □

IX.5.3. The Theories VNC^1, $\widehat{VNC^1}$, and $\overline{VNC}^1$. We define the theories VNC^1, $\widehat{VNC^1}$ and $\overline{VNC}^1$ as in Section IX.2 using the formula δ_{MFV} above. In Section IX.5.5 we will define $VNC^1 V$, another universal conservative extension of VNC^1 using number recursion.

DEFINITION IX.5.4 (VNC^1). Let

$$MFV \equiv \exists Y \leq 2a + 1 \delta_{MFV}(a, G, I, Y). \tag{275}$$

The theory VNC^1 has vocabulary $\mathcal{L}_A^2$ and is axiomatized by MFV and the axioms of V^0.

DEFINITION IX.5.5. The function $Fval(a, G, I)$ is defined as follows:

$$Fval(a, G, I) = Y \leftrightarrow |Y| \leq 2a \wedge \delta_{MFV}(a, G, I, Y). \tag{276}$$

Note that $Fval$ is Σ_1^B-definable in VNC^1.

The next result is an immediate consequence of Theorem IX.5.3 and Lemma IX.1.5.

COROLLARY IX.5.6. *FNC^1 is the closure of $Fval$ under AC^0 reducibility.*

Details of the proof of the next result are easier to carry out after later technical developments (see Exercise IX.5.15). Here we outline the proof.

LEMMA IX.5.7. *The aggregate $Fval^*$ of $Fval$ is Σ_1^B-definable in VNC^1, and $VNC^1(Fval, Fval^*)$ proves*

$$\forall i < b, Fval^*(b, A, G, I)^{[i]} = Fval((A)^i, G^{[i]}, I^{[i]}).$$

PROOF SKETCH. This is an exercise in circuit design similar to the proof of Lemma VIII.1.10, which shows a similar result for the theory VP. Since the set of Σ_1^B-definable functions in any theory extending V^0 is closed under composition it suffices to find suitable FAC^0 functions EXT, a_0, G_0, I_0 such that

$$Fval^*(b, A, G, I) = EXT(Fval(a_0(*), G_0(*), I_0(*)), *)$$

where $*$ stands for b, A, G, I. Thus the tree circuit C described by (a_0, G_0, I_0) has, for each $i < b$, a subtree C_i which computes $Fval((A)^i, G^{[i]}, I^{[i]})$. We fix things so that each subtree C_i is a padded version of the circuit described by $((A)^i, G^{[i]}, I^{[i]})$ that it is simulating, so that all circuits C_i have the same size, and the number of leaves of each is a power of 2. This makes it easy to define the function EXT so that $EXT(Y, *)^{[i]}$ extracts the gate values of C_i when Y is the string of gate values for C.   □

To define $\widehat{VNC^1}$ we use the following quantifier-free defining axiom for $Fval$, where x occurs as a free variable. It is easy to see that this defining axiom is equivalent to (276).

$$Fval(a, G, I) = Y \supset (|Y| \le 2a) \wedge Y(0) \wedge (x < a \supset (Y(x + a) \leftrightarrow$$
$$I(x))) \wedge (0 < x \wedge x < a) \supset (Y(x) \leftrightarrow ((G(x) \wedge Y(2x) \wedge$$
$$Y(2x + 1)) \vee (\neg G(x) \wedge (Y(2x) \vee Y(2x + 1)))))). \quad (277)$$

DEFINITION IX.5.8. $\widehat{VNC^1}$ is the universal theory over the vocabulary $\mathcal{L}_{\widehat{VNC^1}} = \mathcal{L}_{FAC^0} \cup \{Fval\}$ with axioms those of $\overline{V}^0$ and the defining axiom (277) for $Fval$.

DEFINITION IX.5.9. $\mathcal{L}_{FNC^1}$ is the smallest set that contains $\mathcal{L}_{\widehat{VNC^1}}$ such that for every $\mathcal{L}_A^2$-term t and every quantifier-free $\mathcal{L}_{FNC^1}$-formula φ there is a function $F_{\varphi(z),t}$ in $\mathcal{L}_{FNC^1}$ with defining axiom (86) (page 125).

$\overline{VNC}^1$ is the universal theory over $\mathcal{L}_{FNC^1}$ that is axiomatized by the axioms of $\widehat{VNC^1}$ and (86) for each function $F_{\varphi(z),t}$ of $\mathcal{L}_{FNC^1}$.

The next theorem follows from results in Section IX.2 concerning the general treatment of the theories VC, $\widehat{VC}$, and $\overline{VC}$.

THEOREM IX.5.10. *Assume that either $\mathcal{L}$ is $\mathcal{L}_{\widehat{VNC^1}}$ and T is $\widehat{VNC^1}$, or $\mathcal{L}$ is $\mathcal{L}_{FNC^1}$ and T is $\overline{VNC}^1$.*

(a) *A function is in FNC^1 iff it is represented by a term in $\mathcal{L}_{\widehat{VNC^1}}$ (and for a string function) iff it is represented by a function in $\mathcal{L}_{FNC^1}$. A relation is in NC^1 iff it is represented by an open (or Σ_0^B) formula of $\mathcal{L}$.*

(b) *Every $\Sigma_1^B(\mathcal{L})$ formula is equivalent in T to a $\Sigma_1^B(\mathcal{L}_A^2)$ formula.*

(c) *T proves $\Sigma_0^B(\mathcal{L})$-COMP, $\Sigma_0^B(\mathcal{L})$-IND, and $\Sigma_0^B(\mathcal{L})$-MIN.*

(d) *$\overline{VNC}^1$ is a universal conservative extension of $\widehat{VNC^1}$, which is in turn a universal conservative extension of VNC^1.*

(e) *A function is in FNC^1 iff it is Σ_1^B-definable in VNC^1 iff it is Σ_1^B-definable in T.*

(f) *A relation is in NC^1 iff it is Δ_1^B-definable in VNC^1 iff it is Δ_1^B-definable in T.*

The original definition of VNC^1 [45] uses the axiom scheme Σ_0^B-$TreeRec$ instead of the axiom MFV.

DEFINITION IX.5.11 (Σ_0^B-$TreeRec$). Σ_0^B-$TreeRec$ is the set of axioms of the form

$$\exists Y \forall x < a \big(Y(x+a) \leftrightarrow \psi(x)) \wedge$$
$$(0 < x \supset (Y(x) \leftrightarrow \varphi(x)[Y(2x), Y(2x+1)])\big)) \quad (278)$$

where $\psi(x)$ is a Σ_0^B formula, $\varphi(x)[p,q]$ is a Σ_0^B formula which contains two Boolean variables p and q, and Y does not occur in ψ and φ.

We will show that our definition of VNC^1 here is equivalent to the original definition. Since MFV is an instance of the Σ_0^B-$TreeRec$ axiom scheme, we need only to show that Σ_0^B-$TreeRec$ is provable in VNC^1. This is proved in Theorem IX.5.12 below. In Section IX.5.4 below we will show that VNC^1 proves several generalizations of Σ_0^B-$TreeRec$ (Theorems IX.5.13 and IX.5.14).

THEOREM IX.5.12. *The Σ_0^B-$TreeRec$ axiom scheme is provable in VNC^1.*

PROOF. Given a, ψ and φ, the idea is to construct a (large) treelike circuit (b, G) and inputs I so that from $Fval(b, G, I)$ we can extract Y (using Σ_0^B-$COMP$) that satisfies (278).

Notice the "gates" $\varphi(x)[p,q]$ in (278) can be any of the sixteen Boolean functions in two variables p, q. We will (uniformly) construct binary treelike $\wedge$-$\vee$ circuits of constant depth that compute $\varphi(x)[p,q]$.

Let

$$\beta_1, \ldots, \beta_8, \beta_9 \equiv \neg\beta_1, \ldots, \beta_{16} \equiv \neg\beta_8$$

be the sixteen Boolean functions in two variables p, q. Each β_i can be computed by a binary treelike and-or circuit of depth 2 with inputs among $0, 1, p, q, \neg p, \neg q$. For $1 \le i \le 16$, let X_i be defined by

$$X_i(x) \leftrightarrow \big(x < a \wedge (\varphi(x)[p,q] \leftrightarrow \beta_i(p,q))\big).$$

Then,

$$\varphi(x)[p,q] \leftrightarrow \bigvee_{i=1}^{16} (X_i(x) \wedge \beta_i(p,q)).$$

Consequently, $\varphi(x)[p,q]$ can be computed by a binary and-or tree T_x of depth 7 whose inputs are $0, 1, p, \neg p, q, \neg q, X_i(x)$. Similarly, $\neg\varphi(x)[p,q]$ is computed by a binary and-or tree T_x' having the same depth and set of inputs. Our large tree G has one copy of T_1, and in general for each copy of T_x or T_x', there are multiple copies of $T_{2x}, T_{2x+1}, T_{2x}', T_{2x+1}'$ that supply the inputs $Y(2x), Y(2x+1), \neg Y(2x), \neg Y(2x+1)$, and other trivial treelike circuits that provide inputs $0, 1, X_i(x)$ $(1 \le i \le 16)$.

Finally, I is defined as follows: $I(x) \leftrightarrow (x < a \wedge \psi(x))$. $\square$

IX.5.4. $VTC^0 \subseteq VNC^1$. It is known that $TC^0 \subseteq NC^1$ (although it is unknown whether the inclusion is proper). Here we will show, informally, that VNC^1 proves this inclusion. In particular, we will show that VNC^1 extends VTC^0. Note that for this it suffices to show that VNC^1 proves the axiom *NUMONES*. Our proof is by formalizing in VNC^1 the construction of NC^1 circuits that compute *numones* and prove in VNC^1 the correctness of this construction. Here we formalize the construction by Buss [22].

The next two theorems show that VNC^1 proves some generalizations of Σ_0^B-*TreeRec*. They are useful in formalizing the construction of the counting circuits. They are also useful in proving that the function *Fval** is Σ_1^B-definable in VNC^1 (see Exercise IX.5.15), a result that we need for Theorem IX.5.10 stated earlier.

First, Theorem IX.5.13 asserts, informally, that we can evaluate in VNC^1 formulas whose underlying trees have an arbitrary constant branching factor (as opposed to binary trees).

THEOREM IX.5.13. *Suppose that $2 \leq k \in \mathbb{N}$, $\psi(x)$ is a Σ_0^B formula, and $\varphi(x)[p_0, \ldots, p_{k-1}]$ is a Σ_0^B formula that contains also Boolean variables p_i. Then VNC^1 proves*

$$\exists Y, \forall x < ka, \, a \leq x \supset (Y(x) \leftrightarrow \psi(x)) \wedge$$
$$\forall x < a, \, Y(x) \leftrightarrow \varphi(x)[Y(kx), \ldots, Y(kx + k - 1)]. \quad (279)$$

PROOF. We prove for the case $k = 4$; similar arguments work for other cases.

Using Theorem IX.5.12 we will define a', ψ', φ' so that from Y' that satisfies the Σ_0^B-*TreeRec* axiom (278) for a', ψ' and φ' we can obtain Y that satisfies (279) above.

Intuitively, consider Y in (279) as a forest of three trees whose nodes are labeled with $Y(x)$, $x < |Y|$. Then Y has branching factor of 4 (since $k = 4$), and the three trees are rooted at $Y(1)$, $Y(2)$ and $Y(3)$. (See Figure 10.) Note also that each layer in Y corresponds to two layers in the binary tree Y'.

We will define an injective map f so that $Y(x) \leftrightarrow Y'(f(x))$. Since the trees rooted at $Y(1)$, $Y(2)$ and $Y(3)$ are disjoint, f is defined so that these trees are the images of disjoint subtrees in the tree Y'. For example, we can choose the subtrees rooted at $Y'(4)$, $Y'(5)$ and $Y'(6)$. Thus,

$$f(1) = 4, f(2) = 5, f(3) = 6.$$

In general, consider the function f defined by:

$$f(4^m + y) = 4^{m+1} + y \qquad \text{for } 0 \leq y < 3 \cdot 4^m.$$

(By the results in Chapter III, f is provably total in $I\Delta_0$, and hence also in V^0.)

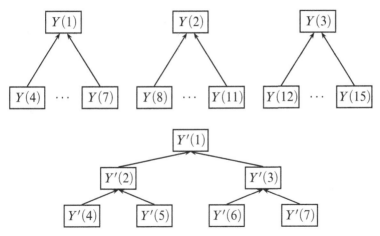

FIGURE 10. The forest Y in Theorem IX.5.13 when $k = 4$. Trees rooted at $Y(1)$, $Y(2)$ and $Y(3)$ are simulated by the sub-trees $Y'(4)$, $Y'(5)$ and $Y'(6)$, respectively.

Now we need ψ' such that for $a \leq x < 4a$

$$\psi'(f(x)) \leftrightarrow \psi(x).$$

So define ψ' as follows: for $y < 3 \cdot 4^m$ and $a \leq 4^m + y < 4a$,

$$\psi'(4^{m+1} + y) \leftrightarrow \psi(4^m + y).$$

To obtain φ', write $\varphi(x)[p_0, p_1, p_2, p_3]$ in the form

$$\varphi_1(x)[\varphi_2(x)[p_0, p_1], \varphi_3(x)[p_2, p_3]]$$

where φ_i is Σ_0^B with at most 2 Boolean variables, for $1 \leq i \leq 3$. Define φ' as follows:

$$\varphi'(4^{m+1} + y)[p, q] \leftrightarrow \varphi_1(4^m + y)[p, q] \qquad \text{for } y < 3 \cdot 4^m,$$

$$\varphi'(2 \cdot 4^{m+1} + 2y)[p, q] \leftrightarrow \varphi_2(4^m + y)[p, q] \qquad \text{for } y < 3 \cdot 4^m/2,$$

$$\varphi'(2 \cdot 4^{m+1} + 2y + 1)[p, q] \leftrightarrow \varphi_3(4^m + y)[p, q] \qquad \text{for } y < 3 \cdot 4^m/2.$$

Finally, let $a' = f(a)$. Let Y' satisfies (278) for a', ψ' and φ', and let Y be such that

$$Y(x) \leftrightarrow Y'(f(x)).$$

It is straightforward to verify that Y satisfies (279). $\qquad \square$

The next theorem shows that in VNC^1 we can evaluate multiple interconnected Boolean circuits where each has logarithmic depth and constant fan-in.

THEOREM IX.5.14. *Suppose that* $1 \leq m, \ell \in \mathbb{N}$, *and for* $1 \leq i \leq m$, $\psi_i(x, y)$ *and* $\varphi_i(x, y)[p_1, q_1, \ldots, p_{m\ell}, q_{m\ell}]$ *are* Σ_0^B *formulas where* $\vec{p}, \vec{q}$ *are* the Boolean variables. Then $\mathbf{VNC}^1$ proves the existence of $Z_1, \ldots, Z_m$ such that

$$\forall z < c \forall x < a \bigwedge_{i=1}^{m} \left((Z_i^{[z]}(x+a) \leftrightarrow \psi_i(z,x)) \wedge 0 < x \supset \left(Z_i^{[z]}(x) \leftrightarrow \right. \right.$$

$$\varphi_i(z,x)[Z_1^{[z]}(2x), Z_1^{[z]}(2x+1), \ldots, Z_m^{[z+\ell-1]}(2x), Z_m^{[z+\ell-1]}(2x+1)]) \bigg).$$

PROOF. Using Theorem IX.5.13 above, the idea is to construct a constant k, a number a' and Σ_0^B formulas $\psi'(c,x)$ and $\varphi'(c,x)[p_0, \ldots, p_{k-1}]$ so that from the set Y that satisfies (279) (for k, a', ψ' and φ') we can obtain $Z_1, \ldots, Z_m$.

Consider for example $m = 2, \ell = 2$. W.l.o.g., assume that $c \geq 1$. The (overlapping) subtrees

$$Z_1^{[0]}, Z_2^{[0]}, \ldots, Z_1^{[c-1]}, Z_2^{[c-1]} \tag{280}$$

have branching factor 8 (i.e., $2m\ell$). So let $k = 8$ (i.e., $k = 2m\ell$). We will construct Y (with branching factor 8) so that the disjoint subtrees rooted at

$$Y(c), \ldots, Y(3c-1) \tag{281}$$

are exactly the subtrees listed in (280).

We will define an 1-1, into map

$$s : \{1,2\} \times \mathbb{N}^2 \to \mathbb{N}$$

so that

$$Z_i^{[z]}(x) \leftrightarrow Y(s(i,z,x)).$$

The map s must be defined in such a way that the nodes of the trees listed in (280) match with those whose roots are listed in (281). For example, for the root level we need

$$s(1,0,1) = c, \ s(2,0,1) = c+1, \ s(1,1,1) = c+2, \ s(2,1,1) = c+3, \ldots.$$

For other levels we need: If $s(i,z,x) = y$, then

$$s(1,z,2x) = 8y, \ s(1,z,2x+1) = 8y+1, \ldots, s(2,z+1,2x+1) = 8y+7.$$

To define s we define partial, onto maps $f, g : \mathbb{N} \to \mathbb{N}$ and $h : \mathbb{N} \to \{1,2\}$ so that

$$s(h(y), g(y), f(y)) = y.$$

In other words,

$$Y(y) \leftrightarrow Z_{h(y)}^{[g(y)]}(f(y)).$$

For example, for $0 \leq z < 2c$:

$$f(c+z) = 1, \qquad g(c+z) = \lfloor z/2 \rfloor, \qquad h(c+z) = 1 + (z \mod 2).$$

In general, we need to define f, g, h only for values of x of the form $8^r c + z$ for $0 \leq z < 2 \cdot 8^r c$. The definitions of f, g, h at $8^r c + z$ are straightforward using the base 8 notation for z, where $0 \leq z < 2 \cdot 8^r c$.

Once f, g, h are defined, the formula ψ' and φ' are defined by

$$\psi'(c, x) \leftrightarrow \psi_{h(x)}(g(x), f(x))$$

and

$$\varphi'(c, x)[\ldots] \leftrightarrow \varphi_{h(x)}(g(x), f(x))[\ldots]$$

(where $\ldots$ is the list of $2m\ell$ Boolean variables). $\qquad\square$

EXERCISE IX.5.15. Use Theorem IX.5.14 to give a more detailed proof of Lemma IX.5.7 stating that the function $Fval^\star$ is Σ_1^B-definable in VNC^1.

For the next theorem we use $Sum(a, X)$ for the sum of a rows of X:

$$Sum(a, X) = \begin{cases} \varnothing & \text{if } a = 0, \\ X^{[0]} + X^{[1]} + \cdots + X^{[a-1]} & \text{if } a \geq 1. \end{cases}$$

(We introduced the function $Sum(m, n, X)$ in (250) and (251) on page 294. The two functions $Sum(a, X)$ and $Sum(m, n, X)$ have the same name but different arity, so the exact meaning is clear from context.)

THEOREM IX.5.16. *The function $Sum(a, X)$ with the following defining axiom is Σ_1^B-definable in $\overline{VNC}^1$:*

$$Sum(a, X) = Y \leftrightarrow$$
$$|Y| \leq \langle a, |X| \rangle \wedge Y^{[0]} = \varnothing \wedge \forall x < a (Y^{[x+1]} = Y^{[x]} + X^{[x]}). \quad (282)$$

The fact that $VTC^0 \subseteq VNC^1$ follows easily:

COROLLARY IX.5.17. $VTC^0 \subseteq VNC^1$.

PROOF OF THEOREM IX.5.16. Informally we need to construct a circuit that adds all rows

$$X^{[0]}, X^{[1]}, \ldots, X^{[a-1]}.$$

The idea is to use the divide-and-conquer technique. We will construct a balanced binary tree Z that has $(2a - 1)$ nodes (see Figure 11 for an example):

- a leaves $Z^{[a]}, Z^{[a+1]}, \ldots, Z^{[2a-1]}$ such that

$$Z^{[a+x]} = X^{[x]} \qquad \text{for } 0 \leq x < a.$$

- $(a - 1)$ inner nodes $Z^{[1]}, Z^{[2]}, \ldots, Z^{[a-1]}$; the two children of node $Z^{[x]}$ are $Z^{[2x]}$ and $Z^{[2x+1]}$, so that

$$Z^{[x]} = Z^{[2x]} + Z^{[2x+1]} \qquad \text{for } 1 \leq x < a.$$

Lemma IX.5.18. *Let $DaCAdd(a, I, Z)$ be the formula*

$$\forall x < a, \; Z^{[a+x]} = I^{[x]} \wedge x > 0 \supset Z^{[x]} = Z^{[2x]} + Z^{[2x+1]}. \qquad (283)$$

($DaCAdd$ stands for "divide-and-conquer addition".) Then

$$\overline{\mathbf{VNC}}^1 \vdash \forall a \forall I \exists Z \, DaCAdd(a, I, Z).$$

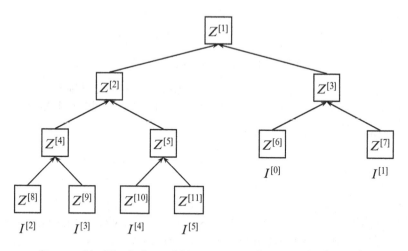

Figure 11. The balanced binary tree Z for $DaCAdd(6, I, Z)$.

Proof. We show how to compute Z by an $\mathbf{NC}^1$ circuit. Note that if for each $x < a$ we simply construct an $\mathbf{AC}^0$ circuit that performs string addition to compute $Z^{[x]}$ from $Z^{[2x]}$ and $Z^{[2x+1]}$ (i.e. $Z^{[x]} = Z^{[2x]} + Z^{[2x+1]}$) and stack them together, the resulting circuit has depth $\mathcal{O}(\log n)$ (where n is the number of input bits) but unbounded fan-in.

Here we use the fact that

$$X + Y + Z = G(X, Y, Z) + H(X, Y, Z) \qquad (284)$$

where $G(X, Y, Z)$ is the string of bit-wise sums, and $H(X, Y, Z)$ is the string of carries:

$$G(X, Y, Z)(z) \leftrightarrow X(z) \oplus Y(z) \oplus Z(z),$$

$$H(X, Y, Z)(0) \leftrightarrow \bot,$$

$$H(X, Y, Z)(z + 1) \leftrightarrow \big((X(z) \wedge Y(z)) \vee (X(z) \wedge Z(z)) \vee$$
$$(Y(z) \wedge Z(z))\big).$$

Exercise IX.5.19. Show that $V^0(G, H)$ proves the equation (284).

Thus, for each $Z^{[x]}$ we have a pair of strings $(S^{[x]}, C^{[x]})$ where $S^{[x]}$ is the string of bit-wise sums and $C^{[x]}$ is the string of carries; and for $1 \le x < a$,

$$Z^{[x]} = S^{[x]} + C^{[x]}.$$

(For $a \leq x < 2a$, we will take $S^{[x]} = I^{[x \dot{-} a]}$ and $C^{[x]} = \varnothing$.)

We need for $1 \leq x < a$,

$$S^{[x]} + C^{[x]} = S^{[2x]} + C^{[2x]} + S^{[2x+1]} + C^{[2x+1]}.$$

So

$$S^{[x]} = G(C^{[2x+1]}, U, V), \qquad C^{[x]} = H(C^{[2x+1]}, U, V)$$

where

$$U = G(S^{[2x]}, C^{[2x]}, S^{[2x+1]}), \qquad V = H(S^{[2x]}, C^{[2x]}, S^{[2x+1]}).$$

In other words, let F_1, F_2 be the AC^0 functions:

$$F_1(X, Y, Z, W) = G(W, G(X, Y, Z), H(X, Y, Z)),$$
$$F_2(X, Y, Z, W) = H(W, G(X, Y, Z), H(X, Y, Z)).$$

Then

$$S^{[x]} = F_1(S^{[2x]}, C^{[2x]}, S^{[2x+1]}, C^{[2x+1]})$$

and

$$C^{[x]} = F_2(S^{[2x]}, C^{[2x]}, S^{[2x+1]}, C^{[2x+1]}).$$

In summary we need to prove in VNC^1 the existence of S and C such that

$$\forall x < a, S^{[x+a]} = I^{[x]} \wedge C^{[x+a]} = \varnothing \wedge$$
$$(0 < x \supset (S^{[x]} = F_1(S^{[2x]}, C^{[2x]}, S^{[2x+1]}, C^{[2x+1]}) \wedge$$
$$C^{[x]} = F_2(S^{[2x]}, C^{[2x]}, S^{[2x+1]}, C^{[2x+1]}))).$$

Notice that for each z, the bits $S^{[x]}(z)$, $C^{[x]}(z)$ are computed from the bits

$$\{S^{[2x]}(y), S^{[2x+1]}(y), C^{[2x]}(y), C^{[2x+1]}(y) : z \dot{-} 2 \leq y \leq z\}$$

(where we define $S^{[2x]}(y) \equiv \bot$ if $y < 0$, etc.). This is not in the form of the hypothesis of Theorem IX.5.14, but we can put it in the required form by transposing S and C. Recall the function *Transpose* from (252) (page 295). We will first compute $S_t = Transpose(b, b, S)$ and $C_t = Transpose(b, b, C)$ where $b = |I|$ is a sufficiently large bound.

Thus $S_t^{[z]}(x)$ and $C_t^{[z]}(x)$ are computed from

$$\{S_t^{[y]}(2x), S_t^{[y]}(2x + 1), C_t^{[y]}(2x), C_t^{[y]}(2x + 1) : z - 2 \leq y \leq z\}$$

by a Σ_0^B formulas. Therefore by Theorem IX.5.14, VNC^1 proves the existence of S_t and C_t.   □

Notice that V^0 proves the uniqueness of Z in (283). Define $Sum(a, X)$ as follows. First,

$$Sum(0, X) = \varnothing.$$

For $a \geq 1$ we apply the above lemma for a *full* binary tree. Thus let a_1 be the smallest power of 2 that is $\geq a$, and define X_1 such that

$$X_1^{[x]} = X^{[x]} \text{ for } x < a \text{ and } X_1^{[x]} = \varnothing \text{ for } a \leq x < a_1. \qquad (285)$$

Let Z be the string that satisfies $DaCAdd(a_1, X_1, Z)$ as in Lemma IX.5.18 (see Figure 12). Define

$$Sum(a, X) = Z^{[1]}.$$

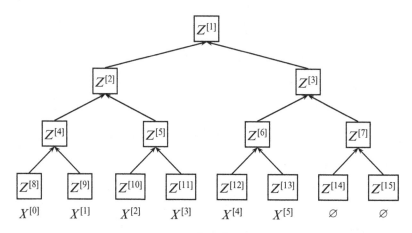

FIGURE 12. Defining $Sum(6, X)$ using a full binary tree.

It remains to show that $\overline{VNC}^1(Sum)$ proves (282). For this, it suffices to show that

$$Sum(a + 1, X) = Sum(a, X) + X^{[a]}. \qquad (286)$$

When $a = 0$ this is straightforward. So first consider the case where $a \geq 1$ and a is not a power of 2. Let X_1 be as in (285), and let X_2 be such that

$$X_2^{[x]} = X^{[x]} \text{ for } x \leq a \text{ and } X_2^{[x]} = \varnothing \text{ for } a < x < a_1.$$

Let Z_1 and Z_2 be such that $DaCAdd(a_1, X_1, Z_1)$ and $DaCAdd(a_1, X_2, Z_2)$ hold. By definition,

$$Sum(a, X) = Z_1^{[1]} \qquad \text{and} \qquad Sum(a + 1, X) = Z_2^{[1]}.$$

So we have to show that

$$Z_2^{[1]} = Z_1[1] + X[a].$$

The trees Z_1 and Z_2 have the same height $h = \lceil \log a_1 \rceil$. Note that $h + 1$ is the length of the binary representation of $(a_1 + a)$. Also, h is definable in $I\Delta_0$ (see Section III.3.3). Let

$$d_0 = 1, d_1 = 3, d_2, \ldots, d_h = (a_1 + a)$$

be all initial segments of the binary representation of $(a_1 + a)$. Then

$$Z_2^{[d_0]}, Z_2^{[d_1]}, \ldots, Z_2^{[d_h]}$$

are all nodes in the tree Z_2 on the path from the root to the leaf $Z^{[a_1 + a]} = X^{[a]}$.

It can be proved by reverse induction on i that

$$(Z_2^{[d_i]} = Z_1^{[d_i]} + X^{[a]}) \wedge \forall x < a_1(|x| = |d_i| \wedge x < d_i \supset Z_2^{[x]} = Z_1^{[x]}).$$

For $i = 0$ we obtain $Z_2^{[1]} = Z_1^{[1]} + X^{[a]}$ as required.

The case where a is a power of 2 is left as an exercise. $\qquad\square$

EXERCISE IX.5.20. Finish the proof of the theorem by showing that (286) is true when a is a power of 2.

IX.5.5. The Theory VNC^1V. In this section we will define VNC^1V using 5-BNR, a bounded number recursion scheme that characterizes FNC^1. This recursion theoretic characterization is based on Barrington's Theorem that asserts that NC^1 is the class of relations computable by width 5 branching programs, or equivalently the word problem for the permutation group S_5 is complete for NC^1.

Here the vocabulary $\mathcal{L}_{VNC^1V}$ consists of symbols for all FNC^1 functions, but their defining axioms are based on 5-BNR rather than on AC^0-reductions to the function $Fval$ (that are used to define $\overline{VNC}^1$). Recall the bounded number recursion (BNR) scheme in Section IX.3.3.

THEOREM IX.5.21 (Barrington). *A function is in FNC^1 iff it can be obtained from the empty set of functions by finitely many applications of AC^0 reduction and 5-BNR.*

By Theorem IX.1.7 it follows also that FNC^1 is the class of functions obtained from FAC^0 by finitely many applications of composition, string comprehension and 5-BNR.

DEFINITION IX.5.22. The vocabulary $\mathcal{L}_{VNC^1V}$ is the smallest set that contains $\mathcal{L}_{FAC^0}$ such that:

- For each $\mathcal{L}_A^2$-term t and quantifier-free $\mathcal{L}_{VNC^1V}$-formula φ there is a function $F_{\varphi(z),t}$ in $\mathcal{L}_{VNC^1V}$ with defining axiom (86):

$$F_{\varphi(z),t}(\vec{x}, \vec{X})(z) \leftrightarrow z < t(\vec{x}, \vec{X}) \wedge \varphi(z, \vec{x}, \vec{X}). \tag{287}$$

- For every string function $F(\vec{x}, \vec{X})$ in $\mathcal{L}_{VNC^1V}$ the number function $f_F(\vec{x}, \vec{X})$ is also in $\mathcal{L}_{VNC^1V}$ with defining axiom

$$f_F(\vec{x}, \vec{X}) = |F(\vec{x}, \vec{X})|. \tag{288}$$

- For any number functions $g(\vec{x}, \vec{X})$ and $h(y, z, \vec{x}, \vec{X})$ in $\mathcal{L}_{VNC^1V}$, there is a number function $f_{g,h}(y, \vec{x}, \vec{X})$ in $\mathcal{L}_{VNC^1V}$ with defining axiom (omitting $\vec{x}, \vec{X}$)

$$(g < 5 \supset f_{g,h}(0) = g) \wedge (g \geq 5 \supset f_{g,h}(0) = 0) \wedge$$
$$(h(y, f_{g,h}(y)) < 5 \supset f_{g,h}(y+1) = h(y, f_{g,h}(y))) \wedge$$
$$(h(y, f_{g,h}(y)) \geq 5 \supset f_{g,h}(y+1) = 0). \quad (289)$$

The next corollary follows from Theorem IX.5.21.

COROLLARY IX.5.23. (a) *A function is in* $\mathbf{FNC}^1$ *iff it is represented by a function in* $\mathcal{L}_{VNC^1V}$.

(b) *A relation is in* $\mathbf{NC}^1$ *iff it is represented by an open (or a* Σ_0^B*) formula of* $\mathcal{L}_{VNC^1V}$.

DEFINITION IX.5.24 ($\mathbf{VNC}^1V$). The theory $\mathbf{VNC}^1V$ has vocabulary $\mathcal{L}_{VNC^1V}$ and axioms those of $\overline{V}^0$ together with (287) for each function $F_{\varphi(z),t}$ and (288) for each function f_F and (289) for each function $f_{g,h}$.

The next exercise can be proved as in Lemma IX.3.16 and Corollary IX.3.21.

EXERCISE IX.5.25. (a) Show that the theory $\mathbf{VNC}^1V$ proves the axiom schemes

$$\Sigma_0^B(\mathcal{L}_{VNC^1V})\text{-}\mathbf{COMP}, \ \Sigma_0^B(\mathcal{L}_{VNC^1V})\text{-}\mathbf{IND}, \ \text{and} \ \Sigma_0^B(\mathcal{L}_{VNC^1V})\text{-}\mathbf{MIN}.$$

(b) Show that for every $\Sigma_0^B(\mathcal{L}_{VNC^1V})$ formula φ^+ there is a Σ_1^B formula φ so that $\mathbf{VNC}^1V \vdash \varphi^+ \leftrightarrow \varphi$.

COROLLARY IX.5.26. (a) *A function is in* $\mathbf{FNC}^1$ *iff it is* Σ_1^B*-definable in* $\mathbf{VNC}^1V$.

(b) *A relation is in* $\mathbf{NC}^1$ *iff it is* Δ_1^B*-definable in* $\mathbf{VNC}^1V$.

PROOF SKETCH. (a) This part follows from Theorem IX.5.21 and Exercise IX.5.25 above. It is proved in the same way as Corollary IX.3.22.

(b) From (a) and Theorem V.4.35. □

THEOREM IX.5.27. $\mathbf{VNC}^1V$ *is a universal conservative extension of* $\mathbf{VNC}^1$.

We outline the proof below. For details see [82].

PROOF IDEA. To show that $\mathbf{VNC}^1V$ extends $\mathbf{VNC}^1$, the main task is to show that $\mathbf{VNC}^1V \vdash \mathbf{MFV}$. The idea is to formalize in $\mathbf{VNC}^1V$ the proof that the Boolean Sentence Value Problem (see page 321) can be computed using width 5 branching programs (the $\Longrightarrow$ direction of Theorem IX.5.21). To show that $\mathbf{VNC}^1V$ is conservative over $\mathbf{VNC}^1$ essentially we need to show that width 5 branching programs can be simulated by families of $\mathbf{NC}^1$ circuits. The proof can be by induction on the definition of $\mathbf{VNC}^1V$. (See also Section VIII.2.2 for the proof that $\mathbf{VPV}$ is a universal conservative extension of $\mathbf{VP}$.) □

IX.5.6. Theories for the NC Hierarchy. We develop the theories for AC^k and NC^{k+1} using the fact that the Circuit Value Problem is complete for the respective classes under appropriate restriction on the given circuits. Consider encoding a layered, monotone Boolean circuit C with $(d + 1)$ layers and n unbounded fan-in ($\wedge$ or $\vee$) gates on each layer. We need to specify the type (either $\wedge$ or $\vee$) of each gate, and the wires between the gates. Suppose that layer 0 contains the inputs which are specified by a string variable I of length $|I| \leq n$. To encode the gates on other layers, there is a string variable G such that for $1 \leq z \leq d$, $G(z, x)$ holds if and only if gate x on layer z is an $\wedge$-gate (otherwise it is an $\vee$-gate). Also, the wires of C are encoded by a 3-dimensional array E: $\langle z, x, y \rangle \in E$ iff the output of gate x on layer z is connected to the input of gate y on layer $z + 1$.

The following algorithm computes the outputs of C using $(d + 1)$ loops: in loop z it identifies all gates on layer z which output 1. It starts by identifying the input gates with the value 1. Then in each subsequent loop $(z + 1)$ the algorithm identifies the following gates on layer $(z + 1)$:

- $\vee$-gates that have at least one input which is identified in loop z;
- $\wedge$-gates all of whose inputs are identified in loop z.

The formula $\delta_{LMCV}(n, d, E, G, I, Y)$ below formalizes this algorithm (LMCV stands for "layered monotone circuit value"). The 2-dimensional array Y stores the result of computation: For $1 \leq z \leq d$, row $Y^{[z]}$ contains the gates on layer z that output 1.

$$\delta_{LMCV}(n, d, E, G, I, Y) \equiv \forall x < n \forall z < d \left((Y(0, x) \leftrightarrow I(x)) \wedge \right.$$
$$\left(Y(z + 1, x) \leftrightarrow (G(z + 1, x) \wedge \forall u < n, E(z, u, x) \supset Y(z, u)) \vee \right.$$
$$\left. (\neg G(z + 1, x) \wedge \exists u < n, E(z, u, x) \wedge Y(z, u)))\right). \quad (290)$$

For NC^k we need the following formula which states that the circuit with underlying graph (n, d, E) has fan-in 2:

$$Fanin2(n, d, E) \equiv \forall z < d \forall x < n \exists u_1 < n \exists u_2 < n \forall v < n \big(E(z, v, x) \supset$$
$$(v = u_1 \vee v = u_2) \big).$$

Recall (Section III.3.3) that the function $|x| = \lceil \log(x + 1) \rceil$ is an AC^0 function with a Δ_0 graph. Define the functions $Lmcv_k$ and $Lmcv_{k,2}$ as follows:

$$Lmcv_k(n, E, G, I) = Y \leftrightarrow \left(|Y| \leq \langle |n|^k + 1, n \rangle \wedge \right.$$
$$\left. \delta_{LMCV}(n, |n|^k, E, G, I, Y) \right).$$

and

$$Lmcv_{k,2}(n, E, G, I) = Y \leftrightarrow \big((\neg Fanin2(n, d, E) \wedge Y = \varnothing) \vee$$

$$(Fanin2(n, d, E) \wedge |Y| \leq \langle |n|^k + 1, n \rangle \wedge \delta_{LMCV}(n, |n|^k, E, G, I, Y))\big).$$

THEOREM IX.5.28. *For* $k \geq 1$, FAC^k *is the closure of* $Lmcv_k$ *under* AC^0 *reductions. For* $k \geq 2$, FNC^k *is the closure of* $Lmcv_{k,2}$ *under* AC^0 *reductions.*

PROOF. It is easy to see that every function in uniform FAC^k (resp. FNC^k) is AC^0 many-one reducible to $Lmcv_k$ (resp. $Lmcv_{k,2}$). Also, the proof that FAC^k and FNC^k are closed under AC^0 reductions is similar to the proof of Theorem IX.5.3 for the class NC^1, where now we use the ATM characterizations (271) and (272) for these classes. It remains to show that the $Lmcv$ functions belong to their respective classes.

We show that $Lmcv_k$ is in FAC^k by using its ATM characterization (271). Thus we describe an ATM M using space $\mathcal{O}(\log n)$ and alternations $\mathcal{O}((\log n)^k)$ which computes the bit graph of $Lmcv_k$.

Let C be the circuit of depth $|n|^k$ and width n described by (n, E, G, I). In order to compute bit i of $Lmcv_k(n, E, G, I)$ the machine M must accept the input (n, E, G, I, i) iff gate i is $\top$. Thus M starts by guessing that gate i is $\top$, and then follows a path down to an input gate. For each gate g on the path, M verifies that g is $\top$ as follows. If g is an $\vee$ gate then M guesses (using existential states) which input g' of g is $\top$, and proceeds to g' as the next gate. If g is an $\wedge$ gate, then M enters universal states and and proceeds to an arbitrary input g'. If g is an input to the circuit, then M accepts iff $I(g)$ holds.

It is easy to see that this computation is correct. Further M needs to keep only two gates written on its tape at once, so the space is $\mathcal{O}(\log n)$. Finally the number of alternations is bounded by the depth of the circuit, which is $\mathcal{O}((\log n)^k)$.

For $k \geq 2$ we show that $Lmcv_{k,2}$ is in NC^k by using (272). Thus we need an ATM M which uses space $\mathcal{O}(\log n)$ and time $\mathcal{O}((\log n)^k)$ which computes the bit graph of $Lmcv_{k,2}$. This proof is more complicated than the one above, and we follow the argument given by Ruzzo [100, Theorem 4].

Let C be a circuit of fanin 2, depth $|n|^k$ and width n described by (n, E, G, I). M determines whether a given gate i in C is $\top$ by following an arbitrary path from i down to an input gate in a manner similar to the AC^k case above, except now M uses a different notation for the current gate g in the path. Initially M sets a variable h to i (in binary), where in general h is the gate at the head of the path segment it is remembering. M also remembers the path p from h to g, where p is a string over the alphabet $\{L, R\}$, where the j-th element of p is L if the path follows the left input out of the j-th gate in the path, and otherwise the j-th element is R (here we use the fanin 2 assumption). At each gate g in the path, M guesses whether g is $\wedge$ or $\vee$, and then (in parallel using universal states) verifies the guess and continues the computation. It verifies the guess by following the path p from h to the current gate, where at each step j along the path it guesses and verifies the j-th gate. The verification can be done in space $\mathcal{O}(\log n)$ and time $\mathcal{O}(m \log n)$, where m is the length of p.

The original computation continues by, depending on whether g is $\vee$ or $\wedge$, either using existential or universal states to determine the next node in the path (there are two choices: L or R for the two possible inputs to g). The path ends when an input gate (at level 0) is reached, and the computation accepts iff that input is $\top$.

The above algorithm needs to be modified, because the depth of the circuit is not $\mathcal{O}(\log n)$, so the path p becomes too long to write down in space $\mathcal{O}(\log n)$ (and the verification time might be too big). So each time the path p reaches length $\log n$ the machine guesses the name (i.e. the pair $\langle z, y \rangle$) of g, and (in parallel using universal states) verifies the guess as explained above. Then p is set to the empty string, and the head h of the path is set to the new gate g.

The space required is $\mathcal{O}(\log n)$ because during the main part of the computation M need only remember the origin h of the current path segment, and a path $p \in \{L, R\}^*$ of length at most $\log n$.

The computation time of M is $\mathcal{O}((\log n)^k)$ because the depth of the circuit C is $\mathcal{O}((\log n)^k)$, and the main part of each computation path of M takes on the average constant time per gate. Each step in which the path head h is reset to a new value requires time $\log n$, but these expensive steps represent only a fraction $1/\mathcal{O}(\log n)$ of the total, so the total time spent by them is still $\mathcal{O}((\log n)^k)$.

Note that the verification algorithms are done in parallel with the main computation path, because they begin with a universal state which forks between the main path and the verification. Hence the total computation time of the algorithm is the maximum of the times of the main algorithm and each of its verification computations. Each verification takes time at most $\mathcal{O}(m \log n) = \mathcal{O}((\log n)^2)$. $\qquad\square$

Note that we do not know whether $Lmcv_{1,2}$ is in NC^1 (or even nonuniform NC^1).

DEFINITION IX.5.29 (VAC^k, VNC^k and VNC). For $k \geq 1$, the theory VAC^k has vocabulary $\mathcal{L}_A^2$ and is axiomatized by V^0 and the axiom

$$\exists Y \leq \langle |n|^k + 1, n \rangle \, \delta_{LMCV}(n, |n|^k, E, G, I, Y).$$

For $k \geq 2$, VNC^k has vocabulary $\mathcal{L}_A^2$ and is axiomatized by V^0 and the axiom

$$(Fanin2(n, |n|^k, E) \supset \exists Y \leq \langle |n|^k + 1, n \rangle \, \delta_{LMCV}(n, |n|^k, E, G, I, Y)).$$

Also,

$$VNC = \bigcup_{k=1}^{\infty} VNC^k.$$

It is straightforward to show that the aggregate functions $Lmcv_k^\star$ for $k \geq 1$ (resp. $Lmcv_{k,2}^\star$, for $k \geq 2$) is Σ_1^B-definable in VAC^k (resp. VNC^k, for $k \geq 2$). Details are left as an exercise.

EXERCISE IX.5.30. Show that for $k \geq 1$ $Lmcv_k^\star$ and $Lmcv_{k+1,2}^\star$ are respectively Σ_1^B-definable in VAC^k and VNC^{k+1}.

The next result follows from the general development in Section IX.2.

COROLLARY IX.5.31. *For* $k \geq 1$:

(a) *The* Σ_1^B-*definable functions of* VAC^k (*resp.* VNC^{k+1}) *are precisely the functions in* FAC^k (*resp.* FNC^{k+1}).

(b) *The* Δ_1^B-*definable functions of* VAC^k (*resp.* VNC^{k+1}) *are precisely the relations in* AC^k (*resp.* NC^{k+1}).

COROLLARY IX.5.32. (a) *A function is in* FNC *iff it is* Σ_1^B-*definable in* VNC *iff it is* Σ_1^B-*definable in* VAC^k *for some* $k \geq 0$.

(b) *A relation is in* NC *iff it is* Δ_1^B-*definable in* VNC *iff it is* Δ_1^B-*definable in* VAC^k *for some* $k \geq 0$.

Now we define U^1, another theory that characterizes NC. Let $|x|$ denote the length of the binary representation of the number x. Using the fact that the predicate BIT is Δ_0-definable (Section III.3.3) we can show that $|x|$ is an AC^0 function (see also Section VIII.8.3). Therefore, by Lemma V.6.7 if $\varphi(z)$ is a Φ formula, where Φ is Σ_i^B or Π_i^B for some $i \geq 0$, then $\varphi(|z|)$ is translates to an equivalent to a Φ formula. Below we will use $\varphi(|z|)$ to denote this translation. (In particular, $\varphi(|z|)$ will be an $\mathcal{L}_A^2$ formula.)

DEFINITION IX.5.33 (Length induction – two-sorted case). For a set Φ of formulas, Φ-*LIND* (length induction for Φ) is the set

$$[\varphi(0) \wedge \forall x, \ \varphi(x) \supset \varphi(x+1)] \supset \forall z \varphi(|z|) \qquad (291)$$

where $\varphi(x)$ is a formula in Φ that may contain variables other than x.

(Note that the single-sorted length induction axiom schemes given in Definition VIII.8.2 (also denoted by *LIND*) roughly corresponds, i.e., via RSUV isomorphism, to our number induction schemes *IND*.)

DEFINITION IX.5.34 (U^1). The theory U^1 has vocabulary $\mathcal{L}_A^2$ and is axiomatized by the axioms of V^0 together with Σ_1^B-*LIND*.

The following theorem follows from results in [35, 108].

THEOREM IX.5.35. *A function is in* FNC *iff it is* Σ_1^B-*definable in* U^1. *A relation is in* NC *iff it is* Δ_1^B-*definable in* U^1.

IX.6. Theories for *NL* and *L*

The class *NL* (resp. *L*) is the class of problems solvable by a nondeterministic (resp. deterministic) Turing machine in space $\mathcal{O}(\log n)$ (see Appendices A.1 and A.2). It is straightforward that $L \subseteq NL$ and both are subclasses of *P*. In fact, it can be shown that $NL \subseteq AC^1$ (see Exercise IX.6.12 below). To see that $NC^1 \subseteq L$ note that the complete problem R_{MFV} (Definition IX.5.1) can be solved in deterministic log space using depth-first search (the path from the root to a node of depth d is specified by a binary string of length d). (Formal arguments will be given in Theorem IX.6.38.) It is also easy to see that *L* is closed under AC^0 reductions, while for *NL* this follows from the important theorem of Immerman and Szelepcsényi which states that *NL* is closed under complementation.

The theory *VNL* is developed using the fact that the *st*-Connectivity (*st*-CONN) problem is AC^0-complete for *NL*. Here the problem is to decide, for a given directed graph G and two designated vertices s and t, whether there is a path from s to t in G.

Krom formulas are propositional formulas in conjunctive normal form where each clause contains at most two literals. The Krom-SAT problem, which is the problem of deciding whether a given Krom formula is satisfiable, is known to be complete for *co-NL* (and hence also for *NL*). It has been used to develop the theory V^1-*KROM* in the same style as V^1-*HORN* (Section VIII.4). We will show that V^1-*KROM* is equivalent to *VNL*.

Now consider a restricted version of the *st*-CONN problem where every vertex in G has out-degree at most one. This is called the PATH problem and it is AC^0-many-one complete for *L*. We will use this fact to develop the triple *VL*, $\widehat{VL}$ and $\overline{VL}$ in the family of theories discussed in Section IX.2.

Finally, the bounded number recursion scheme pBNR (Section IX.3.3) can be used to characterize *FL*. Based on this we will develop a universal theory call *VLV* in the style of *VPV* and VTC^0V. Here the vocabulary of *VLV* contains symbols for every function in *FL*. Their defining axioms are based on pBNR.

This section is organized as follows. We define the theory *VNL* and its universal conservative extensions $\widehat{VNL}$ and $\overline{VNL}$ in Section IX.6.1. We define V^1-*KROM* and show that it is equivalent to *VNL* in Section IX.6.2. In Section IX.6.3 we define *VL*, $\widehat{VL}$ and $\overline{VL}$. Finally, in Section IX.6.4 we define *VLV*.

IX.6.1. The Theories *VNL*, $\widehat{VNL}$, and $\overline{VNL}$. Our theories for *NL* are based on the fact that the *st*-CONN problem is complete for *NL*. We encode a directed graph G by a pair (a, E) as follows:

- a is the number of vertices in G, and the vertices of G are numbered $0, \ldots, (a-1)$, and

- for $x, y < a$, $E(x, y)$ holds if and only if there is a directed edge from x to y in G.

Our designated "source" s is always the vertex 0. Consider the algorithm that solves the st-CONN problem by inductively computing all vertices in G that have distance from s at most $0, 1, \ldots, (a - 1)$. The formula $\delta_{CONN}(a, E, Y)$ below states that $Y^{[z]}$ is the set of all vertices with distance at most z from 0 (recall that $x \in Y^{[z]} \leftrightarrow Y(z, x)$):

$$\delta_{CONN}(a, E, Y) \equiv ROW(a, Y) \wedge Y(0, 0) \wedge$$
$$\forall x < |Y|(x \neq 0 \supset \neg Y(0, x)) \wedge \forall z < a \dotminus 1 \, \forall x < a + |Y|\big(Y(z+1, x) \leftrightarrow$$
$$x < a \wedge (Y(z, x) \vee \exists y < a(Y(z, y) \wedge E(y, x))))\big) \quad (292)$$

where

$$ROW(a, Y) \equiv \forall u < |Y|(Y(u) \supset \exists i < a \exists j < |Y| \, u = \langle i, j \rangle). \quad (293)$$

Here $ROW(a, Y)$ asserts that Y is uniquely determined by its rows $Y^{[0]}$, $\ldots, Y^{[a-1]}$.

EXERCISE IX.6.1. Show that V^0 proves that $\delta_{CONN}(a, E, Y)$ uniquely determines Y for every pair a, E. That is show

$$V^0 \vdash \big(\delta_{CONN}(a, E, Y) \wedge \delta_{CONN}(a, E, Y')\big) \supset Y = Y'.$$

We define the relation R_{CONN} below by assigning the "target" vertex t number 1.

DEFINITION IX.6.2.
$$R_{CONN}(a, E) \leftrightarrow \exists Y \leq \langle a, a \rangle (\delta_{CONN}(a, E, Y) \wedge Y(a, 1)).$$

THEOREM IX.6.3. *The relation R_{CONN} is in **NL**, and for every relation $R(\vec{x}, \vec{X})$ in **NL** there are **AC**^{0} functions a_0, E_0 such that*

$$R(\vec{x}, \vec{X}) \leftrightarrow R_{CONN}(a_0(\vec{x}, \vec{X}), E_0(\vec{x}, \vec{X})).$$

PROOF SKETCH. The fact that R_{CONN} is in **NL** is straightforward: on input (a, E) the nondeterministic Turing machine guesses a path from 0 to 1 by enumerating the edges on the path.

Now let $R(\vec{x}, \vec{X})$ be a relation in **NL**, so R is accepted by a nondeterministic Turing machine M that works in logspace. Suppose without loss of generality that M has a unique accepting configuration. The configurations of M (without the input tape content) can be encoded by numbers less than $t(\vec{x}, \vec{X})$ (for some number term t bounding the running time of M) such that 0 is the initial configuration and 1 is the only accepting configuration (see also Exercise VI.2.9). Consider the directed graph G with vertices the numbers less than $t(\vec{x}, \vec{X})$ such that there is an edge from z_1 to z_2 iff z_2 is a next configuration of z_1. Then M accepts $(\vec{x}, \vec{X})$ iff there is a path from 0 to 1 in G.

The fact that z_1 encodes a next configuration of z_2 can be expressed by a Σ_0^B formula $\varphi(z_1, z_2, \vec{x}, \vec{X})$. Thus there is an AC^0 string function $E_0(\vec{x}, \vec{X})$ whose value is the adjacency matrix of the graph whose nodes are configurations and whose edges indicate one-step transitions. The bit-graph of E_0 satisfies, for $z_1, z_2 < t(\vec{x}, \vec{X})$,

$$E_0(\vec{z}, \vec{X})(z_1, z_2) \leftrightarrow \varphi(z_1, z_2, \vec{x}, \vec{X}).$$

Consequently M accepts $(\vec{x}, \vec{X})$ iff $R_{CONN}(t(\vec{x}, \vec{X}), E_0(\vec{x}, \vec{X}))$ holds. □

Recall the definition of the function class **FNL** (Definition V.2.3) associated with **NL**. Prior to the Immerman–Szelepcsényi Theorem stating that **NL** is closed under complementation, it was not known that **FNL** is closed under composition. In fact it was not known that the characteristic function $C(a, E)$ of $R_{CONN}(a, E)$ is in **FNL**, because in order to verify that $C(a, E) = 0$ a nondeterministic log space Turing machine would have to verify that there is no path from node 0 to node 1 in the graph specified by E. However knowing that **NL** = *co*-**NL** it is easy to show that **FNL** is closed under composition. For example, to verify that bit i of $F(G(X))$ is 0 a nondeterministic log space Turing machine M simulates the machine for F on input $G(X)$, where each time that machine needs a bit j of $G(X)$ M guesses whether the bit is 1 or 0, and in either case can verify the guess by simulating the machine for G or the complement machine.

THEOREM IX.6.4. **FNL** *is closed under* AC^0*-reductions.*

PROOF SKETCH. By Theorem IX.1.7 it suffices to show that **FNL** is closed under composition and string comprehension. We argued the case for composition above. For string comprehension, to determine whether $i = f(y)$ for some $y < b$ a log space machine guesses the answer and verifies it. If the guess is YES, the verification is a nondeterministic log space computation (because b is small and can be written in binary on the work tape), and if the answer is NO the machine verifies it using the complementary machine. □

DEFINITION IX.6.5 (**VNL**). The theory **VNL** has vocabulary $\mathcal{L}_A^2$ and is axiomatized by the axioms of V^0 together with the axiom $CONN$, where

$$CONN \equiv \exists Y \leq \langle a, a \rangle + 1\, \delta_{CONN}(a, E, Y).$$

The function $REACH$ has defining axiom

$$REACH(a, E) = Y \leftrightarrow \delta_{CONN}(a, E, Y). \tag{294}$$

The next result is immediate from Theorems IX.6.3 and IX.6.4.

COROLLARY IX.6.6. *The function REACH is complete for* **FNL** *under* AC^0*-reductions.*

LEMMA IX.6.7. *Both REACH and its aggregate function $REACH^\star$ are* Σ_1^B*-definable in* **VNL**, *and* **VNL**(*Row, REACH, REACH**) *proves* (166):

$$\forall i < b,\ REACH^\star(b, X, E)^{[i]} = REACH((X)^i, E^{[i]}).$$

Proof. That *REACH* is definable is immediate from the axiom *CONN*
for **VNL** and Exercise IX.6.1. That *REACH** is definable is a simple
exercise in coding the disjoint union of a sequence of graphs as a single
graph with a common initial vertex 0.   □

Exercise IX.6.8. Give details for *REACH** in the above Lemma.

Following the method of Section IX.2 we now define the theories $\widehat{VNL}$
and $\overline{VNL}$. Let $\delta'_{CONN}(a, E, Y)$ be the quantifier-free $\mathcal{L}_{FAC^0}$-formula that
is equivalent to $\delta_{CONN}(a, E, Y)$ over $\overline{V}^0$ (see Lemma V.6.3). Let *REACH'*
be defined by

$$REACH'(a, E) = Y \leftrightarrow \delta'_{CONN}(a, E, Y). \qquad (295)$$

Then *REACH* and *REACH'* are semantically equal functions but have
different defining axioms.

Definition IX.6.9 ($\widehat{VNL}$). Let $\mathcal{L}_{\widehat{VNL}} = \mathcal{L}_{FAC^0} \cup \{REACH'\}$. $\widehat{VNL}$ is
the theory with vocabulary $\mathcal{L}_{\widehat{VNL}}$ and is axiomatized by the axioms of $\overline{V}^0$
together with (295).

Definition IX.6.10 ($\overline{VNL}$). The vocabulary $\mathcal{L}_{FNL}$ is the smallest set
that contains $\mathcal{L}_{\widehat{VNL}}$ such that for every $\mathcal{L}_A^2$-term t and every quantifier-free
$\mathcal{L}_{FNL}$-formula φ the function $F_{\varphi(z),t}$ with defining axiom (86):

$$F_{\varphi(z),t}(\vec{x}, \vec{X})(z) \leftrightarrow z < t(\vec{x}, \vec{X}) \wedge \varphi(z, \vec{x}, \vec{X}) \qquad (296)$$

is in $\mathcal{L}_{FNL}$.
    The theory $\overline{VNL}$ has vocabulary $\mathcal{L}_{FNL}$ and is axiomatized by the axioms
of $\widehat{VNL}$ and (296) for each function $F_{\varphi(z),t}$.

The Definability Theorems for our theories here follow from our discussion in Section IX.2.

Corollary IX.6.11. *Here either $\mathcal{L}$ is $\mathcal{L}_{\widehat{VNL}}$ and $\mathcal{T}$ is $\widehat{VNL}$, or $\mathcal{L}$ is $\mathcal{L}_{FNL}$
and $\mathcal{T}$ is $\overline{VNL}$.*

(a) *A function is in **FNL** iff it is represented by a term in $\mathcal{L}_{\widehat{VNL}}$ (and for a
    string function) iff it is represented by a function symbol in $\mathcal{L}_{FNL}$. A
    relation is in **NL** iff it is represented by an open (or a Σ_0^B) formula of
    $\mathcal{L}$.*
(b) *For every $\Sigma_1^B(\mathcal{L})$ formula φ^+ there is a $\Sigma_1^B(\mathcal{L}_A^2)$ formula φ so that
    $\mathcal{T} \vdash \varphi^+ \leftrightarrow \varphi$.*
(c) *$\mathcal{T}$ proves $\Sigma_0^B(\mathcal{L})$-**COMP**, $\Sigma_0^B(\mathcal{L})$-**IND**, and $\Sigma_0^B(\mathcal{L})$-**MIN**.*
(d) *$\overline{VNL}$ is a universal conservative extension of $\widehat{VNL}$ which is in turn a
    universal conservative extension of **VNL**.*
(e) *The Σ_1^B-definable functions of **VNL** (or $\mathcal{T}$) are precisely the functions
    in **FNL**.*

(f) *The Δ_1^B-definable relations of* ***VNL*** *(or $\mathcal{T}$) are precisely the relations in* ***NL***.

Exercise IX.6.12. Recall the theory ***VAC***^{1} from Section IX.5.6. Show that ***VNL*** $\subseteq$ ***VAC***^{1}. The idea is that the transitive closure of a graph on n vertices can be computed by squaring the adjacency matrix of the graph $\log n$ times, and this can be computed by a circuit of depth $\mathcal{O}(\log n)$ with unbounded fanin $\wedge$ and $\vee$ gates.

IX.6.2. The Theory V^1-*KROM*. A Krom formula is a propositional formula in conjunctive normal form where each clause contains at most two literals. The Satisfiability Problem for Krom formulas, Krom-SAT, is complete for *co-NL* (or equivalently *NL*, by the Immerman–Szelepcsényi Theorem). In descriptive complexity theory Grädel's Theorem states that *NL* is the class of finite models of the second-order Krom formulas [53]. This idea was used in [44] to develop the theory V^1-***KROM***.

We start by defining Σ_1^1-***Krom*** formulas, which are Σ_1^1 and resemble propositional Krom formulas. In Theorem IX.6.18 we show that Σ_1^1-***Krom*** formulas represent precisely the *co-NL* relations.

Definition IX.6.13 (Σ_1^1-***Krom*** Formula). A $\Sigma_1^1(\mathcal{L}_A^2)$-formula $\psi(\vec{x}, \vec{X})$ is called a Σ_1^1-***Krom*** formula if it is of the form:

$$\exists P_1 \ldots \exists P_k \forall z_1 \leq t_1(\vec{x}, \vec{X}) \ldots \forall z_m \leq t_m(\vec{x}, \vec{X}) \varphi(\vec{z}, \vec{x}, \vec{X}, \vec{P}) \quad (297)$$

where t_i are $\mathcal{L}_A^2$-terms and $\varphi(\vec{z}, \vec{x}, \vec{X}, \vec{P})$ is a quantifier-free formula in conjunctive normal form. Each clause contains at most two literals of the form $P_j(s(\vec{z}, \vec{x}, \vec{X}))$ or $\neg P_j(s(\vec{z}, \vec{x}, \vec{X}))$ for some number term s, but may contain any number of literals of the form (possibly negated) $X_i(t)$, $t_1 \leq t_2$, and $t_1 =_1 t_2$. No occurrence of $=_2$ or a term of the form $|P_j|$ is allowed.

Notice that $\Sigma_0^B \not\subseteq \Sigma_1^1$-***Krom***. However Corollary IX.6.16 below shows each Σ_0^B-formula is equivalent in the theory V^0 to a Σ_1^1-***Krom*** formula.

Example IX.6.14 (Transitive Closure in Graphs). Suppose that a graph G is coded by (a, E) as before (page 339). The formula $ContainTC(a, E, P)$ below states that P contains the transitive closure of G, i.e., if there is a path from x to y in G, then $P(x, y)$ holds:

$$ContainTC(a, E, P) \equiv \forall x < a \forall y < a \forall z < a,$$
$$(E(x, y) \supset P(x, y)) \wedge (P(x, y) \wedge E(y, z) \supset P(x, z)).$$

The following Σ_1^1-***Krom*** formula states that there is *no* path from x_1 to x_2 in G:

$$\varphi_{\neg Reach}(x_1, x_2, a, E) \equiv \exists P(ContainTC(a, E, P) \wedge \neg P(x_1, x_2)). \quad (298)$$

The set Y that satisfies comprehension for $\varphi_{\neg Reach}$:

$$|Y| \leq a \wedge \forall y < a(Y(y) \leftrightarrow \varphi_{\neg Reach}(x, y, a, E))$$

is the set of all vertices that are *not* reachable from vertex x.

The formula φ in (297) is a quantifier-free formula. In some cases it is convenient to allow the non-P_i part of φ to be a Σ_0^B formula. The next lemma shows that this is possible.

LEMMA IX.6.15. *Suppose that $\psi(\vec{x}, \vec{X})$ is a Σ_1^1 formula*

$$\exists \vec{P} \forall \vec{z} \leq \vec{t} \bigwedge_i \varphi_i(\vec{z}, \vec{x}, \vec{X}, \vec{P}) \tag{299}$$

where each formula φ_i is a disjunction of the form

$$\ell \vee \ell' \vee \rho_i(\vec{z}, \vec{x}, \vec{X})$$

*where ℓ, ℓ' are literals (possibly omitted) of the form $P_j(\vec{s})$ or $\neg P_j(\vec{s})$ (for some number terms $\vec{s}$ not containing any of $\vec{P}$) and ρ_i is a Σ_0^B formula that does not contain any of $\vec{P}$. Then ψ is equivalent in V^0 to a Σ_1^1-**Krom** formula.*

The special case in which $\vec{P}$ and $\vec{z}$ are missing yields:

COROLLARY IX.6.16. *Every Σ_0^B-formula is equivalent in V^0 to a Σ_1^1-**Krom** formula.*

PROOF OF LEMMA IX.6.15. We prove the lemma by structural induction on the formulas ρ_i. Assume w.l.o.g. that they are in prenex form. The base case (all ρ_i are quantifier-free) is obvious. Consider the induction step. First suppose that for some i the formula ρ_i has the form

$$\forall u \leq t \rho_i'(u, \vec{z})$$

where we have suppressed the variables $\vec{x}, \vec{X}$. Let $\varphi'(u, \vec{z}, \vec{P})$ be obtained from $\varphi_i(\vec{z}, \vec{P})$ by replacing ρ_i by ρ_i'. Then

$$\psi \leftrightarrow \exists \vec{P} \forall \vec{z} \leq \vec{t} \forall u \leq t \left(\varphi_i'(u, \vec{z}, \vec{P}) \wedge \bigwedge_{j \neq i} \varphi_j(\vec{z}, \vec{P}) \right).$$

Now consider the case where ρ_i begins with $\exists u \leq t$. Suppose w.l.o.g. that we can write ϕ_i in the form

$$\varphi_i \equiv (P_1(\vec{s}) \wedge \forall u \leq t \rho_i'(u, \vec{z})) \supset P_2(\vec{r}).$$

We introduce a new variable Q and force $Q(v, \vec{z})$ to be true if $P_1(\vec{s}) \wedge \forall u \leq v \rho_i'(u, \vec{z})$ holds. Define φ' by

$$\varphi'(u, \vec{z}, Q, P_1, P_2) \equiv (P_1(\vec{s}) \wedge \rho_i'(0, \vec{z}) \supset Q(0, \vec{z})) \wedge$$
$$(Q(u, \vec{z}) \wedge \rho_i'(u+1, \vec{z}) \supset Q(u+1, \vec{z})) \wedge (Q(t, \vec{z}) \supset P_2(\vec{r})).$$

Note that involving $P_1(\vec{s})$ in the definition of Q allows the last clause in φ' to have only two occurrences of literals from $Q, \vec{P}$. It is straightforward to prove in V^0 that

$$\psi \leftrightarrow \exists \vec{P} \exists Q \forall \vec{z} \leq \vec{t} \forall u \leq t \big(\varphi_i'(u, \vec{z}, Q, P_1, P_2) \wedge \bigwedge_{j \neq i} \varphi_j(\vec{z}, \vec{P})\big).$$

$\square$

COROLLARY IX.6.17. *Suppose that ψ is a formula of the form* (299) *with the same restrictions as in the Lemma except now ρ_i is a $\Sigma_0^B(\mathcal{L}_{FAC^0})$-formula instead of a Σ_0^B-formula and the number terms $\vec{s}$ in $P_i(\vec{s})$ are $\mathcal{L}_{FAC^0}$-terms instead of $\mathcal{L}_A^2$-terms. Then ψ is equivalent in $\overline{V}^0$ to a Σ_1^1-Krom formula.*

PROOF. Replace each occurrence of the form $P_i(\vec{s_i})$ in ψ by $P(\vec{u_i}) \vee \vec{u_i} \neq \vec{s_i}$, for new variables $\vec{u_i}$, and add the bounded quantifiers $\forall \vec{u_i} \leq \vec{t}'$ to the prefix $\forall \vec{z} \leq \vec{t}$ in (299) for suitable bounding terms $\vec{t}'$. Similarly for occurrences $\neg P_i(\vec{s_i})$. Now use Lemma V.6.3 to replace each $\Sigma_0^B(\mathcal{L}_{FAC^0})$-formula by an equivalent Σ_0^B-formula. The Corollary now follows from Lemma IX.6.15.

$\square$

The next result is essentially due to Grädel [53].

THEOREM IX.6.18 (Σ_1^1-*Krom* Representation). *A relation is represented by a Σ_1^1-Krom formula if and only if it is in co-NL.*

PROOF. First we prove the ONLY IF direction. Let $R(\vec{x}, \vec{X})$ be a relation represented by the Σ_1^1-*Krom* formula (297):

$$\exists P_1 \ldots \exists P_k \forall z_1 \leq t_1(\vec{x}, \vec{X}) \ldots \forall z_m \leq t_m(\vec{x}, \vec{X}) \varphi(\vec{z}, \vec{x}, \vec{X}, \vec{P}).$$

For a given input $(\vec{x}, \vec{X})$, let v_i be the value of t_i (for $1 \leq i \leq m$). Now for each $(z_1, z_2, \ldots, z_m)$ where

$$0 \leq z_i \leq v_i \qquad (\text{for } 1 \leq i \leq m)$$

we treat the atoms of the form $P_j(s(\vec{z}, \vec{x}, \vec{X}))$ as propositional variables. Since all terms and other variables in φ can be evaluated, $\varphi(\vec{z}, \vec{x}, \vec{X}, \vec{P})$ can be made into a Krom formula $A_{z_1, \ldots, z_m}$ whose variables are of the form $P_j(s(\vec{z}, \vec{x}, \vec{X}))$. Semantically,

$$\forall z_1 \leq t_1(\vec{x}, \vec{X}) \ldots \forall z_m \leq t_m(\vec{x}, \vec{X}) \varphi(\vec{z}, \vec{x}, \vec{X}, \vec{P})$$

is equivalent to the Krom formula

$$\bigwedge_{z_1=0}^{v_1} \cdots \bigwedge_{z_m=0}^{v_m} A_{z_1, \ldots, z_m}. \tag{300}$$

Therefore $(\vec{x}, \vec{X}) \in R$ iff (300) is satisfiable.

Notice that (300) can be obtained from the formula (297) and $(\vec{x}, \vec{X})$ in deterministic logspace (in fact, AC^0). So the fact that R is in *co-NL* follows from the following result.

LEMMA IX.6.19. *The Satisfiability problem for Krom formulas is in co-NL.*

PROOF. We associate with each Krom formula K a graph G_K whose nodes are the set of literals ℓ such that ℓ or $\bar{\ell}$ occurs in K, and whose edges are pairs $(\bar{\ell_1}, \ell_2)$ such that one of the clauses $(\ell_1 \vee \ell_2)$, $(\ell_2 \vee \ell_1)$ occurs in K. The Lemma follows from the following exercise.   □

EXERCISE IX.6.20. Show that K is unsatisfiable iff the graph G_K has a directed cycle containing both ℓ and $\bar{\ell}$ for some literal ℓ.

Now we prove the IF direction. Suppose that $R(\vec{x}, \vec{X})$ is a *co-NL* relation, we show that R can be represented by a Σ_1^1-*Krom* formula. By Proposition IX.6.3 there are AC^0 functions $a_0(\vec{x}, \vec{X})$ and $E_0(\vec{x}, \vec{X})$ so that

$$R(\vec{x}, \vec{X}) \leftrightarrow \neg R_{CONN}(a_0(\vec{x}, \vec{X}), E_0(\vec{x}, \vec{X}))$$

i.e., $(\vec{x}, \vec{X}) \in R$ iff 1 is not reachable from 0 in the graph $(a_0(\vec{x}, \vec{X}), E_0(\vec{x}, \vec{X}))$. Thus by Example IX.6.14,

$$R(\vec{x}, \vec{X}) \leftrightarrow \varphi_{\neg Reach}(0, 1, a_0(\vec{x}, \vec{X}), E_0(\vec{x}, \vec{X})).$$

By Corollary IX.6.17 $\varphi_{\neg Reach}(0, 1, a_0(\vec{x}, \vec{X}), E_0(\vec{x}, \vec{X}))$ is equivalent in V^0 to a Σ_1^1-*Krom* formula.   □

DEFINITION IX.6.21. The theory V^1-*KROM* has vocabulary $\mathcal{L}_A^2$ and is axiomatized by 2-*BASIC* (Figure 2 on page 96) and the comprehension axiom scheme for all Σ_1^1-*Krom* formulas.

Although $\Sigma_0^B \not\subseteq \Sigma_1^1$-*Krom*, we will show that V^1-*KROM* extends V^0:

LEMMA IX.6.22. $V^0 \subseteq V^1$-*KROM*.

First we prove:

LEMMA IX.6.23. V^1-*KROM* *proves the multiple comprehension axioms* (*see Lemma V.4.25*) *for quantifier-free formulas.*

PROOF. We have to show that V^1-*KROM* proves

$$\exists X \leq \langle y_1, \ldots, y_k \rangle \forall z_1 < y_1 \ldots \forall z_k < y_k (X(z_1, \ldots, z_k) \leftrightarrow \varphi(z_1, \ldots, z_k)) \tag{301}$$

for any quantifier-free formula φ.

A first attempt to prove this lemma might be to show that

$$V^1\text{-}KROM \vdash \exists X \leq \langle \vec{y} \rangle \forall x < \langle \vec{y} \rangle (X(x) \leftrightarrow \exists \vec{z} < \vec{y}(x = \langle \vec{z} \rangle \wedge \varphi(\vec{z}))).$$

However, $\exists \vec{z} < \vec{y}(x = \langle \vec{z} \rangle \wedge \varphi(\vec{z}))$ is not a Σ_1^1-*Krom* formula.

Here we prove (301) using Σ_1^1-*Krom-COMP* as follows. Let X satisfy:

$$\exists X \leq \langle \vec{y} \rangle \forall x < \langle \vec{y} \rangle (X(x) \leftrightarrow \exists P \forall \vec{z} < \langle \vec{y} \rangle ((P(\langle \vec{z} \rangle) \leftrightarrow \varphi(\vec{z})) \wedge P(x))).$$

It is straightforward to verify that such X also satisfies (301).   □

PROOF OF LEMMA IX.6.22. We prove the lemma by showing that V^1-*KROM* proves the multiple comprehension axiom for any Σ_0^B formula φ. The proof is by structural induction on φ. Assume without loss of generality that φ is in prenex form.

The base case, where φ is a quantifier-free formula, follows from the lemma above because a quantifier-free formula is also in Σ_1^1-***Krom***.

For the induction step, suppose that we need to prove

$$V^1\text{-}\textbf{\textit{KROM}} \vdash \exists X \le \langle \vec{a} \rangle \forall \vec{x} < \vec{a},\ X(\vec{x}) \leftrightarrow \varphi(\vec{x}). \tag{302}$$

First consider the case where $\varphi(\vec{x}) \equiv \forall z < a\ \psi(\vec{x}, z)$. By the induction hypothesis for ψ,

$$V^1\text{-}\textbf{\textit{KROM}} \vdash \exists X' \le \langle \vec{a}, a \rangle \forall \vec{x} < \vec{a} \forall z < a,\ X'(\vec{x}, z) \leftrightarrow \psi(\vec{x}, z).$$

Now we can apply the multiple comprehension axiom for the Σ_1^1-***Krom*** formula $\forall z < a\ X'(\vec{x}, z)$:

$$V^1\text{-}\textbf{\textit{KROM}} \vdash \exists X \le \langle \vec{a} \rangle \forall \vec{x} < \vec{a},\ X(\vec{x}) \leftrightarrow \forall z < a\ X'(\vec{x}, z).$$

Such X satisfies (302).

Finally suppose that $\varphi(\vec{x}) \equiv \exists z < a\ \psi(\vec{x}, z)$. Let $\psi'(\vec{x}, z)$ be the prenex formula equivalent to $\neg\psi(\vec{x}, z)$ obtained by pushing the $\neg$ connective through the block of quantifiers using De Morgan's laws. By the previous case

$$V^1\text{-}\textbf{\textit{KROM}} \vdash \exists X' \le \langle \vec{a} \rangle \forall \vec{x} < \vec{a},\ X'(\vec{x}) \leftrightarrow \forall z < a\ \psi'(\vec{x}, z).$$

Let X be such that

$$|X| \le \langle \vec{a} \rangle \wedge \forall \vec{x} < \vec{a},\ X(\vec{x}) \leftrightarrow \neg X'(\vec{x}).$$

Then X satisfies (302). $\square$

Now we prove the main result of this section. The proof ends with Exercise IX.6.27 on page 350.

THEOREM IX.6.24. $V^1\text{-}\textbf{\textit{KROM}} = \textbf{\textit{VNL}}$.

PROOF. First we show that $\textbf{\textit{VNL}} \subseteq V^1\text{-}\textbf{\textit{KROM}}$. By Lemma IX.6.22 above, $V^1\text{-}\textbf{\textit{KROM}}$ is an extension of V^0. It remains to show that $V^1\text{-}\textbf{\textit{KROM}}$ proves the axiom *CONN* (Definition IX.6.5).

The fact that $V^1\text{-}\textbf{\textit{KROM}}$ extends V^0 also gives us:

CLAIM. $V^1\text{-}\textbf{\textit{KROM}}$ proves the multiple comprehension axiom scheme (Lemma V.4.25) for Σ_1^1-***Krom*** formulas. For each Σ_1^1-***Krom*** formula φ, $V^1\text{-}\textbf{\textit{KROM}}$ proves the comprehension for $\neg\varphi$.

Recall that in the formula $\delta_{CONN}(a, E, Y)$ in (292), $Y(z, x)$ holds iff in the graph G coded by (a, E) there is a path from 0 to x of length $\le z$. The Σ_1^1-***Krom*** formula $\varphi_{\neg Dist}(x_1, x_2, z, a, E)$ below states that there is *no* path from x_1 to x_2 in G of length $\le z$. The string variable P codes a superset of the "connectivity to x_1" relation, i.e.,

if there is a path from x_1 to y of length $\le u$ then $P(u, y)$ holds.

$(P(u, y)$ might hold even if there is no $x_1 y$ path of length $\leq u$.)

$$\varphi_{\neg Dist}(x_1, x_2, z, a, E) \equiv \exists P \forall u < z \forall x < a \forall y < a$$
$$\neg P(z, x_2) \wedge P(0, x_1) \wedge (P(u, x) \wedge E(x, y) \supset P(u + 1, y)).$$

By the claim above, V^1-**KROM** proves the existence of Y such that

$$\forall z < a \forall x < a, \ Y(z, x) \leftrightarrow \neg \varphi_{\neg Dist}(0, x, z, a, E).$$

In other words, $Y(z, x)$ holds iff the distance from 0 to x is at most z, i.e., Y satisfies $\delta_{CONN}(a, E, Y)$ (292). The formal argument is left as an exercise.

EXERCISE IX.6.25. Show that

$$V^1\text{-}\boldsymbol{KROM} \vdash \delta_{CONN}(a, E, Y).$$

Now we show that V^1-**KROM** $\subseteq$ **VNL**. Let

$$\psi(y, \vec{x}, \vec{X}) \equiv \exists \vec{P} \forall \vec{z} \leq \vec{t} \varphi(y, \vec{z}, \vec{x}, \vec{X}, \vec{P})$$

be a Σ_1^1-**Krom** formula. We need to show that the comprehension axiom for ψ is provable in **VNL**:

$$\boldsymbol{VNL} \vdash \exists Y \leq b \forall y < b, \ Y(y) \leftrightarrow \exists \vec{P} \forall \vec{z} \leq \vec{t} \varphi(y, \vec{z}, \vec{x}, \vec{X}, \vec{P}). \quad (303)$$

The idea is to formalize in **VNL** the ONLY IF direction in the proof of Theorem IX.6.18. For a fixed set of values for $\vec{x}, \vec{X}$, for each value of $y < b$ we consider the propositional formula (300):

$$\psi_y \equiv \bigwedge_{z_1=0}^{v_1} \cdots \bigwedge_{z_m=0}^{v_m} A_{z_1,\ldots,z_m}. \quad (304)$$

As in the proof of Theorem IX.6.18, $\neg \exists \vec{P} \forall \vec{z} \leq \vec{t} \varphi(y, \vec{z}, \vec{x}, \vec{X}, \vec{P})$ holds iff ψ_y is unsatisfiable. Let G_y be the graph G_K defined in the proof of Lemma IX.6.19, where K is now ψ_y. Thus the vertices of G_y are the literals of ψ_y, and there is an edge from ℓ_1 to ℓ_2 in G_y iff the clause

$$\ell_1 \supset \ell_2$$

is in ψ_y. Note that if the edge (ℓ_1, ℓ_2) is in G_y then so is the edge $(\neg \ell_2, \neg \ell_1)$.

By Exercise IX.6.20 ψ_y is unsatisfiable iff the graph G_y has a directed cycle containing both ℓ and $\neg \ell$ for some literal ℓ, i.e. iff ψ_y contains a set of clauses of the form

$$\ell_0 \supset \ell_1, \ell_1 \supset \ell_2, \ \ldots, \ \ell_k \supset \neg \ell_0, \ \neg \ell_0 \supset \ell_1', \ \ell_1' \supset \ell_2', \ \ldots, \ \ell_n' \supset \ell_0. \quad (305)$$

Here we need to formalize the proof of this exercise in **VNL**.

The encoding of G_y by a pair $(a(y), E^{[y]})$ can be described by a Σ_0^B formula and we omit the details here. It is important that we can check simultaneously in each G_y whether there is a path from any vertex u to a vertex v. For this we use the fact that the aggregate function $REACH^*$ is definable in **VNL** (Lemma IX.6.7).

The fact (303) follows from the next lemma:

LEMMA IX.6.26. *VNL proves that* $\neg \exists \vec{P} \forall \vec{z} \leq \vec{t} \varphi(y, \vec{z}, \vec{x}, \vec{X}, \vec{P})$ *is equivalent to the statement that* G_y *contains a path from* p *to* $\neg p$ *and a path from* $\neg p$ *to* p *for some propositional variable* p *of* ψ_y.

It remains to prove the lemma. Argue in *VNL*: the ($\Longleftarrow$) direction is straightforward, so consider the ($\Longrightarrow$) direction. We prove the contrapositive. Suppose that G_y does not contain simultaneously a path from p to $\neg p$ and a path from $\neg p$ to p for any propositional variable p, we will define a set of values for the string variables $\vec{P}$ that satisfies

$$\forall \vec{z} \leq \vec{t} \varphi(y, \vec{z}, \vec{x}, \vec{X}, \vec{P}).$$

It is easy to define such a set in polytime, however, here we need to define it in *NL*. We will give an *NL* algorithm that assigns values for the propositional variables that satisfies ψ_y. It will be clear that the values for $\vec{P}$ defined accordingly satisfy the requirement. Furthermore, it is straightforward that these arguments can be formalized in *VNL*.

The algorithm works as follows. First, identify all literals ℓ such that there is a path from $\neg \ell$ to ℓ in G_y. Assign $\top$ to all such ℓ and all other literals that are reachable from them. (These are the literals that are forced to be true.) Note that by the hypothesis, no variable gets conflicting truth value.

Now suppose that

$$p_1, p_2, \ldots, p_n$$

are the remaining variables. The main part of the algorithm is to assign truth values to these variables. Let G'_y be the induced subgraph of G_y on the literals

$$p_1, \neg p_1, p_2, \neg p_2, \ldots, p_n, \neg p_n.$$

For each literal ℓ let

$$C(\ell) = \{\ell' \ : \ \text{there is a path from } \ell \text{ to } \ell' \text{ or from } \ell' \text{ to } \ell \text{ in } G'_y\}.$$

Note that for any literal ℓ', at most one of $\ell', \neg \ell'$ is in $C(\ell)$. Also,

$$\ell \in C(\ell') \qquad \text{iff} \qquad \ell' \in C(\ell).$$

Let (see Figure 13 below)

$$C^+(\ell) = \{\ell\} \cup \{\ell' \ : \ \text{there is a path from } \ell \text{ to } \ell' \text{ in } G'_y\},$$
$$C^-(\ell) = C(\ell) - C^+(\ell).$$

Notice that if $\ell_1 \notin C(\ell_2)$ (hence $\ell_2 \notin C(\ell_1)$), then

$$C^+(\ell_1) \cap C^-(\ell_2) = C^-(\ell_1) \cap C^+(\ell_2) = \varnothing. \tag{306}$$

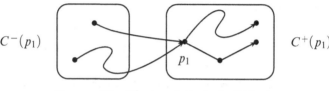

$C^-(p_1)$ $C^+(p_1)$

FIGURE 13. $C(p_1)$ and $C^-(p_1), C^+(p_1)$.

The idea is to select indices $i_1 \leq i_2 \leq \cdots \leq i_n \leq n$ (possibly with repetition) such that

for every variable p, exactly one of $\{p, \neg p\}$ is in $C = \bigcup_j C(p_{i_j})$. (307)

Then we assign $\top$ to every literal in

$$C^+ = \bigcup_j C^+(p_{i_j})$$

and $\bot$ to every literal in

$$C^- = \bigcup_j C^-(p_{i_j}).$$

The condition (307) ensures that every variable get a unique truth value.
   Notice that

$$\ell \in C(\ell') \qquad \text{iff} \qquad \neg\ell \in C(\neg\ell').$$

The indices $i_1, i_2, \ldots, i_n$ are defined (in parallel) as follows:

$$i_j = \min\{t \ : \ t \geq j \text{ and } p_t, \neg p_t \notin \bigcup_{r<j} (C(p_r) \cup C(\neg p_r))\}.$$

Observe that the sequence $i_1, i_2, \ldots, i_n$ is nondecreasing, and if $i_j < i_k$, then both p_{i_k} and $\neg p_{i_k}$ are not in $C(p_{i_j}) \cup C(\neg p_{i_j})$, so the observation (306) guarantees that no literal ℓ belongs to both $C^+(p_{i_j})$ and $C^-(p_{i_k})$, or both $C^-(p_{i_j})$ and $C^+(p_{i_k})$. As a result, no literal ℓ belongs to both C^+ and C^-. Consequently our truth assignment described above is well defined.
   For any t, the truth value of p_t is determined as follows:
   • find the smallest j such that p_t or $\neg p_t$ is in $C(p_{i_j})$;
   • if either $p_t \in C^+(p_{i_j})$ or $\neg p_t \in C^-(p_{i_j})$ then assign p_t the value $\top$, otherwise assign p_t $\bot$.

To complete the proof of Theorem IX.6.24 we need to show that the truth assignment above is correct. This is left as an exercise. □

   EXERCISE IX.6.27. Complete the argument above, i.e.,
   (a) show that $p_{i_1}, p_{i_2}, \ldots, p_{i_n}$ satisfy the condition (307), and
   (b) show that the truth assignment described above satisfies ψ_y.

IX.6.3. The Theories VL, $\widehat{VL}$, and $\overline{VL}$. Given a directed graph G whose vertices have outdegree at most one, and two vertices s, t of G, the PATH problem is to decide whether there is a path in G from s to t. (So PATH is the restriction of the st-CONN problem where the graphs have outdegree at most one.) Here we develop the theories VL, $\widehat{VL}$ and $\overline{VL}$ based on the fact that PATH is a complete problem for L.

Below, Exercise IX.6.37 shows that our definition of VL is equivalent to an earlier definition given in [113]. Then in Theorem IX.6.38 we show that VNC^1 is a subtheory of VL.

First we formalize the PATH problem. As before, our "source" s is always the vertex 0. Recall the function $seq(v, P) = (P)^v$ that encodes a sequence of numbers by P (Definition V.4.31 on page 115). Let $\delta_{PATH}(a, E, P)$ be the Σ_0^B equivalent of

$$(P)^0 = 0 \wedge \forall v < a, \, E((P)^v, (P)^{v+1}) \wedge (P)^{v+1} < a. \qquad (308)$$

Here P codes a path in G starting at 0: $(P)^v$ is the v-th vertex on the path.

The relation R_{PATH} below is AC^0-many-one complete for L. Here the designated "target" vertex t is one.

THEOREM IX.6.28. *Let*

$$R_{PATH}(a, E) \equiv (\forall x < a \exists! y < a E(x, y)) \wedge \exists P(\delta_{PATH}(a, E, P) \wedge (P)^a = 1).$$

The relation R_{PATH} is in L, and for every relation $R(\vec{x}, \vec{X})$ in L there are AC^0 functions $a_0(\vec{x}, \vec{X})$, $E_0(\vec{x}, \vec{X})$ so that

$$R(\vec{x}, \vec{X}) \leftrightarrow R_{PATH}(a_0(\vec{x}, \vec{X}), E_0(\vec{x}, \vec{X})).$$

PROOF SKETCH. The fact that R_{PATH} is in L is straightforward. The second fact can be proved as in Proposition IX.6.3 except for the vertices in the graph G now have outdegree at most one because the Turing machine is deterministic.   □

DEFINITION IX.6.29 (VL). Let $PATH$ be the axiom

$$Unique(a, E) \supset \exists P \leq \langle a, a \rangle \delta_{PATH}(a, E, P) \qquad (309)$$

where

$$Unique(a, E) \equiv a \neq 0 \wedge \forall x < a \exists! y < a E(x, y).$$

VL is the theory over $\mathcal{L}_A^2$ that is axiomatized by $PATH$ and the axioms of V^0.

Now consider the function $Path$ with the following defining axiom (see page 283 for $SEQ(a, P)$):

$$Path(a, E) = P \leftrightarrow SEQ(a, P) \wedge ((Unique(a, E) \wedge \delta_{PATH}(a, E, P)) \vee$$
$$(\neg Unique(a, E) \wedge |P| = 0)). \qquad (310)$$

It is easy to check that V^0 proves that P is uniquely determined by (a, E) in this definition, so by (309) we have

Lemma IX.6.30. *The function Path is Σ_1^B-definable in VL.*

The proof of the next result is an easier version of the proof of Theorem IX.6.4.

Theorem IX.6.31. *FL is closed under AC^0-reductions.*

Corollary IX.6.32. *The function Path is complete for FL under AC^0-reductions.*

Proof. It is easy to see that *Path* is in *FL* and hence by the preceding theorem every function AC^0-reducible to *Path* is in *FL*. The other direction follows from Theorem IX.6.28.   □

Lemma IX.6.33. *The function $Path^\star$ is Σ_1^B-definable in VL, and* (166) *(displayed below) is provable in $VL(Row, Path, Path^\star)$.*

$$\forall i < b, \; Path^\star(b, X, E)^{[i]} = Path((X)^i, E^{[i]}).$$

Proof. The arguments for $Path^\star$ code b graphs $(a_0, E^{[0]}), \dots,$ $(a_{b-1}, E^{[b-1]})$. By modifying these using Σ_0^B-formulas we may assume that each a_u has the same value a (set to the maximum of the original a_u) and $Unique(a, E^{[u]})$ holds for each u. We need to construct simultaneously in *VL* the paths $P^{[0]}, P^{[1]}, \dots, P^{[b-1]}$ so that for $0 \le u < b$, $P^{[u]}$ satisfies $\delta_{PATH}(a, E^{[u]}, P^{[u]})$.

Formally we need to prove that the following is a theorem of *VL*:

$$\forall u < b \forall x < a \exists! y < a \, E^{[u]}(x, y) \supset \exists P \forall u < b \, (P^{[u]})^0 = 0 \wedge$$
$$\forall v < a \, (E^{[u]}((P^{[u]})^v, (P^{[u]})^{v+1}) \wedge (P^{[u]})^{v+1} < a). \quad (311)$$

We will construct a graph G' encoded by (a', E') that contains a path $Q = Path(a', E')$ from which we can define the paths $P^{[0]}, \dots, P^{[b-1]}$. In fact, we will define G' so that Q is just the concatenation of the paths $P^{[u]}$, $0 \le u < b$. More precisely, the nodes of G' are encoded by triples $\langle u, v, x \rangle$ in such a way that if $P^{[u]}$ encodes the path $(0, x_1, \dots, x_a)$, then in Q there is the sub-path of the form

$$\langle u, 0, 0 \rangle, \langle u, 1, x_1 \rangle, \dots, \langle u, a, x_a \rangle.$$

In other words, we will have

$$(P^{[u]})^v = x$$

for all nodes $\langle u, v, x \rangle$ on the path Q.

Thus we have the following edges in G' (for $0 \le u < b$):

$$(\langle u, v, x \rangle, \langle u, v+1, y \rangle) \in E' \quad \text{for } 0 \le v, x, y < a \text{ and } (x, y) \in E^{[u]},$$
$$(\langle u, a, x \rangle, \langle u+1, 0, 0 \rangle) \in E' \quad \text{for } x < a.$$

Let $a' = \langle b, a, a \rangle$, then the graph encoded by (a', E') satisfies the hypothesis of *PATH*. Let Q be the path for this graph. We can prove by induction that the $(u(a + 1) + v)$-th node on the path is of the form $\langle u, v, x \rangle$:

$$(Q)^{u(a+1)+v} = \langle u, v, x \rangle \quad \text{for some } x, 0 \leq x < a.$$

Define P so that

$$(P^{[u]})^v = x \text{ iff } (Q)^{u(a+1)+v} = \langle u, v, x \rangle.$$

It is straightforward to show that each $P^{[u]}$ satisfies $\delta_{PATH}(a, E^{[u]})$. $\square$

Now we define the universal theories $\widehat{VL}$ and $\overline{VL}$. For this we need the function *Path'* which is semantically equal to *Path* but has as its defining equation the quantifier-free formula over $\mathcal{L}_{FAC^0}$ which is equivalent over $\overline{V}^0$ to the RHS of (310).

DEFINITION IX.6.34 ($\widehat{VL}$). The theory $\widehat{VL}$ has vocabulary $\mathcal{L}_{\widehat{VL}}$ $= \mathcal{L}_{FAC^0} \cup \{Path'\}$. The axioms of $\widehat{VL}$ consist of the axioms of $\overline{V}^0$ and the quantifier-free equivalent of (310).

DEFINITION IX.6.35 ($\overline{VL}$). The vocabulary $\mathcal{L}_{FL}$ is the smallest set that contains $\mathcal{L}_{\widehat{VL}}$ such that for every $\mathcal{L}_A^2$-term t and every quantifier-free $\mathcal{L}_{FL}$ formula φ there is a function $F_{\varphi(z),t}$ in $\mathcal{L}_{FL}$ with defining axiom (86):

$$F_{\varphi(z),t}(\vec{x}, \vec{X})(z) \leftrightarrow z < t(\vec{x}, \vec{X}) \wedge \varphi(z, \vec{x}, \vec{X}). \tag{312}$$

The theory $\overline{VL}$ has vocabulary $\mathcal{L}_{FL}$ and axioms those of $\mathcal{L}_{\widehat{VL}}$ and (312) for each function $F_{\varphi(z),t}$.

We have as corollaries of the results from Section IX.2 the Definability Theorems for *VL*, $\widehat{VL}$ and $\overline{VL}$:

COROLLARY IX.6.36. *Here either $\mathcal{L}$ is $\mathcal{L}_{\widehat{VL}}$ and T is $\widehat{VL}$, or $\mathcal{L}$ is $\mathcal{L}_{FL}$ and T is $\overline{VL}$.*

(a) *A function is in **FL** iff it is represented by a term in $\mathcal{L}_{\widehat{VL}}$. A string function is in **FL** iff it is represented by a string function in $\mathcal{L}_{FL}$. A relation is in **L** iff it is represented by an open (or a Σ_1^B) formula of $\mathcal{L}$.*

(b) *Every $\Sigma_1^B(\mathcal{L})$ formula is equivalent in T to a $\Sigma_1^B(\mathcal{L}_A^2)$ formula.*

(c) *T proves $\Sigma_0^B(\mathcal{L})$-**COMP**, $\Sigma_0^B(\mathcal{L})$-**IND**, and $\Sigma_0^B(\mathcal{L})$-**MIN**.*

(d) *$\overline{VL}$ is a conservative extension of $\widehat{VL}$ which is in turn a conservative extension of **VL**.*

(e) *The Σ_1^B-definable functions of **VL** (or $\widehat{VL}$, $\overline{VL}$) are precisely functions in **FL**.*

(f) *The Δ_1^B-definable relations of **VL** (or $\widehat{VL}$, $\overline{VL}$) are precisely relations in **L**.*

In [113] Zambella introduced the theory Σ_0^B-***Rec*** and showed that it characterizes *L*. It can be shown to be equivalent to *VL*. Here Σ_0^B-***Rec*** is defined using the following axiom scheme:

$$\forall w < b \forall x < a \exists y < a \, \varphi(w, x, y) \supset \exists Z, \forall w < b \, \varphi(w, (Z)^w, (Z)^{w+1})$$
(313)

for all Σ_0^B formulas φ not involving Z.

Note that our axiom (309) is provable in V^0 from an instance of (313). So to prove the above equivalence, the main task is to show that (313) is provable in our theory. To do this, given a, b and the formula $\varphi(w, x, y)$ we construct the edge relation E of a graph whose nodes are pairs $\langle w, x \rangle$ for $w < b$ and $x < a$. Then $E(\langle w, x \rangle, \langle w', y \rangle)$ iff $w' = w + 1$ and y is the smallest number satisfying $\varphi(w, x, y)$. If P is the path guaranteed to exist by the axiom (309) for $\textbf{\textit{VL}}$ applied to E, then we can define Z so if $(P)^w = \langle w, x \rangle$ then $(Z)^w = x$, so $\varphi(w, (Z)^w, (Z)^{w+1})$ holds.

EXERCISE IX.6.37. Fill in the details of the above outline to show that $\textbf{\textit{VL}}$ proves the axiom scheme (313).

Finally we prove:

THEOREM IX.6.38. $\textbf{\textit{VNC}}^1 \subseteq \textbf{\textit{VL}}$.

PROOF. Since $\widehat{\textbf{\textit{VL}}}$ is conservative over $\textbf{\textit{VL}}$, it suffices to show that $\textbf{\textit{VNC}}^1 \subseteq \widehat{\textbf{\textit{VL}}}$. Recall the formula MFV (Definition IX.5.4. We need to show that

$$\widehat{\textbf{\textit{VL}}} \vdash MFV.$$

The idea is to formalize in $\widehat{\textbf{\textit{VL}}}$ a logspace algorithm that evaluates a balanced Boolean sentence. The algorithm that we consider here makes a depth-first-search traversal on the tree structure of the sentence, skipping a whole subtree whenever possible (e.g., if A is true then $A \vee B$ is true, so we do not have to examine B).

Thus consider a balanced sentence specified by (a, G, I) as in Section IX.5.2. For each $1 \leq x < a$ we construct a graph encoded by $(a_x, E^{[x]})$ so that the bit $Fval(a, G, I)(x)$ can be obtained from $Path(a_x, E^{[x]})$. Then by Lemma IX.6.33 all bits of $Fval(a, G, I)$ can be obtained simultaneously, and we are done.

We show how to obtain the bit $Fval(a, G, I)(1)$. Other bits can be obtained similarly. The graph $(a_1, E^{[1]})$ describes a depth-first search traversal in the circuit (a, G) to compute the output of the root. Each vertex is a (potential) state of the traversal. There is a starting node (vertex 0), and every other vertex is numbered by

$$\langle x, \mathrm{d}, 0 \rangle \quad \text{or} \quad \langle x, \mathrm{u}, v \rangle, \qquad \text{where } 1 \leq x < 2a, 0 \leq v \leq 1.$$

Here $\mathrm{d} = 1$ (down) and $\mathrm{u} = 2$ (up) indicate the direction of the traversal. A vertex $\langle x, \mathrm{d}, 0 \rangle$ corresponds to the state when the depth-first traversal visits the gate numbered x for the first time (so in general it will go "down"). Similarly, a state $\langle x, \mathrm{u}, v \rangle$ is when the search visits gate x the second time (thus the direction is "up"); by this time the truth value of the gate is known, and v carries this truth value.

The edges of this graph represent the transition between the states of the search. The search starts at the root, thus we have the following edge:

$$(0, \langle 1, \mathsf{d}, 0 \rangle).$$

When the search visits a gate x for the first time, it will travel down along the left-most branch from x:

$$(\langle x, \mathsf{d}, 0 \rangle, \langle 2x, \mathsf{d}, 0 \rangle) \text{ for } 1 \le x < a.$$

And here are the transitions when it reaches the input gates:

$$(\langle x + a, \mathsf{d}, 0 \rangle, \langle x + a, \mathsf{u}, 0 \rangle) \qquad \text{if } \neg I(x), 0 \le x < a,$$
$$(\langle x + a, \mathsf{d}, 0 \rangle, \langle x + a, \mathsf{u}, 1 \rangle) \qquad \text{if } I(x), 0 \le x < a.$$

For an $\vee$-gate x (i.e., if $\neg G(x)$, where $1 \le x < a$) notice that the search in the subtree rooted at x can be completed when (i) either child of x outputs $\top$, or (ii) the right child of x outputs $\bot$. Furthermore, if the left child of x outputs $\bot$, then the search continue at the right child. We have the following transitions:

either child outputs $\top$: $(\langle 2x, \mathsf{u}, 1 \rangle, \langle x, \mathsf{u}, 1 \rangle)$ and $(\langle 2x + 1, \mathsf{u}, 1 \rangle, \langle x, \mathsf{u}, 1 \rangle)$,

the right child outputs $\bot$: $(\langle 2x + 1, \mathsf{u}, 0 \rangle, \langle x, \mathsf{u}, 0 \rangle)$,

the left child outputs $\bot$: $(\langle 2x, \mathsf{u}, 0 \rangle, \langle 2x + 1, \mathsf{d}, 0 \rangle)$.

EXERCISE IX.6.39. Give the transitions for an $\wedge$-gate.

Notice that the graph described so far have outdegree *at most* 1. To make the outdegree *exactly* 1 we can create an extra node and connect all vertices with outdegree 0 to it. Let the resulting graph be encoded by $(a_1, E^{[1]})$. Note that our traversal does not visit all gates of the circuit (a, G, I). But if it does visit a gate, then the gate will be evaluated. In particular, the output of gate 1 (i.e. $Fval(a, G, I)(1)$) is

$$\exists w < a_1 \, (Path(a_1, E^{[1]}))^w = \langle 1, \mathsf{u}, 1 \rangle.$$

Similarly we construct a graph $(a_x, E^{[x]})$ to evaluate each node $x < a$ of the circuit (a, G, I), where now the initial edge is

$$(0, \langle x, \mathsf{d}, 0 \rangle).$$

By $\Sigma_0^B(\widehat{VL})$-*COMP* there is a string Y which uses these values together with the inputs I to evaluate all the nodes in the circuit.

It remains to prove (in $\widehat{VL}$) that $\delta_{MFV}(a, G, I, Y)$ holds. This is left as an exercise.   □

EXERCISE IX.6.40. Complete the proof above by showing that $\widehat{VL}$ proves $\delta_{MFV}(a, G, I, Y)$.

IX.6.4. The Theory *VLV*. Recall the notion of polynomial-bounded number recursion (pBNR) from Section IX.3.3. We develop the universal theory *VLV* using the fact that the function class *FL* can be characterized using pBNR. Thus *VLV* has the same style as *VPV*. Its vocabulary contains symbols for all functions in *FL*. Here their defining axioms are given using the above fact. First we state the characterization of *FL*.

THEOREM IX.6.41 (Lind). *A function is in FL iff it can be obtained by AC^0-reduction and pBNR iff it can be obtained from FAC^0 by finitely many applications of composition, string comprehension, and pBNR.*

PROOF SKETCH. First, it is easy to see that $FAC^0 \subseteq FL$ and that *FL* is closed under composition, string comprehension and pBNR. By Theorem IX.1.7 it remains to show that functions in *FL* can be obtained from FAC^0 by AC^0-reduction and pBNR.

Suppose that $F(\vec{x}, \vec{X})$ is a function in *FL* and let M be a logspace polytime Turing machine that computes F. As in the proof of Propositions IX.6.3 and IX.6.28, the configurations of M (without the input and output tape content) are encoded by numbers $< t(\vec{x}, \vec{X})$ for some number term bounding the running time of M such that 0 and 1 are respectively the initial and (the only) accepting configuration.

Since M is deterministic, there is an AC^0 function $next_M(z, \vec{x}, \vec{X})$ such that for $z < t(\vec{x}, \vec{X})$, $next_M(z, \vec{x}, \vec{X})$ is the next configuration of z if z is a non-final configuration of M, otherwise:

$$next_M(z, \vec{x}, \vec{X}) = \begin{cases} 0 & \text{if } z \text{ does not code a configuration of M}, \\ z & \text{if } z \text{ is a final configuration of M, e.g., 1.} \end{cases}$$

Let $conf_M(y, \vec{x}, \vec{X})$ denote the configuration of M at time y. Then we have

$$conf_M(0, \vec{x}, \vec{X}) = 0,$$
$$conf_M(y + 1, \vec{x}, \vec{X}) = nextM(conf_M(y, \vec{x}, \vec{X}), \vec{x}, \vec{X}).$$

In other words, $conf_M$ can be obtained from AC^0 functions by pBNR.

Now the bits of the string $F(\vec{x}, \vec{X})$ computed by M can be extracted from the numbers

$$conf_M(0, \vec{x}, \vec{X}), conf_M(1, \vec{x}, \vec{X}), \ldots, conf_M(t(\vec{x}, \vec{X}), \vec{x}, \vec{X}).$$

First we need to determine the times at which M writes to its output tape. This can be done using pBNR as well.

EXERCISE IX.6.42. Define using pBNR from $conf_M(y, \vec{x}, \vec{X})$ the function

$$next\_write_M(y, \vec{x}, \vec{X})$$

which is the first time $y' > y$ such that M writes to its output tape at time y'. Use this to define the function $write_M(y, \vec{x}, \vec{X})$ which is the time at which M performs the y-th write.

The bits $F(\vec{x}, \vec{X})(y)$ can be extracted from $conf_M(write_M(y, \vec{x}, \vec{X}), \vec{x}, \vec{X})$ by some AC^0 functions. Consequently, F can be obtained by AC^0-reduction and pBNR.   □

DEFINITION IX.6.43 (*VLV*). The vocabulary $\mathcal{L}_{VLV}$ is the smallest set that contains $\mathcal{L}_{FAC^0}$ such that for every $\mathcal{L}_A^2$-term t, quantifier-free $\mathcal{L}_{VLV}$-formula φ, number functions g, h in $\mathcal{L}_{VLV}$, and string function F in $\mathcal{L}_{VLV}$ there are:

- a string function $F_{\varphi(z),t}$ in $\mathcal{L}_{VLV}$ with defining axiom (86):

$$F_{\varphi(z),t}(\vec{x}, \vec{X})(z) \leftrightarrow z < t(\vec{x}, \vec{X}) \wedge \varphi(z, \vec{x}, \vec{X});$$

- a number function $f_{t,g,h}$ with defining axiom

$$(g < t \supset f_{g,h}(0) = g) \wedge (g \geq t \supset f_{g,h}(0) = 0) \wedge$$
$$(h(y, f_{g,h}(y)) < t \supset f_{g,h}(y+1) = h(y, f_{g,h}(y))) \wedge$$
$$(h(y, f_{g,h}(y)) \geq t \supset f_{g,h}(y+1) = 0); \quad (314)$$

- a number function f_F with defining axiom

$$f_F(\vec{x}, \vec{X}) = |F(\vec{x}, \vec{X})|. \quad (315)$$

VLV is the theory with vocabulary $\mathcal{L}_{VLV}$ and is axiomatized by the axioms of $\overline{V}^0$ and (86) for every function $F_{\varphi(z),t}$, (314) for every function $f_{t,g,h}$, and (315) for every function f_F.

The next corollary follows from Theorem IX.6.41.

COROLLARY IX.6.44. *A function is in **FL** iff it is represented by a term in* $\mathcal{L}_{VLV}$. *A relation is in **L** iff it is represented by an open (or a* Σ_0^B*) formula of* $\mathcal{L}_{VLV}$.

The following facts can be proved as in Section IX.3.4 and we leave the proofs as exercises.

EXERCISE IX.6.45.   (a) Show that *VLV* proves the axiom schemes

$$\Sigma_0^B(\mathcal{L}_{VLV})\text{-}\textbf{COMP}, \ \Sigma_0^B(\mathcal{L}_{VLV})\text{-}\textbf{IND}, \ \Sigma_0^B(\mathcal{L}_{VLV})\text{-}\textbf{MIN}.$$

(b) Show that every $\Sigma_1^B(\mathcal{L}_{VLV})$ formula is equivalent over *VLV* to a $\Sigma_1^B(\mathcal{L}_A^2)$ formula.

EXERCISE IX.6.46. Show that

(a) A function is in *FL* iff it is Σ_1^B-definable in *VLV*.
(b) A relation is in *L* iff it is Δ_1^B-definable in *VLV*.

Finally, the relationship between *VL* and *VLV* is also left as an exercise.

EXERCISE IX.6.47. *VLV* is a universal conservative extension of *VL*.

IX.7. Open Problems

IX.7.1. Proving Cayley–Hamilton in VNC^2. One complexity class that we do not consider here is $\#L$, which can be defined as the set of all functions $count_M$, where M is a nondeterministic log space Turing machine, and $count_M(X)$ is the number of accepting computations of M on input X. This class is especially interesting because the problem of computing the determinant of integer matrices is complete for $GapL$ [79], where a function in $GapL$ has the form $C_1(X) - C_2(X)$ for C_1, C_2 in $\#L$.

If $V\#L$ is a theory for $\#L$ in the style of this chapter then the Σ_1^B-definable functions in $V\#L$ would be those that are AC^0-reducible to $\#L$. The class of relations Δ_1^B-definable in $V\#L$ would be $AC^0(\#L)$, those relations that are AC^0-reducible to $\#L$. According to Allender [7] this class is the $\#L$ hierarchy, and it is the union

$$AC^0(\#L) = L \cup L^{\#L} \cup L^{\#L^{\#L}} \cup \cdots .$$

$AC^0(\#L)$ can also be characterized as the set of relations AC^0-reducible to integer determinants, and since these are in the class FNC^2 it follows that

$$AC^0(\#L) \subseteq NC^2.$$

It turns out that many standard computational problems in linear algebra over the field of rationals are AC^0-reducible to $\#L$, such as computing matrix inverses and solving systems of linear equations. However a major open question is whether the correctness of these algorithms can be proved in $V\#L$ or VNC^2 or even VNC. These questions are discussed in [104] in the more general context of linear algebra over arbitrary fields. There it is proved that in many cases correctness is equivalent to the Cayley–Hamilton Theorem (that a matrix satisfies its characteristic polynomial).

Hence a major open question is whether VNC^2 proves the Cayley–Hamilton Theorem (over the field of rationals).

A simple matrix identity,

$$AB = I \supset BA = I$$

is not known to be provable (over the rationals) in VNC^2, although it does follow from the Cayley–Hamilton Theorem [104]. The propositional translation of this identity over the field of two elements yields a nice tautology family that seems to be hard for most of the proof systems discussed in Chapter X.

IX.7.2. VSL and $VSL \stackrel{?}{=} VL$. The class SL consists of languages that are accepted by a *symmetric* nondeterministic Turing machines working in logspace. A nondeterministic Turing machine M is said to be symmetric iff for any two configurations c_1, c_2 of M:

if c_2 is a next configuration of c_1, then c_1 is a next configuration of c_2.

It can be shown that the st-connectivity problem for undirected graph is AC^0-many-one complete for SL.

A deterministic Turing machine can be seen as a symmetric nondeterministic Turing machine, so

$$L \subseteq SL.$$

Recent breakthrough by Reingold [97] shows that indeed

$$L = SL.$$

Before this was shown, the fact that $SL = co\text{-}SL$ was established in [85].

The Distance Problem for undirected graph (UDP) is to decide, given a undirected graph G and two of its vertices s, t and a positive integer d, whether the distance between s and t is exactly d. It turns out that UDP is complete for NL, so the function $Conn$ (Section IX.6.1) restricted to undirected graphs is complete for NL, hence we cannot use it to define a theory for VSL.

Here we define VSL as follows. Recall the formula $\delta_{PATH}(a, E, P)$ from (308) (on page 351) which asserts that P encodes a path starting at the vertex 0 in the graph specified by (a, E). Let $\delta_{UCONN}(a, E, C, P)$ be the formula given below that states that $C(u)$ holds iff u is in the transitive closure of vertex 0, and in that case $P^{[u]}$ encodes a path from 0 to u.

$$\delta_{UCONN}(a, E, C, P) \equiv C(0) \wedge \forall u < a \forall v < a \big((C(u) \wedge E(u, v)) \supset C(v) \big)$$

$$\wedge \, \forall u < a \big(C(u) \supset (\delta_{PATH}(a, E, P^{[u]}) \wedge (P^{[u]})^a = u) \big).$$

We need the Σ_0^B-formula $Symm(a, E)$, which asserts that the graph (a, E) is undirected and all nodes have self-loops:

$$Symm(a, E) \equiv \forall x < a \forall y < a, \, E(x, x) \wedge (E(x, y) \supset E(y, x)).$$

DEFINITION IX.7.1. VSL is the theory over $\mathcal{L}_A^2$ that is axiomatized by the axioms of V^0 together with the axiom $UCONN$:

$$Symm(a, E) \supset \exists C \leq a \exists P \leq \langle a, a, a \rangle \, \delta_{UCONN}(a, E, C, P).$$

The fact that the functions Σ_1^B-definable of VSL are exactly functions in FSL can be proved using the fact that the relation R_{UCONN} defined below is complete for SL. Let R_{UCONN} be the following relation

$$R_{UCONN}(a, E) \leftrightarrow Symm(a, E) \wedge$$

$$\exists C \leq a \exists P \leq \langle a, a, a \rangle \big(\delta_{UCONN}(a, E, C, P) \wedge C(1) \big).$$

EXERCISE IX.7.2. Show that the relation R_{UCONN} above is complete for SL.

EXERCISE IX.7.3. Develop universal conservative extensions $\overline{VSL}$ and $\widehat{VSL}$ of VSL (in the style of $\overline{VC}$ and $\widehat{VC}$) and show that their Σ_1^B-definable functions are precisely the functions in FSL.

EXERCISE IX.7.4. Show that $VL \subseteq VSL$.

Despite the fact [97] that $SL = L$, it is an open question whether the corresponding theories are the same.

OPEN PROBLEM IX.7.5. Is $VSL = VL$?

IX.7.3. Defining $\lfloor X/Y \rfloor$ in VTC^0. The string division function $\lfloor X/Y \rfloor$ (or also $X \div Y$) is defined so that

$$\lfloor X/Y \rfloor \times Y \leq X < S(\lfloor X/Y \rfloor) \times Y \qquad (316)$$

where S is the string successor function (Example V.4.17). Exercise VI.2.8 shows that $\lfloor X/Y \rfloor$ is Σ_1^B-definable in V^1 by formalizing in V^1 a polytime algorithm that computes $\lfloor X/Y \rfloor$.

A breakthrough result by Hesse et. al. [55] shows that this function is computable in TC^0. However, it has been an open problem whether this algorithm can be formalized and proved correct in the theory VTC^0.

OPEN PROBLEM IX.7.6. Is the function $\lfloor X/Y \rfloor$ with defining axiom (316) Σ_1^B-definable in $\overline{VTC}^0$?

IX.7.4. Proving *PHP* and *Count$_{m'}$* in $V^0(m)$. Recall the Modulo m Counting Principle *Count$_m$* from Sections IX.4.3 and IX.4.7. Exercise IX.4.38 (b) shows that $V^0(m)$ proves *Count$_{m'}$* whenever m and m' share a common nontrivial divisor. However, we do not know whether the same is true if the $\gcd(m, m') = 1$.

OPEN PROBLEM IX.7.7. Let $m, m' \in \mathbb{N}$, $m, m' \geq 2$, $\gcd(m, m') = 1$. Does $V^0(m) \vdash$ *Count$_{m'}(a, X)$*?

We also know that $V^0(m)$ proves *OPHP*, but not whether $V^0(m)$ proves *PHP*:

OPEN PROBLEM IX.7.8. Let $m \in \mathbb{N}$, $m \geq 2$. Does $V^0(m) \vdash$ *PHP*?

IX.8. Notes

The string comprehension operation (Definition IX.1.6) can be seen as a two-sorted version of the *concatenation recursion on notation* (CRN) operation for single-sorted classes [36].

The families *VC* and $\overline{VC}$ (Sections IX.2.1 and IX.2.3) are from [83]. The theories $\widehat{VC}$ are new.

The number recursion operations (in Theorems IX.3.12, IX.4.15, IX.4.42, IX.5.21 and IX.6.41) are from [82] and are based on previous work of Lind [77] (for *FL*) and Clote and Takeuti's [38] (for *FAC^0*(2), *FAC^0*(6) and *FNC^1*). The characterizations in [38] go back to [89] (for

$FAC^0(2)$ and $FAC^0(6)$) and [9] (for FNC^1). The proof of the Theorems IX.4.42 and IX.5.21 can be found in [82].

Various problems computable in TC^0 are discussed in [32, 55].

The descriptive complexity characterizations of TC^0, $AC^0(m)$ (Sections IX.3.1, IX.4.1) are from [10], and Grädel's characterization of NL by second-order logic (Section IX.6.2) is in [53].

Section IX.3.5 (proving the Pigeonhole Principle in VTC^0) formalizes a "folklore" fact that the PHP can be proved using counting, and was inspired by Buss's proof of the PHP in the *Frege* proof system [21]. Section IX.3.6 (defining $X \times Y$ in VTC^0) is based on [18, 32].

Regarding the relationships between *PHP*, *OPHP* and the modulo counting principles *Count*$_m$ (Sections IX.4.3 IX.4.7) it follows from [99] that V^0 does not prove the implications *OPHP* $\supset$ *PHP* and *Count*$_m$ $\supset$ *PHP* (for $2 \leq m \in \mathbb{N}$). These unprovability results come from super-polynomial lower bounds for constant-depth Frege systems augmented with appropriate axioms, as shown in Corollary IX.4.9 (see also [98] for other lower bounds). These superpolynomial lower bounds (as well as that of Theorems IX.4.39, IX.4.11 and IX.4.6) have been improved to exponential lower bounds in [12, 14].

The proof of the Discrete Jordan Curve Theorem (Section IX.4.5) is from [84] which contains also the more complicated proof of the sequence version of JCT in V^0.

The theory VNC^1 (Section IX.5.3) was first defined in [45] and is based on Arai's theory AID [8]. The current axiomatization is from [82]. Both VNC^1 and AID are based on Buss's Theorem that the Boolean Formula Value Problem is complete for NC^1 [22]. The fact that $VTC^0 \subseteq VNC^1$ (Section IX.5.4) is based on [21]. The theory $VNC^1 V$ is called $VALV$ in [82]. It is developed based on Barrington's Theorem which is from [9]. A proof of Theorem IX.5.27 can be found in [82].

The theory VNL is from [83] and V^1-$KROM$ is from [44]. The results in Section IX.6.2 are from [67]. Immerman–Szelepcsényi Theorem (that NL is closed under complement) is from [58] and [106]. The theory VL is from [82]. The equivalent theory Σ_0^B-Rec is from [113].

Chapter X

PROOF SYSTEMS AND THE REFLECTION
PRINCIPLE

An association between V^i and the proof system $G_i^\star$ (for $i \geq 1$) is shown in Chapter VII by the fact that each bounded theorem of the theory V^i translates into a family of tautologies that have polynomial-size $G_i^\star$ proofs. Our theories and their associated proof systems are more deeply connected than as shown by just the propositional translation theorems. In this chapter we will present some more connections between the proof systems, their associated theories and the underlying complexity classes.

In general, for each proof system $\mathcal{F}$ we study the principle that asserts that the system is sound, i.e, that formulas that have $\mathcal{F}$-proofs are valid. This is known as the Reflection Principle (RFN) for $\mathcal{F}$. We will show in this chapter that the theories V^i and TV^i prove the RNF for their associated proof systems when the principles are stated for Σ_i^q formulas. Together with the Propositional Translation Theorems, these show that the systems $G_i^\star$ and G_i are the strongest systems (for proving Σ_i^q formulas) whose RFN are provable in the theories V^i and TV^i, respectively.

A connection between a propositional proof system $\mathcal{F}$ and the complexity class C definable in the theory T associated with $\mathcal{F}$ will be seen by the fact that the Witnessing Problem for $\mathcal{F}$ is complete for C. Recall Theorem VII.4.13 which shows that the Witnessing Problem for $G_1^\star$ (and equivalently for *eFrege*) are solvable by a polytime algorithm. (In fact here we will formalize this algorithm in V^1 in order to show that the Σ_1^q-RFN of $G_1^\star$ is provable in V^1 as mentioned above.) The fact that the Witnessing Problem for $G_1^\star$ is hard for P can be proved by using the Proposition Translation Theorem for V^1 and the fact that theorems of V^1 can be proved using the RFN for $G_1^\star$.

We will also present some connections between subtheories of TV^0 and their associated proof systems. Here VNC^1 is associated with the sequent calculus PK introduced in Chapter II in the same way that V^0 is associated with bounded depth PK or V^1 is associated with *eFrege*. We prove that PK is the strongest propositional proof system whose reflection principle is provable in VNC^1. The theory VTC^0 is associated with bounded depth PTK, the systems that extend bounded depth PK by a new kind of connective corresponding to the counting gates in TC^0 circuits.

This chapter is organized as follows. We start by formalizing propositional proofs in Section X.1. The formalizations are needed for stating the Reflection Principle. They also enable us to state the Propositional Translation Theorems as theorem in our theory VTC^0. In fact we will prove (in Section X.1.3) the Propositional Translation Theorems for TV^i as a theorem of VTC^0, and restate various theorems from Chapter VII this way. The RFN and Witnessing Problems for $G_i^\star$ and G_i will be discussed in Section X.2. Finally, in Sections X.3 and X.4 we discuss the Propositional Translation Theorems for the theories VNC^1 and VTC^0.

X.1. Formalizing Propositional Translations

Recall (Definition VII.1.2) that a proof system is defined to be a polytime, surjective function:

$$F : \{0,1\}^* \longrightarrow TAUT.$$

It turns out that all proof systems that we have discussed are TC^0 functions. This is because for these systems, to compute $F(X)$ the main task is often to verify whether X is a legitimate proof. The verification in turn consists of recognizing (quantified) propositional formulas, sequents and proofs. The recognition can be done using counting gates, for example, to check that parentheses are properly nested in formulas, or that inference rules are properly applied in a proof. Therefore the property of being a legitimate proof (or formula, or sequent) is a TC^0 relation.

Verifying proofs in polytime is often straightforward and therefore omitted. However, to show that it can be done in TC^0 is less straightforward. So in Section X.1.1 below we will carry this out in some detail. Recall (Section IX.3.2) that a relation is in TC^0 iff it is Δ_1^B-definable in VTC^0, iff it is represented by an open $\mathcal{L}_{FTC^0}$ formula, and iff it is represented by a $\Sigma_0^B(\mathcal{L}_{FTC^0})$ formula (Theorem IX.3.7).

The propositional translations of LK^2 proofs from Chapter VII produce uniform propositional proofs. In fact, in Section X.1.2 we will show that these translations are computable by TC^0 functions, and the propositional translation theorems are theorems of VTC^0.

Then in Section X.1.3 we will prove the Propositional Translation Theorem for TV^i. Following the discussion from the previous section, we will show that this is also a theorem of VTC^0.

X.1.1. Verifying Proofs in TC^0. We will consider proofs of G. Other systems can be handled in similar way or with minor modifications. Recall the definition of G from Section VII.3. First we present a simple encoding of proofs in our two-sorted vocabulary $\mathcal{L}_A^2$. The pairing function (Example V.4.20) can be used to avoid using deliminators for sequents in a proof or formulas in a cedent.

Let

$$\pi = \mathcal{S}_\ell, \mathcal{S}_{\ell-1}, \ldots, \mathcal{S}_1, \mathcal{S}_0$$

be a proof in G where $\mathcal{S}_0$ is the end sequent. A simple way of present π in our two-sorted vocabulary is to view it as an array whose rows $\pi^{[i]}$ encode the sequents $\mathcal{S}_i$. To simplify our verification $\pi^{[i]}$ will also contain the indices of all parents of $\mathcal{S}_i$. Thus, we let

$$\pi^{[i]} = \langle\langle j, k \rangle, \mathcal{S}_i \rangle \tag{317}$$

where either $j = k = 0$ or $j > i \wedge k = 0$ or $j > i \wedge k > i$. Here j, k are indices of the parents of $\mathcal{S}_i$ (if a parent is not present, the corresponding index is 0). Also, $\langle j, k \rangle$ is the pairing function from Example V.4.20, and $\langle x, Y \rangle$ is the pairing function from Definition VIII.7.2.

The number of sequents in π can be easily extracted from π. We will require that every sequent except for the end sequent is used at least once. This can be checked by stating that for all j, where $0 < j \leq \ell$, there exists $i < j$ such that $\pi^{[i]}$ has the form

$$\langle\langle j, k \rangle, \mathcal{S}_i \rangle \qquad \text{or} \qquad \langle\langle k, j \rangle, \mathcal{S}_i \rangle$$

(for some k).

Verifying that the rules are properly applied in π will be discussed in the proof of Lemma X.1.5 below. Now we briefly discuss the other ingredients of a proof, i.e., sequents and formulas. A sequent S is encoded as two arrays $\mathcal{S}^{[0]}$ and $\mathcal{S}^{[1]}$ that encode its antecedent and succedent, respectively. For example, $\mathcal{S}^{[0]}$ is $\varnothing$ if the antecedent is empty; otherwise $\mathcal{S}^{[0.0]}$ encodes the first formula of the antecedent, etc.

Next, assume that all propositional variables are either x_k (bound variables) or p_k (free variables), for $k \geq 0$. Using the letters x and p, these can be written respectively as $x 11 \ldots 1$ and $p 11 \ldots 1$ with k 1's in each string (when $k = 0$ the strings are just x and p, respectively). Thus quantified propositional formulas are written as strings over the alphabet

$$\{\top, \bot, p, x, 1, (,), \wedge, \vee, \neg, \exists, \forall\}. \tag{318}$$

(Note that we use unary notation for writing the indices of variables. For a propositional formula of size n there are at most n variables and their indices are at most n.)

We will encode a string S over (318) by a binary string X in such a way that the i-th symbol $X[i]$ of S can be easily (in AC^0) extracted from X. The exact encoding is not important; for example each symbol in (318) can be encoded by a five-bit string with high-order bit 1, and X could be the concatenation of the codes for the symbols in S. Formal proofs for most of the results below are straightforward but at the same time tedious.

So our arguments will often be informal or sketched. Interested readers are encouraged to carry out the proofs in detail themselves.

Our TC^0 algorithm for recognizing formulas requires the following notion.

NOTATION. The outside pair of parentheses in a string "(Z)" over the vocabulary (318) is said to *match* if Z contains the same number of left and right parentheses, and every initial segment of Z contains at least as many left parentheses as right parentheses.

NOTATION. A string over the vocabulary (318) is called a *pseudo formula* if it has the form

$$\underbrace{\neg \ldots \neg}_{n_1} Q_1 x \underbrace{11 \ldots 1}_{i_1} \underbrace{\neg \ldots \neg}_{n_2} Q_2 x \underbrace{11 \ldots 1}_{i_2} \ldots Q_k x \underbrace{11 \ldots 1}_{i_k} \underbrace{\neg \ldots \neg}_{n_{k+1}} Y \quad (319)$$

where $Q_j \in \{\exists, \forall\}$ for $1 \leq j \leq k$; $k, n_1, n_2, \ldots, n_{k+1}, i_1, i_2, \ldots, i_k \geq 0$ (i.e., the string preceding Y might be empty); and the substring Y satisfies the condition:

1) either Y is one of the following strings:

$$\top, \qquad \bot, \qquad p \underbrace{11 \ldots 1}_{\ell}, \qquad x \underbrace{11 \ldots 1}_{\ell}$$

(for some $\ell \geq 0$),

2) or Y has the form "(Z)" where the indicated pair of parentheses match.

Note that any formula is also a pseudo formula.

NOTATION. For a string $s_0 s_1 \ldots s_n$ over the vocabulary in (318) that is encoded by an $\mathcal{L}_A^2$ string (i.e., set) X, we use $X[i]$ for the symbols s_i, for $0 \leq i \leq n$. Also for $i \leq j$, let $X[i, j]$ denotes the substring of X that consists of the symbols $X[i], X[i + 1], \ldots, X[j]$.

The next lemma implies that there is a TC^0 algorithm that accepts precisely proper encodings of formulas.

LEMMA X.1.1. *A string X over the alphabet* (318) *is a formula iff it is a pseudo formula, and*

- *for every substring of X the form "(S)" where the indicated parentheses are matched, S has the form $Y \wedge Z$ or $Y \vee Z$, where Y and Z are pseudo formulas; and*
- *every maximal substring of X of the form $x 11 \ldots 1$ (with s 1's) is contained in a pseudo formula of the form* (319) *where $i_j = s$ for some j.*

PROOF. See the definition of formula given in Section VII.3. The second condition in the lemma ensures that every variable beginning with x is quantified. That every formula satisfies the two conditions is proved by induction on the length of the formula. The converse is proved by induction on the length of X.   □

DEFINITION X.1.2. For a proof system $\mathcal{F}$ let $FLA_{\mathcal{F}}(X)$ denote the property that X encodes a formula in $\mathcal{F}$. Let $PRF_{\mathcal{F}}(\pi, X)$ hold iff the string π encodes an $\mathcal{F}$-proof of the formula X. (We will omit the subscript $\mathcal{F}$ when it is clear from context.)

It follows from Lemma X.1.5 that for $\boldsymbol{G}$ and its subsystems the above predicates FLA and $PRF_{\mathcal{F}}$ are in $\boldsymbol{TC^0}$. The more general treatment below will be useful for later results.

In general, a proof system is a polytime function, and hence it is Σ_1^B-definable in $\boldsymbol{TV^0}$ and its graph $PRF_{\mathcal{F}}(\pi, X)$ is Δ_1^B-definable in $\boldsymbol{TV^0}$ (see Chapter VIII). Here we are interested in formulas of special forms that represent $PRF_{\mathcal{F}}(\pi, X)$ and $FLA_{\mathcal{F}}(X)$. These forms will be useful for our proof of Lemma X.1.7 and several results in Section X.2.2. Recall (Section VIII.3.2) that for each Σ_0^B formula $\varphi(y, \vec{x}, \vec{X}, Y)$ the $\boldsymbol{BIT\text{-}REC}$ axiom for φ is defined using the Σ_0^B formula

$$\varphi^{rec}(y, \vec{x}, \vec{X}, Y) \equiv \forall i < y(Y(i) \leftrightarrow \varphi(i, \vec{x}, \vec{X}, Y^{<i})).$$

(The notation $Y^{<i}$ stands for $Cut(i, Y)$ and is defined in (97) on page 139.) For the exercise below, the idea is that the formula

$$\varphi^{rec}(t + 1, \vec{x}, \vec{X}, Y)$$

states that Y encodes a polytime computation for the relation $R(\vec{x}, \vec{X})$, and the bit $Y(t)$ is the "check bit": $Y(t)$ is true iff $R(\vec{x}, \vec{X})$ holds. (Note also that for strings Y of length $|Y| \leq t + 1$, $Y(t) \leftrightarrow |Y| = t + 1$.)

EXERCISE X.1.3. Show that a relation $R(\vec{x}, \vec{X})$ is in $\boldsymbol{P}$ iff there are a Σ_0^B formula $\varphi(y, \vec{x}, \vec{X}, Z)$ and an $\mathcal{L}_A^2$ term $t(\vec{x}, \vec{X})$ so that both formulas below represent R:

$$\varphi_1(\vec{x}, \vec{X}) \equiv \exists Y \leq t + 1\big(\varphi^{rec}(t + 1, \vec{x}, \vec{X}, Y) \wedge Y(t)\big), \qquad (320)$$

$$\varphi_2(\vec{x}, \vec{X}) \equiv \forall Y \leq t + 1\big(\varphi^{rec}(t + 1, \vec{x}, \vec{X}, Y) \supset Y(t)\big) \qquad (321)$$

and that

$$\boldsymbol{TV^0} \vdash \varphi_1(\vec{x}, \vec{X}) \leftrightarrow \varphi_2(\vec{x}, \vec{X}).$$

Thus for each proof system $\mathcal{F}$ there are a Σ_0^B formula $\varphi_{\mathcal{F}}(y, \pi, X, Y)$ and a term $t_{\mathcal{F}}(\pi, X)$ such that

$$Prf_{\mathcal{F}}^{\Sigma}(\pi, X) \equiv \exists Y \leq t_{\mathcal{F}} + 1\big(\varphi_{\mathcal{F}}^{rec}(t_{\mathcal{F}} + 1, \pi, X, Y) \wedge Y(t_{\mathcal{F}})\big), \qquad (322)$$

$$Prf_{\mathcal{F}}^{\Pi}(\pi, X) \equiv \forall Y \leq t_{\mathcal{F}} + 1\big(\varphi_{\mathcal{F}}^{rec}(t_{\mathcal{F}} + 1, \pi, X, Y) \supset Y(t_{\mathcal{F}})\big) \qquad (323)$$

both represent $PRF_{\mathcal{F}}(\pi, X)$ and

$$\boldsymbol{TV^0} \vdash Prf_{\mathcal{F}}^{\Sigma}(\pi, X) \leftrightarrow Prf_{\mathcal{F}}^{\Pi}(\pi, X). \qquad (324)$$

(Lemma X.1.5 below shows that for the case of G and its subsystems, the theory TV^0 can be replaced by VTC^0. This fact will be useful, for example, for Theorem X.2.23.)

We are also interested in similar formulas that represent the FLA relation.

COROLLARY X.1.4. *There is an open* $\mathcal{L}_{FTC^0}$ *formula* $\psi(X)$ *that represents* $FLA_G(X)$. *The relation* $FLA_G(X)$ *is* Δ_1^B*-definable in* VTC^0. *Moreover, there are a* Σ_0^B *formula* $\varphi_{FLA}(y, X, Y)$ *and an* $\mathcal{L}_A^2$ *term* t_{FLA} *such that both formulas* $Fla^\Sigma(X)$ *and* $Fla^\Pi(X)$ *below represent* $FLA_G(X)$ *and* $VTC^0 \vdash Fla^\Sigma(X) \leftrightarrow Fla^\Pi(X)$:

$$Fla^\Sigma(X) \equiv \exists Y \leq t_{FLA} + 1\big(\varphi_{FLA}^{rec}(t_{FLA} + 1, X, Y) \wedge Y(t_{FLA})\big), \quad (325)$$

$$Fla^\Pi(X) \equiv \forall Y \leq t_{FLA} + 1\big(\varphi_{FLA}^{rec}(t_{FLA} + 1, X, Y) \supset Y(t_{FLA})\big). \quad (326)$$

PROOF SKETCH. The open $\mathcal{L}_{FTC^0}$ formula $\psi(X)$ expresses the conditions listed in Lemma X.1.1.

To prove the existences of the formulas Fla^Σ and Fla^Π as required, it is easier to start with a $\Sigma_0^B(\mathcal{L}_{\widetilde{VTC^0}})$ (i.e., $\Sigma_0^B(Numones')$) formula $\eta(X)$ that is equivalent to the above open $\mathcal{L}_{FTC^0}$ formula $\psi(X)$ (see Theorem IX.3.7). The idea is to successively remove the occurrences of $Numones'$ in η using the axiom $NUMONES$ (which can be used as a defining axiom for $Numones'$). Note that $NUMONES$ is already an instance of Σ_0^B-*BIT-REC*.   □

Now we show that for the subsystems of G in (324) we can use VTC^0 instead of TV^0.

LEMMA X.1.5. *For each proof system* $\mathcal{F}$ *that we have discussed (e.g.,* G, $G_i^\star$, G_i, *eFrege, etc.) there are a* Σ_0^B *formula* $\varphi_{\mathcal{F}}$ *and an* $\mathcal{L}_A^2$ *term* $t_{\mathcal{F}}$ *so that the formulas* $Prf_{\mathcal{F}}^\Sigma(\pi, X)$ *and* $Prf_{\mathcal{F}}^\Pi(\pi, X)$ *as in* (322) *and* (323) *both represent the relation* $PRF_{\mathcal{F}}(\pi, X)$, *and such that*

$$VTC^0 \vdash Prf_{\mathcal{F}}^\Sigma(\pi, X) \leftrightarrow Prf_{\mathcal{F}}^\Pi(\pi, X).$$

PROOF SKETCH OF LEMMA X.1.5. We will argue for G. The arguments for other proof systems F are similar.

First we sketch a TC^0 algorithm that verifies that (i) π properly encodes a proof, and (ii) the last sequent in π is

$$\longrightarrow X.$$

In fact, it can be shown that there is a $\Sigma_0^B(\mathcal{L}_{FTC^0})$ formula $\psi'(\pi, X)$ that is true iff the algorithm accepts (π, X). Therefore, as in Corollary X.1.4, it is straightforward to obtain $Prf_{\mathcal{F}}^\Sigma$ and $Prf_{\mathcal{F}}^\Pi$ as desired.

Verifying (ii) is straightforward. For (i) note that by our encoding of proofs, the i-th sequent in π can be easily extracted from π, see (317). So first we check that each row $\pi^{[i]}$ of π is of the form (317) where $\mathcal{S}_i$ consists of two lists of formulas $\mathcal{S}_i^{[0]}$ and $\mathcal{S}_i^{[1]}$. Next, it remains to verify locally that for each i, either $\mathcal{S}_i$ is an axiom and $j = k = 0$, or $\mathcal{S}_i$ follows from sequent(s) $\mathcal{S}_j$ (and $\mathcal{S}_k$) by an inference rule.

Verifying that $\mathcal{S}_i$ is an axiom is straightforward. Now consider the case where $\mathcal{S}_i$ follows from $\mathcal{S}_j$ by the $\exists$-left rule; other rules are similar or easier. This case can be checked by

1) verifying that $\mathcal{S}_j$ and $\mathcal{S}_i$ are identical except for $\mathcal{S}_j$ contains a formula of the form $A(p_k)$ and $\mathcal{S}_i$ contains a formula of the form $\exists x_t A(x_t)$ at the same location in the antecedent,

2) verifying that p_k does not occur in $\mathcal{S}_i$.

It is easy to see that task 2) can be done by an $\boldsymbol{AC^0}$ algorithm. For (1) we first need to identify the scope of the existential quantifier $\exists x_t$: this is the smallest pseudo formula that contain $\exists x_t$ (see Lemma X.1.1). Then we need to check that all occurrences of p_k have been properly replaced by x_t. For this the counting gates are used, for example, to count the number of occurrences of p_k and x_t in subformulas of $A(p_k)$ and $A(x_t)$, respectively.

For treelike proofs (e.g., $\boldsymbol{G}_i^\star$) we have to verify in addition that every sequent is used at most once. For this we simply check that all nonzero indices j, k as in (317) appear at most once.   □

Now we show that the polytime algorithms for recognizing formulas and proofs translate into polytime algorithms for generating $\boldsymbol{G}_0^\star$ proofs verifying that formulas are formulas and proofs are proofs. We need the following notation.

DEFINITION X.1.6. For an $\mathcal{L}_A^2$ formula $\varphi(X)$ that might contain other free variables and a constant string X_0 we use

$$\varphi(X_0)[\,]$$

to denote the propositional formula that is obtained from the translation $\varphi(X)[n]$, where $n = |X_0|$, by plugging the values $(\top, \bot)$ of the bits $X_0(j)$ for p_j^X.

(Thus if X is the only free variable in $\varphi(X)$, then $\varphi(X_0)[\,]$ is a sentence.) Recall the propositional translation given in Sections VII.2 and VII.5.

LEMMA X.1.7. *Let $\mathcal{F}$ be a proof system with defining formulas as in (322) and (323). Then there is a polytime algorithm that, given a string X_0 that encodes a formula A and a string π_0 that encodes an $\mathcal{F}$-proof of A, outputs $G_0^\star$ proofs of the following sequents:*

$$\longrightarrow Fla^\Sigma(X_0)[\,], \tag{327}$$

$$\longrightarrow Fla^\Pi(X_0)[\,], \tag{328}$$

$$\longrightarrow Prf_\mathcal{F}^\Sigma(\pi_0, X_0)[\,], \tag{329}$$

$$\longrightarrow Prf_\mathcal{F}^\Pi(\pi_0, X_0)[\,]. \tag{330}$$

For the proof we need the following:

EXERCISE X.1.8. Show that there is a polytime algorithm that, given a Boolean sentence A, outputs a cut-free $\boldsymbol{PK^\star}$ proof of $\longrightarrow A$ if A is true and $A \longrightarrow$ if A is false. Use induction on A to show the existence of the proofs.

PROOF OF LEMMA X.1.7. First we give a $\boldsymbol{G_0^\star}$ proof of (327). The idea is to use a polytime algorithm that, given X_0, computes the (unique) string Y_0 of length $n = val(t_{FLA} + 1)$ that witnesses Y in $Fla^\Sigma(X_0)$ (see (325)). Then by Exercise X.1.8 we can generate a cut-free $\boldsymbol{PK^\star}$ of

$$\longrightarrow \left(|Y_0| \leq t_{FLA} + 1 \wedge (\varphi_{FLA}^{rec}(t_{FLA} + 1, X_0, Y_0) \wedge Y_0(t_{FLA}))\right)[\,].$$

Finally (327) can be derived by a series of applications of the $\exists$-right rule.

Formally, let $m = |X_0|$. By definition $Fla^\Sigma(X)[m]$ is the translation of

$$\exists Y \leq t_{FLA} + 1(\varphi_{FLA}^{rec}(t_{FLA} + 1, X, Y) \wedge Y(t_{FLA})).$$

Let $r = t_{FLA}(m)$. First we translate

$$\varphi_{FLA}^{rec}(t_{FLA} + 1, X, Y) \wedge Y(t_{FLA}) \tag{331}$$

for X of length m and each Y of length $k \leq r + 1$. Note that when $k = r + 1$, $Y(t_{FLA})$ translates into $\top$ and hence (331) translates into

$$\varphi_{FLA}^{rec}(t_{FLA} + 1, X, Y)[m, r + 1].$$

On the other hand, for $k \leq r$, $Y(t_{FLA})$ translates into $\bot$ and therefore (331) also translates into $\bot$. As a result, $Fla^\Sigma(X)[m]$ is

$$\exists p_0^Y \exists p_1^Y \ldots \exists p_{r-1}^Y \left(\varphi_{FLA}^{rec}(t_{FLA} + 1, X, Y)[m, r + 1]\right).$$

So, as outlined above, first we compute in polytime the unique Y_0 that satisfies $|Y_0| \leq r + 1$ and $\varphi_{FLA}^{rec}(t_{FLA} + 1, X_0, Y_0) \wedge Y_0(t_{FLA})$. Then we derive the following sequent as in Exercise X.1.8:

$$\varphi_{FLA}^{rec}(t_{FLA} + 1, X_0, Y_0)[\,].$$

The sequent $\longrightarrow Fla^\Sigma(X_0)[\,]$ can now be obtained by applying the $\exists$ right rule r times.

Now we construct a $\boldsymbol{G_0^\star}$ proof of (328). It will be clear that our construction can be done in polynomial time. As above let $m = |X_0|$ and $r = t_{FLA}(m)$. By definition $Fla^\Pi(X_0)[\,]$ is

$$\forall p_0^Y \forall p_1^Y \ldots \forall p_{r-1}^Y \bigwedge_{k=0}^{r+1} \left(\varphi_{FLA}^{rec}(t_{FLA} + 1, X_0, Y) \supset Y(t_{FLA})\right)[k].$$

So we will give a $\boldsymbol{PK^\star}$ proof of the following sequent

$$\longrightarrow \bigwedge_{k=0}^{r+1} \left(\varphi_{FLA}^{rec}(t_{FLA} + 1, X_0, Y) \supset Y(t_{FLA})\right)[k]. \tag{332}$$

The $\boldsymbol{G_0^\star}$ proof of (328) is then obtained by applying the $\forall$-right rule r times.

When $k = r + 1$ the atom $Y(t_{FLA})$ translates into $\top$, so

$$\left(\varphi_{FLA}^{rec}(t_{FLA} + 1, X_0, Y) \supset Y(t_{FLA})\right)[r + 1]$$

is $\top$ and hence is deleted from the conjunction. For $k \leq r$ the atom $Y(t_{FLA})$ translates into $\bot$ and hence

$$\left(\varphi_{FLA}^{rec}(t_{FLA} + 1, X_0, Y) \supset Y(t_{FLA})\right)[k]$$

is

$$\neg\left(\varphi_{FLA}^{rec}(t_{FLA} + 1, X_0, Y)[k]\right).$$

Consequently (332) is

$$\longrightarrow \bigwedge_{k=0}^{r} \neg\left(\varphi_{FLA}^{rec}(t_{FLA} + 1, X_0, Y)[k]\right).$$

Intuitively the above sequent is valid because any string Y that satisfies $\varphi_{FLA}^{rec}(t_{FLA}+1, X_0, Y)$ must have length exactly $r+1$. To derive the sequent we need to derive the following sequents (for $k \leq r$):

$$\varphi_{FLA}^{rec}(t_{FLA} + 1, X_0, Y)[k] \longrightarrow .$$

Recall that

$$\varphi_{FLA}^{rec}(t_{FLA} + 1, X_0, Y) \equiv \forall i \leq t_{FLA}(Y(i) \leftrightarrow \varphi_{FLA}(i, X_0, Y^{<i})).$$

Therefore

$$\varphi_{FLA}^{rec}(t_{FLA} + 1, X_0, Y)[k] \equiv \bigwedge_{i=0}^{r} \left((Y(i) \leftrightarrow \varphi_{FLA}(i, X_0, Y^{<i}))[k]\right).$$

It suffices to construct, for each $k \leq r$, a $\boldsymbol{PK^{\star}}$ proof of the following sequent:

$$(Y(0) \leftrightarrow \varphi_{FLA}(i, X_0, Y^{<0}))[k], (Y(1) \leftrightarrow \varphi_{FLA}(i, X_0, Y^{<1}))[k], \dots,$$
$$(Y(r) \leftrightarrow \varphi_{FLA}(i, X_0, Y^{<r}))[k] \longrightarrow . \quad (333)$$

Suppose that $k = 0$, then $Y(i)[0]$ is $\bot$, and $\varphi_{FLA}(i, X_0, Y^{<i}))[0]$ is a sentence B_i, for $0 \leq i \leq r$. Therefore the sequent (333) becomes

$$\neg B_0, \neg B_1, \dots, \neg B_r \longrightarrow .$$

Since X_0 encodes a formula, at least one of the sentences B_i must be true (otherwise $Y = \varnothing$ will satisfy $\varphi_{FLA}^{rec}(t_{FLA} + 1, X_0, Y)$). Moreover, such a B_i can be found in polytime. It follows from Exercise X.1.8 that the above sequent has a polynomial size $\boldsymbol{PK^{\star}}$ proof that is computable in polynomial time.

The case $k = 1$ is similar. Now consider the case $1 < k \leq r$. We will write p_i for the variables p_i^Y, for $i \geq 0$. Recall that

$$Y(i)[k] =_{\text{def}} \begin{cases} p_i & \text{if } i < k - 1, \\ \top & \text{if } i = k - 1, \\ \bot & \text{if } i > k - 1 \end{cases}$$

and $\varphi_{FLA}(i, X_0, Y^{<i>}))[k]$ has the form

$$\varphi_{FLA}(i, X_0, Y^{<i>}))[k] = \begin{cases} B_0 & \text{if } i = 0, \\ B_i(p_0, \ldots, p_{i-1}) & \text{if } 1 \leq i < k, \\ B_i(p_0, \ldots, p_{k-2}) & \text{if } k \leq i \leq r \end{cases}$$

for some formulas B_i with all free variables displayed (in particular, B_0 is a sentence). The sequent (333) becomes

$$p_0 \leftrightarrow B_0, p_1 \leftrightarrow B_1(p_0), \ldots, p_{k-2} \leftrightarrow B_{k-2}(p_0, \ldots, p_{k-3}),$$
$$B_{k-1}(\vec{p}), \neg B_k(\vec{p}), \ldots, \neg B_r(\vec{p}) \longrightarrow \quad (334)$$

(here $\vec{p} = p_0, p_1, \ldots, p_{k-2}$). Compute inductively in polytime the Boolean values b_0 of B_0, b_1 of $B_1(b_0)$, ..., b_{k-2} of $B_{k-2}(b_0, \ldots, b_{k-3})$. Then we can construct $\boldsymbol{PK}^\star$ proofs of the following sequents:

$$p_0 \leftrightarrow B_0 \longrightarrow p_0 \leftrightarrow b_0,$$
$$p_1 \leftrightarrow B_1(p_0), p_0 \leftrightarrow b_0 \longrightarrow p_1 \leftrightarrow b_1,$$

$$\cdots$$

$$p_{k-2} \leftrightarrow B_{k-2}(p_0, \ldots, p_{k-3}), p_0 \leftrightarrow b_0, \ldots, p_{k-3} \leftrightarrow b_{k-3} \longrightarrow p_{k-2} \leftrightarrow b_{k-2},$$
$$B_{k-1}(\vec{p}), p_0 \leftrightarrow b_0, \ldots, p_{k-2} \leftrightarrow b_{k-2} \longrightarrow B_{k-1}(\vec{b}),$$
$$\neg B_k(\vec{p}), p_0 \leftrightarrow b_0, \ldots, p_{k-2} \leftrightarrow b_{k-2} \longrightarrow \neg B_k(\vec{b}),$$

$$\cdots$$

$$\neg B_r(\vec{p}), p_0 \leftrightarrow b_0, \ldots, p_{k-2} \leftrightarrow b_{k-2} \longrightarrow \neg B_r(\vec{b}).$$

Argue as in previous cases, at least one of the sentences

$$B_{k-1}(\vec{b}), \neg B_k(\vec{b}), \ldots, \neg B_r(\vec{b})$$

must have value $\bot$, and such a sentence can be found in polynomial time. Using Exercise X.1.8 we can now compute a $\boldsymbol{PK}^\star$ proof of the sequent (334). This completes our argument for the sequent (328).

Deriving (329) and (330) is similar.   □

The formulas Fla^Σ, Fla^Π, $Prf_{\mathcal{F}}^\Sigma$ and $Prf_{\mathcal{F}}^\Pi$ are important for this chapter. They will be used to define the Reflection Principle (Definition X.2.11). Note that

$$V^0 \vdash \forall X (Fla^\Sigma(X) \supset Fla^\Pi(X))$$

and (by *Y-IND* on the string Y in (323)):

$$V^0 \vdash \forall \pi \forall X (Prf^\Sigma_{\mathcal{F}}(\pi, X) \supset Prf^\Pi_{\mathcal{F}}(\pi, X)).$$

We also use Fla^Σ and $Prf^\Sigma_{\mathcal{F}}$ for the following notions.

DEFINITION X.1.9. Let $F(\vec{n})$ be a function in the vocabulary $\mathcal{L}$ of a theory $\mathcal{T}$. Suppose that $F(\vec{n})$ is the encoding of a formula $A_{\vec{n}}$, for all $\vec{n}$. We say that F *provably in* $\mathcal{T}$ computes $A_{\vec{n}}$ if

$$\mathcal{T} \vdash \forall \vec{n} Fla^\Sigma(F(\vec{n})). \tag{335}$$

Similarly we say that a function $G(\vec{n})$ *provably in* $\mathcal{T}$ computes an $\mathcal{F}$-proof $\pi_{\vec{n}}$ of a formula $A_{\vec{n}}$ (encoded by $\widehat{A_{\vec{n}}}$) if $G(\vec{n}) = \pi_{\vec{n}}$ for all $\vec{n}$, and

$$\mathcal{T} \vdash \forall \vec{n} Prf^\Sigma_{\mathcal{F}}(G(\vec{n}), \widehat{A_{\vec{n}}}). \tag{336}$$

In these cases we also say that the formulas $A_{\vec{n}}$ (resp. proofs $\pi_{\vec{n}}$) are *provably in* $\mathcal{T}$ *computable by* F (*resp. G*), or just *provably computable in* $\mathcal{T}$.

We will often view a formula A as a tree whose leaves are labeled with the constants $\top$, $\bot$ or atomic subformulas p_k, x_k, and whose inner nodes are labeled with the Boolean connectives or quantifiers. Then all paths from the root to the leaves can be identified as follows. For each leaf B of the tree we can identify all pseudo formulas that contain B. Then it can be shown that these pseudo formulas are indeed all subformulas of A that contain B, and hence they form the path from the root of A to B. For this path, using the counting gates we can compute, for example, the alternation depths of quantifiers or connectives. It follows in particular that there is a TC^0 number function, called *qdepth*, that computes the maximum alternation depth of quantifiers in X.

Some basic properties of proofs can be proved as theorems of our theories. We leave these as exercises.

EXERCISE X.1.10. Show that VTC^0 proves the subformula property of G_i proofs: If π is a G_i proof of a formula A, then all formulas in π are either in $(\Sigma^q_i \cup \Pi^q_i)$ or a subformula of A.

EXERCISE X.1.11. Recall the notion of free variable normal form proofs from Section II.2.4. Show that there is a polytime function G so that for every treelike proof π, $G(\pi)$ is provably in VPV (Definition VIII.2.2) a treelike proof in free variable normal form of the same endsequent. (Hint: we need to find all paths in π.)

X.1.2. Computing Propositional Translations in TC^0. Recall from Chapter VII (Sections VII.2 and VII.5) that each bounded $\mathcal{L}^2_A$ formula $\varphi(\vec{x}, \vec{X})$ is translated into a family $\|\varphi\|$ of propositional formulas $\varphi(\vec{x}, \vec{X})[\vec{m}; \vec{n}]$, for $\vec{m}, \vec{n} \in \mathbb{N}$. Each formula $\varphi(\vec{x}, \vec{X})[\vec{m}; \vec{n}]$ is obtained from $\varphi(\vec{x}, \vec{X})$ by substituting the numerals $\vec{m}$ for $\vec{x}$ and introducing for each string variable

X of intended length n the propositional variables p_i^X that represent the bits $X(i)$ of X (for $0 \le i < n - 1$).

In general the family $\|\varphi\|$ involves bound variables

$$p_0^X, p_1^X, \ldots \qquad p_0^Y, p_1^Y, \ldots \qquad \text{etc.}$$

and free variables

$$p_0^\alpha, p_1^\alpha, \ldots \qquad p_0^\beta, p_1^\beta, \ldots \qquad \text{etc.}$$

Encoding and verifying the formulas in $\|\varphi\|$ will be as described in Section X.1.1. We assume that the original string variables $X, Y, \ldots$ and $\alpha, \beta, \ldots$ have been assigned distinct numbers, and we represent a bound variable p_i^X by $x1^\ell$ (x followed by a string of 1's of length ℓ) where $\ell = \langle i, j \rangle$ and X has number j. Similarly we represent a free variable p_i^α by $p1^\ell$.

Following the inductive definition of the propositional translations

$$\varphi(\vec{x}, \vec{X})[\vec{m}; \vec{n}]$$

from Sections VII.2.1 and VII.5 we can show that the encoding Y of $\varphi(\vec{x}, \vec{X})[\vec{m}; \vec{n}]$ can be described by a $\Sigma_0^B(\mathcal{L}_{FTC^0})$ formula and its length $|Y|$ can be expressed by some $\mathcal{L}_{FTC^0}$ function $t_\varphi(\vec{m}, \vec{n})$. For example, suppose that φ is $\exists y \le t \psi(\vec{x}, y, \vec{X})$. Then

$$\varphi(\vec{x}, \vec{X})[\vec{m}; \vec{n}] \equiv \bigvee_{i=0}^{v} \psi(\vec{x}, y, \vec{X})[\vec{m}, i; \vec{n}]$$

(where $v = val(t)$). Thus,

$$t_\varphi(\vec{m}, \vec{n}) = 3v + \sum_{i=0}^{v} t_\psi(\vec{m}, i, \vec{n}).$$

($3v$ is the number of parentheses plus the number of occurrences of $\vee$.)

LEMMA X.1.12. *For every bounded $\mathcal{L}_A^2$ formula $\varphi(\vec{x}, \vec{X})$ there is a $\Sigma_0^B(\mathcal{L}_{FTC^0})$ formula $\psi(\vec{m}, \vec{n}, Y)$ and an $\mathcal{L}_{FTC^0}$ function $t_\varphi(\vec{m}, \vec{n})$ such that for all $\vec{m}$, $\vec{n}$ and Y, $\psi(\vec{m}, \vec{n}, Y)$ is true iff Y encodes $\varphi(\vec{x}, \vec{X})[\vec{m}; \vec{n}]$, and $|Y| = t_\varphi(\vec{m}, \vec{n})$ when $\psi(\vec{m}, \vec{n}, Y)$ holds.*

PROOF IDEA. We can prove by structural induction on φ the existence of both $\psi(\vec{m}, \vec{n}, Y)$ and $t_\varphi(\vec{m}, \vec{n})$, as illustrated above.   □

NOTATION. For a (quantified) propositional formula A, we use $\widehat{A}$ to denote the string that encodes A.

The next corollary follows easily. (Recall Definition X.1.9.)

COROLLARY X.1.13. *For every bounded $\mathcal{L}_A^2$ formula $\varphi(\vec{x}, \vec{X})$ there is an FTC^0 function $T_\varphi(\vec{m}, \vec{n})$ that provably in $\overline{VTC}^0$ computes $\widehat{\varphi(\vec{x}, \vec{X})[\vec{m}; \vec{n}]}$.*

Moreover, $\overline{VTC}^0$ proves the definitions of the translation given in Sections VII.2.1 and VII.5, such as

$$T_{\exists y < t\varphi(y,X)}(n) = \widehat{A}$$

where

$$A = \bigvee_{i=0}^{v-1} B_i, \qquad \text{for } B_i \text{ such that } T_{\varphi(\underline{i},X)}(n) = \widehat{B_i}$$

where $v = val(t(\underline{n}))$.

PROOF IDEA. Using the formula ψ and the function t_φ from Lemma X.1.12, T_φ can be defined as follows:

$$T_{\varphi(\vec{x},\vec{X})}(\vec{m},\vec{n}) = Y \leftrightarrow \left(|Y| \le t_\varphi(\vec{m},\vec{n}) \wedge \psi(\vec{m},\vec{n},Y) \right).$$

It is easy to see that T_φ is in $\mathcal{L}_{FTC^0}$.

The fact that

$$\overline{VTC}^0 \vdash Fla^\Sigma(T_\varphi(\vec{m},\vec{n}))$$

and that $\overline{VTC}^0$ proves the definitions of the translation as required are straightforward.   □

In Chapter VII we proved a number of Propositional Translation Theorems of the following form for a theory T and an associated proof system P: for certain theorems $\varphi(\vec{x}, \vec{X})$ of T, the families $\|\varphi\|$ of propositional tautologies $\varphi(\vec{x}, \vec{X})[\vec{m}; \vec{n}]$ have polynomial-size proofs in P. Here we will strengthen these theorems by showing that the P-proofs of $\varphi(\vec{x}, \vec{X})[\vec{m}; \vec{n}]$ are in fact provably in $\overline{VTC}^0$ computable by some FTC^0 function $F_\varphi(\vec{m}, \vec{n})$ that depends on φ. In Section X.1.3 will prove one more such theorem for the theories TV^i and the proof systems G_i (where $i \ge 1$).

Theorem X.1.14 below strengthens Theorem VII.5.6 in the way mentioned above. Here our propositional proofs of $\varphi(\vec{x}, \vec{X})[\vec{m}; \vec{n}]$ are computable in TC^0 because they consist of disjoint components each of which can be computed by a TC^0 function. For example, suppose that S is a first-order sequent that is obtained from the sequent(s) S_1 (and S_2). Then the propositional proof of $S[\vec{m}; \vec{n}]$ is obtained from the propositional proof(s) of $S_1[\vec{m}; \vec{n}]$ (and $S_2[\vec{m}; \vec{n}]$) by adding some derivations that can also be computed in TC^0.

THEOREM X.1.14. *Suppose that $\varphi(\vec{x}, \vec{X})$ is a bounded theorem of V^0. Then there is a constant d and an FTC^0 function F_φ so that provably in $\overline{VTC}^0$, $F_\varphi(\vec{m}, \vec{n})$ is a d-$G_0^\star$ proof of $\varphi(\vec{a}, \vec{\alpha})[\vec{m}; \vec{n}]$, for all $\vec{m}, \vec{n}$.*

PROOF SKETCH. The constant d will be the same as in Theorem VII.5.6, and we will follow the proof of Theorem VII.5.6 to construct F_φ. Let π be the LK^2-V^0 proof of φ as in the proof of Theorem VII.5.6. For each sequent S in π we will construct an FTC^0 function $F_{S,\pi}(\vec{m},\vec{n})$ that computes the d-$G_0^\star$ proofs of the translations $S[\vec{m};\vec{n}]$. Then $F_\varphi = F_{S_0,\pi}$ for the last sequent S_0 of π.

For each sequent S, the function $F_{S,\pi}(\vec{m},\vec{n})$ is obtained by composition from earlier functions $F_{S_1,\pi}$, $F_{S_2,\pi}$ (for parents S_1, S_2 of S) and some FTC^0 functions. Therefore it will be straightforward that $F_{S,\pi}$ are in FTC^0. Moreover, the fact (336):

$$\overline{VTC}^0 \vdash \forall \vec{m} \forall \vec{n} Prf^\Sigma(F_\varphi(\vec{m},\vec{n}))$$

can be proved by verifying at each step that

$$\overline{VTC}^0 \vdash \forall \vec{m} \forall \vec{n} Prf^\Sigma(F_{S,\pi}(\vec{m},\vec{n})).$$

Exercise VII.5.7 can be strengthened to show that the translations of formulas in π are provably in VTC^0 computable by FTC^0 functions that depend only on π. Details are left as an exercise (see also Corollary X.1.13 above).

EXERCISE X.1.15. Show that for each Σ_0^B formula $\psi(\vec{x}, \vec{X})$ in π there is an FTC^0 function $G_{\psi,\pi}(\vec{m},\vec{n})$ that depends only on π and that, provably in $\overline{VTC}^0$, computes the translation $\psi(\vec{x},\vec{X})[\vec{m};\vec{n}]$ of ψ.

Following the proof of Theorem VII.5.6, it can be shown that there are TC^0 computable proofs of the tautology (156). This is left as an exercise (see also Exercise VII.5.3 and Theorem VII.1.8).

*EXERCISE X.1.16. Suppose that $T_\varphi(i)$ is an FTC^0 function that provably in VTC^0 computes the translation $\varphi(x)[i]$ for an Σ_0^B formula $\varphi(x)$ as in Corollary X.1.13. Show that there is an FTC^0 function $H(\ell)$ that provably in $\overline{VTC}^0$ computes a $PK^\star$ proof of the sequent

$$\longrightarrow \bigwedge_{i=0}^{\ell} \neg A_i, A_0 \wedge \bigwedge_{i=1}^{\ell} \neg A_i, A_1 \wedge \bigwedge_{i=2}^{\ell} \neg A_i, \ldots, A_{\ell-1} \wedge \neg A_\ell, A_\ell$$

where A_i denotes $T_\varphi(i)$. Hint: first describe using a $\Sigma_0^B(\mathcal{L}_{FTC^0})$ formula a $PK^\star$ proof of

$$\longrightarrow \bigwedge_{i=0}^{\ell} \neg p_i, p_0 \wedge \bigwedge_{i=1}^{\ell} \neg p_i, p_1 \wedge \bigwedge_{i=2}^{\ell} \neg p_i, \ldots, p_{\ell-1} \wedge \neg p_\ell, p_\ell$$

then substitute $T_\varphi(i)$ for p_i.

Now we proceed inductively as in the proof of Theorem VII.5.6. Here we can show that if S is derived from S_1 (and S_2), then the proof $F_{S,\pi}$ of $S[\vec{m};\vec{n}]$ can be obtained by compositions from $F_{S_1,\pi}$ (and $F_{S_2,\pi}$) and some other FTC^0 functions.   □

Formalizing the V^i Translation Theorem (Theorem VII.5.2) is similar and is left as an exercise.

*EXERCISE X.1.17. Show that for each bounded theorem $\varphi(\vec{x}, \vec{X})$ of V^i there is an $\boldsymbol{FTC}^0$ function $F_\varphi(\vec{m}, \vec{n})$ that, provably in $\overline{\boldsymbol{VTC}}^0$, computes a $G_i^\star$-proof of $\varphi(\vec{x}, \vec{X})[\vec{m}; \vec{n}]$.

X.1.3. The Propositional Translation Theorem for $\boldsymbol{TV}^i$. Recall the theories $\boldsymbol{TV}^i$ from Section VIII.3. Analogous to the V^i Translation Theorem (Theorem VII.5.2) we will show here that theorems of $\boldsymbol{TV}^i$ translate into families of tautologies that have polynomial size G_i proofs. In fact, we will show that provably in $\overline{\boldsymbol{VTC}}^0$ these G_i proofs can be computed by $\boldsymbol{FTC}^0$ functions that depend only on the theorems of $\boldsymbol{TV}^i$. First we need the following facts whose proofs are left as exercises. (Recall that $\widehat{A}$ is the encoding of a propositional formula A.)

EXERCISE X.1.18. Let the functions $T_{\varphi(Z)}(n)$ and $T_{\psi(x)}(m)$ be as in Corollary X.1.13, i.e., provably in $\overline{\boldsymbol{VTC}}^0$, $T_{\varphi(Z)}(n)$ computes

$$A(p_0^Z, p_1^Z, \ldots, p_{n-2}^Z) =_{\text{def}} \varphi(Z)[n]$$

and $T_{\psi(x)}(m)$ computes

$$B_m =_{\text{def}} \psi(x)[m].$$

Then the formula $A(B_0, B_1, \ldots, B_{n-2})$ is also provably computable in $\overline{\boldsymbol{VTC}}^0$ by some $\boldsymbol{FTC}^0$ function of the form $T_{\varphi'(y)}(n)$.

Below, Exercise X.1.19 formalizes a generalization of Lemma VII.4.11 and Exercise X.1.20 formalizes Lemma VII.4.10 (for our translation formulas $\varphi[\vec{m}; \vec{n}]$).

EXERCISE X.1.19. Suppose that the $\boldsymbol{FTC}^0$ functions $T_{\varphi(Z)}(n)$, $T_{\psi_1(x)}(i)$ and $T_{\psi_2(x)}(i)$ provably in $\overline{\boldsymbol{VTC}}^0$ compute a (quantified) formula $A(\vec{p})$ and quantifier-free formulas $\overrightarrow{B_i^1}$ and $\overrightarrow{B_i^2}$. Then there is an $\boldsymbol{FTC}^0$ function that provably in $\overline{\boldsymbol{VTC}}^0$ computes $G_0^\star$ proofs of

$$A(\overrightarrow{B^1}), B_0^1 \leftrightarrow B_0^2, \ldots, B_{n-2}^1 \leftrightarrow B_{n-2}^2 \longrightarrow A(\overrightarrow{B^2}).$$

Hint: first describe a proof of the sequent by structural induction on A, then argue that such a proof can actually be computed in parallel.

EXERCISE X.1.20. Consider a sequent of formulas in $(\Sigma_i^q \cup \Pi_i^q)$:

$$\Gamma(\vec{p}), \Gamma' \longrightarrow \Delta(\vec{p}), \Delta'.$$

Suppose that all formulas in this sequent are provably in $\overline{\boldsymbol{VTC}}^0$ computable by $\boldsymbol{FTC}^0$ functions of the form $T_{\varphi(Z)}(n)$. Suppose also that B_j are quantifier-free formulas that are provably in $\overline{\boldsymbol{VTC}}^0$ computable by an

FTC^0 function $T_{\psi(x)}(j)$. Then provably in $\overline{VTC}^0$ there is a G_i^* derivation of the form

$$\frac{\Gamma(\vec{p}), \Gamma' \longrightarrow \Delta(\vec{p}), \Delta'}{\Gamma(\vec{B}), \Gamma' \longrightarrow \Delta(\vec{B}), \Delta'}$$

Now we prove the main theorem of this section.

THEOREM X.1.21 (TV^i Propositional Translation). *Let $i \geq 1$. For each bounded theorem $\varphi(\vec{x}, \vec{X})$ of TV^i there is an FTC^0 function $F(\vec{m}, \vec{n})$ that, provably in $\overline{VTC}^0$, computes a G_i proof of $\varphi(\vec{x}, \vec{X})[\vec{m}; \vec{n}]$, for all $\vec{m}, \vec{n} \in \mathbb{N}$.*

PROOF. First we will translate first-order proofs of theorems of TV^i into propositional proofs as in Theorems VII.2.3 and VII.5.2. Then we will argue that the propositional proofs can be provably in $\overline{VTC}^0$ computed by FTC^0 functions that depend on the theorems of TV^i. We will consider the case where $i = 1$; other cases are similar.

Recall that $LK^2\text{-}TV^1$ (Definition VIII.5.14) is a complete system for TV^1. To simplify our translation we modify the string induction rule $SIND$ as follows. Let $S(X, Y)$ be a formula representing the graph of the string successor function, i.e., the "successor relation" (we redefine the symbol S used in Example V.4.17 where it denotes the successor function; the exact meaning is easily understood from context):

$$S(X, Y) \equiv \forall i \leq |X| + |Y| Y(i) \leftrightarrow \big(i \leq |X| \wedge$$
$$\big((X(i) \wedge \exists j < i \neg X(j)) \vee (\neg X(i) \wedge \forall j < i X(j))\big)\big).$$

Now let $\Sigma_1^B\text{-}SIND'$ be the rule:

$$\frac{S_1}{S_2} = \frac{\Gamma, A(\alpha), S(\alpha, \beta) \longrightarrow A(\beta), \Delta}{\Gamma, A(\varnothing) \longrightarrow A(\gamma), \Delta} \tag{337}$$

In this rule, A is a Σ_1^B formula, and α and β do not appear in Γ, Δ.

It is straightforward to verify that the modified $LK^2\text{-}TV^1$ system is also complete for TV^1. (See the discussion for $LK^2\text{-}TV^1$ following Definition VIII.5.14 and also the arguments for $LK^2\text{-}\widetilde{V}^1$ in Section VI.4.1.) In other words, a formula is a theorem of TV^1 if and only if it has an anchored LK^2 proof where all nonlogical axioms are instances of axioms of V^0 and instances of the $\Sigma_1^B\text{-}SIND'$ rule are allowed. (Here a proof is anchored if the cut formulas are instances of axioms of V^0 or instances of $A(\varnothing)$ or $A(\gamma)$ in the bottom sequent of (337).)

Let π be an anchored LK^2 proof of $\varphi(\vec{x}, \vec{X})$ where the rule (337) is allowed. For each sequent $S(\vec{x}, \vec{X})$ in π we will define the propositional proofs $F_S(\vec{m}, \vec{n})$ for the tautologies $S[\vec{m}; \vec{n}]$. This is done inductively for all sequents in π, starting with the axioms. The base case (where S is an axiom) and most of the induction step have been dealt with in the proof of Theorem VII.5.2 (see also Exercise X.1.17). The only remaining case for the induction step is the case of the Σ_1^B-***SIND'*** rule above.

Thus consider an instance of the rule Σ_1^B-***SIND'***. Suppressing other free variables in $\mathcal{S}_1$ and $\mathcal{S}_2$, for the lengths ℓ, m, n of α, β, γ we have

$$\mathcal{S}_1[\ell, m, n] \equiv \Gamma[n], A(\alpha)[\ell], S(\alpha, \beta)[\ell, m] \longrightarrow A(\beta)[m], \Delta[n],$$

$$\mathcal{S}_2[n] \equiv \Gamma[n], A(\varnothing)[] \longrightarrow A(\gamma)[n], \Delta[n].$$

We need to show that for each n, $\mathcal{S}_2[n]$ can be derived from $\mathcal{S}_1[\ell, m, n]$ (for polynomially many values of ℓ and m) by some polynomial size G_1 derivation. (Furthermore, the derivation is computable by an ***FTC***^{0} function.)

At first sight such a derivation might seem impossible. Informally, assuming that both Γ and Δ are empty, then $\mathcal{S}_1$ allows us to obtain $A(\alpha + 1)$ from $A(\alpha)$ (here 1 is really the set $\{0\}$ and $+$ is the string addition function). So it appears that in order to get $A(\gamma)$ from $A(0)$ (i.e., $A(\varnothing)$) we need to use $\mathcal{S}_1$ γ times, i.e., we need exponentially many cuts.

Lemma X.1.22 below shows that the number of cuts can be effectively reduced to just a polynomial in $|\gamma|$, by showing roughly that using $\mathcal{S}_1$ we can obtain $A(\beta)$ from $A(\alpha)$ for any β of length $|\beta| = |\alpha| + 1$.

Formally we use the following notation:

NOTATION. Let $S_k(X, Y)$ be the Σ_0^B formula

$$X \le Y \wedge Y \le X + \{k\}.$$

(Recall that $\{k\}$, or also $POW2(k)$, is an ***AC***^{0} function defined by $\{k\}(x) \leftrightarrow x = k$. See Example VIII.3.12.)

LEMMA X.1.22. *For each Σ_1^B formula $A(X)$ and distinct string variables $\alpha, \beta, \sigma, \delta$ there is a **FTC**^{0} function $H(k, d)$ which is provably in $\overline{VTC}^0$ a G_1 derivation whose nonlogical axioms are from the set*

$$\{A(\alpha)[\ell], S(\alpha, \beta)[\ell, m] \longrightarrow A(\beta)[m] : \ell, m \le d\}$$

and that contains all sequents in the set

$$\{A(\sigma)[s], S_k(\sigma, \delta)[s, n] \longrightarrow A(\delta)[n] : s, n \le d\}.$$

Lemma X.1.22 completes the induction step for describing the proofs $F_S(\vec{m}, \vec{n})$ of the translations $S[\vec{m}; \vec{n}]$ of sequents S in π. It can be verified that F_S is in ***FTC***^{0} when S is an axiom in π. When S is derived from $\mathcal{S}_1$ (and $\mathcal{S}_2$) then as in Theorem X.1.14 and Exercise X.1.17 it can be shown that F_S is obtained by composition from $F_{\mathcal{S}_1}$ (and $F_{\mathcal{S}_2}$) and some other ***FTC***^{0} functions (here we need also the ***FTC***^{0} functions from Lemma X.1.22). Thus F_S are in ***FTC***^{0} for all S in π. The fact that $\overline{VTC}^0$ proves that $F_S(\vec{m}, \vec{n})$ are proofs of $S[\vec{m}; \vec{n}]$ can be proved by induction on the sequent S. $\square$

Proof of lemma X.1.22. First we will describe $H(k, d)$ simply as a polynomial-size derivation. The definition is by induction on k. Then we will argue that H is in fact a FTC^0 function that provably in $\overline{VTC}^0$ computes the desired derivation. From now on we will denote the desired derivation by

$$\frac{\{A(\alpha)[\ell], S(\alpha, \beta)[\ell, m] \longrightarrow A(\beta)[m] \; : \; \ell, m \le d\}}{\{A(\sigma)[s], S_k(\sigma, \delta)[s, n] \longrightarrow A(\delta)[n] \; : \; s, n \le d\}} \tag{338}$$

Consider the base case, $k = 0$. Note that $S_0(\sigma, \delta)[s, n]$ is false if $n < s$ or $n > s + 1$, and in these cases

$$A(\sigma)[s], S_0(\sigma, \delta)[s, n] \longrightarrow A(\delta)[n] \tag{339}$$

can easily be shown to have polynomial size $G_0^\star$ proofs. So we focus on the cases $n = s$ or $n = s + 1$.

By Exercise X.1.19 there is provably in $\overline{VTC}^0$ an FTC^0-computable $G_0^\star$ derivation of

$$A(\sigma)[s], (\sigma = \delta)[s, s] \longrightarrow A(\delta)[s]. \tag{340}$$

Also, note that for $n \ne s$, the formula $(\sigma = \delta)[s, n]$ is false, and the sequent

$$A(\sigma)[s], (\sigma = \delta)[s, n] \longrightarrow A(\delta)[n] \tag{341}$$

can be shown to have polynomial size proof in $G_0^\star$.

By Exercise X.1.20 there are (provably in $\overline{VTC}^0$) FTC^0-computable $G_1^\star$ derivations

$$\frac{A(\alpha)[s], S(\alpha, \beta)[s, n] \longrightarrow A(\beta)[n]}{A(\sigma)[s], S(\sigma, \delta)[s, n] \longrightarrow A(\delta)[n]}$$

for $n = s$ and $n = s + 1$. Combine these derivations we obtain $G_1^\star$ derivations

$$\frac{\{A(\alpha)[s], S(\alpha, \beta)[s, m] \longrightarrow A(\beta)[m] \; : \; m \in \{s, s + 1\}\}}{\{A(\sigma)[s], (\sigma = \delta)[s, n] \vee S(\sigma, \delta)[s, n] \longrightarrow A(\delta)[n] \; : \; n \in \{s, s + 1\}\}} \tag{342}$$

Now note that V^0 proves

$$S_0(X, Y) \leftrightarrow (X = Y \vee S(X, Y)).$$

So by Theorem X.1.14 there is provably in $\overline{VTC}^0$ an FTC^0-computable $G_0^\star$ derivation of

$$S_0(\sigma, \delta)[s, n] \longrightarrow (\sigma = \delta)[s, n] \vee S(\sigma, \delta)[s, n]. \tag{343}$$

From this and (342) above we obtain a $G_1^\star$ derivation

$$\frac{\{A(\alpha)[s], S(\alpha, \beta)[s, m] \longrightarrow A(\beta)[m] \; : \; s \le d, \text{ and } m = s \text{ or } m = s + 1\}}{\{A(\sigma)[s], S_0(\sigma, \delta)[s, n] \longrightarrow A(\delta)[n] \; : \; s \le d, \text{ and } n = s \text{ or } n = s + 1\}}$$

Combine this and the derivations in (339) we obtain the derivation for the base case.

For the induction step, suppose that there is a polynomial size G_1 derivations of the form (338):

$$\frac{\{A(\alpha)[\ell], S(\alpha, \beta)[\ell, m] \longrightarrow A(\beta)[m] \ : \ \ell, m \le d\}}{\{A(\sigma)[s], S_k(\sigma, \delta)[s, n] \longrightarrow A(\delta)[n] \ : \ s, n \le d\}} \tag{344}$$

We will augment this derivation with additional derivations in order to obtain one that contains also all sequents in the set

$$\{A(\sigma)[s], S_{k+1}(\sigma, \delta)[s, n] \longrightarrow A(\delta)[n] \ : \ s, n \le d\}. \tag{345}$$

By Exercise X.1.20 there are $\pmb{FTC}^0$ functions that provably in $\overline{\pmb{VTC}}^0$ compute some derivations of the following sequents from the bottom sequents in (344):

$$\{A(\gamma)[n], S_k(\gamma, \delta)[n, p] \longrightarrow A(\delta)[p] \ : \ n, p \le d\}. \tag{346}$$

From the sequents in (346) and the sequents at the bottom of (344) we obtain

$$\{A(\sigma)[s], S_k(\sigma, \gamma)[s, n] \wedge S_k(\gamma, \delta)[n, p] \longrightarrow A(\delta)[p] \ : \ s, n, p \le d\}.$$

For each pair (s, p) $(s, p \le d)$, from the above sequents with $n = 0, 1, \ldots, p$ using the $\vee$-left and $\exists$-left rules we obtain

$$A(\sigma)[s], \big(\exists Z \le |\delta|(S_k(\sigma, Z) \wedge S_k(Z, \delta))\big)[s, p] \longrightarrow A(\delta)[p]. \tag{347}$$

Notice that $\{k\} + \{k\} = \{k + 1\}$, and

$$V^0 \vdash S_{k+1}(\sigma, \delta) \longrightarrow \exists Z \le |\delta|(S_k(\sigma, Z) \wedge S_k(Z, \delta)).$$

Therefore by Theorem X.1.14 there is provably in $\overline{\pmb{VTC}}^0$ an $\pmb{FTC}^0$-computable $G_0^\star$ proof of

$$S_{k+1}(\sigma, \delta)[s, p] \longrightarrow \big(\exists Z \le |\delta|(S_k(\sigma, Z) \wedge S_k(Z, \delta))\big)[s, p]. \tag{348}$$

From this and (347) we obtain the following member of (345):

$$A(\sigma)[s], S_{k+1}(\sigma, \delta)[s, p] \longrightarrow A(\delta)[p].$$

This completes the description of the polynomial size derivation (338). Observe that the top sequents in (344) are used more than once, so the resulting derivation is daglike.

Now we show that $H \in \pmb{FTC}^0$; the fact that provably in $\overline{\pmb{VTC}}^0$ the function $H(k, d)$ computes the desired derivations is straightforward. That $H \in \pmb{FTC}^0$ can be seen by observing that (i) $H(0, d)$ is a $\pmb{FTC}^0$ function, and (ii) $H(k + 1, d)$ is obtained from $H(k, d)$ by augmenting additional derivations that are computed by functions in $\pmb{FTC}^0$. In other words, the string $H(k, d)$ consists of disjoint fragments that can be defined independently by $\pmb{FTC}^0$ functions. Therefore $H(k, d)$ is in $\pmb{FTC}^0$. $\quad\square$

Recall that V^1 is Σ_1^B-conservative over TV^0 (Theorem VIII.3.10 and Corollary VIII.2.18) and $G_1^\star$ is equivalent to ePK for proving prenex Σ_1^q formulas (Theorem VII.4.16). Thus the V^i Translation Theorem (Theorem VII.5.2) shows that Σ_1^B theorems of TV^0 translate into families of propositional tautologies that have polynomial-size ePK proofs. The next exercise is to formalize in $\overline{VTC}^0$ a more direct proof of this fact.

*Exercise X.1.23 (Propositional Translation Theorem for TV^0).
Show, by translating the axiom MCV (Definition VIII.1.1) and using Theorem X.1.14, that for each Σ_1^B theorem $\varphi(\vec{x}, \vec{X})$ of TV^0 there is a function F_φ in FTC^0 that provably in $\overline{VTC}^0$ computes an ePK proof of $\varphi(\vec{x}, \vec{X})[\vec{m}; \vec{n}]$. (Hint: consider a free variable normal form LK^2-TV^0 proof π and treat the bits of the existentially quantified string variable in MCV that appear in antecedents in π as extension variables.)

X.2. The Reflection Principle

The Reflection Principle (RFN) for a proof system $\mathcal{F}$ states that $\mathcal{F}$ is sound, i.e., the endsequent of any $\mathcal{F}$-proof is a valid sequent. In order to state the principle we need to formalize the notion of truth definitions, i.e. the relation $(Z \models X)$ that holds iff the truth assignment Z satisfies a propositional formula X. It is straightforward that for $i \geq 1$ the relation $(Z \models X)$ is in Σ_i^P (resp. Π_i^P) whenever X is a Σ_i^q (resp. Π_i^q) formula. When X is a quantifier-free propositional formula, it is also straightforward that $(Z \models X)$ is a polytime relation and is Δ_1^B-definable in TV^0. (A difficult result, due to Buss, states that $(Z \models X)$ is an NC^1 relation when X is a quantifier-free. See Section X.3.2.) Formulas that represent the relations $(Z \models X)$ (for different classes of X) are presented in Section X.2.1.

Using the formulas expressing $(Z \models X)$ we can state and prove the following "back and forth" properties. On the one hand, let $\widehat{A}$ denote the string (of $\mathcal{L}_A^2$) that encodes a propositional formula $A(\vec{p})$. Then for all truth assignments Z, intuitively the propositional translations of $(Z \models \widehat{A})$ are equivalent to $A(\overrightarrow{p^Z})$, where $\overrightarrow{p^Z}$ are the values of $\vec{p}$ under Z. These equivalences will be stated as propositional tautologies, and we will give polytime algorithms that compute $G_0^\star$ proofs for them.

On the other hand, let $\varphi(Z)$ be a formula of $\mathcal{L}_A^2$ and $\widehat{A}$ be the string encoding the propositional translation $\varphi(Z)[n]$ of φ. Then for Z of length $|Z| = n$ we must have

$$(Z \models \widehat{A}) \leftrightarrow \varphi(Z).$$

We will show that this equivalence is a theorem of $\overline{VTC}^0$. Detailed discussions are given in Section X.2.2.

The Φ-RFN for $\mathcal{F}$ will be defined in Section X.2.3, where Φ is a class of formulas and $\mathcal{F}$ is a proof system. There we will show that the Φ-RFN for each system $\boldsymbol{G}_i^\star$ and $\boldsymbol{G}_i$ is provable in the associated theory, where Φ includes (at least) $\Sigma_i^q \cup \Pi_i^q$. In Section X.2.4 we will show that for $i \geq 1$ the Σ_{i+1}^q-RFN for $\boldsymbol{G}_i^\star$ (resp. $\boldsymbol{G}_i$) can be used to axiomatize the associated theories $\boldsymbol{V}^i$ (resp. $\boldsymbol{TV}^i$). Then in Section X.2.5 we will show that $\boldsymbol{G}_i^\star$ and $\boldsymbol{G}_i$ are the strongest (w.r.t. p-simulation) proof systems whose RFN can be proved in $\boldsymbol{V}^i$ and $\boldsymbol{TV}^i$, respectively.

Recall the Witnessing Theorem for $\boldsymbol{V}^1$ (Theorem VII.4.13). In Section X.2.6 we consider generally the problem of finding witness for a Σ_j^q formula

$$A(\vec{p}) \equiv \exists \vec{x} B(\vec{p}, \vec{x})$$

given a truth assignment to $\vec{p}$ and a $\boldsymbol{G}_i$ (or $\boldsymbol{G}_i^\star$) proof π of A. The Witnessing Problem is closely related to the RFN. Indeed, our proof of the fact that $\boldsymbol{V}^1$ proves the Σ_1^q-RFN for $\boldsymbol{G}_1^\star$ is by formalizing the proof of the Witnessing Theorem for $\boldsymbol{V}^1$ (Theorem VII.4.13). We will show in Section X.2.6 that the Witnessing Problems for the systems $\boldsymbol{G}_i$ and $\boldsymbol{G}_i^\star$ are complete for the classes that are definable in the associated theories.

X.2.1. Truth Definitions. Suppose that X encodes a (quantified) propositional formula. Then each string Z specifies a truth assignment to the variables p_i in X as follows:

$$p_i \text{ is assigned the value of } Z(i).$$

Thus all possible truth assignment to variables in X can be specified by strings Z of length $|Z| \leq |X|$.

Here we present $\mathcal{L}_A^2$-formulas that represent the relation

$$(Z \models X)$$

which holds for a truth assignment Z and a formula X iff Z satisfies X. We will consider separate cases depending on whether X is quantifier-free or X belongs to Σ_i^q or Π_i^q where $i \geq 1$.

First let

$$(Z \models_0 X)$$

hold iff X encodes a quantifier-free formula, and Z is a satisfying truth assignment to X. Lemma X.2.1 below follows from Exercise X.1.3 and the fact that $(Z \models_0 X)$ is in $\boldsymbol{P}$. (Recall the axiom $\Sigma_0^B\text{-}\boldsymbol{BIT}\text{-}\boldsymbol{REC}$ from Section VIII.3.2.) Nevertheless, we will give some details describing the formula $\varphi_0(y, X, Z, E)$ for the Lemma, since we will need them later. In Section X.3.2 we will show that $(Z \models_0 X)$ is indeed in $\boldsymbol{NC}^1$ and Δ_1^B-definable in $\boldsymbol{VNC}^1$.

LEMMA X.2.1. *There are a Σ_0^B formula $\varphi_0(y, X, E)$ and an $\mathcal{L}_A^2$ term $t_0(y, X)$ so that both*

$$(Z \models_0^\Sigma X) \equiv \exists E \le t_0 + 1 \big(\varphi_0^{rec}(t_0 + 1, X, Z, E) \wedge E(t_0) \big),$$

$$(Z \models_0^\Pi X) \equiv \forall E \le t_0 + 1 \big(\varphi_0^{rec}(t_0 + 1, X, Z, E) \supset E(t_0) \big)$$

represent $(Z \models_0 X)$ and such that

$$\boldsymbol{TV^0} \vdash (Z \models_0^\Sigma X) \leftrightarrow (Z \models_0^\Pi X).$$

PROOF IDEA. The formula $\varphi_0^{rec}(t_0 + 1, X, Z, E)$ asserts that E encodes a polytime algorithm that consists of two stages: first it verifies that X is a formula, then it evaluates X in a bottom up fashion. If the first stage rejects then the algorithm rejects, otherwise its output is the value of the evaluation and is stored in $E(t_0)$.

For the first stage in the algorithm we use the formula $\varphi_{FLA}(y, X, Y)$ from Corollary X.1.4 that essentially states that Y encodes a computation of the relation $FLA(X)$. The "check bit" $Y(t_{FLA})$ indicates whether the computation accepts. Thus, for $i \le t_{FLA}$ we have

$$\varphi_0(i, X, Z, E^{<i}) \leftrightarrow \varphi_{FLA}(i, X, E^{<i}).$$

If the first stage rejects (i.e., $E(t_{FLA})$ is false) then the algorithm rejects, i.e., for all $t_{FLA} < i \le t_0$ we have

$$\varphi_0(i, X, Z, E^{<i}) \leftrightarrow \bot.$$

Suppose now that $E(t_{FLA})$ is true. Let $n = |X|$. To encode the second stage we will store the value of each subformula $X[i, j]$ of X (for some $0 \le i \le j < n$) in the bit $E(a_{i,j})$, for distinct terms $a_{i,j} > t_{FLA}$ defined below.

In order to conform with the axiom scheme **BIT-REC**, where the bits $E(z)$ is computed from $E^{<z}$, and since the subformulas of X are evaluated bottom up, we can order the subformulas of X in nondecreasing order of their lengths, i.e., we want

$$a_{i,j} < a_{i',j'}$$

whenever $j' - i' > j - i$. Thus, let

$$a_{i,j} = t_{FLA} + 1 + (j - i + 1)\langle n, n \rangle + \langle i, j \rangle.$$

Now, for example if $X[i, j]$ is an atom p_s, then we have

$$\varphi_0(a_{i,j}, X, Z, E^{<a_{i,j}}) \leftrightarrow Z(s).$$

For another example, suppose that $X[i, j]$ is the formula

$$(C \wedge D)$$

where $C = X[i + 1, \ell]$ and $D = X[\ell + 2, j - 1]$ for some ℓ, then

$$\varphi_0(a_{i,j}, X, Z, E^{<a_{i,j}}) \leftrightarrow (E(a_{i+1,\ell}) \wedge E(a_{\ell+2,j-1})).$$

The "check bit" for E is $E(t_0)$, where $t_0 = a_{0,n-1}$. The value of this check bit is the value of the formula X. Also, for all other bits $E(r)$, where $t_{FLA} < r < t_0$ and $r \neq a_{i,j}$ for all subformulas $X[i, j]$ of X, we set $E(r)$ to $\top$ by having

$$\varphi_0(r, X, Z, E^{<r}) \leftrightarrow \top.$$

This completes the description of the formula $\varphi_0(i, X, Z, E)$. It is easy to see that

$$TV^0 \vdash (Z \models^{\Sigma}_0 X) \leftrightarrow (Z \models^{\Pi}_0 X). \qquad \square$$

Recall that $\widehat{A}$ denotes the $\mathcal{L}^2_A$ string encoding a propositional formula A.

Exercise X.2.2. Show that the theory V^0 proves

1) $(Z \models^{\Sigma}_0 \widehat{A \wedge B}) \leftrightarrow ((Z \models^{\Sigma}_0 \widehat{A}) \wedge (Z \models^{\Sigma}_0 \widehat{B}))$,
2) $(Z \models^{\Sigma}_0 \widehat{A \vee B}) \leftrightarrow ((Z \models^{\Sigma}_0 \widehat{A}) \vee (Z \models^{\Sigma}_0 \widehat{B}))$,
3) $(Z \models^{\Sigma}_0 \widehat{\neg A}) \leftrightarrow \neg(Z \models^{\Pi}_0 \widehat{A})$.

Show also that V^0 proves similar theorems with $\models^{\Pi}_0$ instead of $\models^{\Sigma}_0$.

Now we consider the classes of formulas Σ^q_i and Π^q_i (for $i \geq 1$). Here it can be seen that evaluating Σ^q_i (resp. Π^q_i) sentences can be done in Σ^P_i (resp. in Π^P_i). So in this case formulas that represent $(Z \models X)$ belong to Σ^B_i (resp. Π^B_i).

Lemma X.2.3. Let $1 \leq i \in \mathbb{N}$. There is a Σ^B_i formula $(Z \models_{\Sigma^q_i} X)$ that is true iff X encodes a Σ^q_i formula and the truth assignment Z satisfies X. Similarly, there is a Π^B_i formula $(Z \models_{\Pi^q_i} X)$ that is true iff X encodes a Π^q_i formula and the truth assignment Z satisfies X.

Proof Idea. We show how to construct $(Z \models_{\Sigma^q_i} X)$; the formula $(Z \models_{\Pi^q_i} X)$ is constructed in the same way. The idea is to encode quantified propositional variables by the bits of quantified string variables. Let A be the Σ^q_i formula encoded by X.

First suppose that A is a prenex formula of the form

$$\exists \vec{x}_i \forall \vec{x}_{i-1} \ldots Q\vec{x}_1 B$$

where B is a quantifier-free formula, and $Q \in \{\exists, \forall\}$: if i is odd then Q is $\exists$, otherwise Q is $\forall$. Then $(Z \models_{\Sigma^q_i} X)$ has the form

$$\exists X_i \leq n \, \forall X_{i-1} \leq n \ldots QX_1 \leq n\psi(\widehat{B}, X_1, \ldots, X_i, Z) \qquad (349)$$

where $n = |X|$, ψ is in Σ^B_1 if i is odd, and ψ is in Π^B_1 if i is even (so the whole formula is Σ^B_i in either case). Here ψ is obtained as in Lemma X.2.1 to express the fact that the truth assignment defined by Z and $X_1, X_2, \ldots, X_i$ satisfies the formula B.

Now suppose that A is not in prenex form. Note that by definition no string quantifier in a Σ^B_i formula is in the scope of a number quantifier or a Boolean connective. So first we have to put A into prenex form.

The procedure described in Theorem II.5.12 is sequential. A parallel procedure is as follows.

TC^0 *prenexification.* First we compute the quantifier depth of each quantified variable using the function *qdepth* mentioned on page 373. (Here we can assume that A has an outer most existential quantifier.) After renaming the quantified variables (so that they are distinct) we can safely move the quantifiers into their proper block in the prefix. Consider a quantifier $\exists x_i$ or $\forall x_i$ that occurs in X at position t. We will simply rename simultaneously all occurrences of x_i that are caught by this quantifier to x_{n+t}, where n is the length of the original formula A. Note that in A all variables have index at most n. Also, all variables (including both bound and free variables) in the new formula will have distinct indices.

Consider for example the following scenario:

$$\ldots \exists x_2 (\ldots \forall x_2 (\ldots x_2 \ldots) \ldots x_2 \ldots) \ldots$$

where the $\exists$ is at position 7 and the $\forall$ is at position 20, and n is 100. Then the first and the fourth occurrences of x_2 are renamed to x_{107}, while the other two occurrences of x_2 are renamed to x_{120}.

Finally we must determine (in TC^0) whether each (original) quantifier is in the scope of an odd number of $\neg$'s, and if so change it from $\forall$ to $\exists$ or from $\exists$ to $\forall$.

It can be seen that the length of the resulting formula is at most n^2. It can be seen that the transformation can be done by a TC^0 algorithm. In fact, it can be shown that there are a Σ_1^B formula $\varphi_1(X, X')$ and a Π_1^B formula $\varphi_2(X, X')$ that are true iff X' is the result of the transformation of X described above, and such that

$$VTC^0 \vdash \varphi_1(X, X') \leftrightarrow \varphi_2(X, X')$$

and

$$VTC^0 \vdash \exists X' \le n^2 \varphi_1(X, X').$$

Now the Σ_i^B formula $Z \models_{\Sigma_i^q} X$ has the form

$$\exists X_i \le n^2 \, \forall X_{i-1} \le n^2 \, \ldots QX_1 \le n^2 \, QX' \le n^2 \, \psi(X', \vec{X}, Z) \qquad (350)$$

where Q is $\exists$ if i is odd and Q is $\forall$ otherwise. Suppose that i is odd. Then ψ is a Σ_1^B formula; it is is obtained from φ_1 and the Σ_1^B formula (obtained as in Lemma X.2.1) that expresses the fact that the truth assignment defined by Z and $X_1, X_2, \ldots, X_i$ satisfies the formula coded by X'. The case where i is even is similar.   □

EXERCISE X.2.4. Let A, B be Σ_i^q formulas. Show that the following are theorems of V^0:

1) $(Z \models_{\Sigma_i^q} \widehat{A \wedge B}) \leftrightarrow ((Z \models_{\Sigma_i^q} \widehat{A}) \wedge (Z \models_{\Sigma_i^q} \widehat{B}))$.
2) $(Z \models_{\Sigma_i^q} \widehat{A \vee B}) \leftrightarrow ((Z \models_{\Sigma_i^q} \widehat{A}) \vee (Z \models_{\Sigma_i^q} \widehat{B}))$.
3) $(Z \models_{\Sigma_i^q} \widehat{A}) \leftrightarrow \neg(Z \models_{\Pi_i^q} \widehat{\neg A})$.

4) $(Z \models_{\Sigma_i^q} \widehat{\exists x A(x)}) \leftrightarrow ((Z \models_{\Sigma_i^q} \widehat{A(\bot)}) \vee (Z \models_{\Sigma_i^q} \widehat{A(\top)}))$.

5) $(Z \models_{\Sigma_i^q} \widehat{\forall x A(x)}) \leftrightarrow ((Z \models_{\Sigma_i^q} \widehat{A(\bot)}) \wedge (Z \models_{\Sigma_i^q} \widehat{A(\top)}))$ (if $\forall x A(x)$ is a Σ_i^q formula).

Give similar theorems of V^0 that involve $(Z \models_{\Pi_i^q} X)$.

X.2.2. Truth Definitions vs Propositional Translations. In this section we consider a kind of back and forth relationship between propositional translation (from first-order theories to proof systems) and the formalization of propositional proofs in our theories.

Consider for example a quantifier-free propositional formula

$$A(p_0, p_1, \ldots, p_{n-1}).$$

As before let $\widehat{A}$ be the encoding of A. Recall that for Z of length $|Z| = n + 1$, the intended meaning of $(Z \models_0 \widehat{A})$ defined in Section X.2.1 is

$$A(p_0^Z, p_1^Z, \ldots, p_{n-1}^Z).$$

Therefore, intuitively, the propositional formulas

$$(Z \models_0^\Pi \widehat{A})[n+1] \qquad \text{and} \qquad (Z \models_0^\Sigma \widehat{A})[n+1]$$

should both be equivalent to $A(p_0^Z, p_1^Z, \ldots, p_{n-1}^Z)$.

We will show that there are polytime algorithms that compute $G_0^\star$ proofs of these equivalences. In addition, if A is a Φ formula (where $\Phi \in \{\Sigma_i^q, \Pi_i^q\}$ for $i \geq 1$) then there is a polytime algorithm that computes a $G_i^\star$ proof of the equivalence between $A(\overrightarrow{p^Z})$ and $(Z \models_\Phi \widehat{A})[n+1]$. In other words, the systems $G_i^\star$ prove the correctness of the composition of our truth definitions and translation (and the $G_i^\star$ proofs can be computed in polytime).

In Theorem X.2.10 we will turn the above observation around and show that the theory VTC^0 proves the correctness of the composition of propositional translation and truth definition.

For the next theorem recall (Definition X.1.6) that for a constant string $\widehat{A_0}$ of length m, the notation

$$(Z \models_0^\Sigma \widehat{A_0})[n+1]$$

denotes the propositional formula with variables $\overrightarrow{p^Z}$ that is obtained from the translation $(Z \models_0^\Sigma X)[m, n+1]$ (where $m = |\widehat{A_0}|$) by plugging the values of the bits $\widehat{A_0}(j)$ for p_j^X. Since the truth values of the variables $\overrightarrow{p^Z}$ are intended to assign truth values to the free variables $\vec{p}$ of $A_0(\vec{p})$ we generally assume that n is greater than or equal to the number of free variables in A_0.

THEOREM X.2.5. *There are polytime algorithms that, given a quantifier-free formula $A_0(\vec{p})$ that has no more than n free variables $\vec{p}$, compute $G_0^\star$ proofs of the sequents:*

(a) $(Z \models_0^\Pi \widehat{A_0})[n+1] \longrightarrow A_0(\overrightarrow{p^Z})$.

(b) $(Z \models_0^\Sigma \widehat{A_0})[n+1] \longrightarrow A_0(\overrightarrow{p^Z})$.

(c) $A_0(\overrightarrow{p^Z}) \longrightarrow (Z \models_0^\Sigma \widehat{A_0})[n+1]$.

(d) $A_0(\overrightarrow{p^Z}) \longrightarrow (Z \models_0^\Pi \widehat{A_0})[n+1]$.

PROOF. We will prove (a) and (b) and leave the proofs of (c) and (d) as an exercise. Refer to the proof of Lemma X.2.1 for detailed description of the formula $\varphi_0^{rec}(t_0 + 1, \widehat{A_0}, Z, E)$. Let $m = |\widehat{A_0}|$ and $r = t_0(m)$.

(a) We will construct a G_0^* proof of the sequent, and it can be verified that the construction is in polytime. Recall (Lemma X.2.1) that $(Z \models_0^\Pi \widehat{A_0})$ is the Π_1^B formula

$$(Z \models_0^\Pi \widehat{A_0}) \equiv \forall E \le t_0 + 1\left(\varphi_0^{rec}(t_0 + 1, \widehat{A_0}, Z, E) \supset E(t_0)\right)$$

so $(Z \models_0^\Pi \widehat{A_0})[n+1]$ is

$$\forall p_0^E \forall p_1^E \ldots \forall p_{r-1}^E \bigwedge_{k=0}^{r+1} \left(\varphi_0^{rec}(t_0 + 1, \widehat{A_0}, Z, E) \supset E(t_0)\right)[n+1, k].$$

The idea is to prove an instance of the following sequent

$$\bigwedge_{k=0}^{r+1} \left(\varphi_0^{rec}(t_0 + 1, \widehat{A_0}, Z, E) \supset E(t_0)\right)[n+1, k] \longrightarrow A_0(\overrightarrow{p^Z}) \qquad (351)$$

where the variables $\overrightarrow{p^E}$ have the right values, and then apply the $\forall$-left rule repeatedly. In particular, note that the first $(t_{FLA} + 1)$ bits of E parse $\widehat{A_0}$ and the remaining bits evaluate A_0 bottom up: if the substring $\widehat{A_0}[u, v]$ of $\widehat{A_0}$ encodes a subformula $A_{u,v}$ of A_0, then

$$E(a_{u,v}) \leftrightarrow A_{u,v}$$

for the term $a_{u,v}$ described in the proof of Lemma X.2.1. Thus, as in Lemma X.1.7 we can compute the "parsing" bits of E (and compute a cut-free PK^* proof for their correctness) in polytime. For the "evaluating" bits in E, the only relevant bits are bits of the form $E(a_{u,v})$ as above, and here we substitute the subformulas $A_{u,v}$ for them.

More precisely, consider the antecedent of (351). For $k = r + 1$, $E(t_0)$ translates into $\top$, so

$$\left(\varphi_0^{rec}(t_0 + 1, \widehat{A_0}, Z, E) \supset E(t_0)\right)[n+1, r+1]$$

is $\top$ and is deleted from the conjunction. For $k \le r$, $E(t_0)$ translates into $\bot$, so

$$\left(\varphi_0^{rec}(t_0 + 1, \widehat{A_0}, Z, E) \supset E(t_0)\right)[n+1, k] \equiv$$
$$\neg\varphi_0^{rec}(t_0 + 1, \widehat{A_0}, Z, E)[n+1, k].$$

Therefore to derive (351) it suffices to derive an instance of the following sequent (where the free variables $\overrightarrow{p^E}$ are replaced by appropriate formulas):

$$\longrightarrow \varphi_0^{rec}(t_0+1, \widehat{A_0}, Z, E)[n+1, 0], \varphi_0^{rec}(t_0+1, \widehat{A_0}, Z, E)[n+1, 1], \ldots,$$
$$\varphi_0^{rec}(t_0+1, \widehat{A_0}, Z, E)[n+1, r], A_0(\overrightarrow{p^Z}).$$

In fact, we will give a $PK^\star$ proof of an instance of the following sequent and then apply the weakening rule:

$$\longrightarrow \varphi_0^{rec}(t_0+1, \widehat{A_0}, Z, E)[n+1, r], A_0(\overrightarrow{p^Z}). \tag{352}$$

It remains to describe a substitution for the free variables $\overrightarrow{p^E}$ so that (352) has a polynomial size $PK^\star$ proof.

Recall that

$$\varphi_0^{rec}(t_0+1, \widehat{A_0}, Z, E) \equiv \forall i \leq t_0(E(i) \leftrightarrow \varphi_0(i, \widehat{A_0}, Z, E^{<i})).$$

So $\varphi_0^{rec}(t_0+1, \widehat{A_0}, Z, E)[n+1, r]$ is

$$\bigwedge_{i=0}^{r-2} ((E(i) \leftrightarrow \varphi_0(i, \widehat{A_0}, Z, E^{<i}))[n+1, r]) \wedge$$
$$\varphi_0(r-1, \widehat{A_0}, Z, E^{<r-1})[n+1, r] \wedge \neg(\varphi_0(r, \widehat{A_0}, Z, E^{<r})[n+1, r]).$$

Also recall that the formula $\varphi_0(i, \widehat{A_0}, Z, E)$ is defined so that:

- bits $E(0), E(1), \ldots, E(t_{FLA})$ "parse" $\widehat{A_0}$;
- if $\widehat{A_0}[u, v]$ is a subformula of A_0 of the form an atom p_s, then

$$\varphi_0(a_{u,v}, \widehat{A_0}, Z, E^{<a_{u,v}}) \leftrightarrow Z(s);$$

- if $\widehat{A_0}[u, v]$ is a subformula of A_0 of the form

$$(A_0[u+1, w] \wedge A_0[w+2, v-1])$$

then

$$\varphi_0(a_{u,v}, \widehat{A_0}, Z, E^{<a_{u,v}}) \leftrightarrow E(a_{u+1,w}) \wedge E(a_{w+2,v-1})$$

and similarly for other kinds of connectives;
- if $t_{FLA} < i < t_0$ ($t_0 = a_{0,m-1}$) and $u \neq a_{u,v}$ for all $0 \leq u \leq v < m$, then

$$\varphi_0(i, \widehat{A_0}, Z, E^{<i}) \leftrightarrow \top.$$

In polytime we can compute the right Boolean values $b_0, b_1, \ldots, b_{t_{FLA}}$ for the bits $E(0), E(1), \ldots, E(t_{FLA})$ and $PK^\star$ proofs of their correctness, i.e., $PK^\star$ proofs of

$$\longrightarrow b_i \leftrightarrow B_i \tag{353}$$

for $0 \le i \le t_{FLA}$, where B_i is the sentence obtained from

$$\varphi_0(i, \widehat{A_0}, Z, E^{<i})[n+1, r]$$

by substituting b_j for p_j^E, for $j < i$.

From now on we will assume that $b_0, b_1, \ldots, b_{t_{FLA}}$ have been substituted for the bits $E(0), E(1), \ldots, E(t_{FLA})$. Suppose that $\widehat{A_0}[u, v]$ is an atomic subformula p_s of A_0, then the following sequent is valid and has a short $PK^\star$ proof:

$$\longrightarrow \varphi_0(a_{u,v}, \widehat{A_0}, Z, E^{<a_{u,v}})[n+1, r] \leftrightarrow p_s^Z.$$

In addition, if $\widehat{A_0}[u, v]$ is a subformula of A_0 of the form

$$(A_0[u+1, w] \wedge A_0[w+2, v-1])$$

then we have a short $PK^\star$ proof of the sequent

$$\longrightarrow \varphi_0(a_{u,v}, \widehat{A_0}, Z, E^{<a_{u,v}})[n+1, r] \leftrightarrow (p_{a_{u+1,w}}^E \wedge p_{a_{w+2,v-1}}^E).$$

Similarly for other subformulas of A_0. Also, for $t_{FLA} < i < r$ and $i \ne a_{u,v}$ for all subformulas $\widehat{A_0}[u, v]$ of A_0 (for $0 \le u \le v < m$), then there are short $PK^\star$ proof of

$$\longrightarrow \varphi_0(i, \widehat{A_0}, Z, E^{<i})[n+1, r]. \tag{354}$$

(In particular, it can be verified that $r - 1 \ne a_{u,v}$ for $0 \le u \le v < m$, so

$$\longrightarrow \varphi_0(r-1, \widehat{A_0}, Z, E^{<r-1})[n+1, r] \tag{355}$$

has a short $PK^\star$ proof.)

Thus, if $\widehat{A_0}[u, v]$ is a proper subformula C of A_0, we will substitute C for $p_{a_{u,v}}^E$. Also, for i such that $t_{FLA} < i < r$ and $i \ne a_{u,v}$ for all subformulas $\widehat{A_0}[u, v]$ of A_0, we will substitute $\top$ for p_i^E. The above argument shows that for any subformula $C = \widehat{A_0}[u, v]$ of A_0 we can derive

$$\longrightarrow C \leftrightarrow \varphi_0(a_{u,v}, \widehat{A_0}, Z, E^{<a_{u,v}})[n+1, r]. \tag{356}$$

In particular we can derive

$$\longrightarrow A_0 \leftrightarrow \varphi_0(r, \widehat{A_0}, Z, E^{<r})[n+1, r]. \tag{357}$$

Now it can be seen that under the described substitution, the sequent (352) can be derived from the sequents (353), (354), (355), (356) and (357).

(b) As in (a) we will construct a $G_0^\star$ proof of the given sequent and leave it to the reader to verify that the construction is in polytime.

Recall that

$$(Z \models_0^\Sigma \widehat{A_0}) \equiv \exists E \le t_0 + 1 \left(\varphi_0^{rec}(t_0 + 1, \widehat{A_0}, Z, E) \wedge E(t_0) \right).$$

So by definition, $(Z \vDash_0^\Sigma \widehat{A_0})[n + 1]$ is (the simplification of)

$$\exists p_0^E \exists p_1^E \ldots \exists p_{r-1}^E \bigvee_{k=0}^{r+1} ((\varphi_0^{rec}(t_0 + 1, \widehat{A_0}, Z, E) \wedge E(t_0))[n + 1, k]).$$

We have $E(t_0)[r + 1] =_{\text{def}} \top$, and $E(t_0)[k] =_{\text{def}} \bot$ for $k \le r$. Therefore

$$(Z \vDash_0^\Sigma \widehat{A_0})[n + 1] =_{\text{def}}$$
$$\exists p_0^E \exists p_1^E \ldots \exists p_{r-1}^E (\varphi_0^{rec}(t_0 + 1, \widehat{A_0}, Z, E)[n + 1, r + 1]).$$

Thus, to prove the given sequent we will give a **PK*** proof of the sequent (358) in Lemma X.2.6 below and then apply repeatedly the rule ∃-left. We conclude the proof by proving Lemma X.2.6.   □

Lemma X.2.6. *Let A_0 and n, r be as in Theorem X.2.5 and its proof. Then there is a polytime algorithm that computes a* **PK*** *proof of the following sequent:*

$$\varphi_0^{rec}(t_0 + 1, \widehat{A_0}, Z, E)[n + 1, r + 1] \longrightarrow A_0(\overrightarrow{p^Z}). \tag{358}$$

Proof. We have

$$\varphi_0^{rec}(t_0 + 1, \widehat{A_0}, Z, E) \equiv \forall i \le t_0(E(i) \leftrightarrow \varphi_0(i, \widehat{A_0}, Z, E^{<i})).$$

So by definition (recall $E(t_0)[r + 1] \equiv \top$):

$$\varphi_0^{rec}(t_0 + 1, \widehat{A_0}, Z, E)[n + 1, r + 1] \equiv$$
$$\bigwedge_{i=0}^{r-1} (p_i^E \leftrightarrow \varphi_0(i, \widehat{A_0}, Z, E^{<i})[n + 1, r + 1]) \wedge$$
$$\varphi_0(t_0, \widehat{A_0}, Z, E^{<t_0})[n + 1, r + 1].$$

Let $B_0, B_1, \ldots, B_{r-1}$ be the correct values of $\overrightarrow{p^E}$, i.e.,

- $B_0, B_1, \ldots, B_{t_{FLA}}$ are the (only) Boolean values of the bits

$$E(0), E(1), \ldots, E(t_{FLA})$$

  that correctly parse the formula A_0;
- for a subformula $C_{u,v}$ of A_0 that is encoded by $\widehat{A_0}[u, v]$ (for $0 \le u \le v < m$), $B_{a_{u,v}}$ is $C_{u,v}$;
- for $t_{FLA} < i < r$ such that $i \ne a_{u,v}$ for all $0 \le u \le v < m$, $B_i \equiv \top$.

For $1 \le i \le r$ let

$$\Lambda_i = p_0^E \leftrightarrow B_0, \ldots, p_{i-1}^E \leftrightarrow B_{i-1}.$$

Now we can prove by induction that there are polynomial size $PK^\star$ proofs of the following sequents:

$$p_0^E \leftrightarrow \varphi_0(0, \widehat{A_0}, Z, E^{<0})[n+1, r+1] \longrightarrow p_0^E \leftrightarrow B_0,$$

$$p_1^E \leftrightarrow \varphi_0(1, \widehat{A_0}, Z, E^{<1})[n+1, r+1], \Lambda_1 \longrightarrow p_1^E \leftrightarrow B_1,$$

$$\dots$$

$$p_{r-1}^E \leftrightarrow \varphi_0(r-1, \widehat{A_0}, Z, E^{<r-1})[n+1, r+1], \Lambda_{r-1} \longrightarrow p_{r-1}^E \leftrightarrow B_{r-1},$$

$$\varphi_0(r, \widehat{A_0}, Z, E^{<r})[n+1, r+1], \Lambda_r \longrightarrow A_0(\overrightarrow{p^Z}).$$

From these we can obtain a polynomial size $PK^\star$ proof of (358).   □

EXERCISE X.2.7. Prove parts (c) and (d) of Theorem X.2.5.

THEOREM X.2.8. *Let $i \geq 1$ and $\Phi \in \{\Sigma_i^q, \Pi_i^q\}$. There are polytime algorithms that on input a prenex Φ formula $A_0(\vec{p})$, with no more than n free variables $\vec{p}$, compute $G_0^\star$ proofs of the following sequents:*

$$(Z \models_\Phi \widehat{A_0})[n+1] \longrightarrow A_0(\overrightarrow{p^Z}) \tag{359}$$

and

$$A_0(\overrightarrow{p^Z}) \longrightarrow (Z \models_\Phi \widehat{A_0})[n+1].$$

If A_0 is is not a prenex formula, then the proofs are in $G_i^\star$.

PROOF SKETCH. We consider the first sequent; the argument for the second is similar. Suppose that A_0 is Σ_i^q, so A_0 has the form

$$\exists \vec{x}_i \forall \vec{x}_{i-1} \dots Q\vec{x}_1 B_0(\vec{x}_1, \dots, \vec{x}_i, \vec{p}).$$

Then by (349) $(Z \models_{\Sigma_i^q} \widehat{A_0})$ has the form (for $m = |\widehat{A_0}|$)

$$\exists X_i \leq m \, \forall X_{i-1} \leq m \, \dots QX_1 \leq m \psi(\widehat{B_0}, X_1, \dots, X_i, Z).$$

Suppose that i is even, so ψ is Π_1^B and Q is $\forall$. By a slight generalization of part (a) of Theorem X.2.5 we can compute in polytime a $G_0^\star$ proof of

$$\psi(\widehat{B_0}, X_1, X_2, \dots, X_i, Z)[m+1, m+1, \dots, m+1, n+1]$$
$$\longrightarrow B_0(\overrightarrow{p^{X_1}}, \dots, \overrightarrow{p^{X_i}}, \overrightarrow{p^Z}). \tag{360}$$

To turn this proof into a $G_0^\star$ proof of (359) involves applying the quantifier introduction rules together with $\vee$ and $\wedge$ introduction. To see how this is done, refer to Section VII.5 on propositional translations, formulas (149) and (150). Since the innermost quantifier of A_0 is $\forall X_1$ we refer to (150). Starting with (360) we apply successive weakenings and $\wedge$-left to obtain the required conjunction on the left side, and then apply $\forall$-left repeatedly to quantify the variables of $\overrightarrow{p^{X_1}}$, and finally $\forall$-right repeatedly to quantify the same variables on the right.

The next quantifier of A_0 is $\exists X_2$. According to (149) the translation on the left has a disjunction over all lengths k of X_2 from $k = 0$ to $m+1$.

Here we need the fact that assigning a length $k < m + 1$ to X_2 implicitly assigns some high-order bits of X_2 to $\perp$. Thus (360) continues to hold with the second $m + 1$ on the left replaced by k, provided the high-order variables for $\overrightarrow{p^{X_2}}$ on the right are replaced by $\perp$. Hence for all values of k the sequent holds when on the right all variables in $\overrightarrow{p^{X_2}}$ are existentially quantified (after universal quantifiers have been applied to the variables $\overrightarrow{p^{X_1}}$). Now we can put these derivations together and successively apply $\vee$-left, followed by repeated applications of $\exists$-left to quantify the variables of $\overrightarrow{p^{X_2}}$. We continue in this way until all variables associated with the X_js have been quantified, to obtain a $G_0^\star$ proof of (359).

For the second statement, first let A_0' be the prenex formula equivalent to A_0 as output by the TC^0 prenexification procedure described in the proof of Lemma X.2.3. We leave the proofs of the following facts as an exercise:

EXERCISE X.2.9. Show that there are polytime algorithms that compute $G_i^\star$ proofs of the following sequents:

$$A_0' \longrightarrow A_0 \qquad \text{and} \qquad A_0 \longrightarrow A_0'.$$

(Hint: construct the proofs by structural induction on A_0.)

Consider the first sequent:

$$(Z \models_\Phi \widehat{A_0})[n + 1] \longrightarrow A_0(\overrightarrow{p^Z}).$$

By the first statement there is a polynomial size $G_0^\star$ proof of

$$(Z \models_\Phi \widehat{A_0'})[n + 1] \longrightarrow A_0'(\overrightarrow{p^Z}).$$

The desired sequent can now be derived from this and $A_0' \longrightarrow A_0$ using cut on the Σ_i^q prenex formula A_0'.

The second sequent is derived similarly.   □

Now we prove the category-theoretic reverse direction of Theorems X.2.5 and X.2.8. Let $\varphi(Z)$ be a Σ_i^B formula whose only free variable is Z (for some $i \geq 0$). By Corollary X.1.13 there is an FTC^0 function $T_\varphi(n)$ that provably in $\overline{VTC}^0$ computes the encoding of $\varphi(Z)[n]$, for all n. Formally,

$$T_\varphi(n) = \widehat{\varphi(Z)[n]}$$

and

$$\overline{VTC}^0 \vdash \forall n \, Fla^\Sigma(T_\varphi(n))$$

where $n = |Z|$. Now intuitively it should be clear that

$$(Z \models_{\Sigma_i^q} \widehat{\varphi(Z)[n]}) \iff \varphi(Z).$$

We will show that this equivalence is indeed provable in our theory $\overline{VTC}^0$.

NOTATION. $(Z \models_{\Sigma_0^q} X)$ and $(Z \models_{\Pi_0^q} X)$ are defined to be $Z \models_0^\Sigma X$ and $Z \models_0^\Pi X$, respectively.

THEOREM X.2.10. *Let $i \geq 0$ and $\varphi(Z)$ be a Σ_i^B formula with a single free variable Z as shown. Then*

$$\overline{VTC}^0 \vdash n = |Z| \supset ((Z \models_{\Sigma_i^q} \widehat{\varphi(Z)[n]}) \leftrightarrow \varphi(Z)).$$

Similarly, if $\varphi(Z)$ is Π_i^B, then

$$\overline{VTC}^0 \vdash n = |Z| \supset ((Z \models_{\Pi_i^q} \widehat{\varphi(Z)[n]}) \leftrightarrow \varphi(Z)).$$

PROOF IDEA. First consider the case $i = 0$. Suppose that φ is a Σ_0^B formula. Reasoning in $\overline{VTC}^0$. Let $n = |Z|$. We will show that

$$(Z \models_0^\Sigma \widehat{\varphi(Z)[n]}) \leftrightarrow \varphi(Z).$$

The fact that

$$(Z \models_0^\Pi \widehat{\varphi(Z)[n]}) \leftrightarrow \varphi(Z)$$

is similar.

Let A denote $\varphi(Z)[n]$. Recall from the proof of Lemma X.2.1 that $(Z \models_0^\Sigma \widehat{A})$ has the form:

$$\exists E \leq t_0 + 1(\varphi_0^{rec}(t_0 + 1, \widehat{A}, Z, E) \wedge E(t_0))$$

where the first $(t_{FLA} + 1)$ bits of E encode a computation that parses the formula $\widehat{A}$, and the remaining bits in E evaluate A in a bottom up fashion, where the value of a subformula $A_{i,j}$ (encoded by $\widehat{A}[i, j]$) is stored as the bit $E(a_{i,j})$ for the term $a_{i,j}$ as in the proof of Lemma X.2.1 (note that $a_{0,m-1} = t_0$, where $m = |\widehat{A}|$).

First we show that

$$(Z \models_0^\Sigma \widehat{A}) \supset \varphi(Z).$$

Let E satisfy $\varphi_0^{rec}(t_0 + 1, \widehat{A}, Z, E) \wedge E(t_0)$. Then we can show by structural induction on the (constant number of) subformulas φ_k of φ that

$$\varphi_k \leftrightarrow E(a_{\ell_k, r_k}) \tag{361}$$

where ℓ_k, r_k are the indices so that $\widehat{A}[\ell_k, r_k]$ encodes the translation of φ_k. As a result, from $E(t_0)$ (i.e., $E(a_{0,m-1})$) we conclude $\varphi(Z)$.

Now we show that

$$\varphi(Z) \supset (Z \models_0^\Sigma \widehat{A}).$$

Here we need to prove the existence of the string E that parses and then evaluates $\widehat{A}$. Recall Corollary X.1.13 that $\widehat{A} = \widehat{\varphi(Z)[n]}$ is provably computable in $\overline{VTC}^0$ by an FTC^0 function. So, informally, the "parsing" part in E (i.e., up to bit $E(t_{FLA})$) exists because

$$\overline{VTC}^0 \vdash Fla^\Sigma(\widehat{A}).$$

(This part of E can be extracted from the string Y that satisfies $\varphi_{FLA}^{rec}(t_{FLA}+1, \widehat{A}, Y) \wedge Y(t_{FLA})$.)

The "evaluating" part in E can be proved to exist by $\Sigma_0^B(\mathcal{L}_{FTC^0})$-**COMP** using the observation (361). The fact that these bits satisfy φ_0^{rec} is straightforward. Note that by assuming that $\varphi(Z)$ is true we also have that $E(a_{0,m-1})$ (i.e., $E(t_0)$) is true.

Now consider the case $i = 2$; the cases for other values $i \geq 1$ are similar. First, suppose that $\varphi(Z)$ is a Σ_2^B formula. Let $n = |Z|$ as before. Without loss of generality, suppose that $\varphi(Z)$ has the form

$$\exists X \leq t(|Z|) \forall Y \leq t(|Z|) \psi(X, Y, Z)$$

where ψ is a Σ_0^B formula. Then $\varphi(Z)[n]$ has the form (recall (149) and (150), page 191):

$$\exists p_0^X \ldots \exists p_{r-2}^X \bigvee_{\ell=0}^{r} \forall p_0^Y \ldots \forall p_{r-2}^Y \bigwedge_{m=0}^{r} \psi(X, Y, Z)[\ell, m, n]$$

where $r = t(n)$.

In defining $(Z \models_{\Sigma_2^q} \varphi(Z)[n])$ (see Lemma X.2.3) we first get the following prenex form of $\varphi(Z)[n]$ (with bound variables renamed):

$$\exists \overrightarrow{p^{X'}} \forall \overrightarrow{p^{Y'}} \bigvee_{\ell=0}^{r} \bigwedge_{m=0}^{r} A_{\ell,m}(\overrightarrow{p^{X'}}, \overrightarrow{p^{Y'}}, \overrightarrow{p^{Z}})$$

where $\overrightarrow{p^{X'}}$ is a sequence of $(r-1)$ distinct variables obtained by renaming $\overrightarrow{p^X}$ and $\overrightarrow{p^{Y'}}$ contains $(r+1)(r-1)$ distinct variables resulting from renaming $\overrightarrow{p^Y}$. Also, $A_{\ell,m}(\overrightarrow{p^{X'}}, \overrightarrow{p^{Y'}}, \overrightarrow{p^{Z}})$ is $\psi(X, Y, Z)[\ell, m, n]$ with the bound variables renamed. Note that the renaming of variables is performed by a TC^0 function. Note also that for $0 \leq \ell \neq \ell' \leq r$ and $0 \leq m, m' \leq r$, $A_{\ell,m}(\overrightarrow{p^{X'}}, \overrightarrow{p^{Y'}}, \overrightarrow{p^{Z}})$ and $A_{\ell',m'}(\overrightarrow{p^{X'}}, \overrightarrow{p^{Y'}}, \overrightarrow{p^{Z}})$ contain disjoint subsets of $\overrightarrow{p^{Y'}}$ corresponding to disjoint substrings of Y'.

Now the formula $(Z \models_{\Sigma_2^q} \varphi(Z)[n])$ has the form

$$\exists X' \leq s \forall Y' \leq s (X', Y', Z \models_0^\Pi \widehat{A})$$

for some term s, where

$$A \equiv \bigvee_{\ell=0}^{r} \bigwedge_{m=0}^{r} A_{\ell,m}(\overrightarrow{p^{X'}}, \overrightarrow{p^{Y'}}, \overrightarrow{p^{Z}})$$

and $(X', Y', Z \models_0^\Pi \widehat{A})$ is the Π_1^B formula obtained as in Lemma X.2.1 expressing the fact that the truth assignment specified by X', Y', Z satisfies A.

We mentioned above that the renaming functions are in TC^0. It can also be seen that their inverses are also in TC^0. Thus, there are TC^0

functions $F(\ell, X')$, $G(\ell, m, Y')$ and $F'(X)$, $G'(\ell, Y)$ where

$$|F(\ell, X')| = \ell, \quad |G(\ell, m, Y')| = m, \quad |F'(X)| \leq s, \quad |G'(\ell, Y)| \leq s$$

such that (using the case $i = 0$ above, and write $A_{\ell,m}$ for $A_{\ell,m}(\overrightarrow{p^{X'}}, \overrightarrow{p^{Y'}}, \overrightarrow{p^Z})$):

- for $|X| = \ell$:

$$\overrightarrow{VTC}^0 \vdash \psi(X, G(\ell, m, Y'), Z) \leftrightarrow \left(F'(X), Y', Z \models_0^\Pi \widehat{A_{\ell,m}}\right)$$

and

- for $|Y| = m$ and any string Y' that shares with $G'(\ell, Y)$ the substrings corresponding to the $\overrightarrow{p^{Y'}}$-variables in $A_{\ell,m}$:

$$\overrightarrow{VTC}^0 \vdash \psi(F(\ell, X'), Y, Z) \leftrightarrow \left(X', Y', Z \models_0^\Pi \widehat{A_{\ell,m}}\right).$$

From this it can be shown that

$$\left(\exists X' \leq s \forall Y' \leq s (X', Y', Z \models_0^\Pi \widehat{A})\right) \leftrightarrow \exists X \leq t \forall Y \leq t \psi(X, Y, Z)$$

so we obtain

$$\overrightarrow{VTC}^0 \vdash \left(Z \models_{\Sigma_2^q} \widehat{\varphi(Z)}[n]\right) \leftrightarrow \varphi(Z)$$

as required.

The second statement (i.e., for a Π_2^B formula φ) is proved similarly.   □

X.2.3. RFN and Consistency for Subsystems of G. For a class Φ of formulas and a (quantified) propositional proof system $\mathcal{F}$, the Φ-Reflection Principle for $\mathcal{F}$, denoted by Φ-$RFN_\mathcal{F}$, asserts that every formula of Φ that has an $\mathcal{F}$-proof is valid. Here we will show that for $i \geq 1$, the Π_{i+1}^q-RFN for $G_i^\star$ (resp. G_i) is provable in the associated theory V^i (resp. TV^i). In Section X.2.4 we will show that indeed the theories can be axiomatized using the RFN of the associated proof systems.

To state the principle we need the formulas $(Z \models_{\Sigma_i^q} X)$ and $(Z \models_{\Pi_i^q} X)$ from Section X.2.2. Recall that $(Z \models_{\Sigma_0^q} X)$ and $(Z \models_{\Pi_0^q} X)$ stands for $(Z \models_0^\Sigma X)$ and $(Z \models_0^\Pi X)$, respectively. Recall also the formulas Fla^Σ, Fla^Π, $Prf_\mathcal{F}^\Sigma$, and $Prf_\mathcal{F}^\Pi$ (see Corollary X.1.4 and Lemma X.1.5).

NOTATION. For $i \geq 0$ and $\Phi \in \{\Sigma_i^q, \Pi_i^q\}$, let $Fla_\Phi^\Pi(X)$ (resp. $Fla_\Phi^\Sigma(X)$) be the Π_1^B (resp. Σ_1^B) formula that represents the relation $FLA(X)$ for formulas X in Φ.

DEFINITION X.2.11 (The Reflection Principle). For a proof system $\mathcal{F}$ and $\Phi \in \{\Sigma_i^q, \Pi_i^q\}$ $(i \geq 0)$ the Φ-Reflection Principle for $\mathcal{F}$, denoted Φ-$RFN_\mathcal{F}$, is the $\mathcal{L}_A^2$ sentence defined as follows:

$$\Sigma_i^q\text{-}RFN_\mathcal{F} \equiv \forall \pi \forall X \forall Z \left((Fla_{\Sigma_i^q}^\Pi(X) \wedge Prf_\mathcal{F}^\Pi(\pi, X)) \supset (Z \models_{\Sigma_i^q} X)\right),$$

$$\Pi_i^q\text{-}RFN_\mathcal{F} \equiv \forall \pi \forall X \forall Z \left((Fla_{\Pi_i^q}^\Sigma(X) \wedge Prf_\mathcal{F}^\Sigma(\pi, X)) \supset (Z \models_{\Pi_i^q} X)\right).$$

Also,

$$i\text{-}RFN_{\mathcal{F}} \equiv \Sigma_i^q\text{-}RFN_{\mathcal{F}} \wedge \Pi_i^q\text{-}RFN_{\mathcal{F}}.$$

Note that for $i \geq 1$, $\Sigma_i^q\text{-}RFN_{\mathcal{F}}$ is equivalent to a $\forall\Sigma_i^B$ sentence, and $\Pi_i^q\text{-}RFN_{\mathcal{F}}$ is equivalent to a $\forall\Pi_i^B$ sentence. Also, $\Sigma_0^q\text{-}RFN_{\mathcal{F}}$ is equivalent to a $\forall\Sigma_1^B$ sentence, while $\Pi_0^q\text{-}RFN_{\mathcal{F}}$ is equivalent to a $\forall\Sigma_0^B$ sentence.

The principle $\Pi_0^q\text{-}RFN_{\mathcal{F}}$ is equivalent to the *consistency* statement which asserts that the proof system $\mathcal{F}$ does not prove a contradiction:

DEFINITION X.2.12 (Consistency). For a proof system $\mathcal{F}$, the sentence $CON_{\mathcal{F}}$ is defined to be

$$\forall\pi\neg Prf_{\mathcal{F}}^{\Sigma}(\pi, \bot).$$

Note that $CON_{\mathcal{F}}$ is equivalent to a $\forall\Sigma_0^B$ sentence.

LEMMA X.2.13. *Let $\mathcal{F}$ be the system $\boldsymbol{G}_i$ or $\boldsymbol{G}_i^{\star}$ (where $i \geq 0$). Then*

$$\boldsymbol{TV}^0 \vdash CON_{\mathcal{F}} \leftrightarrow \Pi_0^q\text{-}RFN_{\mathcal{F}}.$$

PROOF. The direction

$$\Pi_0^q\text{-}RFN_{\mathcal{F}} \supset CON_{\mathcal{F}}$$

is obvious. So consider proving

$$CON_{\mathcal{F}} \supset \Pi_0^q\text{-}RFN_{\mathcal{F}}.$$

Reason in $\boldsymbol{TV}^0$. Assume for a contradiction that $\neg\Pi_0^q\text{-}RFN_{\mathcal{F}}$. That is, there are a quantifier-free formula A with an $\mathcal{F}$-proof π and a truth assignment Z such that $\neg(Z \models_{\Pi_0^q} \widehat{A})$. (Recall that $\widehat{A}$ denotes the encoding of formula A.)

By Exercise X.2.4 we have $(Z \models_{\Pi_0^q} \widehat{\neg A})$. Let A_0 be the formula A with the bits of Z substituted for the free variables in A. Then we have $\neg A_0$, and by Exercise X.1.8 there is a $\boldsymbol{PK}^{\star}$-proof π' of $\neg A_0$. Also, by substituting the bits of Z for the parameter variables in π we obtain a $\mathcal{F}$-proof π'' of A_0. Combine π' and π'' by a cut we obtain an $\mathcal{F}$-proof of $\bot$, and this violates $CON_{\mathcal{F}}$. $\square$

Observe that asserting that a formula $A(\vec{p})$ of the form

$$\forall\vec{x}B(\vec{p}, \vec{x})$$

is valid is essentially equivalent to asserting that $B(\vec{p}, \vec{q})$ is valid. So, if an $\mathcal{F}$-proof of any Π_{i+1}^q formula $A(\vec{p})$ can be transformed (in a theory $\mathcal{T}$) into a proof of B (or some other Σ_i^q formula), then $\mathcal{T}$ proves

$$\Sigma_i^q\text{-}RFN_{\mathcal{F}} \supset \Pi_{i+1}^q\text{-}RFN_{\mathcal{F}}.$$

We illustrate this in the next lemma where we prove the implication for treelike proof systems. The transformation in this case can be computed by a polytime function, and this explains why the implication is provable in $\boldsymbol{TV}^0$.

Lemma X.2.14. *For* $i, j \geq 0$,

$$TV^0 \vdash \Sigma_i^q\text{-}RFN_{G_j^*} \supset \Pi_{i+1}^q\text{-}RFN_{G_j^*}.$$

It follows immediately that:

Corollary X.2.15. *For* $i \geq 1$, $j \geq 0$,

$$TV^0 \vdash \Sigma_i^q\text{-}RFN_{G_j^*} \leftrightarrow i\text{-}RFN_{G_j^*}.$$

Proof sketch of Lemma X.2.14. Assuming $\Sigma_i^q\text{-}RFN_{G_j^*}$ we need to prove $\Pi_{i+1}^q\text{-}RFN_{G_j^*}$. Thus let π be a G_j^* proof of a Π_{i+1}^q formula $A(\vec{p})$. Informally we need to show that A is valid.

If A is in Σ_i^q then we can use $\Sigma_i^q\text{-}RFN_{G_j^*}$ and the conclusion is trivial. So suppose that A is in $(\Pi_{i+1}^q - \Sigma_i^q)$. Using the fact that π is a treelike proof, we will transform π into a proof of a Σ_i^q formula $A'(\vec{p}, \vec{q})$ so that

$$\forall A' \supset \forall A \qquad (362)$$

and such that the transformation is in polytime. Then by $\Sigma_i^q\text{-}RFN_{G_j^*}$ we have that A' is valid, and hence A is valid.

Below we will describe the transformation and the formula A'. They can be computed from π and A in polytime, and (362) can be formalized and proved in TV^0, so we are done.

There are two cases depending on whether A is a prenex formula or not. We consider the simpler case first.

Case I. A is a prenex formula. Here A has the form

$$\forall x_m \ldots \forall x_1 B(\vec{p}, x_1, \ldots, x_m) \qquad (363)$$

where B is a prenex Σ_i^q formula.

If $j > i$ then there is an easy argument as follows. There is an obvious G_0^* proof of

$$A \longrightarrow B(\vec{p}, \vec{q}).$$

This together with the proof π of A and cut gives a G_j^* proof of $B(\vec{p}, \vec{q})$, so we can take A' to be B. Thus we may assume $j \leq i$, although the argument below can be made to work for any $j \geq 0$.

Since π is treelike, we can assume that π is in free variable normal form (recall Section II.2.4 and see Exercise X.1.11).

We use the idea of Gentzen's Midsequent theorem, and transform π into a proof of a sequent of the form

$$\longrightarrow B(\vec{p}, \vec{q^1}), B(\vec{p}, \vec{q^2}), \ldots, B(\vec{p}, \vec{q^k}) \qquad (364)$$

for some k. Here $\vec{q^l}$ are all eigenvariables that introduce the universal variables $\vec{x}$ shown in (363). Intuitively, we retain these eigenvariables by ignoring the $\forall$-right rule.

Formally, suppose that C is a $(\Pi_{i+1}^q - \Sigma_i^q)$ ancestor of A in the succedent of a sequent S in π. Then C has the form

$$\forall x_t \ldots \forall x_1 B(\vec{p}, x_1, \ldots, x_t, q_{t+1}, \ldots, q_m)$$

(for some t, $1 \leq t \leq m$). Note that C can only be in the succedent of S. We transform S by replacing C by a list of formulas as in (364) that contains all ancestors of C of the type $B(\vec{p}, \vec{q})$. (In case C occurs in an axiom $C \longrightarrow C$ we may assume C has the form $B(\vec{p}, \vec{q})$ by using the axiom $B \longrightarrow B$ and adding universal quantifiers to both sides.)

The replacement above is performed for all such C. Let S' denote the transformed sequent, and π' denote the transformed proof. We can easily turn π' into a legitimate proof by (i) deleting the $\forall$-right that introduces the variables $\forall \vec{x}$ of A as shown in (363) as well as contraction right involving C, and (ii) inserting necessary weakenings.

Finally, from a proof of the sequent of the form (364) using the $\vee$-right we obtain a proof of $A'(\vec{p}, \vec{q})$ where A' has the form

$$\bigvee B(\vec{p}, \overrightarrow{q^\ell}). \tag{365}$$

Case II. A is not a prenex formula. The description of A' in this case is more complicated, so we only outline the arguments here. For illustration, consider a $(\Pi_{i+1}^q - \Sigma_i^q)$ subformula A_1 of A of the form (363) where here B is in Σ_i^q but is not necessarily in prenex form. Then following the above procedure, A_1 is replaced by a formula A_1' of the form (365).

We need to extend the above transformation to other $(\Pi_{i+1}^q - \Sigma_i^q)$ subformulas of A. The transformation will be done in a top-down fashion. Thus, for example, a superformula of A_1 may be replaced by several different copies all containing A_1. These copies of A_1 can then be replaced by different disjunctions of the form (365).

This motivates the following definition. For simplicity, assume that in A all $\neg$ connective occur only in front of atoms.

Definition X.2.16. For a formula A in $(\Pi_{i+1}^q - \Sigma_i^q)$, a Σ_i^q-expansion of A is any Σ_i^q formula that can be obtained from A by finitely many repeated applications of the operations that consist of the following steps:

1) let A_1 be a non-Σ_i^q subformula of A;
2) replace A_1 by a formula as follows:
   - if A_1 has the form $\forall x B(x)$ then let $q_1, q_2, \ldots, q_r$ be a list of new free variables (q_t need not be distinct), and replace A_1 by the disjunction
   $$\bigvee_{1 \leq t \leq r} B(q_t),$$
   - otherwise A_1 is replaced by $(A_1 \vee A_1)$.

For example, the formula in (365) is a Σ_i^q-expansion of (363). For another example, suppose

$$A \equiv \forall x_1 \left(\exists y_1 B(x_1, y_1) \wedge \forall x_2 \exists y_2 C(x_1, x_2, y_2) \right)$$

where B, C are quantifier-free formulas. Then the following formula is a Σ_1^q-expansion of A:

$$\exists y_1 B(x_1, q_1) \wedge \left(\exists y_2 C(q_1, q_2, y_2) \vee \exists y_2 C(q_1, q_3, y_2) \right).$$

Now, the $G_j^\star$ proof π of $A(\vec{p})$ can be transformed into a $G_j^\star$ proof π' of an Σ_i^q-expansion $A'(\vec{p}, \vec{q})$ of A. The transformation can be seen to be computable by a polytime function, and the formalization of (362) can be shown to be provable in TV^0.   □

Now we prove the RFN of the systems $G_i^\star$ and G_i in our theories. We take the following approaches to show that the endsequent of given proof π (in $G_i^\star$ or G_i) is valid. The first (see part (a) of the theorem below) is to proceed by induction on the length of π to show that all sequents in π are valid. Notice that if $A(\vec{p})$ is a Σ_i^q or Π_{i+1}^q formula and $\widehat{A}$ encodes A, then the statement asserting A is valid:

$$\forall Z \le |\widehat{A}|(Z \models_{\Sigma_i^q} \widehat{A})$$

is in Π_{i+1}^B. Thus, informally, to prove Π_{i+1}^q-RFN_{G_i} we need Π_{i+1}^B-IND (so V^{i+1} suffices).

Another approach for proving the RFN for $G_1^\star$ is to formalize the proof of the Witnessing Theorem for $G_1^\star$ (Theorem VII.4.13). In general, suppose that $A(\vec{p})$ is a Σ_i^q formula of the form

$$\exists \vec{x} B(\vec{p}, \vec{x})$$

where B is a Π_{i-1}^q formula (here $i \ge 1$). We wish to define a witnessing function for A that, given the values for $\vec{p}$, computes $\vec{x}$ that satisfy $B(\vec{p}, \vec{x})$. The graph of the function is Π_{i-1}^q, so this suggests that the witnessing function is in $\mathcal{L}_{FP^i}$ (see Definition VIII.7.6). For part (b) of Theorem X.2.17 below we will outline the formalization of the proof of the Witnessing Theorem for $G_1^\star$. The proof of the general case is similar and will be left as an exercise.

For part (c) of Theorem X.2.17 (due to Perron) we will need to formalize a more complicated witnessing argument. We refer to [91] for a proof of this part.

THEOREM X.2.17. *For $i \ge 1$:*

(a) $V^i \vdash \Pi_i^q$-$RFN_{G_{i-1}}$;
(b) $V^i \vdash \Pi_{i+1}^q$-$RFN_{G_i^\star}$;
(c) [Perron] $V^i \vdash \Pi_{i+2}^q$-$RFN_{G_i^\star}$.

PROOF. (a) Reasoning in V^i. Let π be a G_{i-1} proof of a Π_i^q formula. By Exercise X.1.10 all formulas in π are Π_i^q. Moreover, it can be shown that all formulas in the antecedents of sequents in π are in $(\Sigma_{i-1}^q \cup \Pi_{i-1}^q)$.

The idea is to prove by induction on t that the t-th sequent in π is valid. Suppose first that $i \geq 2$. Let

$$S_t = A_0, \ldots, A_n \longrightarrow B_0, \ldots, B_m \qquad (366)$$

be the t-th sequent in π. Here all A_j are in $(\Sigma_{i-1}^q \cup \Pi_{i-1}^q)$ and all B_k are in Π_i^q.

DEFINITION X.2.18. For $i \geq 1$ define $(Z \models_i X)$ to be the formula

$$((Z \models_{\Sigma_i^q} X) \vee (Z \models_{\Pi_i^q} X)).$$

Thus $(Z \models_i X)$ iff X is in $\Sigma_i^q \cup \Pi_i^q$ and Z satisfies X. Note that $(Z \models_i X)$ is in $(\Sigma_{i+1}^B \cap \Pi_{i+1}^B)$.

Formally we will prove the following formula:

$$\forall Z \leq |\pi|(\forall j \leq n(Z \models_{i-1} A_j) \supset (Z \models_{\Pi_i^q} \bigvee_k B_k)). \qquad (367)$$

Since $(Z \models_{i-1} X)$ is Σ_i^B and $Z \models_{\Pi_i^q} X)$ is Π_i^B, by Corollary VI.3.8 we can prove (367) by induction on t. Both the base case and the induction step are straightforward.

Now we prove

$$V^1 \vdash \Pi_1^q\text{-}RFN_{G_0}.$$

Let π be a G_0-proof of a Π_1^q formula. Let S_t as in (366) be the t-th sequent in π, here all A_j are quantifier-free and all B_k are Π_1^q formulas. We prove in V^1 the following formula:

$$\forall Z \leq |\pi|((Z \models_0^\Sigma \bigwedge_j A_j) \supset (Z \models_{\Pi_1^q} \bigvee_k B_k)). \qquad (368)$$

Because $(Z \models_0^\Sigma X)$ is a Σ_1^B formula and $(Z \models_{\Pi_1^q} X)$ is a Π_1^B formula, (368) is equivalent in V^1 to a Π_1^B formula. Therefore (368) can be proved in V^1 by induction on t (using Π_1^B-*IND*, see by Corollary VI.1.4).

(b) By Lemma X.2.14 it suffices to show that

$$V^i \vdash \Sigma_i^q\text{-}RFN_{G_i^*}.$$

Let π be a $G_i^\star$ proof of a Σ_i^q formula A. First consider the case $i = 1$, and consider the interesting case where A is in $(\Sigma_1^q - \Sigma_0^q)$. Note that by the subformula property (see Exercise X.1.10) all formulas in π are Σ_1^q.

To show that this formula is valid, the idea is to prove the Witnessing Theorem for $G_1^\star$ (Theorem VII.4.13) that there is a polytime function that produces the witnesses for the existentially quantified variables. Recall that this requires Theorem VII.4.7 and the second half of Theorem VII.4.16. It is straightforward to formalize in TV^0 the proof of both theorems, and hence also the proof of Theorem VII.4.13.

The proof for the case where $i > 1$ is similar. Here the outermost existentially quantified variables in A can be witnessed by some $\textbf{\textit{FP}}^{\Sigma^p_{i-1}}$ functions. These witnessing functions can in fact be defined by examining π directly (without introducing an analogue of $\textbf{\textit{ePK}}$). Details are left as an exercise.

(c) By Lemma X.2.14 it suffices to show that

$$V^i \vdash \Sigma^q_{i+1}\text{-}RFN_{G^*_i}.$$

This is Theorem 5.1.2 in [91]. □

*Exercise X.2.19. Prove part (b) above for the case where $i > 1$.

Corollary X.2.20. *For $i \geq 0$:*

(a) $\textbf{\textit{TV}}^i \vdash \Pi^q_{i+2}\text{-}RFN_{G^*_{i+1}}$;

(b) $\textbf{\textit{TV}}^i \vdash \Pi^q_{i+1}\text{-}RFN_{G_i}$.

Proof. (a) $\Pi^q_{i+2}\text{-}RFN_{G^*_{i+1}}$ is equivalent to a $\forall\Sigma^B_{i+1}$ sentence and by Theorem X.2.17 (b) it is provable in V^{i+1}. By Theorem VIII.7.13 V^{i+1} is Σ^B_{i+1}-conservative over $\textbf{\textit{TV}}^i$, hence $\textbf{\textit{TV}}^i$ also proves $\Pi^q_{i+2}\text{-}RFN_{G^*_{i+1}}$.

(b) Similar to part (a) here for $i \geq 1$ the sentence $\Pi^q_{i+1}\text{-}RFN_{G_i}$ is $\forall\Pi^B_{i+1}$, which is the same as $\forall\Sigma^B_i$. So the fact that $\textbf{\textit{TV}}^i$ proves $\Pi^q_{i+1}\text{-}RFN_{G_i}$ follows from the fact that V^{i+1} proves $\Pi^q_{i+1}\text{-}RFN_{G_i}$ (Theorem X.2.17 (a)) and the fact that V^{i+1} is Σ^B_{i+1}-conservative over $\textbf{\textit{TV}}^i$.

For the case where $i = 0$, $\Pi^q_1\text{-}RFN_{G_0}$ is a $\forall\Pi^B_1$ sentence that is provable in V^1. Since V^1 is Σ^B_1-conservative over $\textbf{\textit{TV}}^0$, $\Sigma^q_0\text{-}RFN_{G_0}$ is also provable in $\textbf{\textit{TV}}^0$. □

Exercise X.2.21. (a) For $i \geq j \geq 0$, show that

$$\textbf{\textit{TV}}^0 \vdash \Sigma^q_j\text{-}RFN_{G^*_{i+1}} \supset \Sigma^q_j\text{-}RFN_{G_i}.$$

(b) For $i \geq 0$ and $j \geq 0$, show that

$$\textbf{\textit{TV}}^0 \vdash \Sigma^q_j\text{-}RFN_{G_i} \supset \Sigma^q_j\text{-}RFN_{G^*_{i+1}}.$$

(Hint: formalize the p-simulations given in the proofs of Theorems VII.4.3 and VII.4.8.)

*Exercise X.2.22. Let Fla^{Π}_{PK} and Prf^{Π}_{ePK} be Π^B_1 formulas that represent the relations FLA_{PK} and PRF_{ePK}, respectively. The Reflection Principle for $\textbf{\textit{ePK}}$ is defined as follows:

$$RFN_{ePK} \equiv \forall\pi\forall X\forall Z\big((Fla^{\Pi}_{PK}(X) \wedge Prf^{\Pi}_{ePK}(\pi, X)) \supset (Z \models^{\Sigma}_0 X)\big).$$

Note that RFN_{ePK} is equivalent to a $\forall\Sigma^B_1$ sentence. Show that

$$\textbf{\textit{TV}}^0 \vdash RFN_{ePK}.$$

(Hint: show that $V^1 \vdash RFN_{ePK}$ and then use the fact that V^1 is Σ^B_1-conservative over $\textbf{\textit{TV}}^0$.)

X.2.4. Axiomatizations Using RFN. In this section we present results of the following type. We will show that the RFN of a proof system $\mathcal{F}$ ($G_i^\star$ or G_i) can be used together with a base theory (e.g., VTC^0) to axiomatize the associated theory T (V^i or TV^i). In Section X.2.3 above we have shown one direction, i.e., the RFN of $\mathcal{F}$ is provable in T. For the other direction we need to show that all theorems of T are provable from the base theory and the RFN of $\mathcal{F}$. Informally, this can be seen as follows. First, the propositional version of each theorem of T have been shown to have proofs in $\mathcal{F}$ that are definable in VTC^0 (Section X.1.2). So by the RFN for $\mathcal{F}$ these propositional translations are valid. Next, $\overline{VTC}^0$ proves that the validity of the propositional translations implies the validity of the first-order formulas (Theorem X.2.10 in Section X.2.2). As the result, the theorem of T can be proved using VTC^0 and the RFN of $\mathcal{F}$ (because $\overline{VTC}^0$ is a conservative extension of VTC^0).

First, we prove:

THEOREM X.2.23. (a) *Let* $i \geq j \geq 1$. *Then* $\Sigma_j^B(V^i)$, *the* Σ_j^B *consequences of* V^i, *can be axiomatized by the axioms of* VTC^0 *together with* $\Sigma_j^q\text{-}RFN_{G_i^\star}$.
(b) *For* $j \geq 1$ *and* $i \geq j+1$, $\Sigma_j^B(V^i)$ *can also be axiomatized by the axioms of* VTC^0 *together with* $\Sigma_j^q\text{-}RFN_{G_{i-1}}$.
(c) *For* $i \geq 1$, *the theory* V^i *can be axiomatized by the axioms of* VTC^0 *and* $\Sigma_{i+1}^q\text{-}RFN_{G_i^\star}$. V^i *can also be axiomatized by the axioms of* VTC^0 *and* $\Sigma_{i+1}^q\text{-}RFN_{cut\text{-}free\ G^\star}$.

PROOF. (a) Note that by Theorem X.2.17 (b) V^i proves $\Sigma_j^q\text{-}RFN_{G_i^\star}$. Therefore $\Sigma_j^q\text{-}RFN_{G_i^\star}$ is a Σ_j^B consequence of V^i. Since $j \geq 1$, all axioms of VTC^0 are also in $\Sigma_j^B(V^i)$. Thus it remains to show that every Σ_j^B consequence of V^i can be proved in $VTC^0 + \Sigma_j^q\text{-}RFN_{G_i^\star}$. We prove this for $i = j = 1$, since other cases are similar. Here we have to show that all Σ_1^B theorems of V^1 are provable using $VTC^0 + \Sigma_1^q\text{-}RFN_{G_1^\star}$.

Suppose that φ is a Σ_1^B theorem of V^1. Assume without loss of generality that it has a single free variable Z. Let $T_{\varphi(Z)}(n)$ be the FTC^0 function as in Corollary X.1.13 that provably in $\overline{VTC}^0$ computes the encoding of $\varphi(Z)[n]$. Thus we have

$$T_{\varphi(Z)}(n) = \widehat{\varphi(Z)[n]}$$

and

$$\overline{VTC}^0 \vdash Fla_{\Sigma_1^q}^{\Pi}(T_{\varphi(Z)}(n)). \tag{369}$$

Now by Exercise X.1.17 there is an FTC^0 function $F_{\varphi(Z)}(n)$ that provably in $\overline{VTC}^0$ computes a $G_1^\star$ proof of $\varphi(Z)[n]$. In other words, we have

$$\overline{VTC}^0 \vdash Prf_{G_1^\star}^\Pi(F_{\varphi(Z)}(n), T_{\varphi(Z)}(n)). \qquad (370)$$

From (369) and (370), using $\Sigma_1^q\text{-}RFN_{G_1^\star}$ we obtain

$$\forall Z(Z \models_{\Sigma_1^q} T_{\varphi(Z)}(n)).$$

From this and Theorem X.2.10 we obtain $\forall Z \varphi(Z)$.

(b) This is suggested by (a) since by Corollary VII.4.9, $G_i^\star$ and G_{i-1} are p-equivalent for proving Σ_j^q formulas for $j \leq i - 1$. However the p-equivalence seems to require TV^0 rather than VTC^0 to prove, so we prove (b) as follows.

Note that V^i is Σ_i^B-conservative over TV^{i-1} (Theorem VIII.7.13). So the proof here similar to (a). Here Theorem X.2.17 (a) gives us that

$$\Sigma_j^B(V^i) \vdash \Sigma_j^q\text{-}RFN_{G_{i-1}}.$$

Therefore all consequences of $VTC^0 + \Sigma_j^q\text{-}RFN_{G_{i-1}}$ are also in $\Sigma_j^B(V^i)$.

For the other direction, let φ be a Σ_j^B theorem φ of V^i. Since V^i is Σ_i^B-conservative over TV^{i-1}, φ is also a theorem of TV^{i-1} and hence has a G_{i-1} proof which is provably computable in $\overline{VTC}^0$ (by Theorem X.1.21). The argument is similar to (a).

(c) For the first sentence, the fact that all theorems of V^i are provable from

$$VTC^0 + \Sigma_{i+1}^q\text{-}RFN_{G_i^\star}$$

can be proved as in part (a).

The proof of the other direction, i.e., that V^i proves $\Sigma_{i+1}^q\text{-}RFN_{G_i^\star}$, is part (c) of Theorem X.2.17.

The second sentence follows from the first and Theorem X.2.27 below. □

As a corollary, we obtain a finite axiomatization of TV^i as follows.

COROLLARY X.2.24. (a) *For $i \geq 0$, the theory TV^i can be axiomatized by the axioms of VTC^0 together with $\Sigma_{i+1}^q\text{-}RFN_{G_{i+1}^\star}$.*

(b) *For $i \geq 1$, the theory TV^i can be axiomatized by the axioms of TV^0 and $\Sigma_{i+1}^q\text{-}RFN_{G_i}$.*

PROOF. Part (a) follows from Theorem X.2.23 (a) applied to $\Sigma_{i+1}^B(V^{i+1})$, and the fact that TV^i can be axiomatized by the Σ_{i+1}^B consequences of V^{i+1}, because TV^i have Σ_{i+1}^B axioms and V^{i+1} is Σ_{i+1}^B-conservative over TV^i (Theorem VIII.7.13).

Part (b) follows from (a) and Exercise X.2.21. □

We obtain an alternative proof for the finite axiomatizability of the theories TV^i (see Theorem VIII.7.3):

Corollary X.2.25. *For $i \geq 0$ the theories V^{i+1} and TV^i are finitely axiomatizable. For $i \geq j \geq 1$, the Σ_j^B consequences of V^i are finitely axiomatizable.*

Proof. The conclusion follows from Corollary X.2.24 and Theorem X.2.23, and the fact that VTC^0 is finitely axiomatizable.   □

Exercise X.2.26. Show that TV^0 can be axiomatized by the axioms of VTC^0 together with RFN_{ePK} defined in Exercise X.2.22. (Hint: one direction follows from Exercise X.2.22. For the other direction, use Exercise X.1.23.)

In Theorem X.2.23 above we have considered $\Sigma_j^q\text{-}RFN_{G_i^\star}$ only for the values of j such that $1 \leq j \leq i$. Now we consider $\Sigma_j^q\text{-}RFN_{G_i^\star}$ for $j > i$. It turns out that in this case $\Sigma_j^q\text{-}RFN_{G_i^\star}$ is equivalent to $\Sigma_j^q\text{-}RFN_{cut\text{-}free\ G^\star}$, because any $G_i^\star$ proof of a Σ_j^q formula can be transformed into a cut free $G^\star$ proof of an equivalent Σ_j^q formula A'. This observation is due to Perron [91]. Here we need VTC^0 to prove the equivalence, essentially because the transformation given in the proof below is computable in TC^0 (while the equivalence between A and A' is provable in V^0).

Theorem X.2.27. *Let $i \geq 0$. The theory VTC^0 proves that the following are all equivalent*:

$$\Sigma_{i+1}^q\text{-}RFN_{cut\text{-}free\ G^\star},\ \Sigma_{i+1}^q\text{-}RFN_{G_0^\star},\ \ldots,\ and\ \Sigma_{i+1}^q\text{-}RFN_{G_i^\star}.$$

Proof. Since *cut-free* $G^\star$ is a subsystem of $G_0^\star$, which in turn is a subsystem of $G_1^\star$, etc., and since $\overline{VTC}^0$ is a conservative extension of VTC^0, it suffices to show that

$$\overline{VTC}^0 + \Sigma_{i+1}^q\text{-}RFN_{cut\text{-}free\ G^\star} \vdash \Sigma_{i+1}^q\text{-}RFN_{G_i^\star}.$$

The idea is as follows. Let π be a $G_i^\star$ proof of a Σ_{i+1}^q formula A. We will transform π into a cut free $G^\star$ proof of a Σ_{i+1}^q formula A' of the form (372) below. Our transformation can be seen to be in TC^0. So, formally, the transformed prove is provably in $\overline{VTC}^0$ computable by some FTC^0 function F. Then using $\Sigma_j^q\text{-}RFN_{cut\text{-}free\ G^\star}$ we have that A' is valid. Finally, since V^0 proves that A and A' are equivalent we conclude that A is valid.

We will first transform π into a *cut-free* $G^\star$ proof of the following sequent:

$$\longrightarrow A, \exists(C_1 \wedge \neg C_1), \exists(C_2 \wedge \neg C_2), \ldots, \exists(C_k \wedge \neg C_k) \qquad (371)$$

where C_t (for $1 \leq t \leq k$) are all cuts formulas in π, and $\exists(C_t \wedge \neg C_t)$ is the sentence obtained from $(C_t \wedge \neg C_t)$ by existentially quantifying all free variables. Then, by using the $\vee$-right we get a proof of

$$A' =_{\text{def}} A \vee \bigvee_{1 \leq t \leq k} (\exists(C_t \wedge \neg C_t)). \qquad (372)$$

Notice that each C_i is a Σ_i^q formula, so A' is in Σ_{i+1}^q.

The derivation from (371) to

$$\longrightarrow A'$$

is obvious, so we will focus on the derivation of (371).

Let Δ denote the sequence

$$\exists(C_1 \wedge \neg C_1), \exists(C_2 \wedge \neg C_2), \ldots, \exists(C_k \wedge \neg C_k)$$

as in (371). We transform π as follows. First, add Δ to the succedent of every sequent in π. For each sequent S in π let S' be the result of this addition.

To obtain a legitimate derivation, note that if S is derived from S_1 (and S_2) by an inference of G, then S' can be derived from S_1' (and S_2') by the same inference with possibly some applications of the exchange rule. In addition, each axiom $B \longrightarrow B$ now becomes $B \longrightarrow B, \Delta$ so we also add the following derivation (using weakening)

$$\frac{B \longrightarrow B}{B \longrightarrow B, \Delta}$$

Thus, the result, called π_1, is a $G_i^\star$ proof.

Next, consider an instance of the cut rule in π_1:

$$\frac{\Lambda \longrightarrow \Gamma, \Delta, C \qquad C, \Lambda \longrightarrow \Gamma, \Delta}{\Lambda \longrightarrow \Gamma, \Delta}$$

We insert the following derivation

$$\frac{\Lambda \longrightarrow \Gamma, \Delta, C \quad \dfrac{C, \Lambda \longrightarrow \Gamma, \Delta}{\Lambda \longrightarrow \Gamma, \Delta, \neg C}}{\dfrac{\dfrac{\Lambda \longrightarrow \Gamma, \Delta, (C \wedge \neg C)}{\Lambda \longrightarrow \Gamma, \Delta, \exists(C \wedge \neg C)}}{\Lambda \longrightarrow \Gamma, \Delta}}$$

Here the bottom double line represents applications of exchange and contraction right. The double line above it represents a series of $\exists$-right.

It can be seen that the result is a cut-free $G^\star$ proof of (371) as desired. We will briefly verify that the above transformation is computable in TC^0. The fact that it can be formalized in VTC^0 is straightforward.

For the transformation, first we need to identify all cuts and cut formulas in π (here we need TC^0 circuits to recognize formulas). Once this has been done, it is easy to see that the cut free $G^\star$ proof of (371) described above can be computed by some TC^0 circuit (here we need the counting gates, e.g., to put $\exists(C_t \wedge \neg C_t)$ into the list Δ). The last step is to obtain a derivation of A' from (371), and this can also be computed by a TC^0 circuit.   □

X.2.5. Proving p-Simulations Using RFN. In this section we will show how to use the Propositional Translation Theorems (e.g., Theorems VII.5.2, X.1.21) to prove p-simulations between proof systems. Informally, the result is as follows. Suppose that $\mathcal{G}$ (such as $\mathbf{G}_1^\star$) is a proof system associated with a theory $\mathcal{T}$ (such as V^1) (where the association is by propositional translation). Then any proof system $\mathcal{F}$ (such as *eFrege*) which is definable in $\mathcal{T}$ and whose RFN is provable in $\mathcal{T}$ is p-simulated by $\mathcal{G}$. (See the precise statement in Theorem X.2.29 below.)

Intuitively the reason why $\mathcal{G}$ p-simulates such $\mathcal{F}$ is as follows. Because $\mathcal{T}$ proves the soundness (i.e., the RFN) of $\mathcal{F}$, and $\mathcal{G}$ is a nonuniform version of $\mathcal{T}$, there are short $\mathcal{G}$ derivations of the fact that $\mathcal{F}$ is sound. Therefore, given a proof of $\mathcal{F}$ of A we can derive a short $\mathcal{G}$ derivation of A (we will need the derivations from Theorems X.2.5 and X.2.8).

First we need to prove in $\mathbf{G}_i^\star$ the "honesty" of our encoding of the RFN. Here we translate the RFN for a system $\mathcal{F}$ by treating the variables π, X and Z (Definition X.2.11) as free variables and introducing the bits

$$p_0^\pi, p_1^\pi, \ldots; \qquad p_0^X, p_1^X, \ldots; \qquad p_0^Z, p_1^Z \ldots$$

as usual.

Recall Definition X.1.6 that for some constant values of π_0 and X_0 of length n and m respectively, we use

$$\Sigma_i^q\text{-}RFN_{\mathcal{F}}(\pi_0, X_0, Z)[k] \tag{373}$$

to denote the result of substituting the values ($\top$ or $\bot$) of bits $X_0(t)$, $\pi_0(t)$ for the variables p_t^X and p_t^π in the propositional translation

$$\Sigma_i^q\text{-}RFN_{\mathcal{F}}(\pi, X, Z)[n, m, k]$$

where $m = |X_0|$ and $n = |\pi_0|$. (Thus the only free variables in (373) are $p_0^Z, p_1^Z, \ldots$.) For a Σ_i^q formula X_0, note that the translation (373) is

$$(Fla_{\Sigma_i^q}^\Pi(X_0) \wedge Prf_{\mathcal{F}}^\Pi(\pi_0, X_0))[\,] \supset (Z \models_{\Sigma_i^q} X_0)[k].$$

For the proof of the theorem below, it is helpful to review Theorems X.2.5 and X.2.8 and their proofs.

THEOREM X.2.28. *Let $i \geq 1$ and $\mathcal{F}$ be a proof system with defining formulas $Prf_{\mathcal{F}}^\Sigma$ and $Prf_{\mathcal{F}}^\Pi$ as in (322) and (323) (on page 367). Then there is a polytime algorithm that computes a $\mathbf{G}_i^\star$ proof of the sequent below for each Σ_i^q formula A_0 of length m with at most $k-1$ free variables, and $\mathcal{F}$-proof π_0 of A_0 of length n:*

$$\Sigma_i^q\text{-}RFN_{\mathcal{F}}(\pi_0, \widehat{A_0}, Z)[k] \longrightarrow A_0(\overrightarrow{p^Z}).$$

PROOF. The desired sequent is obtain from the sequent in Theorem X.2.8 and the following sequent by a Σ_i^q cut:

$$\Sigma_i^q\text{-}RFN_{\mathcal{F}}(\pi_0, \widehat{A_0}, Z)[k] \longrightarrow (Z \models_{\Sigma_i^q} \widehat{A_0})[k].$$

In turns, the sequent above can be derived from

$$\longrightarrow \left(Fla^{\Pi}(\widehat{A_0}) \wedge Prf^{\Pi}_{\mathcal{F}}(\pi_0, \widehat{A_0}) \right)[\ ].$$

A $G_0^\star$ proof of this sequent can be computed in polytime as follows from Lemma X.1.7. $\square$

THEOREM X.2.29. *Let $i \geq j \geq 1$.*

(a) *Suppose that $\mathcal{F}$ is a proof system such that*

$$V^i \vdash \Sigma^q_j\text{-}RFN_{\mathcal{F}}.$$

Then $G_i^\star$ p-simulates $\mathcal{F}$ w.r.t. Σ^q_j formulas.

(b) *The same is true for TV^i and G_i in place of V^i and $G_i^\star$.*

PROOF. (a) Let $A_0(\vec{p})$ be a Σ^q_j formula and π_0 be an $\mathcal{F}$-proof of $A_0(\vec{p})$. As before, let $\widehat{A_0}$ denote the string encoding of $A_0(\vec{p})$. Let $n = |\pi_0|$, $m = |\widehat{A_0}|$. From the hypothesis and the Propositional Translation Theorem for V^i (Theorem VII.5.2, see also Exercise X.1.17), there are polytime computable $G_i^\star$ proofs of the translations

$$\Sigma^q_j\text{-}RFN_F(\pi_0, \widehat{A_0}, Z)[k].$$

Now using the $G_j^\star$ proof from Theorem X.2.28 we obtain a $G_i^\star$ proof of $A_0(\overrightarrow{p^Z})$.

Part (b) is proved similarly. $\square$

*EXERCISE X.2.30. Let $\mathcal{F}$ be a proof system for (unquantified) propositional formulas. Suppose that $RFN_{\mathcal{F}}$ (i.e., $\Sigma^q_0\text{-}RFN_{\mathcal{F}}$, see Definition X.2.11) is provable in TV^0. Show that ePK p-simulates $\mathcal{F}$. (Hint: use Theorem X.2.29 (a) for $i = 1$ and Theorem VII.4.16.)

We obtain as corollaries some results proved earlier in Chapter VII (Corollary VII.4.9 and Theorem VII.4.16).

COROLLARY X.2.31. *For $i \geq 1$, $G_{i+1}^\star$ and G_i are p-equivalent when proving Σ^q_i formulas. ePK and $G_1^\star$ are p-equivalent for proving prenex Σ^q_1 formulas.*

X.2.6. The Witnessing Problems for G. Recall the notion of search problems defined in Section VIII.5 (see Definitions VIII.5.1 and VIII.5.11). Recall also the witnessing problem given in Theorem VII.4.13. In general, the witnessing problems for (subsystems of) G are search problems that are motivated by the following observation. Let $i \in \mathbb{N}$, $i \geq 1$, and consider a Σ^q_i tautology $A(\vec{p})$ of the form

$$A(\vec{p}) \equiv \exists \vec{x} B(\vec{x}, \vec{p})$$

where B is a Π^q_{i-1} formula. Given A and the values for $\vec{p}$ we wish to find a truth assignment for the existentially quantified variables $\vec{x}$ that satisfies $B(\vec{x}, \vec{p})$. Note that this problem is polytime complete for $FP^{\Sigma^p_i}$. However, a given G-proof π of $A(\vec{p})$ may help us find $\vec{x}$, and it becomes interesting to study the problems when different proofs π are given.

Formally, the problems are defined as follows.

Definition X.2.32 (Witnessing Problem). For a quantified propositional proof system $\mathcal{F}$ and $1 \leq i \in \mathbb{N}$, the Σ_i^q *Witnessing Problem for* $\mathcal{F}$, denoted by $\Sigma_i^q\text{-}WIT_{\mathcal{F}}$, is, given an $\mathcal{F}$-proof π of a Σ_i^q formula $A(\vec{p})$ of the form

$$A(\vec{p}) \equiv \exists \vec{x} B(\vec{x}, \vec{p})$$

where B is a Π_{i-1}^q formula, and a truth assignment to $\vec{p}$, find a truth assignment for $\vec{x}$ that satisfies $B(\vec{x}, \vec{p})$.

Not surprisingly, the Witnessing Problems for G_i and $G_i^{\star}$ are closely related to the classes of definable search problems in the associated theories. For the next theorem it is useful to refer to the the summary table on page 250.

Recall the notion of many-one reduction between search problems (Definition VIII.5.2). For the next theorem we use the notion of TC^0 many-one reduction between search problems. This is defined just as in Definition VIII.5.2 with the exception that the functions $\vec{f}, \vec{F}, G$ are now in FTC^0 (as opposed to FAC^0). The reason that we need TC^0 reductions here is basically because our translation functions (such as in Exercise X.1.17) are TC^0 functions. Theorem X.2.33 is about the witnessing problems for G_i and $G_i^{\star}$ for $i \geq 1$. In Theorem X.3.12 we will state the results for G_0 and $G_0^{\star}$.

Theorem X.2.33. *For $i \geq 1$:*

(a) $\Sigma_i^q\text{-}WIT_{G_i^{\star}}$ *is TC^0-complete for* $FP^{\Sigma_{i-1}^P}$.

(b) $\Sigma_i^q\text{-}WIT_{G_i}$ *is TC^0-complete for* $CC(PLS)^{\Sigma_{i-1}^P}$.

(c) $\Sigma_{i+1}^q\text{-}WIT_{G_i^{\star}}$ *is TC^0-complete for* $FP^{\Sigma_i^P}[wit, \mathcal{O}(\log n)]$.

Proof. (a) First we show that $\Sigma_i^q\text{-}WIT_{G_i^{\star}}$ is in $FP^{\Sigma_{i-1}^P}$. Consider the case $i = 1$. Here the Witnessing Theorem for $G_1^{\star}$ (Theorem VII.4.13) already shows that $\Sigma_1^q\text{-}WIT_{G_1^{\star}}$ is in P. The case where $i > 1$ is similar. In fact, we have pointed out in the proof of Theorem X.2.17 (b) that by analyzing π, a witnessing function can be defined in V^i by a Σ_i^B formula.

Now we show that $\Sigma_i^q\text{-}WIT_{G_i^{\star}}$ is hard for $FP^{\Sigma_{i-1}^P}$. Thus let $Q(\vec{x}, \vec{X})$ be a search problem in $FP^{\Sigma_{i-1}^P}$ with graph $R(\vec{x}, \vec{X}, Z)$. We show that Q is reducible to $\Sigma_i^q\text{-}WIT_{G_i^{\star}}$. By Theorem VIII.7.12 Q is Σ_i^B-definable in V^i by a Σ_i^B formula $\varphi(\vec{x}, \vec{X}, Z)$, i.e.,

$$\varphi(\vec{x}, \vec{X}, Z) \supset R(\vec{x}, \vec{X}, Z)$$

and

$$V^i \vdash \exists Z\, \varphi(\vec{x}, \vec{X}, Z).$$

By the V^i Translation Theorem (Theorem VII.5.2) the Σ_i^B theorem $\exists Z \, \varphi(\vec{x}, \vec{X}, Z)$ of V^i translates into a family of tautologies that have polynomial-size $G_i^\star$ proofs. In fact, by Exercise X.1.17 there is a TC^0 function $F_\varphi(\vec{x}, \vec{X})$ that provably in $\overline{VTC}^0$ computes a $G_i^\star$ proof of the translation of φ. Thus, given $(\vec{x}, \vec{X})$, a value for Z that satisfies $\varphi(\vec{x}, \vec{X}, Z)$ can be easily obtained from the solution of the witnessing problem given by $F_\varphi(\vec{x}, \vec{X})$ and $(\vec{x}, \vec{X})$.

(b) The fact that $\Sigma_i^q\text{-}WIT_{G_i}$ is complete for $CC(PLS)^{\Sigma_{i-1}^P}$ is proved similarly using Corollary X.2.20 (b), the TV^i Translation Theorem X.1.21 and Theorem VIII.7.14.

(c) Similar to part (a). Here $\Sigma_{i+1}^q\text{-}WIT_{G_i^\star}$ is in $FP^{\Sigma_i^P}[wit, \mathcal{O}(\log n)]$ because $\Sigma_{i+1}^q\text{-}RFN_{G_i^\star}$ is provable in V^i (Theorem X.2.23 (c)) and Σ_{i+1}^B-definable search problems in V^i are in $FP^{\Sigma_i^P}[wit, \mathcal{O}(\log n)]$ (Theorem VIII.7.17). The fact that $\Sigma_{i+1}^q\text{-}WIT_{G_i^\star}$ is hard for $FP^{\Sigma_i^P}[wit, \mathcal{O}(\log n)]$ also follows from Theorem VIII.7.17 and the V^i Translation Theorem VII.5.2 as in (a). $\qquad\qquad\qquad\qquad\qquad\qquad\qquad\qquad\qquad\qquad\qquad\square$

The Witnessing Problems for G_0 and $G_0^\star$ are discussed in the next section.

X.3. VNC^1 and $G_0^\star$

Recall the theory VNC^1 from Section IX.5.3. Here we will show that bounded theorems of VNC^1 translate into tautologies with polynomial size $G_0^\star$-proofs. (Section X.3.1.) It will follow that Σ_0^B theorems of VNC^1 translate into families of propositional tautologies that have PK proofs which are provably in $\overline{VTC}^0$ computable by FTC^0 functions.

In order to prove the RFN for PK in VNC^1 we need to show that the relation $(Z \models_0 X)$ (that the truth assignment Z satisfies a formula X, see Section X.2.1) is Δ_1^B-definable in VNC^1. For this we will formalize in VNC^1 an NC^1 algorithm, due to Buss, that computes the Boolean Sentence Value Problem (BSVP). We will present the algorithm in Section X.3.2 and its formalization in VNC^1 in Section X.3.3.

X.3.1. Propositional Translation for VNC^1. Recall (Section VII.4) that $G_0^\star$ is the subsystem of $G^\star$ in which all cut formulas are quantifier-free. Note also that $G_0^\star$ is p-equivalent to G_0 with respect to prenex Σ_1^q formulas (Theorem VII.4.5).

Recall (Section IX.5.3) that the theory VNC^1 is axiomatized by the axioms of V^0 together with the axiom MFV (for *monotone formula value*) that asserts the existence of a string Y which evaluates all subformulas of a balanced formula (encoded by (a, G, I)):

$$\exists Y \leq 2a\delta_{MFV}(a, G, I, Y)$$

where

$$\delta_{MFV}(a, G, I, Y) \equiv \forall x < a\big((Y(x + a) \leftrightarrow I(x)) \wedge Y(0) \wedge$$
$$0 < x \supset \big(Y(x) \leftrightarrow \big((G(x) \wedge (Y(2x) \wedge Y(2x + 1))) \vee$$
$$(\neg G(x) \wedge (Y(2x) \vee Y(2x + 1)))\big)\big)\big). \quad (374)$$

Our goal in this section is to prove the following theorem. Recall Definition X.1.9.

THEOREM X.3.1 (Translation Theorem for VNC^1). *Let $\varphi(\vec{x}, \vec{X})$ be a bounded theorem of VNC^1. Then there is an FTC^0 function $F_\varphi(\vec{m}, \vec{n})$ that provably in $\overline{VTC}^0$ computes a $G_0^\star$ proof of the propositional formulas $\varphi(\vec{x}, \vec{X})[\vec{m}; \vec{n}]$ for all $\vec{m}, \vec{n}$ in $\mathbb{N}$.*

Our translation of an anchored $LK^2\text{-}VNC^1$ proof (in free variable normal form) of a bounded theorem of VNC^1 extends the translation of $LK^2\text{-}V^0$ proofs discussed in Section VII.5.1. Here the new type of cut formulas are instances of the formula $\exists Y \delta_{MFV}(a, G, I, Y)$ (see (374)). Note that the length $|Y|$ in MFV is bounded by $2a$. To make the translation easier we will fix $|Y|$. Thus we will use another axiom $\delta'_{MFV}(a, G, I, Y)$ defined below, where now the length $|Y|$ of Y is required to be exactly $2a + 1$. Informally, this is easily obtained by adding a fixed leading bit (bit $2a$) to the string Y.

The fact that the new axiom is equivalent to MFV over V^0 is easy and is left as an exercise.

EXERCISE X.3.2. Let $\delta'_{MFV}(a, G, I, Y)$ denote

$$|Y| = 2a + 1 \wedge Y(0) \wedge \forall x < a\big((Y(x + a) \leftrightarrow I(x)) \wedge 0 < x \supset$$
$$\big(Y(x) \leftrightarrow \big((G(x) \wedge Y(2x) \wedge Y(2x + 1)) \vee$$
$$(\neg G(x) \wedge (Y(2x) \vee Y(2x + 1)))\big)\big)\big). \quad (375)$$

Then VNC^1 can be axiomatized by V^0 together with

$$MFV' \equiv \exists Y \, \delta'_{MFV}(a, G, I, Y).$$

As a result, every theorem of VNC^1 has an anchored LK^2 proof in which cut formulas are instances of the axioms of V^0 or the axiom MFV' above. To prove the Theorem X.3.1 we will first translate LK^2 proofs of this type, and then argue that the translation is indeed provably computable in $\overline{VTC}^0$.

Translating the proof of a bounded theorem of VNC^1 is done by extending the translation of $LK^2\text{-}V^0$ proofs as described in Section VII.5.1. Here we have to consider in addition instances of the axiom MFV' above. Like the translations of the $\Sigma_0^B\text{-}COMP$ cut formulas, here the translations of the cut MFV' formulas will be tautologies that have PK proofs which are provably in $\overline{VTC}^0$ computable. Recall the notions such as *comprehension variables* from Section VII.5.1.

Proof of Theorem X.3.1. By Exercise X.3.2 above, there is an anchored LK^2-proof π of φ where all cut formulas are instances of the axioms of V^0 or instances of the axiom MFV'. In addition, we can assume that π is in free variable normal form.

The translations of cut $\Sigma_0^B\text{-}COMP$ formulas as described in the proof of Theorem VII.5.6 can be extended here easily. So we will focus on the instances of the cut MFV' axiom. Similar to the notion of comprehension variables, we define:

Notation. A free string variable γ in π is called an *MFV variable* if it is used as the eigenvariable for the string-$\exists$-left rule whose principal formula is an ancestor of a cut formula of the form $\exists Y\, \delta'_{MFV}(t, \alpha, \beta, Y)$. In this case, we also say that (t, α, β) is the *defining triple* of γ.

For example, consider an instance of the string-$\exists$-left rule:

$$\frac{S_1}{S_2} = \frac{\delta'_{MFV}(t, \alpha, \beta, \gamma), \Gamma \longrightarrow \Delta}{\exists Y\delta'_{MFV}(t, \alpha, \beta, Y), \Gamma \longrightarrow \Delta} \qquad (376)$$

Suppose that the formula $\exists Y\delta'_{MFV}(t, \alpha, \beta, Y)$ in S_2 is an ancestor of a cut formula. Then γ is an MFV variable.

In our translation below, if two MFV variables have the same defining triple, they will have identical translations.

Next, we extend the definition of the dependence relation defined in the proof of Theorem VII.5.6 to include MFV variables.

Notation. We say that an MFV variable γ depends on a variable β (or b) if β (or b) occurs in the defining triple of γ.

The *dependence degree* of a variable is defined as before by taking into account the fact that now there are also MFV variables. Formally, all variables that are not comprehension variables nor MFV variables have dependence degree 0. The dependence degree of a comprehension variable (resp. an MFV variable) γ is one plus the maximum dependence degree of all variables occurring in its defining pair (resp. defining triple).

Before translating formulas in π we will remove from π the right branch of a cut rule if the cut formula is an MFV' instance or a $\Sigma_0^B\text{-}COMP$ instance, and remove all remaining MFV' instances that are ancestors of a cut formula (i.e., the MFV' instance in (376) and all its descendants) as well as instances of $\Sigma_0^B\text{-}COMP$ that are ancestors of a cut formula. The reason for doing this is that $\delta'_{MFV}(t, \alpha, \beta, \gamma)$ will translate into tautologies that have short $PK^\star$ proofs, and hence the translations can be cut. The same is true for the Σ_0^B matrix of $\Sigma_0^B\text{-}COMP$ (as shown in the proof of Theorem VII.5.6).

Now the formulas in π are translated in stages just as described in the proof of Theorem VII.5.6. Here the only new case to be handled is the case of a atomic formula of the form $\gamma(s)$ for an MFV variable γ. Thus, let γ be an MFV variable, and suppose that (t, α, β) is the defining triple of γ as in (376). Note that the length of γ is $(2t + 1)$. Let $\vec{n}$ be the list of values/lengths of all non-MFV variables in π. The translation $\gamma(s)[\vec{n}]$ is defined by (reverse) induction on $val(s)$ as follows.

Let $m = val(t)$. First, suppose that $m \leq val(s) < 2m$, then

$$\gamma(s)[\vec{n}] =_{\text{def}} \beta(r)[\vec{n}]$$

where r is the numeral $\underline{val(s) - m}$. Next, suppose that $1 \leq val(s) < m$, then

$$\gamma(s)[\vec{n}] =_{\text{def}} (A \wedge (B_0 \wedge B_1)) \vee (A \wedge (B_0 \vee B_1)) \qquad (377)$$

where $A \equiv \alpha(s)[\vec{n}]$, $B_0 \equiv \gamma(2s)[\vec{n}]$, and $B_1 \equiv \gamma(2s + 1)[\vec{n}]$. Finally,

$$\gamma(s)[\vec{n}] =_{\text{def}} \begin{cases} \bot & \text{if } val(s) > 2m, \\ \top & \text{if } val(s) = 2m \vee val(s) = 0. \end{cases}$$

We show that the translations above are provably in $\overline{VTC}^0$ computable.

LEMMA X.3.3. *For each $\mathcal{L}_A^2$ formula $\varphi(\vec{x}, \vec{X})$ in π there is an FTC^0 function $F_{\varphi,\pi}(\vec{k}, \vec{n})$ that provably in $\overline{VTC}^0$ computes the translation $\varphi(\vec{x}, \vec{X})[\vec{k}; \vec{n}]$ as described above.*

PROOF SKETCH. Following our translation of formulas in π, $F_{\varphi,\pi}$ will be defined in stages: for $i \geq 0$, in stage i we define the functions for all formulas φ that contain some variable of dependence degree i but none of higher degree.

In each stage the construction is by structural induction on the formula φ. Stage 0 is exactly the same as in Exercise X.1.15; and in general, for each stage i (where $i \geq 1$) except for the base case of the formulas $\gamma(s)$ for MFV variables γ, the arguments are the same as in Exercise X.1.15.

So now consider the base case of stage i where $i \geq 1$. Let γ be an MFV variable of dependence degree i, and let (t, α, β) be the defining triple for γ. We need to define the function $F_{\gamma(s),\pi}$. As before, let $m = val(t)$. Then note that the length of γ is understood to be $2m + 1$.

Let $b = val(s)$. If $b = 0$ or $b \geq m$ then by definition $\gamma(s)[\vec{n}]$ is a constant $\top$ or $\bot$ or is a function that has been defined in the previous stage. So consider the case where $1 \leq b < m$.

By the definition (377), $\gamma(s)[\vec{n}]$ can be seen as a binary tree whose leaves are labeled with $\alpha(s)[\vec{n}]$, $\beta(s)[\vec{n}]$ and $\gamma(2s)[\vec{n}]$, $\gamma(2s + 1)[\vec{n}]$. Intuitively we will expand this tree repeatedly at the leaves $\gamma(2s)[\vec{n}]$, $\gamma(2s + 1)[\vec{n}]$ until all leaves are labeled with either $\alpha(r)[\vec{n}]$ or $\beta(r)[\vec{n}]$ for some r.

The depth of the final tree can be shown to be computable by some AC^0 functions (recall Section III.3.3, for example, that the function $\log(x)$ is in AC^0, where $\log(x)$ is the length of the binary representation of x). Additionally, the labels of the nodes on all paths from the root to the leaves of the tree can be identified by AC^0 functions. From these facts, using the counting gates we can compute the string that concatenates all labels on the nodes with parentheses properly inserted. □

Now we verify that for every sequent S in π that is not on the right branches of the cut Σ_0^B-*COMP* or cut MFV' instances, there are provably in VTC^0 computable $G_0^\star$ proofs of the translation of S. The proof is by induction on the sequent S. Except for the case of the string $\exists$-left that introduces a cut MFV' instance, all cases are the same as in the proof of Theorem X.1.14.

So consider an instance of the string $\exists$-left that introduces a cut MFV' as in (376). Consider the interesting case where the translation of S_1 is not simplified to an axiom, and thus has the form

$$S_1[\bar{n}] = \delta'_{MFV}(t, \alpha, \beta, \gamma)[\bar{n}], \Gamma' \longrightarrow \Delta'.$$

Then S translates into

$$S[\bar{n}] = \Gamma' \longrightarrow \Delta'.$$

In order to obtain $S[\bar{n}]$ from $S_1[\bar{n}]$, we need to derive the tautology $\delta'_{MFV}(t, \alpha, \beta, \gamma)[\bar{n}]$ and then apply the cut rule. Here note that $\delta'_{MFV}(t, \alpha, \beta, \gamma)[\bar{n}]$ is just a conjunction of the form

$$\bigwedge (B_i \leftrightarrow B_i)$$

where B_i is the translation of $\gamma(s)$ when $val(s) = i$, for $1 \leq i < 2val(t)$. Hence $\delta'_{MFV}(t, \alpha, \beta, \gamma)[\bar{n}]$ can be easily derived from the axioms

$$B_i \longrightarrow B_i.$$

Given the FTC^0 functions that compute the translations of formulas in S_1, it is straightforward to obtain an FTC^0 function that computes the above derivation of $S[\bar{n}]$ from $S_1[\bar{n}]$. By the same arguments as in the proof of Theorem X.1.14, it follows that the $G_0^\star$ proofs can be provably in $\overline{VTC}^0$ computed by some FTC^0 functions. □

The next corollary follows easily.

Corollary X.3.4. *For any Σ_0^B theorem φ of VNC^1, there is an FTC^0 function that provably in $\overline{VTC}^0$ computes $PK^\star$ proofs of the family of tautologies $\|\varphi\|$.*

X.3.2. The Boolean Sentence Value Problem. Recall (page 321) that the Boolean Sentence Value Problem (BSVP) is to determine the truth value of a Boolean sentence. Here the sentence is given as a string over the alphabet:

$$\{\top, \bot, (,), \wedge, \vee, \neg\}. \tag{378}$$

The sentence is viewed as a tree whose leaves are labeled with constants $\top, \bot$ and whose inner nodes are labeled with connectives. Note that when the tree representing a sentence A is a balanced binary tree then it is straightforward to show that there is an *ALogTime* algorithm that computes the value of A. In fact, by using the axiom *MFV* (Definition IX.5.4) we can easily formalize in VNC^1 such an algorithm. However designing an *ALogTime* algorithm for a general tree structure is more difficult. The algorithm given in the proof of Theorem X.3.5 below is a slight modification of Buss's algorithm given in [25].

THEOREM X.3.5 (Buss). *The Boolean Sentence Value Problem is in ALogTime*.

PROOF. We give an algorithm in terms of a game between two players: one is called the Pebbler and the other is called the Challenger. The game is defined so that the Pebbler has a winning strategy if and only if the given Boolean sentence is true. The actual algorithm works by first playing the game and then determining the winner.

By using De Morgan's laws we can remove all occurrences of $\neg$ from the sentence. This transformation requires counting the number of occurrences of the $\neg$ connective along each path in the tree and therefore can be done in TC^0. Thus we can assume that the underlying tree is a binary tree whose inner nodes are labeled with $\vee$ or $\wedge$ and whose leaves are labeled with $\top$ and $\bot$. By padding the sentence with $\wedge\top$ we can assume that the tree has exactly $(2^{d+1} - 1)$ leaves, for some $d \geq 1$. We number the leaves of the tree from left to right with

$$1, 2, \ldots, 2^{d+1} - 1.$$

(We do not number inner nodes of the tree.)

The pebbling game. The game will be played in at most d rounds; each round consists of a move by the Pebbler followed by a move by the Challenger. In each round the Pebbler will assert the values of some nodes in the tree (by pebbling them with Boolean values) and the Challenger must deny the Pebbler's assertion by challenging one of the pebbled nodes. The Challenger is required not to challenge any node that has been pebbled but unchallenged in a previous round. In effect, the Challenger implicitly agrees with the Pebbler on all pebbled descendants of the currently challenged node. (Following a play of the game, the challenged nodes lie on a path from the root to some agreed node.) The idea is that at the end of at most d rounds the value of the challenged node is easily computed from an agreed node and some leaves, thus revealing the winner of the game. Intuitively, a winning strategy for the Pebbler is to pebble the nodes with their correct values, and if the Pebbler fails to do so, the Challenger can win by challenging some incorrectly pebbled node. So by having the Pebbler start with pebbling the root with $\top$, the sentence is true iff the Pebbler has a winning strategy.

The i-th round of the game involves the following nodes: c_i (c for *challenged*), a_i (a for *agreed*), u_i, v_i, u_i^1, u_i^2, v_i^1, v_i^2 (u_i^j and v_i^j are children of u_i and v_i, respectively) and leaves ℓ_i (ℓ for *left*) and r_i (r for *right*). In general, $\ell_i < r_i$, and ℓ_i (resp. r_i) is never to the right (resp. left) of the subtree rooted at a_i. Moreover, a_i is always a descendant of c_i, u_i as well as v_i. As i increases, the challenged nodes c_i move down a path from the root. Also, all leaf descendants of c_i that are not descendants of a_i are numbered in the range

$$\{\ell_i - 2^{d-i} + 1, \ldots, \ell_i, \ldots, \ell_i + 2^{d-i} - 1\} \cup$$
$$\{r_i - 2^{d-i} + 1, \ldots, r_i, \ldots, r_i + 2^{d-i} - 1\}.$$

Therefore after d rounds the Pebbler's asserted value of the challenged node c_d can be compared with the appropriate combination of a_d and the leaves ℓ_d, r_d, allowing us to determine the winner of the game. A possible configuration of the nodes is given in Figure 14. Here we orient the tree so that the leaves are at the bottom of the diagram.

Each move by a player will be specified by a constant number of bits $0, 1$. Essentially, the moves by the Pebbler (resp. the Challenger) can be interpreted as the existential (resp. universal) states, and playing the game can therefore be seen as running an alternating Turing machine in logtime.

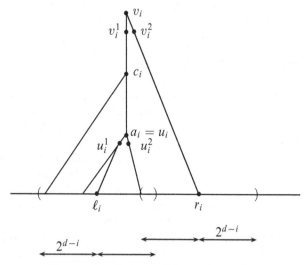

FIGURE 14. One possible configuration.

In the i-th round the Pebbler pebbles nodes

$$u_i, v_i, u_i^1, u_i^2, v_i^1, v_i^2$$

with some Boolean values, and the Challenger must either challenge one of these nodes or rechallenge a node it has challenged in the previous round. The Pebbler needs six bits for this task, and the Challenger needs three.

Later we will summarize the conditions for ending the game in less than d rounds. For instance, the game will end if the Challenger challenges either u_i or v_i. This is because, for example, the asserted value of u_i can be compared with the asserted values of u_i^1 and u_i^2, or if u_i is a leaf its true value is readily available.

The nodes of the i-th round are determined as follows. First, c_i is the challenged node from the previous round (c_1 is understood to be the root). Also,

$$a_1 = 2^d, \qquad \ell_1 = 2^{d-1}, \qquad r_1 = 2^d + 2^{d-1}.$$

For $i \geq 1$:

$$u_i = lca(\ell_i, a_i), \qquad v_i = lca(a_i, r_i)$$

where $lca(n_1, n_2)$ denotes the least common ancestor of nodes n_1 and n_2. The nodes u_i^1 and u_i^2 are the left and right children of u_i, respectively. (If u_i is a leaf, then $u_i^1 = u_i^2 = u_i$.) Similarly for v_i^1 and v_i^2.

Next, for the $(i + 1)$-st round (where $1 \leq i < d$) a_{i+1}, ℓ_{i+1}, r_{i+1} are determined based on the relative positions of c_i, u_i and v_i. For this purpose we need the following notation. Let

$$n_1 \rhd n_2$$

denote the fact that node n_1 is a proper ancestor of n_2, and

$$n_1 \unrhd n_2$$

stand for $n_1 \rhd n_2$ or $n_1 = n_2$. It will be true in general that

$$c_i \unrhd a_i, \qquad u_i \unrhd a_i, \qquad v_i \unrhd a_i.$$

As a result, the only possible relative positions for c_i, u_i, v_i are listed in Table 4 below. Will refer to the cases by their number later.

Case 1	Case 2	Case 3	Case 4
$u_i = v_i$	$c_i \unrhd u_i \rhd v_i$	$c_i \unrhd v_i \rhd u_i$	$u_i \rhd c_i \unrhd v_i$

Case 5	Case 6	Case 7
$v_i \rhd c_i \unrhd u_i$	$u_i \rhd v_i \rhd c_i$	$v_i \rhd u_i \rhd c_i$

TABLE 4. Possible relative positions of c_i, u_i, v_i.

Note that exactly one of these cases will hold, and the Pebbler will use three bits to specify which one holds. Also, $u_i = v_i$ only if $u_i = v_i = a_i$.

The game ends in round i if the Challenger challenges u_i or v_i. So we can assume that these two nodes are not challenged. First, suppose that in round i the Challenger challenges u_i^1 (i.e., $c_{i+1} = u_i^1$). The game ends in round i if $u_i \triangleright c_i$ (Cases 4, 6, 7) or $u_i = v_i$ (Case 1), because in this case the Challenger does not challenge a descendant of the currently challenged node. For other cases,

$$a_{i+1} = \ell_i, \qquad \ell_{i+1} = \ell_i - 2^{d-i-1}, \qquad r_{i+1} = \ell_i + 2^{d-i-1}. \tag{379}$$

See an illustration in Figure 15.

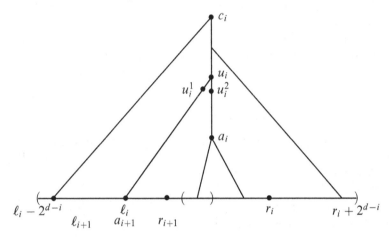

FIGURE 15. $c_{i+1} = u_i^1$ (v_i, v_i^1, v_i^2 are not shown).

Now suppose that the Challenger challenges u_i^2 in the i-th round. The game ends if $u_i = v_i$ (Case 1). If $u_i \triangleright v_i$ (Cases 2, 4, 6) then (see Figure 16 for an illustration):

$$a_{i+1} = v_i, \qquad \ell_{i+1} = \ell_i + 2^{d-i-1}, \qquad r_{i+1} = r_i + 2^{d-i-1}. \tag{380}$$

Otherwise, $v_i \triangleright u_i$ (Cases 3, 5, 7), and

$$a_{i+1} = a_i, \qquad \ell_{i+1} = \ell_i + 2^{d-i-1}, \qquad r_{i+1} = r_i - 2^{d-i-1}.$$

Note that if c_i is a proper descendant of u_i^2 then the Challenger will lose (see below).

The cases where the Challenger challenges v_i^1 or v_i^2 are similar. So suppose now that in the i-th round the Challenger rechallenges c_i. The nodes a_{i+1}, ℓ_{i+1} and r_{i+1} are set as as specified in Table 5. Figure 17 illustrates Case 1.

In summary, in each round the Pebbler gives nine bits specifying the truth values of $u_i, v_i, u_i^1, u_i^2, v_i^1, v_i^2$ and the relative positions of c_i, u_i, v_i given in Table 4. (It is understood that the Pebbler also pebbles the root with $\top$ in the first round.) Each move by the Challenger consists of giving three bits specifying the challenged node.

The following moves cause the Pebbler to lose the game:

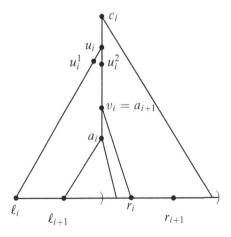

FIGURE 16. $c_{i+1} = u_i^2$ and $c_i \trianglerighteq u_i \rhd v_i$.

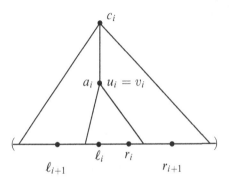

FIGURE 17. $c_{i+1} = c_i$ and $u_i = v_i$.

	$u_i = v_i$	$c_i \trianglerighteq u_i \rhd v_i$	$c_i \trianglerighteq v_i \rhd u_i$	$u_i \rhd c_i \trianglerighteq v_i$	$v_i \rhd c_i \trianglerighteq u_i$	$u_i, v_i \rhd c_i$
Case	(1)	(2)	(3)	(4)	(5)	(6,7)
a_{i+1}	a_i	u_i	v_i	v_i	u_i	a_i
ℓ_{i+1}	$\ell_i - t$	$\ell_i - t$	$\ell_i - t$	$\ell_i + t$	$\ell_i - t$	$\ell_i + t$
r_{i+1}	$r_i + t$	$r_i + t$	$r_i + t$	$r_i + t$	$r_i - t$	$r_i - t$

TABLE 5. The Challenger rechallenges c_i (i.e., $c_{i+1} = c_i$).
Here $t = 2^{d-i-1}$.

1) Pebble a leaf with the wrong value, or pebble incompatible values
   for u_i, u_i^1, u_i^2, v_i, v_i^1, v_i^2. For example, u_1 is an $\wedge$ node and u_1 is
   pebbled with $\bot$ while both u_i^1 and u_i^2 are pebbled with $\top$.
2) Pebble a node with both $\top$ and $\bot$.

3) Make a wrong assertion about the relative positions of c_i, u_i and v_i.

The Challenger loses if he

1) challenges a correctly pebbled leaf;
2) challenges u_i or v_i when they are pebbled compatibly with u_i^1, u_i^2, v_i^1, v_i^2;
3) in round i does not challenge a descendant of the currently challenged node c_i;
4) in round i challenges a descendant of the currently agreed node a_i.

The game is played in at most d rounds. It may end in less than d rounds if a player obviously makes a mistake listed above and therefore loses the game. Thus the game ends as soon as

1) The Challenger challenges either u_i or v_i,
2) The Challenger challenges u_i^1 when the Pebbler says $u_i \rhd c_i$ (Cases 4, 6, 7),
3) The Challenger challenges v_i^2 when the Pebbler says $v_i \rhd c_i$ (Cases 5, 6, 7),
4) The Challenger challenges u_i^j or v_i^j when the Pebbler says $u_i = v_i$ (Case 1).

CLAIM. The Pebbler has a winning strategy iff the given sentence is true.

The Claim is straightforward: If the sentence is true, the Pebbler can always win by pebbling the nodes with their correct values and stating the correct relative positions of c_i, u_i, v_i. If the sentence is false then the Challenger can win by always challenging the lowest node that is incorrectly pebbled.

Determining the winner. We finish the proof of Theorem X.3.5 by showing that the winner of the Pebbling game above can be determined from the plays in *ALogTime*. The task is, given a sequence of moves of the players (represented as a binary string), to determine which player is the first to violate the conditions above. We will indeed show that this can be done in TC^0.

We will first compute all ℓ_i, r_i, then all a_i. From these we can get u_i, v_i, u_i^1, u_i^2, v_i^1, v_i^2 easily. Then it is straightforward to find out the winner. Below we briefly show how to compute ℓ_i, r_i and a_i, for $1 \leq i \leq d$.

For simplicity, assume that the game lasts in exactly d rounds. Notice that ℓ_i and r_i have the form

$$\ell_i = x_d^i 2^d + x_{d-1}^i 2^{d-1} + \cdots + x_{d-i}^i 2^{d-i},$$
$$r_i = y_d^i 2^d + y_{d-1}^i 2^{d-1} + \cdots + y_{d-i}^i 2^{d-i}$$

where $x_j^i, y_j^i \in \{-1, 0, 1\}$. For example,

$$\ell_1 = 0 \times 2^d + 1 \times 2^{d-1}, \qquad r_1 = 1 \times 2^d + 1 \times 2^{d-1}.$$

Also, if the Challenger challenges u_i^1 as in (379) then $x_{d-i-1}^{i+1} = -1$, $y_{d-i-1}^{i+1} = 1$, and for $d - i \leq j \leq d$:

$$x_j^{i+1} = y_j^{i+1} = x_j^i.$$

On the other hand, if the Challenger challenges u_i^2 as in (380) then $x_{d-i-1}^{i+1} = y_{d-i-1}^{i+1} = 1$, and for $d - i \leq j \leq d$:

$$x_j^{i+1} = x_j^i, \qquad y_j^{i+1} = y_j^i.$$

Generally, x_{d-i}^i and y_{d-i}^i can be easily extracted from the moves in round i; and for $0 \leq j < i$, x_{d-j}^i and x_{d-j}^i can be computed from x_{d-j}^j and x_{d-j}^j by counting the number of "jumps" as in (379) where both ℓ_{j+1} and r_{j+1} are computed from only ℓ_j (or only r_j). From this we can conclude that ℓ_i and r_i can be computed in TC^0 from the moves of both players.

Next, notice that we can compute simultaneously all i such that a_i is a leaf. For example a_1 is a leaf; a_{i+1} as in (379) is also a leaf. For each other value of i, let $j < i$ be largest so that a_j is a leaf and hence has been determined. Then a_i is the least common ancestor of a_j and a certain subset S_i of

$$\{\ell_j, r_j, \ell_{j+1}, r_{j+1}, \ldots, \ell_{i-1}, r_{i-1}\}.$$

For example, for $j \leq k < i$, $r_k \in S_i$ if $a_{k+1} = v_k$ (e.g., as in (380)). The set S_i can be computed by an AC^0 function from the moves of the players. Hence all a_i can be computed in TC^0.   □

X.3.3. Reflection Principle for PK. Recall that $(Z \models_0 X)$ holds iff the truth assignment Z satisfies the quantifier-free formula X, and $(Z \models_0^\Sigma X)$ and $(Z \models_0^\Pi X)$ are Σ_1^B and Π_1^B formulas that represent $(Z \models_0 X)$, respectively. These formulas are expressed in terms of a Σ_0^B formula $\varphi_0(y, X, E)$ described in the statement and proof of Lemma X.2.1. By formalizing the algorithm given in the proof of Theorem X.3.5 we can strengthen Lemma X.2.1 by proving the following.

LEMMA X.3.6. $VNC^1 \vdash \exists E \leq t_0 + 1\varphi_0^{rec}(t_0 + 1, X, Z, E)$.

COROLLARY X.3.7. $VNC^1 \vdash (Z \models_0^\Sigma X) \leftrightarrow (Z \models_0^\Pi X)$.

PROOF. The direction

$$(Z \models_0^\Sigma X) \supset (Z \models_0^\Pi X)$$

can in fact be proved in V^0, and the other direction follows immediately from the Lemma.   □

PROOF OF LEMMA X.3.6. In essence we have to prove the existence of the array E' where $E'(i, j)$ is the truth value of the subformula encoded by $X[i, j]$, for all $1 \leq i \leq j \leq n$, where n is the length of X (as a string over the alphabet (378) on page 414). We will evaluate the subformulas $X[i, j]$ in parallel. Using the function $Fval$ (Definition IX.5.5), the idea is that each subformula $X[i, j]$ will be evaluated by constructing a suitable balanced tree encoded by some (a, G, I) (as in Sections IX.5.1 and IX.5.3) such that $Fval(a, G, I)(1)$ is the value of $X[i, j]$. The fact that all subformulas $X[i, j]$ can be evaluated simultaneously in VNC^1 will follow from the fact that $Fval^*$ is provably total in VNC^1 (Exercise IX.5.15).

We will construct a tuple (a, G, I) so that $Fval(a, G, I)(1)$ is the value of the sentence X; the constructions for other subformulas of X are similar. Since $\overline{VNC}^1$ is a conservative extension of VNC^1, we will actually work in $\overline{VNC}^1$. Recall that the *ALogTime* algorithm from the proof of Theorem X.3.5 is obtained by first playing the pebbling game and then determining the winner of the game. The balanced tree (a, G) will encode the game playing part of the algorithm: each path from the root of the tree to a leaf corresponds to a possible play of the game. Each input bit $I(x)$ specifies the winner of the play corresponding to the path ending with that leaf. The value of $I(x)$ is computed by the algorithm that determines the winner of the play.

We will in fact specify a balanced bounded fan-in tree T. Conversion from this tree to a balanced binary tree (a, G) as required for the arguments of *Fval* is straightforward and will be omitted. Let n be the number of constants in X (i.e., the number of leaves in the tree representing X). Let d be such that $2^d \leq n < 2^{d+1}$. As in the *ALogTime* algorithm for BSVP, we will pad the formula X with necessary $\wedge \top$ in order to make X a sentence with exactly $2^{d+1} - 1$ constants $\top, \bot$ (i.e., the underlying tree for X has exactly $2^{d+1} - 1$ leaves).

The tree T has $2d$ alternating layers of nodes corresponding to d rounds of the game. We number the layers starting at the root with number 1. The root is an $\vee$ node; generally, $\vee$ nodes are on layers $2j - 1$ (for $1 \leq j \leq d$) and correspond to the Pebbler's moves. They all have fan-in

$$7 \times 2^6$$

that represents 7×2^6 possibilities for a move by the Pebbler (2^6 different choices of the values for u_i, v_i, u_i^1, u_i^2, v_i^1, v_i^2, and 7 possible relative positions of c_i, u_i, v_i as in Table 4). Each child of an $\vee$ node is a $\wedge$ node that corresponds to a move by the Challenger. Thus all $\wedge$ nodes are on layers $2j$ (for $1 \leq j \leq d$) and have branching factor of 7 which encodes 7 possible choices for the Challenger. The children of an $\wedge$ node on layer $2j$ where $j < d$ correspond to the Pebbler's responses in round $(j + 1)$, and the children of the $\wedge$ nodes on layer $2d$ are inputs that are specified below.

Note that here we make T a balanced tree by having each play of the game end in exactly d rounds. (If some play ends in less than d rounds, simply add arbitrary moves to it.) Using the fact that the relation $BIT(i, x)$ (Section III.3.3) is Δ_0-definable, the (binary form of the) tree T can be defined in V^0.

Now, the "determining the winner" part in the proof of Theorem X.3.5 has been shown to be computable in TC^0; it is straightforward to formalize this part in VTC^0 (and hence in VNC^1). This implies that in $\overline{VNC}^1$ we can define a string of inputs I to the tree T so that $I(x)$ is true iff the path from the root of T to the leaf x corresponds to a play of the pebbling game where the Pebbler wins.

We finish the proof by showing the correctness of our formalization. Simply write $Fval(T, I)$ for $Fval(a, G, I)$, where (a, G) is a balanced binary formula equivalent to T. Then to prove (in $\overline{VNC}^1$) that our formalization is correct, we need to prove:

CLAIM. Suppose that $(A \odot B)$ is a subformula of X, where $\odot \in \{\wedge, \vee\}$, and suppose that $(T_{A \odot B}, I_{A \odot B})$, (T_A, I_A) and (T_B, I_B) are the result of our constructions for the sentences $A \odot B$, A and B, respectively. Then $\overline{VNC}^1$ proves

$$Fval(T_{A \odot B}, I_{A \odot B})(1) = Fval(T_A, I_A)(1) \odot Fval(T_B, I_B)(1). \qquad (381)$$

We prove the claim by structural induction on the subformula $(A \odot B)$. The base case (where A, B are both constants) is obvious. For the induction step consider the following cases:

Case I. $(A \odot B) = (A \wedge B)$.

Case Ia. First suppose that

$$Fval(T_A, I_A)(1) = Fval(T_B, I_B)(1) = \top.$$

We show that $Fval(T_{A \wedge B}, I_{A \wedge B})(1) = \top$ (i.e., that the Pebbler has a winning strategy for the game played on $(A \wedge B)$).

Consider playing the game on $(A \wedge B)$. Intuitively, the Pebbler wins by always giving the correct values for the nodes u_i, v_i, u_i^1, u_i^2, v_i^1 and v_i^2. Formally, we will show that all nodes on the "winning paths" of the tree $T_{A \wedge B}$ are true. Here winning paths are defined in favor of the Pebbler: they are the paths from the root of $T_{A \wedge B}$ that follow the Pebbler's correct move at every $\vee$ node (or any branch of the $\vee$ node if the game has ended earlier with the Pebbler being the winner). To find the winning paths we use the induction hypothesis, i.e., the value of a subformula C of $(A \wedge B)$ is $Fval(T_C, I_C)(1)$. Thus, a path from the root of $T_{A \wedge B}$ to a leaf x is a winning path if at every $\vee$ node the path follows the edge that is specified by (i) the correct relative positions of c_i, u_i and v_i (as in Table 4) and (ii) the values of u_i, v_i, u_i^1, u_i^2, v_i^1 and v_i^2 as given by

$$Fval(T_{U_i}, I_{U_i})(1), \qquad Fval(T_{V_i}, I_{V_i})(1), \qquad \text{etc.}$$

(here U_i denotes the subformula whose root is u_i, etc.).

We will prove by reverse induction on j, $0 \le j \le d$, that all nodes on layer $(2j + 1)$ of all winning paths are true. For $j = 0$ we will have that the root of $T_{A \wedge B}$ is true, i.e., $Fval(T_{A \wedge B}, I_{A \wedge B})(1) = \top$ and we will be done with *Case Ia*.

For the base case, $j = d$, and all nodes on layer $(2d + 1)$ are leaves of $T_{A \wedge B}$. Using the induction hypothesis (of the claim that (381) holds for all subformulas of $(A \wedge B)$) and using the fact that both $Fval(T_A, I_A)(1)$ and $Fval(T_B, I_B)(1)$ are true, by definition the inputs to $T_{A \wedge B}$ at the end of all winning paths are true.

The induction step is straightforward: suppose that node w is on a winning path and w is on layer $(2j + 1)$. Let t be the child of w that corresponds to a correct move by the Pebbler (or any child of w if the Pebbler has won before round $(j + 1)$). Then by the induction hypothesis all children of t are true. Hence both t and w are true, because w is an $\vee$ node and t is an $\wedge$ node.

Case Ib.   At least one of $Fval(T_A, I_A)(1)$ and $Fval(T_B, I_B)(1)$ is $\bot$. The proof in this case is similar to *Case Ia*. Define "losing paths" to be paths from the root of $T_{A \wedge B}$ where the Challenger always challenges the lowest node that is wrongly pebbled (recall that the trees are oriented with the roots at the top). Then by similar arguments as in *Case Ia*, it can be shown that all nodes on the losing paths are false. In particular, the root of $T_{A \wedge B}$ is false.

Case II. $(A \odot B) = (A \vee B)$. This case can be handled similarly to *Case I*.   □

Recall the formulas Fla^{Π} and $Prf_{\mathcal{F}}^{\Pi}$ from Corollary X.1.4 and Lemma X.1.5.

DEFINITION X.3.8.   The Reflection Principle for **PK**, denoted by RFN_{PK}, is the $\forall \Sigma_1^B$ sentence:

$$\forall \pi \forall X \forall Z \big((Fla^{\Pi}(X) \wedge Prf_{PK}^{\Pi}(\pi, X)) \supset (Z \models_0^{\Sigma} X) \big).$$

THEOREM X.3.9. VNC^1 *proves* RFN_{PK}.

PROOF. VNC^1 can extract a sequence $X^{[0]}, \ldots, X^{[r]}$ of strings representing the successive sequents in the proof π, and prove $(Z \models_0^{\Sigma} \widehat{D_i})$ by induction on i, using Lemma X.3.6, where D_i is the formula expressing the semantics of the sequent $X^{[i]}$ (as in (8) on page 10). Details are left as an exercise.   □

EXERCISE X.3.10.   Give details for the argument above.

Analogous to results in Section X.2.5 we prove that **PK** is the strongest propositional proof system whose reflection principle is provable in VNC^1.

THEOREM X.3.11 (*PK* Simulation).   *Let $\mathcal{F}$ be a propositional proof system and let $RFN_{\mathcal{F}}$ denote the reflection principle for $\mathcal{F}$ (defined as Σ_0^q-$RFN_{\mathcal{F}}$ in Definition X.2.11). Suppose that $VNC^1 \vdash RFN_{\mathcal{F}}$. Then **PK** p-simulates $\mathcal{F}$.*

Proof. Let $A_0(\vec{p})$ be a tautology and π_0 be an $\mathcal{F}$-proof of $A_0(\vec{p})$. We need to give a polytime algorithm that on input π_0 (and hence also $A_0(\vec{p})$) computes a PK proof of $A_0(\vec{p})$. We will show how to derive $\longrightarrow A_0(\vec{p})$ in PK, and it can be verified that the derivation can be constructed in time polynomial in $|\pi_0|$.

As usual we use $\widehat{A_0}$ and $\widehat{\pi_0}$ to denote the strings that encode $A_0(\vec{p})$ and π_0, respectively. The idea is to first translate a VNC^1 proof of $RFN_\mathcal{F}$ into a $PK^\star$ proof and then substitute the bits of $\widehat{A_0}$ and $\widehat{\pi_0}$ for the corresponding propositional variables. This is basically a $PK^\star$ proof of the fact that $A_0(\overrightarrow{p})$ is valid. From this we will be able to derive $A_0(\overrightarrow{p})$.

Note that the Translation Theorem for VNC^1 X.3.1 already gives us a $G_0^\star$ proof of the translation of $RFN_\mathcal{F}$. Also, $RFN_\mathcal{F}$ is Σ_1^B so it translates into Σ_1^q formulas. Here we need a $PK^\star$ proof (in particular, no quantifier is allowed). So we have to go further and instantiate the existentially quantified Boolean variables in order to obtain a quantifier-free tautology expressing the same fact (that is, $A_0(\vec{p})$ is valid) and a $PK^\star$ proof of it. Intuitively, this step can be done by bypassing the $\exists$-right and $\forall$-left introduction rules in the $G_0^\star$ proof, and thus retaining all the target formulas as witnesses for the existentially quantified variables in (the translation of) $RFN_\mathcal{F}$. Formal arguments are as follows.

By hypothesis VNC^1 proves $RFN_\mathcal{F}$, so by definition of $RFN_\mathcal{F}$

$$VNC^1 \vdash \forall\pi\forall X\forall Z\big((Fla^\Pi(X) \wedge Prf_\mathcal{F}^\Pi(\pi, X)) \supset (Z \models_0^\Sigma X)\big).$$

Therefore the following sequent is a theorem of VNC^1:

$$Fla^\Pi(X), Prf_\mathcal{F}^\Pi(\pi, X) \longrightarrow Z \models_0^\Sigma X.$$

Let $m = |\widehat{A_0}|$, $\ell = |\widehat{\pi_0}|$ and let n be the number of free variables in A_0. By the Translation Theorem for VNC^1 X.3.1 there is, provably computable in $\overline{VTC}^0$, a polynomial size $G_0^\star$ proof of of the sequent

$$Fla^\Pi(X)[m], Prf_\mathcal{F}^\Pi(\pi, X)[\ell, m] \longrightarrow (Z \models_0^\Sigma X)[m, n+1].$$

By substituting the bits of $\widehat{A_0}$ and $\widehat{\pi_0}$ for the free variables $\overrightarrow{p^X}$ and $\overrightarrow{p^\pi}$ respectively we obtain a $G_0^\star$ proof P of the following sequent (recall the notation from Definition X.1.6)

$$Fla^\Pi(\widehat{A_0})[\,] , Prf_\mathcal{F}^\Pi(\widehat{\pi_0}, \widehat{A_0})[\,] \longrightarrow (Z \models_0^\Sigma \widehat{A_0})[n+1]. \qquad (382)$$

Because the cut formulas in this proof are quantifier-free, it follows that the rules $\forall$-right and $\exists$-left are not used. Therefore we can turn this into a $PK^\star$ proof of a quantifier-free instantiation of (383) as follows.

First we remove the $\exists$ quantifiers in the succedents in P. This can be done as in the proof of Morioka's Theorem VII.4.5. For convenience we reproduce the argument here. Note that all Σ_1^q formula in P are ancestors of $(Z \models^{\Sigma}_0 \widehat{A_0})[n+1]$ and appear in the succedents in P. Recall (Lemma X.2.1 and its proof)

$$(Z \models^{\Sigma}_0 X) \equiv \exists E \leq t_0 + 1(\varphi^{rec}_0(t_0, X, Z, E) \wedge E(t_0)).$$

Therefore $(Z \models^{\Sigma}_0 \widehat{A_0})[n+1]$ is (the simplification of) the Σ^q_1 formula (here $r = t_0(m)$):

$$\exists p^E_0 \exists p^E_1 \ldots \exists p^E_{r-1} \bigvee_{k=0}^{r+1} ((\varphi^{rec}_0(t_0, \widehat{A_0}, Z, E) \wedge E(t_0))[n+1, k]).$$

Since $E(t_0)[k] \equiv \perp$ for $k \leq r$ and $E(t_0)[r+1] \equiv \top$, we have

$$(Z \models^{\Sigma}_0 \widehat{A_0})[n+1] \equiv \exists \overrightarrow{p^E} ((\varphi^{rec}_0(t_0, \widehat{A_0}, Z, E) \wedge E(t_0))[n+1, r+1]).$$

Let $\lambda(p^E_0, p^E_1, \ldots, p^E_{r-1}, p^Z_0, p^Z_1, \ldots, p^Z_{n-1})$ be the quantifier-free matrix of $(Z \models^{\Sigma}_0 \widehat{A_0})[n+1]$:

$$\lambda(\overrightarrow{p^E}, \overrightarrow{p^Z}) \equiv (\varphi^{rec}_0(t_0, \widehat{A_0}, Z, E) \wedge E(t_0))[n+1, r+1].$$

We modify the $G^{\star}_0$ proof P as follows. Let S be a sequent in P that contains a Σ^q_1 ancestor of $(Z \models^{\Sigma}_0 \widehat{A_0})[n+1]$. This Σ^q_1 formula is of the form

$$\exists p^E_i \ldots \exists p^E_{r-1} \lambda(D_0, \ldots, D_{i-1}, p^E_i, \ldots, p^E_{r-1}, \overrightarrow{p^Z}) \qquad (383)$$

for some $0 \leq i < r$ and formulas $D_0, D_1, \ldots, D_{i-1}$ (when $i = 0$ the list $\vec{D}$ is empty). Each maximal quantifier-free ancestor of this Σ^q_1 formula (called a P-prototype) has the form

$$\lambda(D_0, \ldots, D_{i-1}, D'_i, \ldots, D'_{r-1}, \overrightarrow{p^Z})$$

for some formulas $D'_i, \ldots, D'_{r-1}$. We replace each occurrence of (383) in S by the list of all its maximal quantifier-free ancestors:

$$\lambda(D_0, \ldots, D_{i-1}, D^1_i, \ldots, D^1_{r-1}, \overrightarrow{p^Z}), \ldots,$$

$$\lambda(D_0, \ldots, D_{i-1}, D^s_i, \ldots, D^s_{r-1}, \overrightarrow{p^Z})$$

(for some $s \geq 1$).

We apply this procedure to every sequent in P. Then, by removing possible duplicated sequents and inserting necessary weakenings, it can be seen that the result is a $G^{\star}_0$ proof of a sequent of the form

$$Fla^{\Pi}(\widehat{A_0})[\,], Prf^{\Pi}_{\mathcal{F}}(\widehat{\pi_0}, \widehat{A_0})[\,] \longrightarrow \lambda(\overrightarrow{D^1}, \overrightarrow{p^Z}), \lambda(\overrightarrow{D^1}, \overrightarrow{p^Z}), \ldots, \lambda(\overrightarrow{D^s}, \overrightarrow{p^Z})$$

for some $s \geq 1$.

Similarly we can eliminate all $\forall$ quantifiers in the antecedents in P. As a result we arrive at a $\boldsymbol{PK}^\star$ proof of a sequent of the form

$$\psi_1(\overrightarrow{G^1}), \ldots, \psi_1(\overrightarrow{G^{s_1}}), \psi_2(\overrightarrow{H^1}), \ldots, \psi_2(\overrightarrow{H^{s_2}})$$

$$\longrightarrow \lambda(\overrightarrow{D^1}, \overrightarrow{p^Z}), \lambda(\overrightarrow{D^1}, \overrightarrow{p^Z}), \ldots, \lambda(\overrightarrow{D^s}, \overrightarrow{p^Z}) \quad (384)$$

for some $s_1, s_2 \geq 1$ and formulas $\overrightarrow{G^i}$, $\overrightarrow{H^i}$ and $\overrightarrow{D^i}$ that may contain the free variables $\overrightarrow{p^Z}$. Here ψ_1 and ψ_2 are the quantifier-free matrices of $Fla^{\Pi}(\widehat{A_0})[\ ]$ and $Prf_{\mathcal{F}}^{\Pi}(\widehat{\pi}_0, \widehat{A_0})[\ ]$, respectively, i.e., (for $r_1 = t_{FLA}(m)$, $r_2 = t_{\mathcal{F}}(\ell, m)$)

$$\psi_1(\overrightarrow{p^Y}) \equiv \bigwedge_{k=0}^{r_1+1} \left((\varphi_{FLA}^{rec}(t_{FLA}, \widehat{A_0}, Y) \supset Y(t_{FLA}))[k] \right),$$

$$\psi_2(\overrightarrow{p^Y}) \equiv \bigwedge_{k=0}^{r_2+1} \left((\varphi_{\mathcal{F}}^{rec}(t_{\mathcal{F}}, \widehat{\pi}_0, \widehat{A_0}, Y) \supset Y(t_{\mathcal{F}}))[k] \right).$$

(Recall Fla^{Π} from (326) and $Prf_{\mathcal{F}}^{\Pi}$ from (323), pages 368 and 367.)

Now we will show how to derive the following sequents:

$$\longrightarrow \psi_1(\overrightarrow{G^i}) \qquad \text{for } 1 \leq i \leq s_1, \quad (385)$$

$$\longrightarrow \psi_2(\overrightarrow{H^i}) \qquad \text{for } 1 \leq i \leq s_2, \quad (386)$$

$$\lambda(\overrightarrow{D^i}, \overrightarrow{p^Z}) \longrightarrow A_0(\overrightarrow{p^Z}) \qquad \text{for } 1 \leq i \leq s. \quad (387)$$

Then it is straightforward to combine these with (384) to obtain

$$\longrightarrow A_0(\overrightarrow{p^Z}).$$

Consider (385) for some $1 \leq i \leq s_1$. Recall the proof of Lemma X.1.7, in particular we have shown that there are (computable in polytime) a $\boldsymbol{PK}^\star$ proof of the sequent (332) (here r_1 plays the role of r):

$$\longrightarrow \bigwedge_{k=0}^{r_1+1} \left(\varphi_{FLA}^{rec}(t_{FLA} + 1, X_0, Y) \supset Y(t_{FLA}) \right)[k].$$

By simultaneously replacing each p_j^Y by the formula G_j^i we obtain a proof of (385).

The argument for (386) is similar. Consider now (387) for some $1 \leq i \leq s$. This sequent is obtained from the following sequent by simultaneously substituting D_j^i for p_j^E (for $0 \leq j < r$):

$$(\varphi_0^{rec}(t_0, \widehat{A_0}, Z, E) \wedge E(t_0))[n + 1, r + 1] \longrightarrow A_0(\overrightarrow{p^Z}).$$

Recall (Lemma X.2.6) that there is a polytime algorithm that computes a $\boldsymbol{PK}^\star$ proof of the above sequent. A proof of (387) can then be obtained by simultaneously substituting D_j^i for p_j^E. $\qquad\qquad\qquad\qquad \square$

The procedure for transforming the $G_0^\star$ proof of (382) to a $PK^\star$ proof can be used for showing that the witnessing problem $\Sigma_1^q\text{-}WIT_{G_0^\star}$ can be solved in NC^1 (see also [80]). (And if the correctness of such a algorithm could be proved in VNC^1, then we would have $VNC^1 \vdash \Sigma_1^q\text{-}RFN_{G_0^\star}$. However we do not know whether this is true, even when the Reflection Principle for $G_0^\star$ is restricted to prenex Σ_1^q formulas.) Furthermore, similar to Theorem X.2.33 we can show that $\Sigma_1^q\text{-}WIT_{G_0}$ is TC^0-complete for NC^1. Here we state a stronger result from [45] and [80, Section 6.2].

THEOREM X.3.12. *Both* $\Sigma_1^q\text{-}WIT_{G_0}$ *and* $\Sigma_1^q\text{-}WIT_{G_0^\star}$ *are complete for* NC^1 *under* AC^0 *many-one reduction.*

X.4. VTC^0 and Threshold Logic

In this section we introduce PTK, an extension of the sequent calculus PK that has a new kind of connective called a *threshold connective*. This plays the same role in proof systems as the counting function *numones* plays in theories. We are interested in *bounded depth* PTK, the subsystems of PTK obtained by limiting the depth of the cut formulas to constants in $\mathbb{N}$. We will associate the theory VTC^0 with these subsystems. (The full version of PTK is p-equivalent to PK.)

The section is organized as follows. The sequent calculus PTK and its subsystems are introduced in Section X.4.1. The reflection principle for bounded depth PTK is introduced ins Section X.4.2. In Section X.4.3 we show that the families of tautologies translated from Σ_0^B theorems of VTC^0 have polynomial-size bounded-depth PTK proofs. In Section X.4.4 we show that the translation can be extended to quantified tautologies that correspond to any bounded theorem of VTC^0 by using $bGTC_0$, a quantified version of PTK.

X.4.1. The Sequent Calculus PTK. The sequent calculus PTK is defined similarly to PK, but instead of the binary connectives ($\wedge$ and $\vee$) PTK contains *threshold connectives* Th_k (for $1 \le k \in N$) that have unbounded arity. The semantics of Th_k is that

$$\mathsf{Th}_k(A_1, A_2, \ldots, A_n)$$

is true if and only if at least k of the formulas A_i are true. For example,

$$\mathsf{Th}_2(p, q, r) \Leftrightarrow (p \wedge q) \vee (q \wedge r) \vee (r \wedge p).$$

Also,

$$\mathsf{Th}_1(A_1, A_2, \ldots, A_n) \Leftrightarrow \bigvee_{i=1}^{n} A_i; \quad \mathsf{Th}_n(A_1, A_2, \ldots, A_n) \Leftrightarrow \bigwedge_{i=1}^{n} A_i. \quad (388)$$

For readability we will sometimes use $\bigwedge$ and $\bigvee$ in *PTK* formulas in place of Th_1 and Th_n.

Formally *PTK* formulas (or threshold formulas, or just formulas) are built from

- propositional constants $\top, \bot$,
- propositional variables $p, q, r, \ldots$,
- connectives $\neg, \mathsf{Th}_k$,
- parenthesis $(,)$

using the rules:

(a) $\top, \bot$, and p are *atomic* formulas, for any propositional variable p;
(b) if A is a formula, so is $\neg A$;
(c) for $n \geq 2, 1 \leq k \leq n$, if $A_1, A_2, \ldots, A_n$ are formulas, so is $\mathsf{Th}_k(A_1, A_2, \ldots, A_n)$.

Moreover,

$$\mathsf{Th}_0(A_1, A_2, \ldots, A_n) =_{\text{def}} \top, \quad \mathsf{Th}_k(A_1, A_2, \ldots, A_n) =_{\text{def}} \bot \text{ for } k > n.$$

The sequent calculus *PTK* is defined similarly to *PK* (Definition II.1.2). Here the logical axioms are of the form

$$A \longrightarrow A, \qquad \bot \longrightarrow, \qquad \longrightarrow \top$$

where A is any *PTK* formula. The weakening, exchange, contraction, cut and $\neg$ introduction rules are the same as on page 11. The other rules of *PTK* are listed below.

The left and right all-introduction rules (all-left and all-right) are as follows:

$$\frac{A_1, \ldots, A_n, \Lambda \longrightarrow \Gamma}{\mathsf{Th}_n(A_1, \ldots, A_n), \Lambda \longrightarrow \Gamma} \qquad \frac{\Lambda \longrightarrow A_1, \Gamma \quad \cdots \quad \Lambda \longrightarrow A_n, \Gamma}{\Lambda \longrightarrow \mathsf{Th}_n(A_1, \ldots, A_n), \Gamma}$$

Left and right one-introduction rules (one-left and one-right) are:

$$\frac{A_1, \Lambda \longrightarrow \Gamma \quad \cdots \quad A_n, \Lambda \longrightarrow \Gamma}{\mathsf{Th}_1(A_1, \ldots, A_n), \Lambda \longrightarrow \Gamma} \qquad \frac{\Lambda \longrightarrow A_1, \ldots, A_n, \Gamma}{\Lambda \longrightarrow \mathsf{Th}_1(A_1, \ldots, A_n), \Gamma}$$

Th_k-introduction rules (for $2 \leq k \leq n-1$):

$$\frac{\mathsf{Th}_k(A_2, \ldots, A_n), \Lambda \longrightarrow \Gamma \quad A_1, \mathsf{Th}_{k-1}(A_2, \ldots, A_n), \Lambda \longrightarrow \Gamma}{\mathsf{Th}_k(A_1, \ldots, A_n), \Lambda \longrightarrow \Gamma} \mathsf{Th}_k\text{-left}$$

$$\frac{\Lambda \longrightarrow A_1, \mathsf{Th}_k(A_2, \ldots, A_n), \Gamma \quad \Lambda \longrightarrow \mathsf{Th}_{k-1}(A_2, \ldots, A_n), \Gamma}{\Lambda \longrightarrow \mathsf{Th}_k(A_1, \ldots, A_n), \Gamma} \mathsf{Th}_k\text{-right}$$

Like *PK*, *PTK* is sound and complete and does not require the cut rule for completeness.

EXERCISE X.4.1. Show that for each of the above rules, the bottom sequent is a logical consequence of the top sequent(s). Also show that if the bottom sequent is valid, then each top sequent is valid.

THEOREM X.4.2 (Soundness and Completeness of **PTK**). *Any sequent provable in* **PTK** *is valid, and valid sequents have cut-free* **PTK** *proofs.*

PROOF. The proof is an extension of the proof of soundness and completeness of **PK** (Theorems II.1.7 and II.1.8). Soundness follows from the first sentence in the above exercise. To prove completeness, it follows from the second sentence in that exercise that the Inversion Principle for **PK** (Lemma II.1.9) applies more generally to **PTK**. Hence a proof of a valid sequent S can be obtained by successively simplifying the formulas in S by applying appropriate rules in reverse.   □

We will prove in Theorem X.4.6 that **PTK** and **PK** are p-equivalent. So we are mainly interested in subsystems of **PTK** where the cut formulas have bounded depths.

DEFINITION X.4.3 (Depth of a **PTK** Formula). The depth of a **PTK** formula A is the nesting depth of the connectives in A.

So, for example, the atomic **PTK** formulas have depth 0.

DEFINITION X.4.4 (Bounded Depth **PTK**). For each constant $d \in \mathbb{N}$, a d-**PTK** proof is a **PTK** proof in which all cut formulas have depth at most d. A *bounded depth* **PTK** system (or just **bPTK**) is any system d-**PTK** for $d \in \mathbb{N}$.

The treelike versions of **PTK**, d-**PTK** and **bPTK** are denoted by **PTK***, d-**PTK*** and b-**PTK***, respectively.

Recall (Definition VII.1.13) that the depth of a formula A of **PK** is the maximal number of times the connective changes in any path in the tree form of A. When we write a formula $(A_1 \vee A_2 \vee \cdots \vee A_n)$ or $(A_1 \wedge A_2 \wedge \cdots \wedge A_n)$ of the system **PK** we ignore the fact that officially $\vee$ and $\wedge$ are binary connectives and the formula must be fully parenthesized. The justification is that the parenthesization does not affect the depth of a formula, and also according to Lemma VII.1.15 there are simple **PK** proofs converting from one parenthesization to any other.

Thus using the formulas (388) there is a natural way to translate formulas of the system **PK** to formulas of **PTK** which preserves depth. The statement of the next result makes sense when we use this translation.

THEOREM X.4.5. *For any* $d \in \mathbb{N}$, d-**PTK** p-simulates d-**PK** *for proving formulas of depth* d.

PROOF. Let π be a d-**PK** derivation whose end sequent has depth at most d. Note that all formulas in π have depth at most d. We translate each **PK** formula into a **PTK** formula using (388). The result is a **PTK** formula of the same depth. Thus each formula A in π is translated into a **PTK** formula A' of depth at most d. For each sequent S in π, let S' be the translation of S. We prove by induction on the length of π that there is a d-**PTK** proof π' of size polynomial in the size of π that contains all translations S' of sequents S in π.

The base case is obvious because axioms of **PK** are translated into axioms of **PTK** of the same depth. For the induction step, suppose that $\pi = (\pi_1, S)$ where S is the end sequent of π. Consider, for example, the case where S is derived from two sequents S_1 and S_2 in π_1 as follows:

$$\frac{S_1 \qquad S_2}{S} = \frac{\bigvee A_i, \Gamma \longrightarrow \Delta \qquad \bigvee B_j, \Gamma \longrightarrow \Delta}{\bigvee A_i \vee \bigvee B_j, \Gamma \longrightarrow \Delta}$$

Here $\bigvee A_i$ is any parenthesizing of $A_1 \vee A_2 \vee \cdots \vee A_n$, and similarly for $\bigvee B_j$. Note that

$$S_1' = \mathsf{Th}_1(\vec{A}'), \Gamma' \longrightarrow \Delta', \qquad S_2' = \mathsf{Th}_1(\vec{B}'), \Gamma' \longrightarrow \Delta',$$

$$S' = \mathsf{Th}_1(\vec{A}', \vec{B}'), \Gamma' \longrightarrow \Delta'.$$

Using the one-left rule we can derive

$$\mathsf{Th}_1(\vec{A}', \vec{B}') \longrightarrow \mathsf{Th}_1(\vec{A}'), \mathsf{Th}_1(\vec{B}').$$

From this and S_1', S_2', using the cut rule (with cut formulas $\mathsf{Th}_1(\vec{A}')$ and $\mathsf{Th}_1(\vec{B}')$) we derive S'. The derivation π' is obtained from π_1' and the above derivation.

It is easy to see that π' as described above has size bounded by a polynomial in the size of π. $\qquad\square$

The reverse direction of Theorem X.4.5 does not hold, since the pigeon-hole tautologies have polysize **bPTK** proofs (Corollary X.4.20) but do not have polysize **bPK** proofs (Theorem VII.1.16).

However the reverse simulation does hold when we compare the unbounded depth proof systems **PTK** and **PK**, and this is stated in the next theorem. To make sense of this statement we use the fact that there are standard polynomial time translations back and forth from formulas of **PK** to formulas of of **PTK** such that formulas are translated to equivalent formulas. As mentioned before the translation from **PK** formulas to **PTK** formulas uses the equivalences (388). The translation in the other direction is described in the proof below.

THEOREM X.4.6. **PK** *is p-equivalent to* **PTK**.

PROOF SKETCH. The fact that **PTK** *p*-simulates **PK** follows from the proof of Theorem X.4.5 above. It remains to show that **PK** *p*-simulates **PTK**.

In Section IX.5.4 we show that the function *numones* is provably total in VNC^1. In essence we construct a uniform family of formulas that compute *numones*. In other words, there are **PK** formulas $F_{n,k}(p_1, p_2, \ldots, p_n)$ so that

$$F_{n,k}(p_1, p_2, \ldots, p_n) \Leftrightarrow \text{the number of } \top \text{ in } p_1, p_2, \ldots, p_n \text{ is } k.$$

Moreover in the same way that $VNC^1 \vdash NUMONES$ (Theorem IX.5.17) we can prove:

Proposition X.4.7. *There are polynomial-size **PK**-proofs of the following sequents*:

1) $F_{n,n}(A_1, A_2, \ldots, A_n) \longrightarrow A_i$ *(for $1 \le i \le n$)*;
2) $A_1, A_2, \ldots, A_n \longrightarrow F_{n,n}(A_1, A_2, \ldots, A_n)$;
3) $A_i \longrightarrow F_{n,1}(A_1, A_2, \ldots, A_n)$ *(for $1 \le i \le n$)*;
4) $F_{n,1}(A_1, A_2, \ldots, A_n) \longrightarrow A_1, A_2, \ldots, A_n$;
5) $F_{n,\ell}(A_1, A_2, \ldots, A_n) \longrightarrow A_1, F_{n-1,\ell}(A_2, \ldots, A_n)$ *(for $1 \le \ell \le n-1$)*;
6) $A_1, F_{n,\ell}(A_1, A_2, \ldots, A_n) \longrightarrow F_{n-1,\ell-1}(A_2, \ldots, A_n)$ *(for $2 \le \ell \le n$)*;
7) $F_{n-1,\ell}(A_2, \ldots, A_n) \longrightarrow A_1, F_{n,\ell}(A_1, A_2, \ldots, A_n)$ *(for $1 \le \ell \le n-1$)*;
8) $A_1, F_{n-1,\ell-1}(A_2, \ldots, A_n) \longrightarrow F_{n,\ell}(A_1, A_2, \ldots, A_n)$ *(for $2 \le \ell \le n$)*.

Now **PTK** formulas are translated into **PK** formulas (of unbounded depth) inductively using the formulas $F_{n,k}$ as follows. No translation is required for atomic formulas, because they are the same in **PK** and **PTK**. For the inductive step, suppose that A_i has been translated to A_i', for $1 \le i \le n$. Then $\mathsf{Th}_k(A_1, A_2, \ldots, A_n)$ is translated into

$$\bigvee_{\ell=k}^{n} F_{n,\ell}(A_1', A_2', \ldots, A_n').$$

To show that **PK** p-simulates **PTK** it suffices to show that the translations of rules of **PTK** have polynomial-size **PK** proofs.

First, consider the rule all-left. We need to show that there is a polynomial-size **PK** derivation of the form

$$\frac{A_1, A_2, \ldots, A_n, \Lambda \longrightarrow \Gamma}{F_{n,n}(A_1, A_2, \ldots, A_n), \Lambda \longrightarrow \Gamma}$$

For this we can use Proposition X.4.7 (1) with successive cuts on the formulas A_i.

Similarly, the rule one-left can be simulated using Proposition X.4.7 (4).

Consider now the rule Th_k-left. Suppose that $2 \le k \le n - 1$, we need to give polynomial-size **PK**-derivations of the form

$$\frac{\bigvee_{\ell=k}^{n-1} F_{n-1,\ell}(A_2, \ldots, A_n), \Lambda \longrightarrow \Gamma \quad A_1, \bigvee_{\ell=k-1}^{n-1} F_{n-1,\ell}(A_2, \ldots, A_n), \Lambda \longrightarrow \Gamma}{\bigvee_{\ell=k}^{n} F_{n,\ell}(A_1, A_2, \ldots, A_n), \Lambda \longrightarrow \Gamma}$$

It suffices to derive for each ℓ (where $k \le \ell \le n$) the sequent

$$F_{n,\ell}(A_1, A_2, \ldots, A_n), \Lambda \longrightarrow \Gamma. \tag{389}$$

From Proposition X.4.7 (5) we derive (by weakening and $\vee$-right):

$$F_{n,\ell}(A_1, A_2, \ldots, A_n) \longrightarrow A_1, \bigvee_{\ell=k}^{n-1} F_{n-1,\ell}(A_2, \ldots, A_n).$$

Using the cut rule for this sequent and

$$\bigvee_{\ell=k}^{n-1} F_{n-1,\ell}(A_2, \ldots, A_n), \Lambda \longrightarrow \Gamma$$

we obtain

$$F_{n,\ell}(A_1, A_2, \ldots, A_n), \Lambda \longrightarrow A_1, \Gamma. \tag{390}$$

From Proposition X.4.7 (6) for $\ell = k$ we obtain (using weakening and $\vee$-right):

$$A_1, F_{n,k}(A_1, A_2, \ldots, A_n) \longrightarrow \bigvee_{\ell=k-1}^{n-1} F_{n-1,\ell}(A_2, \ldots, A_n).$$

From this and

$$A_1, \bigvee_{\ell=k-1}^{n-1} F_{n-1,\ell}(A_2, \ldots, A_n), \Lambda \longrightarrow \Gamma$$

we derive

$$A_1, F_{n,k}(A_1, A_2, \ldots, A_n), \Lambda \longrightarrow \Gamma. \tag{391}$$

Combined (390) with (391) we obtain the desired derivation.

It is easy to verify that the above derivations have size polynomial in the size of the end sequents. Simulating the other rules of *PTK* is left as an exercise. ☐

EXERCISE X.4.8. Complete the proof of Theorem X.4.6 by showing that the translations of the rules all-right, one-right and Th-right have polynomial-size *PK*-proofs.

X.4.2. Reflection Principles for Bounded Depth *PTK*. We will show that for each depth $d \in \mathbb{N}$, the soundness of d-*PTK* is provable in *VTC*⁰. Let $Z \models_{d\text{-}PTK} X$ hold iff the truth assignment Z satisfies the depth-d *PTK* formula X. It can be shown by induction on d that this relation is a *TC*⁰ relation. We leave the details as an exercise.

EXERCISE X.4.9. Show that for each $d \in \mathbb{N}$ there is a Σ_1^B formula $(Z \models_{d\text{-}PTK}^{\Sigma} X)$ and a Π_1^B formula $(Z \models_{d\text{-}PTK}^{\Pi} X)$ that both represent the relation $Z \models_{d\text{-}PTK} X$, and such that

$$\textit{VTC}^0 \vdash (Z \models_{d\text{-}PTK}^{\Sigma} X) \leftrightarrow (Z \models_{d\text{-}PTK}^{\Pi} X).$$

For each d, the Reflection Principle for d-*PTK*, denoted by d-*RFN*$_{PTK}$, is the sentence:

$$\forall \pi \forall X \forall Z \big((Fla_d^\Pi(X) \wedge Prf_d^\Pi(\pi, X)) \supset (Z \models_{d\text{-}PTK}^{\Sigma} X) \big).$$

Here $Fla_d^\Pi(X)$ is the Π_1^B formula that is true iff X is a formula of depth at most d, and $Prf_d^\Pi(\pi, X)$ is the Π_1^B formula that is true iff π is a d-**PTK** proof of X. (See Corollary X.1.4 and Lemma X.1.5.)

The following result is left as an exercise.

*EXERCISE X.4.10. Show that for each $d \in \mathbb{N}$, d-RFN_{PTK} is a theorem of **VTC**^{0}. (Hint: first show that all formulas in a d-**PTK** proof of a formula of depth d must have depth at most d (see also Exercise X.1.10). Then prove by induction on i that the i-th sequent in the proof is valid.)

X.4.3. Propositional Translation for VTC^0. Our goal in this section is to translate **VTC**^{0}-proofs of Σ_0^B formulas into families of polynomial-size **bPTK**-proofs. One way would be to translate directly all instances of the axiom *NUMONES* (227) into **PTK** formulas (the bits of the "counting sequence" Y in *NUMONES* are translated using the Th_k connectives). Here we take another approach, based on Lemma X.4.12 below.

Recall that $numones'(z, X)$ has the same value as $numones(z, X)$ (which is the number of elements in X that are less than z). For convenience, we list below the defining axioms of $numones'$ ((231), (232) and (233), page 286):

$$numones'(0, X) = 0, \tag{392}$$

$$X(z) \supset numones'(z + 1, X) = numones'(z, X) + 1, \tag{393}$$

$$\neg X(z) \supset numones'(z + 1, X) = numones'(z, X). \tag{394}$$

DEFINITION X.4.11 ($V^0(numones')$). The theory $V^0(numones')$ has vocabulary $\mathcal{L}_A^2 \cup \{numones'\}$ and is axiomatized by 2-**BASIC**, the axioms (392), (393), (394) and the $\Sigma_0^B(numones')$-**COMP** axiom scheme.

LEMMA X.4.12. $V^0(numones')$ *is a conservative extension of* **VTC**^{0}.

PROOF. First, *NUMONES* is provable in $V^0(numones')$ because the counting sequence Y in *NUMONES* can be defined by $\Sigma_0^B(numones')$-**COMP** as follows:

$$(Y)^z = y \leftrightarrow numones'(z, X) = y.$$

Hence $V^0(numones')$ extends **VTC**^{0}.

Also, $\overline{\textbf{VTC}}^0$ is an extension of $V^0(numones')$, so the fact that $\overline{\textbf{VTC}}^0$ is conservative over **VTC**^{0} (Theorem IX.3.7) implies that $V^0(numones')$ is conservative over **VTC**^{0}. □

Suppose that φ is a Σ_0^B theorem of **VTC**^{0}. It follows from Lemma X.4.12 that φ has a $V^0(numones')$-proof π. All formulas in π are $\Sigma_0^B(numones')$, and we will show that π can be translated into a family of polynomial-size bounded-depth **PTK**-proofs for the translation of φ.

We will describe the translation of atomic formulas. The translations of other $\Sigma_0^B(numones')$ formulas build up inductively as in Section VII.2.1 using appropriate connectives Th_k for $\wedge$ and $\vee$.

Thus let $\varphi(\vec{x}, \vec{X})$ be an atomic formula. If φ does not contain *numones'* then the translation $\varphi[\vec{m}, \vec{n}]$ is defined as in Section VII.2.1 (using Th_k instead of $\land, \lor$). So suppose that φ contains *numones'*. Now if φ is of the form $X(t)$, where t contains *numones'*, then we can use the equivalence

$$X(t) \leftrightarrow \exists z < |X|(z = t \land X(z))$$

to translate φ using the translations of other atomic formulas $z = t$ and $X(z)$ (the latter does not contain *numones'*). Thus we only need to focus on atomic formulas φ of the form $s = t$ or $s \leq t$.

Let

$$numones'(t_1, X_1), numones'(t_2, X_2), \ldots, numones'(t_\ell, X_\ell)$$

be all occurrences of *numones'* in $\varphi(\vec{x}, \vec{X})$ (some t_i may contain terms of the form *numones'*(t_j, X_j)). Thus $\varphi(\vec{x}, \vec{X})$ has the form

$$\varphi'(\vec{x}, |\vec{X}|, numones'(t_1, X_1), \ldots, numones'(t_\ell, X_\ell))$$

where φ' is an atomic formula of the form $s' = t'$ or $s' \leq t'$. The truth value of $\varphi(\vec{x}, \vec{X})$ can be determined from the values $\vec{m}$ of $\vec{x}$, the length $\vec{n}$ of $\vec{X}$, and the values of *numones'*(t_i, X_i). So for a fixed sequences $\vec{m}, \vec{n}$, let $S = S_{\varphi, \vec{m}, \vec{n}}$ be the following set (recall *val* on page 166)

$$\{(k_1, k_2, \ldots, k_\ell) : k_i \leq val(t_i(\vec{m}, \vec{n})), \text{ and}$$
$$\varphi \text{ is true when } numones'(t_i, X_i) = k_i, \text{ for } 1 \leq i \leq \ell\}.$$

Recall that for each string variable X_i and a length $n_i \geq 2$ we introduce the propositional variables

$$\vec{p}^{X_i} = p_0^{X_i}, p_1^{X_i}, \ldots, p_{n_i-2}^{X_i}.$$

If the set S is empty, then define $\varphi[\vec{m}; \vec{n}] = \bot$; otherwise $\varphi[\vec{m}; \vec{n}]$ is defined to be the simplification (explained below) of (395). Note that for readability we here use $(A_1 \land A_2 \land \cdots \land A_k)$ for $\mathsf{Th}_k(A_1, A_2, \ldots, A_k)$ and $(A_1 \lor A_2 \lor \cdots \lor A_k)$ for $\mathsf{Th}_1(A_1, A_2, \ldots, A_k)$. The translation of φ is obtained by simplifying the following formula:

$$\bigvee_{\vec{k} \in S}^{\ell} \bigwedge_{i=1} (\mathsf{Th}_{k_i}(p_0^{X_i}, p_1^{X_i}, \ldots, p_{s_i-1}^{X_i}) \land \neg \mathsf{Th}_{k_i+1}(p_0^{X_i}, p_1^{X_i}, \ldots, p_{s_i-1}^{X_i})) \quad (395)$$

where $s_i = val(t_i(\vec{m}, \vec{n}))$, $p_{n_i-1}^{X_i} = \top$, and $p_j^{X_i} = \bot$ for $j \geq n_i$.

The simplification of (395) is performed inductively, starting with the atomic formulas $\mathsf{Th}_{k_i}(p_0^{X_i}, p_1^{X_i}, \ldots, p_{s_i-1}^{X_i})$ and $\mathsf{Th}_{k_i+1}(p_0^{X_i}, p_1^{X_i}, \ldots, p_{s_i-1}^{X_i})$. Each formula is simplified by applying the following procedure repeatedly.

Recall that

$$\mathsf{Th}_0(A_1, A_2, \ldots, A_n) =_{\text{def}} \top$$

and

$$\mathsf{Th}_k(A_1, A_2, \ldots, A_n) =_{\text{def}} \bot, k > n.$$

Simplification Procedure. Whenever possible

- $\mathsf{Th}_1(A)$ is simplified to A,
- $\neg\bot$ is simplified to $\top$,
- $\neg\top$ is simplified to $\bot$,
- $\mathsf{Th}_1(A, A, A_1, A_2, \ldots, A_n)$ is simplified to $\mathsf{Th}_1(A, A_1, A_2, \ldots, A_n)$,
- $\mathsf{Th}_{n+1}(A, A, A_1, \ldots, A_{n-1})$ is simplified to $\mathsf{Th}_n(A, A_1, \ldots, A_{n-1})$,
- $\mathsf{Th}_k(\bot, A_1, A_2, \ldots, A_n)$ is simplified to $\mathsf{Th}_k(A_1, A_2, \ldots, A_n)$,
- $\mathsf{Th}_k(\top, A_1, A_2, \ldots, A_n)$ is simplified to $\mathsf{Th}_{k-1}(A_1, A_2, \ldots, A_n)$.

EXAMPLE X.4.13. Recall the defining axioms (392), (393) and (394) for *numones'*. They are translated as follows.

(a) (392) is translated into $\top$.

(b) To translate (393), first we translate the atomic formula

$$\varphi(z, X) \equiv \textit{numones}'(z + 1, X) = \textit{numones}'(z, X) + 1.$$

Here $\ell = 2$, $t_1 = z + 1$, $t_2 = z$, $X_1 = X_2 = X$. For $m, n \in \mathbb{N}$, $n \geq 2$, we have

$$S_{\varphi, m, n} = \{(k + 1, k) : k \leq m\}.$$

We omit the superscript X for the variables p_i^X, and let $\vec{p}$ denote $p_0, p_1, \ldots, p_{m-1}$. Then $\varphi[m; n]$ is the simplification of

$$\bigvee_{k=0}^{m} \left((\mathsf{Th}_{k+1}(\vec{p}, p_m) \wedge \neg\mathsf{Th}_{k+2}(\vec{p}, p_m)) \wedge (\mathsf{Th}_k(\vec{p}) \wedge \neg\mathsf{Th}_{k+1}(\vec{p})) \right).$$

As a result, (393) translates into

$$\begin{cases} \neg p_m \vee \varphi[m; n] \ \ (\text{see } (396) \text{ below}) & \text{if } m \leq n - 2, \\ \varphi[n - 1; n] \ \ (\text{see } (397) \text{ below}) & \text{if } m = n - 1, \\ \top & \text{if } m \geq n. \end{cases}$$

Note that for $m \leq n - 2$,

$$\varphi[m; n] \equiv (\mathsf{Th}_1(\vec{p}, p_m) \wedge \neg\mathsf{Th}_2(\vec{p}, p_m) \wedge \neg\mathsf{Th}_1(\vec{p})) \vee$$
$$\left(\bigvee_{k=1}^{m-1} \left(\mathsf{Th}_{k+1}(\vec{p}, p_m) \wedge \neg\mathsf{Th}_{k+2}(\vec{p}, p_m) \wedge \mathsf{Th}_k(\vec{p}) \wedge \neg\mathsf{Th}_{k+1}(\vec{p}) \right) \right) \vee$$
$$\left(\mathsf{Th}_{m+1}(\vec{p}, p_m) \wedge \mathsf{Th}_m(\vec{p}) \right) \quad (396)$$

where $\vec{p}$ stands for $p_0, p_1, \ldots, p_{m-1}$. Also,

$$\varphi[n-1;n] \equiv \neg\mathsf{Th}_1(\vec{p}) \vee \left(\bigvee_{k=1}^{n-2} (\mathsf{Th}_k(\vec{p}) \wedge \neg\mathsf{Th}_{k+1}(\vec{p})) \right) \vee \mathsf{Th}_{n-1}(\vec{p})$$
$$(397)$$

where $\vec{p} = p_0, p_1, \ldots, p_{n-2}$.

(c) For (394), consider the atomic formula

$$\psi(z, X) \equiv numones'(z+1, X) = numones'(z, X).$$

Here ℓ, t_1, t_2, X_1, X_2 are as in (b) and

$$S_{\psi,m,n} = \{(k,k) : k \leq m\}.$$

Again drop mention of the superscript X, and let $\vec{p} = p_0, p_1, \ldots, p_{m-1}$. The formula $\psi[m;n]$ is (the simplification of)

$$\bigvee_{k=0}^{m} ((\mathsf{Th}_k(\vec{p}, p_m) \wedge \neg\mathsf{Th}_{k+1}(\vec{p}, p_m)) \wedge$$
$$(\mathsf{Th}_k(\vec{p}) \wedge \neg\mathsf{Th}_{k+1}(\vec{p}))). \quad (398)$$

For $m \geq n$, the simplification of (398) is just $\varphi[n-1;n]$ in (397). Hence, (394) translates into

$$\begin{cases} p_m \vee \psi[m;n] \text{ (see (396))} & \text{if } m \leq n-2, \\ \top & \text{if } m = n-1, \\ \varphi[n-1;n] \text{ (see (397))} & \text{if } m \geq n. \end{cases}$$

We will show that the translations of (392), (393) and (394) described above have d-$\boldsymbol{GTC_0^*}$ proofs of size polynomial in m, n, for some constant $d \in \mathbb{N}$. We need the following lemma.

LEMMA X.4.14.   (a) *The sequents (397) have polynomial size (in n) cut-free* ***PTK*** *proofs.*

(b) *Let $\vec{p}$ denote $p_0, \ldots, p_{m-1}$. The following sequents have polynomial-size (in m) cut-free* ***PTK*** *proofs:*

$$p_m \longrightarrow \neg\mathsf{Th}_2(\vec{p}, p_m) \vee \left(\bigvee_{k=1}^{m-1} (\mathsf{Th}_k(\vec{p}) \wedge \neg\mathsf{Th}_{k+2}(\vec{p}, p_m)) \right) \vee \mathsf{Th}_m(\vec{p}).$$
$$(399)$$

Proof. (a) The cut-free **PTK** proof is as follows:

$$
\cfrac{
\cfrac{
\cfrac{\mathsf{Th}_{n-1}(\vec{p}) \longrightarrow \mathsf{Th}_{n-1}(\vec{p})}{\longrightarrow \neg\mathsf{Th}_{n-1}(\vec{p}), \mathsf{Th}_{n-1}(\vec{p})}\ (7)
}{\mathsf{Th}_{n-2}(\vec{p}) \longrightarrow \mathsf{Th}_{n-2}(\vec{p}) \wedge \neg\mathsf{Th}_{n-1}(\vec{p}), \mathsf{Th}_{n-1}(\vec{p})}\ (6)
}{\vdots}
$$

$$
\cfrac{
\cfrac{
\cfrac{
\cfrac{
\cfrac{\longrightarrow \neg\mathsf{Th}_3(\vec{p}), \mathsf{Th}_3(\vec{p}) \wedge \neg\mathsf{Th}_4(\vec{p}), \ldots, \mathsf{Th}_{n-2}(\vec{p}) \wedge \neg\mathsf{Th}_{n-1}(\vec{p}), \mathsf{Th}_{n-1}(\vec{p})}{\mathsf{Th}_2(\vec{p}) \longrightarrow \mathsf{Th}_2(\vec{p}) \wedge \neg\mathsf{Th}_3(\vec{p}), \ldots, \mathsf{Th}_{n-2}(\vec{p}) \wedge \neg\mathsf{Th}_{n-1}(\vec{p}), \mathsf{Th}_{n-1}(\vec{p})}\ (5)
}{\longrightarrow \neg\mathsf{Th}_2(\vec{p}), \mathsf{Th}_2(\vec{p}) \wedge \neg\mathsf{Th}_3(\vec{p}), \ldots, \mathsf{Th}_{n-2}(\vec{p}) \wedge \neg\mathsf{Th}_{n-1}(\vec{p}), \mathsf{Th}_{n-1}(\vec{p})}\ (4)
}{\mathsf{Th}_1(\vec{p}) \longrightarrow \mathsf{Th}_1(\vec{p}) \wedge \neg\mathsf{Th}_2(\vec{p}), \ldots, \mathsf{Th}_{n-2}(\vec{p}) \wedge \neg\mathsf{Th}_{n-1}(\vec{p}), \mathsf{Th}_{n-1}(\vec{p})}\ (3)
}{\longrightarrow \neg\mathsf{Th}_1(\vec{p}), \mathsf{Th}_1(\vec{p}) \wedge \neg\mathsf{Th}_2(\vec{p}), \ldots, \mathsf{Th}_{n-2}(\vec{p}) \wedge \neg\mathsf{Th}_{n-1}(\vec{p}), \mathsf{Th}_{n-1}(\vec{p})}\ (2)
}{\longrightarrow \neg\mathsf{Th}_1(\vec{p}) \vee \left(\bigvee_{k=1}^{n-2} (\mathsf{Th}_k(\vec{p}) \wedge \neg\mathsf{Th}_{k+1}(\vec{p})) \right) \vee \mathsf{Th}_{n-1}(\vec{p})}\ (1)
$$

Here the top sequent is an axiom, (1) is by the rule one-right, (2, 4, 7) are ¬-right, and the derivations (3, 5, 6) consist of the rule all-right and a derivation from the axiom of the form

$$
\mathsf{Th}_i(\vec{p}) \longrightarrow \mathsf{Th}_i(\vec{p}).
$$

(b) The **PTK** proof is presented below. Because of the space limit, we will give one fragment of the proof at a time. There are $(m+1)$ fragments. The bottom fragment is:

$$
\cfrac{
\cfrac{
\cfrac{
\cfrac{\mathcal{S}_1 \quad \cfrac{p_m, \mathsf{Th}_1(\vec{p}) \longrightarrow \{\mathsf{Th}_k(\vec{p}) \wedge \neg\mathsf{Th}_{k+2}(\vec{p}, p_m)\}_{k=1}^{m-1}, \mathsf{Th}_m(\vec{p})}{p_m, p_m, \mathsf{Th}_1(\vec{p}) \longrightarrow \{\mathsf{Th}_k(\vec{p}) \wedge \neg\mathsf{Th}_{k+2}(\vec{p}, p_m)\}_{k=1}^{m-1}, \mathsf{Th}_m(\vec{p})}\ (4)
}{p_m, \mathsf{Th}_2(\vec{p}, p_m) \longrightarrow \{\mathsf{Th}_k(\vec{p}) \wedge \neg\mathsf{Th}_{k+2}(\vec{p}, p_m)\}_{k=1}^{m-1}, \mathsf{Th}_m(\vec{p})}\ (3)
}{p_m \longrightarrow \neg\mathsf{Th}_2(\vec{p}, p_m), \{\mathsf{Th}_k(\vec{p}) \wedge \neg\mathsf{Th}_{k+2}(\vec{p}, p_m)\}_{k=1}^{m-1}, \mathsf{Th}_m(\vec{p})}\ (2)
}{p_m \longrightarrow \neg\mathsf{Th}_2(\vec{p}, p_m) \vee \left(\bigvee_{k=1}^{m-1} (\mathsf{Th}_k(\vec{p}) \wedge \neg\mathsf{Th}_{k+2}(\vec{p}, p_m)) \right) \vee \mathsf{Th}_m(\vec{p})}\ (1)
$$

Here (1) is by the rule one-right, (2) is ¬-right, (3) is Th_2-left, and (4) is contraction left. The sequent $\mathcal{S}_1$ is the top sequent in (8) below (so our proof is a dag-like proof). The top sequent of (4) is derived in the next fragment:

$$\cfrac{\cfrac{\cfrac{\cfrac{\cfrac{p_m, \mathsf{Th}_2(\vec{p}) \longrightarrow \{\mathsf{Th}_k(\vec{p}) \wedge \neg \mathsf{Th}_{k+2}(\vec{p}, p_m)\}_{k=2}^{m-1}, \mathsf{Th}_m(\vec{p})}{p_m, p_m, \mathsf{Th}_2(\vec{p}) \longrightarrow \{\mathsf{Th}_k(\vec{p}) \wedge \neg \mathsf{Th}_{k+2}(\vec{p}, p_m)\}_{k=2}^{m-1}, \mathsf{Th}_m(\vec{p})} \; (8)}{p_m, \mathsf{Th}_3(\vec{p}, p_m) \longrightarrow \{\mathsf{Th}_k(\vec{p}) \wedge \neg \mathsf{Th}_{k+2}(\vec{p}, p_m)\}_{k=2}^{m-1}, \mathsf{Th}_m(\vec{p})} \; (7)}{p_m \longrightarrow \neg \mathsf{Th}_3(\vec{p}, p_m), \{\mathsf{Th}_k(\vec{p}) \wedge \neg \mathsf{Th}_{k+2}(\vec{p}, p_m)\}_{k=2}^{m-1}, \mathsf{Th}_m(\vec{p})} \; (6)}{p_m, \mathsf{Th}_1(\vec{p}) \longrightarrow \{\mathsf{Th}_k(\vec{p}) \wedge \neg \mathsf{Th}_{k+2}(\vec{p}, p_m)\}_{k=1}^{m-1}, \mathsf{Th}_m(\vec{p})} \; (5)$$

The derivation (5) consists of an all-right and a derivation by weakenings from the axiom

$$\mathsf{Th}_1(\vec{p}) \longrightarrow \mathsf{Th}_1(\vec{p}).$$

The steps (6, 7, 8) are similar to (2, 3, 4) above.

The next fragment derives the top sequent of (8) and is similar.

The top fragment is:

$$\cfrac{\cfrac{p_m, \mathsf{Th}_{m+1}(\vec{p}, p_m) \longrightarrow \mathsf{Th}_m(\vec{p})}{p_m, \longrightarrow \neg \mathsf{Th}_{m+1}(\vec{p}, p_m), \mathsf{Th}_m(\vec{p})} \; (10)}{p_m, \mathsf{Th}_{m-1}(\vec{p}) \longrightarrow \mathsf{Th}_{m-1}(\vec{p}) \wedge \neg \mathsf{Th}_{m+1}(\vec{p}, p_m), \mathsf{Th}_m(\vec{p})} \; (9)$$

The top sequent of (10) is obtained from some axioms by the rules all-left and all-right.   □

LEMMA X.4.15. *The translations of the defining axioms* (392), (393) *and* (394) *for numones' (described in Example X.4.13) have polynomial size d-**PTK** proofs, for some constant d.*

PROOF. The translation of (392) is $\top$, so the conclusion is obvious. Consider the translations of the defining axiom (393) in part (b) of Example X.4.13. Recall the formulas $\varphi[m; n]$ and $\varphi[n-1; n]$ from (396) and (397), respectively. We need to show that the following sequents have polynomial size d-**PTK** proofs, for some d:

$$\longrightarrow \neg p_m \vee \varphi[m; n] \qquad \text{and} \qquad \longrightarrow \varphi[n-1; n].$$

By Lemma X.4.14 (a) the latter has a polynomial size cut-free **PTK** proof. To derive the former, by Lemma X.4.14 (b) it suffices to derive

$$p_m, \neg \mathsf{Th}_2(\vec{p}, p_m) \vee \left(\bigvee_{k=1}^{m-1} \left(\mathsf{Th}_k(\vec{p}) \wedge \neg \mathsf{Th}_{k+2}(\vec{p}, p_m) \right) \right) \vee$$
$$\mathsf{Th}_m(\vec{p}) \longrightarrow \varphi[m; n] \quad (400)$$

(where $\vec{p}$ denotes $p_0, p_1, \ldots, p_{m-1}$). This is left as an exercise (see below).

Finally consider the translation of axiom (394) described in Example X.4.13 (c). As mentioned above, the sequents (397) have polynomial size cut-free **PTK** proofs. It remains to show that (recall $\psi[m; n]$ from (398)):

$$\longrightarrow p_m \vee \psi[m; n] \qquad (401)$$

has polynomial size d-**PTK** proof, for some constant d. This is left as an exercise.   □

Exercise X.4.16. Complete the proof of Lemma X.4.15 above by showing that the sequents (400) and (401) have polynomial size d-**PTK** proofs, for some constant d. Hint: first deriving the following sequents, then use Lemma X.4.14:

1) $p_m, \mathsf{Th}_k(\vec{p}) \longrightarrow \mathsf{Th}_{k+1}(\vec{p}, p_m)$ (for $1 \le k \le m$).
2) $p_m, \neg\mathsf{Th}_{k+2}(\vec{p}, p_m) \longrightarrow \neg\mathsf{Th}_{k+1}(\vec{p})$ (for $0 \le k \le m - 1$).
3) $\mathsf{Th}_k(\vec{p}) \longrightarrow p_m, \mathsf{Th}_k(\vec{p}, p_m)$ for $1 \le k \le m$.
4) $\neg\mathsf{Th}_{k+1}(\vec{p}) \longrightarrow p_m, \neg\mathsf{Th}_{k+1}(\vec{p}, p_m)$ for $0 \le k \le m - 1$.

As in Section X.1.1, it can be shown that formulas, sequents and proofs of **PTK** are Δ_1^B-definable in **FTC**^{0}.

Lemma X.4.17. *For every Σ_0^B (numones') formula $\varphi(\vec{x}, \vec{X})$, there is a constant d and a polynomial $p(\vec{m}, \vec{n})$ so that for all sequences $\vec{m}, \vec{n}$, the propositional formula $\varphi(\vec{x}, \vec{X})[\vec{m}; \vec{n}]$ has depth d and size bounded by $p(\vec{m}, \vec{n})$. Moreover, $\varphi(\vec{x}, \vec{X})[\vec{m}; \vec{n}]$ is provably in $\overline{VTC}^0$ computable by an **FTC**^{0} function $G(\vec{m}, \vec{n})$.*

Proof. By structural induction on φ.   □

Theorem X.4.18. *Suppose that $\varphi(\vec{x}, \vec{X})$ is a Σ_0^B (numones') theorem of V^0(numones'). Then there are a constant $d \in \mathbb{N}$ and an **FTC**^{0} function $F(\vec{m}, \vec{n})$ so that, provably in $\overline{VTC}^0$, $F(\vec{m}, \vec{n})$ is a d-**PTK** proof of $\varphi(\vec{x}, \vec{X})[\vec{m}; \vec{n}]$, for all $\vec{m}, \vec{n}$.*

Proof. The theorem can be proved in the same way that Theorem VII.2.3 is proved in Section VII.2.3, i.e., by showing that every Σ_0^B (numones') theorem of V^0(numones') has an **LK**^{2} proof where the inference rule Σ_0^B (numones')-**IND** (Definition VI.4.11) is allowed.   □

Corollary X.4.19. *For every Σ_0^B theorem $\varphi(\vec{x}, \vec{X})$ of **VTC**^{0}, there are a constant $d \in \mathbb{N}$ and an **FTC**^{0} function $F(\vec{m}, \vec{n})$ so that, provably in $\overline{VTC}^0$, $F(\vec{m}, \vec{n})$ is a d-**PTK** proof of $\varphi(\vec{x}, \vec{X})[\vec{m}; \vec{n}]$, for all $\vec{m}, \vec{n}$.*

Proof. The Corollary follows from Theorem X.4.18 because V^0(numones') is a conservative extension of **VTC**^{0}.   □

The next result is immediate from the above corollary and the fact that **VTC**^{0} proves **PHP**(a, X) (Theorem IX.3.23).

Corollary X.4.20. *The pigeonhole tautologies **PHP** (Definition VII.1.12) have polynomial size **bPTK** proofs.*

In the next section we introduce the quantified threshold formulas and a sequent calculus **GTC** for them. Theorem X.4.18 will be generalized to show that the propositional translations of *bounded* theorems of **VTC**^{0} have proofs in **GTC** that are provably in **VTC**^{0} computable by some **FTC**^{0} functions (Theorem X.4.22).

X.4.4. Bounded Depth GTC_0. Now we consider an extension of PTK which allows quantifiers over propositional variables. We do not allow the quantifiers to be inside the scope of arbitrary threshold connectives. We do want to allow conjunctions and disjunctions of quantified formulas, so we require that the quantifiers cannot occur inside the scope of a threshold connective Th_k unless $k = 1$ (so the connective expresses a disjunction) or $k = n$ and occurs in the context $Th_n(A_1, A_2, \ldots, A_n)$ (so the connective expresses a conjunction).

Formally, *quantified threshold formulas* (or QT formulas, or just formulas) are defined as follows:

(a) Any PTK formula is a QT formula;
(b) If $A(p)$ is a QT formula, then so are $\forall x A(x)$ and $\exists x A(x)$, for any free variable p and bound variable x.
(c) If $A_1, A_2, \ldots, A_n$ are QT formulas, then so are $Th_1(A_1, A_2, \ldots, A_n)$, $Th_n(A_1, A_2, \ldots, A_n)$ and $\neg A_1$.

As before, we will often write

$$\bigvee_{i=1}^{n} A_i \qquad \text{and} \qquad \bigwedge_{i=1}^{n} A_i$$

for $Th_1(A_1, A_2, \ldots, A_n)$ and $Th_n(A_1, A_2, \ldots, A_n)$, respectively.

The system GTC is the extension of PTK where the axioms now consist of

$$\longrightarrow \top, \qquad \bot \longrightarrow, \qquad A \longrightarrow A$$

for all QT formulas A. The introduction rules for the threshold connectives are as given in Section X.4.1 but now the rules Th_k-left and Th_k-right are applied only to PTK formulas. The introduction rules for the quantifiers are as for QPC (Section VII.3).

Theorem X.4.6 can be extended to show that GTC and G are p-equivalent. In fact, for $i \geq 0$ define Σ_i^{qt} and Π_i^{qt} of QT formulas in the same way as Σ_i^q and Π_i^q, and let GTC_i be obtained from GTC by restricting the cut formulas to $\Sigma_i^{qt} \cup \Pi_i^{qt}$. Then it can be shown that GTC_i and G_i are p-equivalent for $i \geq 0$. Here we are interested in the following subsystems of GTC_0.

DEFINITION X.4.21 (Bounded Depth GTC_0). For each $d \in \mathbb{N}$, d-GTC_0 is the subsystem of GTC where all cut and target formulas are quantifier-free and have depth at most d. A *bounded depth GTC_0* (or just $bGTC_0$) system is any system d-GTC_0 for $d \in \mathbb{N}$. Treelike d-GTC_0 (resp. treelike $bGTC_0$) is denoted by d-$GTC_0^\star$ (resp. $bGTC_0^\star$).

As in Section X.1.1, it can be shown that formulas, sequents and proofs in GTC are Δ_1^B-definable in VTC^0. It is also straightforward to extend the translation given in Section X.4.3 so that $\Sigma_i^B(numones')$ and $\Pi_i^B(numones')$

formulas (for $i \geq 1$) are translated into quantified threshold formulas in Σ_i^{qt} and Π_i^{qt}, respectively.

THEOREM X.4.22 (Propositional Translation for $V^0(numones')$).   *Let* $\varphi(\vec{x}, \vec{X})$ *be a bounded theorem of* $V^0(numones')$. *There is a constant* $d \in \mathbb{N}$ *and a function* F *in* FTC^0 *so that* $F(\vec{m}, \vec{n})$ *is provably in* $\overline{VTC}^0$ *a* d-GTC_0^* *proof of* $\varphi(\vec{x}, \vec{X})[\vec{m}; \vec{n}]$, *for all* $\vec{m}$, $\vec{n}$.

The proof of Theorem X.4.22 is similar to the proof of Theorem X.1.14. Here we translate cut $\Sigma_0^B(numones)$-$COMP$ formulas in the same way that cut Σ_0^B-$COMP$ formulas are translated in Theorem VII.5.6. Then it can be shown that the translation of formulas in an LK^2-$V^0(numones')$ proof are, provably in $\overline{VTC}^0$, computable by some FTC^0 functions. Furthermore, the PTK version of the sequent in Exercise VII.5.3 can be shown to have d-PTK^* proofs that are provably in $\overline{VTC}^0$ computable by some FTC^0 function, for some constant d. Details are left as an exercise.

*EXERCISE X.4.23. Prove Theorem X.4.22.

COROLLARY X.4.24 (Propositional Translation Theorem for VTC^0). *For every bounded theorem* $\varphi(\vec{x}, \vec{X})$ *of* VTC^0, *there is a constant* $d \in \mathbb{N}$ *and an* FTC^0 *function* F *such that* $\overline{VTC}^0$ *proves that for all* $\vec{m}$ *and* $\vec{n}$, $F(\vec{m}, \vec{n})$ *is a* d-GTC_0^* *proof of* $\varphi(\vec{x}, \vec{X})[\vec{m}; \vec{n}]$.

PROOF. Since φ is a theorem of VTC^0, by Lemma X.4.12 it is also a theorem of $V^0(numones')$. Now apply Theorem X.4.22.   □

X.5. Notes

The results in Section X.1.3 are from [73].

The formulation 0-$RFN(\mathcal{F})$, the RFN for quantifier-free formulas given in [72], is essentially our Π_0^q-RFN_F (Definition X.2.11), and hence is different from our notation 0-$RFN_{\mathcal{F}}$. Lemma X.2.13 is from [72, Lemma 9.3.12 b] where it is stated more generally for any system $\mathcal{F}$ which is closed under substitution and modus ponens. Lemma X.2.14 is new. Definition X.2.16 is from [91]. Theorem X.2.17 and Corollary X.2.20 strengthen results from [72, Theorem 9.3.16]. Parts (a) and (b) of Theorem X.2.23 strengthen results from [73] (here we use VTC^0 instead of S_2^1). The axiomatizations of V^i in part (c) of Theorem X.2.23 are new. The difficult part (that V^i proves Σ_{i+1}^q-$RFN_{G_i^*}$) is based on [91, Theorem 5.1.2], see Theorem X.2.17 (c).

The idea of using the Reflection Principle for p-simulation is from [39] where a variant of Exercises X.2.22 and X.2.30 is proved. Theorem X.2.27 is a strengthening of Lemma 5.2.1 from [91]. The definition of a Σ_i^q Witnessing Problem given in Section X.2.6 is more general than that of

[80]: in [80] the problem is to witness *prenex* formulas. Consequently the membership directions of Theorem X.2.33 strengthen that of [80, Theorems 6.2, 6.9]. For the hardness directions of Theorem X.2.33 note that we are using TC^0 reductions while the reductions in [80, Theorems 6.2, 6.9] are polytime. Following the proofs given in [80], the fact that the weaker problems as defined in [80] are hard for the search classes under polytime reduction (instead of TC^0 reduction) is due to the fact that we need polytime procedures for producing $G_i^\star$ proofs of the equivalence between prenex and non-prenex Σ_i^q formulas, see Exercise X.2.9.

Clote [34] introduced an equational theory **ALV** for NC^1 and defined a polytime translation from theorems of **ALV** to families of Frege proofs. Arai [8] introduced a first order system **AID** for NC^1, defined a polytime translation from Σ_0^b theorems to Frege proofs, and proved the reflection principle for Frege systems. A version of VNC^1 was introduced in [45] and translation of all bounded theorems to polynomial size families of $G_0^\star$ proofs presented. The translation given in Section X.3.1 is new. Several *ALogTime* algorithms for the Boolean Sentence Value Problem have been presented by Buss [22, 24, 25]. The algorithm presented in the proof of Theorem X.3.5 is from [25]. The algorithm from [24] was formalized in [8]. The algorithm from [25] was also formalized in [92] using the string theory T^1.

The sequent calculus *PTK* is from [29]. The propositional translation for VTC^0 given in Section X.4.3 is new. The quantified system GTC_0 in Section X.4.4 is similar to the system **QTC** in [45].

Appendix A

COMPUTATION MODELS

We give definitions for some basic concepts in computational complexity and state some useful results. See [49, 64, 87, 100, 102, 110] for further details.

In this Appendix f and g stand for functions from the natural numbers to $\mathbb{R}_{\geq 0} = \{x \in \mathbb{R} : x \geq 0\}$. We use the following notation.

- $g = \mathcal{O}(f)$ if there is a constant $c > 0$ so that $g(n) \leq cf(n)$ for all but finitely many n.
- $g = \Omega(f)$ if there is a constant $c > 0$ so that $g(n) \geq cf(n)$ for all but finitely many n.
- $g = \Theta(f)$ if $g = \mathcal{O}(f)$ and $g = \Omega(f)$.
- $\log n$ stands for $\log_2 n$. When $\log n$ is required to be an integer, it is understood that it takes the value $\lceil \log_2 n \rceil$.

The variable n usually refers to the length of an input string to a machine or circuit. When n appears in the definition of a resource class such as $ATime(k \log n + k)$ it refers to the argument of the function bounding the resource. For a class $Resource(f)$ that is defined by having a bound f on some resource we will write $Resource(\mathcal{O}(f))$ for the union

$$\bigcup_{k=1}^{\infty} Resource(kf + k).$$

For example (see Section A.4):

$$ATime(\mathcal{O}(\log n)) = \bigcup_{k=1}^{\infty} ATime(k \log n + k).$$

A.1. Deterministic Turing Machines

A k–tape deterministic Turing machine (DTM) consists of k two–way infinite tapes and a finite state control. Each tape is divided into squares, each of which holds a symbol from a finite alphabet Γ. Each tape also has a read/write head that is connected to the control and that scans the squares on the tape. Depending on the state of the control and the symbols scanned, the machine makes a move which consists of

1) printing a symbol on each tape;
2) moving each head left or right one square, or leaving it fixed;
3) assuming a new state.

DEFINITION A.1.1. For a natural number $k \geq 1$, a k–tape DTM M is specified by a tuple $\langle Q, \Sigma, \Gamma, \sigma \rangle$ where

1) Q is the finite set of *states*. There are 3 distinct designated states $q_{initial}$ (the initial state), q_{accept} and q_{reject} (the states in which M halts).
2) Σ is the finite, non-empty set of input symbols.
3) Γ is the finite set of working symbols, $\Sigma \subset \Gamma$. Γ contains a special symbol $\not{b}$ (read "blank"), and $\not{b} \in \Gamma \setminus \Sigma$.
4) σ is the transition function, i.e., a total function:

$$\sigma : ((Q \setminus \{q_{accept}, q_{reject}\}) \times \Gamma^k) \rightarrow (Q \times (\Gamma \times \{L, R, O\})^k).$$

If the current state is q, the current symbols being scanned are $s_1, \ldots, s_k$, and $\sigma(q, \vec{s}) = (q', s_1', h_1, \ldots, s_k', h_k)$, then q' is the new state, $\vec{s'}$ are the symbols printed, and for $1 \leq i \leq k$, the head of the ith tape will move one square to the left or right or not move, depending on whether h_i is L or R or O.

On an input x (a finite string of Σ symbols) the machine M works as follows. Initially, the input is given on tape 1, called the *input tape*, which is completely blank everywhere else. Other tapes (i.e., the *work tapes*) are blank, and their heads point to some squares. Also the input tape head is pointing to the leftmost symbol of the input (if the input is the empty string, then the input tape will be completely blank, and its head will point to some square). The control is initially in state $q_{initial}$. Then M moves according to the transition function σ.

If M enters either q_{accept} or q_{reject} then it halts. If M halts in q_{accept} we say that it accepts the input x, if it halts in q_{reject} then we say that it rejects x. Note that it is possible that M never halts on some input. Let Σ^* denote the set of all finite strings of Σ symbols. We say that M accepts (or decides, or computes) a language $L \subseteq \Sigma^*$ if M accepts input $x \in \Sigma^*$ iff $x \in L$. We let $L(M)$ denote the language accepted by M.

Unless specified otherwise, Turing machines are multi-tape (i.e., $k > 1$). In this case we require that the input tape head is read-only. Also, for a Turing machine M to compute a (partial) function, tape 2 is called the *output tape* and the content of the output tape when the machine halts in q_{accept} is the output of the machine. For machines that compute a function we require that the output tape is write-only.

A *configuration* of M is a tuple $\langle q, u_1, v_1, \ldots, u_k, v_k \rangle \in Q \times (\Gamma^* \times \Gamma^*)^k$. The intuition is that q is the current state of the control, the string $u_i v_i$ is the content of the tape i, and the head of tape i is on the left-most symbol of v_i. If both u_i and v_i are the empty string, then the head points to a blank square. If only v_i is the empty string then the head points to the left-most blank symbol to the right of u_i. We require that for each i, u_i does not start with the blank symbol $\not{b}$, and v_i does not end with $\not{b}$.

The *computation* of M on an input x is the (possibly infinite) sequence of configurations of M, starting with the *initial configuration* $\langle q_{initial}, \varepsilon, x, \varepsilon, \varepsilon, \ldots, \varepsilon, \varepsilon \rangle$, where ε is the empty string, and each subsequent configuration is obtained from the previous one as specified by the transition function σ. Note that the sequence can contain at most one *final configuration*, i.e., a configuration of the form $\langle q_{accept}, \ldots \rangle$ or $\langle q_{reject}, \ldots \rangle$. The sequence contains a final configuration iff it is finite iff M halts on x. The *length* of the computation is the length of the sequence.

A.1.1. *L*, *P*, *PSPACE*, and *EXP*. Suppose that a Turing machine M $= \langle Q, \Sigma, \Gamma, \sigma \rangle$ halts on input x. Then the running time of M on x, denoted by $time_M(x)$, is the number of moves that M makes before halting (i.e., the number of configurations in the computation of M on x, not counting the initial configuration). Otherwise we let $time_M(x) = \infty$.

Recall that $L(M)$ denotes the language accepted by M. We say that M *runs in time* $f(n)$ if for all but finitely many $x \in \Sigma^*$, $time_M(x) \leq f(|x|)$, where $|x|$ denotes the length of x. In this case we also say that M *accepts the language* $L(M)$ *in time* $f(n)$.

DEFINITION A.1.2 (*DTime*). For a function $f(n)$, define

$$DTime(f) = \{L : \text{there is a DTM accepting } L \text{ in time } f(n)\}.$$

In general, if f is at least linear, then the class $DTime(f)$ is robust in the following sense.

THEOREM A.1.3 (Speed-up). *For any $\varepsilon > 0$,*

$$DTime(f) \subseteq DTime((1 + \varepsilon)n + \varepsilon f).$$

The classes of polynomial time and exponential time computable languages are defined as follows.

DEFINITION A.1.4 (*P* and *EXP*).

$$P = \bigcup_{k=1}^{\infty} DTime(n^k + k), \qquad EXP = \bigcup_{k=1}^{\infty} DTime(2^{n^k} + k).$$

The working space of a (multi-tape) DTM M on input x, denoted by $space_M(x)$, is the total number of squares on the *work tapes* (i.e., excluding the input and output tapes) that M visits at least once during the computation. Note that it is possible that $space_M(x) = \infty$, and also that $space_M(x)$ can be finite even if M does not halt on x.

We say that M *runs in space* $f(n)$ if for all but finitely many $x \in \Sigma^*$, $space_M(x) \leq f(|x|)$. In this case we also say that M *accepts the language* $L(M)$ *in space* $f(n)$.

DEFINITION A.1.5 (*DSpace*). For a function $f(n)$, define

$DSpace(f) = \{L : \text{there is a DTM accepting } L \text{ in space } f(n)\}.$

THEOREM A.1.6 (Tape Compression). *For any* $\varepsilon > 0$ *and any function* f,

$$DSpace(max\{\varepsilon f, 1\}) = DSpace(f).$$

The class of languages computable in logarithmic and polynomial space are defined as follows.

DEFINITION A.1.7 (*L* and *PSPACE*).

$$L = \bigcup_{k=1}^{\infty} DSpace(k \log n + k), \qquad PSPACE = \bigcup_{k=1}^{\infty} DSpace(n^k + k).$$

For a single-tape Turing machine, the working space is the total number of squares visited by the tape head during the computation. The classes *P*, *PSPACE* and *EXP* remain the same even if we restrict to single-tape DTMs. This is due to the following theorem.

THEOREM A.1.8 (Multi Tape). *For each multi-tape Turing machine* M *that runs in time* $t(n)$ *and space* $s(n)$, *there is a single–tape Turing machine* M′ *that runs in time* $(t(n))^2$ *and space* $max\{n, s(n)\}$ *and accepts the same language as* M. *There exists also a 2–tape Turing machine* M″ *that works in space* $s(n)$ *and accepts* $L(M)$.

For the Time Hierarchy Theorem below we need the notion of *time constructible function*. A function $f(n)$ is time constructible if there is a Turing machine M such that on input x the running time of M is $\Theta(f(|x|))$. It turns out that common integer-valued functions such as $kn, n\lceil \log n \rceil, n^k, n^{\lceil \log n \rceil}, 2^n$ are time constructible. We will be concerned only with time bounding functions that are constructible.

THEOREM A.1.9 (Time Hierarchy). *Suppose that* $f(n)$ *is a function,* $f(n) \geq n$, *and* $g(n)$ *is a time constructible function so that*

$$\lim_{n\to\infty} \inf \frac{f(n) \log f(n)}{g(n)} = 0.$$

Then

$$DTime(g) \setminus DTime(f) \neq \emptyset.$$

A function $f(n)$ is *space constructible* if there is Turing machine M such that on input x the working space of M is $\Theta(f(|x|))$. The space bounds that we are interested in are all constructible.

THEOREM A.1.10 (Space Hierarchy). *Suppose that* $f(n)$ *is a function and* $g(n)$ *is a space constructible function so that*

$$g(n) = \Omega(\log n) \qquad and \qquad \lim_{n\to\infty} \inf \frac{f(n)}{g(n)} = 0.$$

Then

$$DSpace(g) \setminus DSpace(f) \neq \emptyset.$$

It is easy to see that

$$L \subseteq P \subseteq PSPACE \subseteq EXP. \tag{402}$$

The Time Hierarchy Theorem shows that

$$DTime(n) \subsetneq DTime(n^2) \subsetneq \cdots \quad \text{and} \quad P \subsetneq DTime(2^{\varepsilon n})$$

for any $\varepsilon > 0$. The Space Hierarchy Theorem shows that $L \subsetneq PSPACE$. However none of the immediate inclusions in (402) is known to be proper.

Sublinear time classes are defined using Turing machines that are equipped with an index tape that operates like a work tape, except its content is used for accessing the input in the following way: the machine queries an input bit by writing its position in binary on the index tape and enter some special state. (Dowd, see [22], shows that a deterministic logtime Turing machine can compute the length of its input written in binary.) Define

$$DLogTime = DTime(\mathcal{O}(\log n))$$

and define *NLogTime* and *ALogTime* similarly using nondeterministic and alternating Turing machines given in Sections A.2 and A.4.

A.2. Nondeterministic Turing Machines

DEFINITION A.2.1. A k–tape nondeterministic Turing machine (NTM) is specified by a tuple $\langle Q, \Sigma, \Gamma, \sigma \rangle$ as in Definition A.1.1, but now the transition function σ is of the form

$$\sigma : ((Q \setminus \{q_{accept}, q_{reject}\}) \times \Gamma^k) \to \mathcal{P}(Q \times (\Gamma \times \{L, R, O\})^k)$$

where $\mathcal{P}(S)$ denotes the *power set* of the set S.

Here $\sigma(q, s_1, \ldots, s_k)$ is the (possibly empty) set of possible moves of M, given that the current state is q and the symbols currently being scanned are $\vec{s}$.

A computation of M on an input x is a (possibly infinite) sequence of configurations of M, starting with the initial configuration

$$\langle q_{initial}, \varepsilon, x, \varepsilon, \varepsilon, \ldots, \varepsilon, \varepsilon \rangle$$

and each subsequent configuration is a configuration that can be obtained from the previous one by one of the possible moves specified by σ. By definition, each computation of M may contain at most one configuration of the form $\langle q_{accept}, \ldots \rangle$ or $\langle q_{reject}, \ldots \rangle$. In the former case we say that it is an *accepting computation*, and in the latter case we say that it is a *rejecting computation*.

We say that the NTM M accepts x is there is an accepting computation of M on x. We say that M accepts x *in time* $f(n)$ if there is an accepting computation of length $\leq f(|x|)$, and M accepts x *in space* $f(n)$ if there is an accepting computation such that the number of squares on the work tapes used by M during this computation is $\leq f(|x|)$.

If for all but finitely many $x \in L(\mathsf{M})$ the NTM M accepts x in time/space $f(n)$, we also say that M accepts the language $L(\mathsf{M})$ in time/space $f(n)$.

DEFINITION A.2.2 (*NTime* and *NSpace*). For a function $f(n)$, define

$$NTime(f) = \{L : \text{there is a NTM accepting } L \text{ in time } f(n)\},$$

$$NSpace(f) = \{L : \text{there is a NTM accepting } L \text{ in space } f(n)\}.$$

The Speed-up Theorem (A.1.3) and Tape Compression Theorem (A.1.6) continue to hold for NTMs.

DEFINITION A.2.3 (*NP* and *NL*).

$$NP = \bigcup_{k \geq 1}^{\infty} NTime(n^k + k), \qquad NL = \bigcup_{k=1}^{\infty} NSpace(k \log n + k).$$

The list in (402) is extended as follows:

$$L \subseteq NL \subseteq P \subseteq NP \subseteq PSPACE.$$

However, it is not known whether any of the immediate inclusions is proper.

For a class C of languages, *co-C* is defined to be the class of the complements of the languages in C. For deterministic classes *L*, *P*, *EXP* we have $C = co\text{-}C$. However it is an open problem whether $NP = co\text{-}NP$. For *NL* and *co-NL* we have an affirmative answer, due to Immerman and Szelepcsényi:

THEOREM A.2.4 (Immerman–Szelepcsényi). *For any space constructible function* $f(n) \geq \log(n)$, $NSpace(f) = co\text{-}NSpace(f)$.

It is also easy to see that

$$P \subseteq co\text{-}NP \subseteq PSPACE.$$

But it is unknown whether either inclusion is proper.

The class of languages computable by NTMs in polynomial space is defined similarly, but by Savitch's Theorem this is the same as *PSPACE*.

THEOREM A.2.5 (Savitch's Theorem). *For any space constructible function* $f(n) \geq \log n$,

$$NSpace(f) \subseteq DSpace(f^2).$$

(*Here the superscript in* f^2 *refers to multiplication, rather than composition.*)

It follows that nondeterministic polynomial space is the same as *PSPACE*, and also that $NL \subsetneq PSPACE$.

A.3. Oracle Turing Machines

Let L be a language. An Oracle Turing machine (OTM) M with oracle L is a Turing machine augmented with the ability to ask questions of the form "is $y \in L$?". Formally, M has a designated write-only tape for the queries, called the *query tape*. It also has 3 additional states, namely q_{query}, q_{Yes} and q_{No}. In order to ask the question "is $y \in L$?", the machine writes the string y on the query tape, and enters the state q_{query}. The next state of M is then either q_{Yes} or q_{No}, depending on whether $y \in L$. Also the query tape is blanked out before M makes the next move.

In case the queries are witnessed (e.g., Definition VIII.7.16) or we want a function oracle, i.e., oracles that answer queries of the form

$$F(W)?$$

for some function F, then the OTM will have a read-only *answer tape* that contains oracle replies. The head of the answer tape is positioned to the left-most non-blank square whenever the machine enters the state q_{query}.

The running time of M on an input x is defined as before. Note that the time it takes to write down the queries (and to read the oracle answers/witnesses) are counted. In particular, an OTM running in polynomial time can ask only polynomially long queries.

A nondeterministic oracle Turing machine (NOTM) is a generalization of OTM where the transition function is a many-valued function. For a language L, we denote by P^L the class of languages accepted by some OTM running in polynomial time with L as the oracle, and similarly NP^L the class of languages accepted by some NOTM running in polynomial time with L as the oracle. For a class C of languages, define

$$P^C = \bigcup_{L \in C} P^L, \qquad \text{and} \qquad NP^C = \bigcup_{L \in C} NP^L.$$

Define $NLinTime^C$ similarly, where (see Definition A.2.2)

$$NLinTime = NTime(\mathcal{O}(n)).$$

(Relativizing logspace classes is more complicated, see [1] for details.)

The polynomial time hierarchy (PH) and linear time hierarchy (LTH) are defined in Section III.4.1 as follows.

DEFINITION A.3.1 (PH). $\Delta_0^p = \Sigma_0^p = \Pi_0^p = P$. For $i \geq 0$,

$$\Sigma_{i+1}^p = NP^{\Sigma_i^p}, \qquad \Pi_{i+1}^p = co\text{-}\Sigma_{i+1}^p, \qquad \Delta_{i+1}^p = P^{\Sigma_i^p}.$$

And

$$PH = \bigcup_{i=0}^{\infty} \Sigma_i^p.$$

Thus $NP \subseteq PH$, and it can be shown that $PH \subseteq PSPACE$. However neither inclusion is known to be proper. It is also not known whether the polynomial time hierarchy is proper.

DEFINITION A.3.2 (LTH).

$$\Sigma_1^{lin} = NLinTime, \qquad \Sigma_{i+1}^{lin} = NLinTime^{\Sigma_i^{lin}} \quad \text{for } i \geq 1,$$

and

$$LTH = \bigcup_{i=1}^{\infty} \Sigma_i^{lin}.$$

Thus $LTH \subseteq PH$, and as far as we know, P and LTH are incomparable. Both PH and LTH can be alternatively defined using the notion of *alternating Turing machines*, which we will define in the next section.

A.4. Alternating Turing Machines

An *alternating Turing machine* (ATM) M is defined as in Definition A.2.1 for a nondeterministic Turing machine, but now the finite set Q \ $\{q_{accept}, q_{reject}\}$ is partitioned into 2 disjoint sets of states, namely the set of $\exists$ states and the set of $\forall$ states.

If a configuration c_2 of M can be obtained from c_1 as specified by the transition function σ, we say that it is *a successor configuration* of c_1. An *existential* (resp. *universal*) configuration is a configuration of the form $\langle q, \ldots \rangle$ where q is an $\exists$-state (resp. a $\forall$-state).

We define the set of *accepting configurations* to be the smallest set of configurations that satisfies:

- a final configuration of the form $\langle q_{accept}, \ldots \rangle$ is an accepting configuration (*a final accepting configuration*);
- an existential configuration is accepting iff at least one of its successor configuration is accepting;
- a universal configuration is accepting iff all of its successor configurations are accepting.

We say that M accepts x iff the initial configuration $\langle q_{initial}, \varepsilon, x, \varepsilon, \varepsilon, \ldots, \varepsilon, \varepsilon \rangle$ is an accepting configuration of M.

A computation of M on an input x is viewed as a tree T with leaves labeled with the configurations as follows:

- the root of T is labeled with the initial configuration of M on x;
- if v is an inner node of T labeled with a universal configuration c which has k successor configurations, then v has k children each labeled uniquely by a successor configuration of c;
- if v is an inner node of T labeled with an existential configuration c which has k successor configurations, then v has k' children for

some k', $1 \le k' \le k$, and each child of v is labeled uniquely by a successor configuration of c.

A finite computation of M is called an *accepting computation* if all its leaves are labeled with a final accepting configuration.

We say that an ATM M accepts (or computes, or decides) x in time t if there is an accepting computation of M on input x where the paths from the root to any leaf has length $\le t$. We say that M accepts (or computes, or decides) x in space s if there is an accepting computation of M on input x in which every configuration has size at most s. Also M accepts (or computes, or decides) $L = L(M)$ in time $f(n)$ (resp. space $f(n)$) if for all $x \in L$, M accepts x in time $f(|x|)$ (resp. space $f(|x|)$).

The alternation depth of a computation is the maximum over all paths from root to leaf of one plus the number of changes of state type (i.e. existential or universal) along the path. In particular, the alternation depth of a computation of a nondeterministic Turing machine is one.

DEFINITION A.4.1. For functions $f(n)$, $g(n)$, *ATime*(f) is the class of languages that are accepted by an ATM in time $f(n)$, *ATime-Alt*(f, g) is the class of languages that are accepted by an ATM in time $f(n)$ with at most $g(n)$ alternations, *ASpace-Time*(f, g) is the class of languages that are accepted by an ATM in space $f(n)$ and time $g(n)$, *ASpace-Alt*(f, g) is the class of languages that are accepted by an ATM in space $f(n)$ with at most $g(n)$ alternations.

It can be seen that for $i \ge 1$, Σ_i^p is the class of languages accepted by a polytime ATM with at most i alternations and an existential initial state, and Π_i^p is defined similarly with a universal initial state.

In the next section we define the circuit classes such as NC^k, AC^k. They can be equivalently defined using ATMs.

A.5. Uniform Circuit Families

A *Boolean circuit* (or just *circuit*) C with inputs $x_0, x_1, \ldots, x_{n-1}$ is a directed acyclic graph in which each gate (i.e., node) that has indegree 0 is either an *input gate* and is labeled with some variable x_i or a *constant gate* and is labeled with a Boolean constant (0 for False and 1 for True), and each gate that has indegree $k > 0$ (called an *inner gate*) is labeled with a Boolean function of k variables. Gates with outdegree 0 are called *output gates*. For a gate g with indegree $k > 0$, the inputs to g are those gates g' for which there is an edge from g' to g. Unless otherwise specified, the circuits are assumed to contain inner gates of type $\neg$, $\wedge$ and $\vee$ only.

When the input gates $x_0, x_1, \ldots, x_{n-1}$ are assigned values according to an input string $\vec{a}$ in $\{0,1\}^n$ then the value (or output) of each gate in C is defined inductively as follows. The value of a constant gate is the label assigned to the gate and the value of an input gate labelled x_i is a_i. The value of every other gate g is the value of the labeling function when applied to the values of the inputs to g.

Suppose that C is a circuit with a single output gate. The value (or output) of C on input $\vec{a}$ is the value of the output gate of C. We say that C accepts $\vec{a}$ iff it outputs 1 on input $\vec{a}$.

Note that each circuit accepts strings of a fixed length. So to decide a language we need a family of circuits, one for each length. Thus we often consider a family $\{C_n : n \geq 1\}$ of circuits where for each $n \geq 1$, C_n is a circuit with n inputs. We say that a language $L \subseteq \{0,1\}^*$ is accepted (or decided, or computed) by such a family if L is the set of all w such that w is accepted by C_n, where n is the length of w.

In the same way we can define the function computed by a family of circuits. Here the circuits are allowed to have a sequence of output gates that make up the output string.

Our definition allows the existence of hardwired families of circuits that accept non-computable languages. However for the complexity classes of interest to us we need uniform families of circuits: i.e. circuit families which can be described by languages in weak complexity classes such as **DLogTime** or **FO**.

Following [100] the uniformity of a family of circuits is defined in terms of its *direct connection language* or *extended connection language* given below. Informally, the former describes gate types and edges of the circuits while the latter describes gate types and paths in the circuits.

Consider for example a family of circuits $\{C_n\}$ where for $n \geq 1$ C_n has n inputs and all inner gates in C_n are either $\neg$ or binary $\wedge$, $\vee$ gates. Thus every inner gate g has either two inputs which we call the left and right input of g and denote by $g(L)$ and $g(R)$, respectively, or one input which we denote by $g(L)$. Suppose that each gate g in a circuit C_n is numbered with a unique natural number less than the size of C_n, so that inputs gates of C_n are numbered $0, 1, \ldots, (n-1)$.

DEFINITION A.5.1. The *direct connection language* L_D of $\{C_n\}$ is the set of (the binary string encodings of)

$$\langle n, g, p, y \rangle$$

where $n, g \in \mathbb{N}$, $p \in \{\varepsilon, L, R\}$, $y \in \{\wedge, \vee, \neg, 0, 1, input\} \cup \mathbb{N}$, such that in C_n either

- $p = \varepsilon$, $y \in \{\wedge, \vee, \neg, 0, 1, input\}$ and gate g is an y-gate, or
- $p \in \{L, R\}$, $y \in \mathbb{N}$ and gate $g(p)$ is numbered y.

The *extended connection language* L_E of $\{C_n\}$ is defined in the same way, except now $p \in \{L, R\}^*$ and $g(p)$ is the gate that is reached from gate g by following the path specified by p.

The complexity of recognizing L_D or L_E determines uniformity of the family $\{C_n\}$. For $k \geq 2$ the definition of NC^k does not depend on whether L_D or L_E is used, however for NC^1 it is important that we take L_E.

DEFINITION A.5.2 (NC^k). For $k \geq 1$, a language L is in uniform NC^k if it is accepted by a family of polynomial–size logarithmic–depth circuits $\{C_n\}$ whose extended connection language L_E is in FO.

When $k = 1$ if we only require the direct connection language L_D of $\{C_n\}$ be in FO, the result, denoted here by NC^1_D, is apparently a bigger class: we have $NC^1 \subseteq NC^1_D$, but it is not known whether $NC^1 = NC^1_D$.

THEOREM A.5.3 (Ruzzo [100]). For $k \geq 1$,

$$NC^k = ASpace\text{-}Time(\mathcal{O}(\log n), \mathcal{O}((\log n)^k)).$$

In particular we have:

$$NC^1 = ATime(\mathcal{O}(\log n)) = ALogTime.$$

The classes AC^k, TC^0 and $AC^0(m)$ are defined using circuits whose gates have unbounded fanin. The direct (resp. extended) connection language L_D (resp. L_E) for these circuits can be defined as in Definition A.5.1 but now $p \in \{1, 2, 3, \dots\}$ (resp. $p \in \{1, 2, 3, \dots\}^*$). It is easy to see that for constant depth circuits (i.e., AC^0, $AC^0(m)$, TC^0) it does not matter whether we use L_D or L_E to define uniformity. It turns out that this is also the case for AC^k for $k \geq 1$.

DEFINITION A.5.4 (AC^k). For $k \geq 0$, a language L is in AC^k iff it is accepted by a family of circuits $\{C_n\}$ of size polynomial in n and depth $\mathcal{O}((\log n)^k)$ whose direct connection language L_E is in FO.

The class AC^0 has several equivalent definitions, see Sections IV.1 and IV.3.2. In particular, $AC^0 = LH$ where

$$LH = ATime\text{-}Alt(\mathcal{O}(\log n), \mathcal{O}(1)).$$

For $k \geq 1$ we have (see also the fact that the problem $Lmcv_k$ is in $ASpace\text{-}Time(\log n, (\log n)^k)$, Theorem IX.5.28):

THEOREM A.5.5 ([40, 110]). For $k \geq 1$,

$$AC^k = ASpace\text{-}Alt(\mathcal{O}(\log n), \mathcal{O}((\log n)^k)).$$

Note that by Immerman–Szelepcsényi Theorem A.2.4

$$ASpace\text{-}Alt(\mathcal{O}(\log n), \mathcal{O}(1)) = NL.$$

The classes uniform TC^0 and $AC^0(m)$ are defined similarly using circuits that have (in addition to $\neg$ gates and unbounded fanin $\wedge$, $\vee$ gates) *majority gates* and *modulo m gates*, respectively. A majority gate outputs one iff at least half of its inputs are one, and a modulo m gate outputs one iff the number of one inputs is 1 modulo m.

DEFINITION A.5.6 ($TC^0, AC^0(m)$). A language is in TC^0 iff it is accepted by a family of polynomial–size constant–depth circuits with majority gates whose extended connection language L_E is in FO. For $m \geq 2$ the class $AC^0(m)$ is defined similarly with modulo m gates replace majority gates. Also,

$$ACC = \bigcup_{m=2}^{\infty} AC^0(m).$$

The class TC^0 can be equivalently defined using *threshold gates*, which outputs one iff the number of one inputs exceeds some given threshold k. It can be shown that

$$ACC \subseteq TC^0 \subseteq NC^1$$

and that

$$AC^0(m) \subseteq AC^0(m')$$

for $m, m' \in \mathbb{N}$ such that $m|m'$. Ajtai [3] and independently Furst, Saxe, and Sipser [52] show that the relation $PARITY$ (see Sections IV.1 and V.5.1) is not in AC^0, and since $PARITY \in AC^0(2)$, it follows that

$$AC^0 \subsetneq AC^0(2).$$

Razborov and Smolensky (see [19]) show that $MODULO_m \notin AC^0(p)$ for any prime p and any $m \geq 2$ which is not a power of p. It follows that

$$AC^0(m) \nsubseteq AC^0(p).$$

Consequently we have

$$AC^0 \subsetneq AC^0(p) \subsetneq ACC.$$

However, it is not known whether $AC^0(6) = NP$.

BIBLIOGRAPHY

[1] KLAUS AEHLIG, STEPHEN COOK, and PHUONG NGUYEN, *Relativizing Small Complexity Classes and their Theories*, **16th EACSL Annual Conference on Computer Science and Logic**, 2007, pp. 374–388.

[2] MANINDRA AGRAWAL, NEERAJ KAYAL, and NITIN SAXENA, *PRIMES is in P*, **Annals of Mathematics**, vol. 160 (2004), no. 2, pp. 781–793.

[3] MIKLÓS AJTAI, Σ_1^1-*formulae on finite structures*, **Annals of Pure and Applied Logic**, vol. 24 (1983), no. 1, pp. 1–48.

[4] ——, *Parity and the Pigeonhole Principle*, **Feasible Mathematics** (S. R. Buss and P. J. Scott, editors), Birkhäuser, 1990, pp. 1–24.

[5] ——, *The complexity of the pigeonhole principle*, **Combinatorical**, vol. 14 (1994), no. 4, pp. 417–433.

[6] ——, *The Independence of the Modulo Counting p Principles*, **Proceedings of the 26th Annual ACM Symposium on Theory of Computing**, 1994, pp. 402–411.

[7] ERIC ALLENDER, *Arithmetic Circuits and Counting Complexity Classes*, **Complexity of Computations and Proofs** (Jan Krajíček, editor), Quaderni di Matematica, 2005, pp. 33–72.

[8] TOSHIYASU ARAI, *A bounded arithmetic AID for Frege systems*, **Annals of Pure and Applied Logic**, vol. 103 (2000), no. 1–3, pp. 155–199.

[9] DAVID A. BARRINGTON, *Bounded-Width Polynomial-Size Branching Programs Recognizes Exactly Those Languages in NC^1*, **Journal of Computer and System Sciences**, vol. 38 (1989), no. 1, pp. 150–164.

[10] DAVID A. MIX BARRINGTON, NEIL IMMERMAN, and HOWARD STRAUBING, *On Uniformity within NC^1*, **Journal of Computer and System Sciences**, vol. 41 (1990), no. 3, pp. 274–306.

[11] PAUL BEAME, RUSSELL IMPAGLIAZZO, JAN KRAJÍČEK, TONIANN PITASSI, and PAVEL PUDLÁK, *Exponential lower bounds for the pigeonhole principle*, **Proceedings of the 24th Annual ACM Symposium on Theory of Computing**, 1992, pp. 200–220.

[12] PAUL BEAME and TONIANN PITASSI, *An exponential separation between the parity principle and the pigeonhole principle*, **Annals of Pure and Applied Logic**, vol. 80 (1996), pp. 195–228.

457

[13] ———, *Propositional Proof Complexity*: *Past, Present and Future*, **Current Trends in Computer Science Entering the 21st Century** (G. Paun, G. Rozenberg, and A. Salomaa, editors), World Scientific Publishing, 2001, pp. 42–70.

[14] PAUL BEAME and SØREN RIIS, *More on the Relative Strength of Counting Principles*, **DIMACS Series in Discrete Mathematics and Theoretical Computer Science**, vol. 39 (1998), pp. 13–35.

[15] JAMES BENNETT, *On Spectra*, Ph.D. thesis, Princeton University, Department of Mathematics, 1962.

[16] J. A. BONDY, *Induced subsets*, **Journal of Combinatorial Theory, Series B**, vol. 12 (1972), pp. 201–202.

[17] MARIA LUISA BONET, SAMUEL R. BUSS, and TONIANN PITASSI, *Are there Hard Examples for Frege Systems?*, **Feasible Mathematics II** (P. Clote and J. B. Remmel, editors), Birkhäuser, 1994, pp. 30–56.

[18] MARIA LUISA BONET, TONIANN PITASSI, and RAN RAZ, *On Interpolation and Automatization for Frege Systems*, **SIAM Journal on Computing**, vol. 29 (2000), no. 6, pp. 1939–1967.

[19] RAVI B. BOPPANA and MICHAEL SIPSER, *The Complexity of Finite Functions*, **Handbook of Theoretical Computer Science, Volume A** (J. van Leeuwen, editor), Elsevier, 1990, pp. 757–804.

[20] SAMUEL BUSS, **Bounded Arithmetic**, Bibliopolis, 1986.

[21] ———, *Polynomial size proofs of the propositional pigeonhole principle*, **The Journal of Symbolic Logic**, vol. 52 (1987), pp. 916–927.

[22] ———, *The Boolean formula value problem is in ALOGTIME*, **Proceedings of the 19th Annual ACM Symposium on Theory of Computing**, 1987, pp. 123–131.

[23] ———, *Axiomatizations and Conservation Results for Fragments of Bounded Arithmetic*, **Logic and Computation, Proceedings of a Workshop held at Carnegie Mellon University**, AMS Contemporary Mathematics (106), 1990, pp. 57–84.

[24] ———, *Propositional Consistency Proofs*, **Annals of Pure and Applied Logic**, vol. 52 (1991), pp. 3–29.

[25] ———, *Algorithms for Boolean formula evaluation and for tree-contraction*, **Arithmetic, Proof Theory, and Computational Complexity** (Peter Clote and Jan Krajíček, editors), Oxford, 1993, pp. 95–115.

[26] ———, *Relating the bounded arithmetic and polynomial time hierarchies*, **Annals of Pure and Applied Logic**, vol. 75 (1995), pp. 67–77.

[27] ———, *An Introduction to Proof Theory*, **Handbook of Proof Theory** (S. Buss, editor), Elsevier, 1998, available on-line at www.math.ucsd.edu/~sbuss/ResearchWeb/HandbookProofTheory/, pp. 1–78.

[28] ———, *First-Order Proof Theory of Arithmetic*, **Handbook of Proof Theory** (S. Buss, editor), Elsevier, 1998, available on-line at www.math.ucsd.edu/~sbuss/ResearchWeb/HandbookProofTheory/, pp. 79–147.

[29] SAMUEL BUSS and PETER CLOTE, *Cutting planes, connectivity and threshold logic*, **Archive for Mathematical Logic**, vol. 35 (1996), pp. 33–62.

[30] SAMUEL BUSS and JAN KRAJÍČEK, *An application of Boolean complexity to separation problems in bounded arithmetic*, **Proc. London Math. Soc.**, vol. 69(3) (1994), pp. 1–21.

[31] SAMUEL BUSS, JAN KRAJÍČEK, and GAISI TAKEUTI, *On Provably Total Functions in Bounded Arithmetic Theories R_3^i, U_2^i, and V_2^i*, **Arithmetic, Proof Theory and Computational Complexity** (Peter Clote and Jan Krajíček, editors), Oxford, 1993, pp. 116–161.

[32] ASHOK K. CHANDRA, LARRY STOCKMEYER, and UZI VISHKIN, *Constant Depth Reducibility*, **SIAM Journal on Computing**, vol. 13(2) (1984), pp. 423–439.

[33] MARIO CHIARI and JAN KRAJÍČEK, *Witnessing functions in bounded arithmetic and search problems*, **The Journal of Symbolic Logic**, vol. 63 (1998), pp. 1095–1115.

[34] P. CLOTE, *ALOGTIME and a conjecture of S. A. Cook*, **Ann. Math. Art. Intell.**, vol. 6 (1990), pp. 57–106, extended abstract in Proc. 13th IEEE Symposium on Logic in Computer Science, 1990.

[35] PEPTER CLOTE and GAISI TAKEUTI, *Bounded arithmetic for NC, ALogTIME, L and NL*, **Annals of Pure and Applied Logic**, vol. 56 (1992), pp. 73–117.

[36] PETER CLOTE, *Sequential, Machine-Independent Characterizations of the Parallel Complexity Classes AlogTIME, AC^k, NC^k and NC*, **Feasible Mathematics** (S. R. Buss and P. J. Scott, editors), Birkhäuser, 1990, pp. 49–70.

[37] ———, *On Polynomial Size Frege Proofs of Certain Combinatorial Principles*, **Arithmetic, Proof Theory, and Computational Complexity** (Peter Clote and Jan Krajíček, editors), Oxford, 1993, pp. 162–184.

[38] PETER CLOTE and GAISI TAKEUTI, *First Order Bounded Arithmetic and Small Boolean Circuit Complexity Classes*, **Feasible Mathematics II** (P. Clote and J. B. Remmel, editors), Birkhäuser, 1995.

[39] STEPHEN COOK, *Feasibly constructive proofs and the propositional calculus*, **Proceedings of the 7th Annual ACM Symposium on Theory of Computing**, (1975), pp. 83–97.

[40] ———, *A Taxonomy of Problems with Fast Parallel Algorithms*, **Information and Control**, vol. 64 (1985), pp. 2–22.

[41] ———, *Proof Complexity and Bounded Arithmetic*, Course Notes for CSC 2429S. http://www.cs.toronto.edu/~sacook/, 2002.

[42] ———, *Theories for Complexity Classes and Their Propositional Translations*, **Complexity of computations and proofs** (Jan Krajíček, editor), Quaderni di Matematica, 2005, pp. 175–227.

[43] STEPHEN COOK and ANTONINA KOLOKOLOVA, *A Second-Order System for Polytime Reasoning Based on Grädel's theorem*, **Annals of Pure and Applied Logic**, vol. 124 (2003), pp. 193–231.

[44] ———, *A Second-order Theory for NL*, **Logic in Computer Science (LICS)**, 2004.

[45] STEPHEN COOK and TSUYOSHI MORIOKA, *Quantified Propositional Calculus and a Second-Order Theory for NC^1*, **Archive for Mathematical Logic**, vol. 44 (2005), no. 6, pp. 711–749.

[46] STEPHEN COOK and ROBERT RECKHOW, *The Relative Efficiency of Propositional Proof Systems*, **The Journal of Symbolic Logic**, vol. 44 (1979), no. 1, pp. 36–50.

[47] STEPHEN COOK and NEIL THAPEN, *The Strength of Replacement in Weak Arithmetic*, **ACM Transactions on Computational Logic**, vol. 7 (2006), no. 4, pp. 749–764.

[48] MARTIN DOWD, **Propositional Representation of Arithmetic Proofs**, Ph.D. thesis, Department of Computer Science, University of Toronto, 1979.

[49] DING-ZHU DU and KER-I KO, **Theory of Computational Complexity**, Wiley-Interscience, 2000.

[50] RONALD FAGIN, **Contributions to the Model Theory of Finite Structures**, Ph.D. thesis, U. C. Berkeley, Department of Mathematics, 1973.

[51] PETER FRANKL, *On the Trace of Finite Sets*, **Journal of Combinatorial Theory, Series A**, vol. 34 (1983), pp. 41–45.

[52] MERRICK FURST, JAMES B. SAXE, and MICHAEL SIPSER, *Parity, circuits and the polynomial-time hierarchy*, **Mathematical Systems Theory**, vol. 17 (1984), pp. 13–27.

[53] ERICH GRÄDEL, *Capturing Complexity Classes by Fragments of Second Order Logic*, **Theoretical Computer Science**, vol. 101 (1992), pp. 35–57.

[54] PETR HÁJEK and PAVEL PUDLÁK, **Metamathematics of First-Order Arithmetic**, Springer-Verlag, 1993.

[55] WILLIAM HESS, ERIC ALLENDER, and DAVID A. MIX BARRINGTON, *Uniform Constant-Depth Threshold Circuits for Division and Iterated Multiplication*, **Journal of Computer and System Sciences**, vol. 65 (2002), pp. 695–716.

[56] ALEKSANDAR IGNJATOVIC, *Delineating Classes of Computational Complexity via Second Order Theories with Weak Set Existence Principles*, **The Journal of Symbolic Logic**, vol. 60 (1995), pp. 103–121.

[57] ALEKSANDAR IGNJATOVIC and PHUONG NGUYEN, *Characterizing Polynomial Time Computable Functions Using Theories with Weak Set Existence Principles*, **Computing: The Australasian Theory Symposium**, Electronic Notes in Theoretical Computer Science, Volume 78, 2003.

[58] NEIL IMMERMAN, *Nondeterministic Space is Closed Under Complementation*, **SIAM J. Comput.**, vol. 17 (1988), no. 5, pp. 935–938.

[59] ———, **Descriptive Complexity**, Springer, 1999.

[60] EMIL JEŘÁBEK, *Weak Pigeonhole Principle, and Randomized Computation*, Ph.D. thesis, Charles University in Prague, Faculty of Mathematics and Physics, 2004.

[61] ———, *Approximate counting by hashing in bounded arithmetic*, preprint, 2007.

[62] JAN JOHANNSEN, *Satisfiability problems complete for deterministic logarithmic space*, **STACS 2004, 21st Annual Symposium on Theoretical Aspects of Computer Science, Proceedings** (Volker Diekert and Michel Habib, editor), 2004, pp. 317–325.

[63] JAN JOHANNSEN and CHRIS POLLETT, *On Proofs about Threshold Circuits and Counting Hierarchies*, **Proceedings of the 13th IEEE Symposium on Logic in Computer Science**, 1998, pp. 444–452.

[64] DAVID S. JOHNSON, *A Catalog of Complexity Classes*, **Handbook of Theoretical Computer Science, Volume A** (J. van Leeuwen, editor), Elsevier, 1990, pp. 67–161.

[65] DAVID S. JOHNSON, CHRISTOS H. PAPADIMITRIOU, and MIHALIS YANNAKAKIS, *How easy is local search?*, **Journal of Computer and System Sciences**, vol. 37 (1988), no. 1, pp. 79–100.

[66] RICHARD M. KARP and RICHARD J. LIPTON, *Turing machines that take advice*, **L'Enseignement Mathematique**, vol. 30 (1982), pp. 255–273.

[67] ANTONINA KOLOKOLOVA, *Systems of Bounded Arithmetic from Descriptive Complexity*, Ph.D. thesis, University of Toronto, 2004.

[68] JAN KRAJÍČEK, *On the number of steps in proofs*, **Annals of Pure and Applied Logic**, vol. 41 (1989), pp. 153–178.

[69] ———, *Exponentiation and second-order bounded arithmetic*, **Annals of Pure and Applied Logic**, vol. 48 (1990), pp. 261–276.

[70] ———, *Fragments of bounded arithmetic and bounded query classes*, **Trans. AMS**, vol. 338 (1993), no. 2, pp. 587–98.

[71] ———, *Lower bounds to the size of constant-depth propositional proofs*, **The Journal of Symbolic Logic**, vol. 59 (1994), pp. 73–86.

[72] ———, **Bounded Arithmetic, Propositional Logic and Computational Complexity**, Cambridge University Press, 1995.

[73] JAN KRAJÍČEK and PUDLÁK, *Quantified Propositional Calculi and Fragments of Bounded Arithmetic*, **Zeitschrift f. mathematische Logik u. Grundlagen d. Mathematik**, vol. 36 (1990), pp. 29–46.

[74] JAN KRAJÍČEK, PAVEL PUDLÁK, and JIRI SGALL, *Interactive computations of optimal solutions*, **Mathematical Foundations of Computer Science** (B. Rovan, editor), Lecture Notes in Computer Science, no. 452, Springer-Verlag, 1990, pp. 48–60.

[75] JAN KRAJÍČEK, PAVEL PUDLÁK, and GAISI TAKEUTI, *Bounded Arithmetic and the Polynommial Hierarchy*, **Annals of Pure and Applied Logic**, vol. 52 (1991), pp. 143–153.

[76] JAN KRAJÍČEK, ALAN SKELLEY, and NEIL THAPEN, *NP Search Problems in Low Fragments of Bounded Arithmetic*, **The Journal of Symbolic Logic**, vol. 72(2) (2007), pp. 649–672.

[77] JOHN C. LIND, *Computing in logarithmic space*, Technical Report 52, MAC Technical Memorandum, 1974.

[78] ALEXIS MACIEL, PHUONG NGUYEN, and TONIANN PITASSI, *Lifting Lower Bounds for Tree-like Proofs*, work in progress, 2009.

[79] MEENA MAHAJAN and V. VINAY, *Determinant: Combinatorics, algorithms, and complexity*, **Chicago Journal of Theoretical Computer Science**, vol. 5 (1997).

[80] TSUYOSHI MORIOKA, **Logical Approaches to the Complexity of Search Problems: Proof Complexity, Quantified Propositional Calculus, and Bounded Arithmetic**, Ph.D. thesis, University of Toronto, Department of Computer Science, 2005.

[81] V. A. NEPOMNJAŠČIJ, *Rudimentary predicates and Turing calculations*, **Soviet Math. Dokl.**, vol. 11 (1970), no. 6, pp. 1462–1465.

[82] PHUONG NGUYEN, **Bounded Reverse Mathematics**, Ph.D. thesis, University of Toronto, 2008, http://www.cs.toronto.edu/~pnguyen/.

[83] PHUONG NGUYEN and STEPHEN COOK, *Theories for TC^0 and Other Small Complexity Classes*, **Logical Methods in Computer Science**, (2005).

[84] —————, *The Complexity of Proving Discrete Jordan Curve Theorem*, **Proceedings of the 22nd IEEE Symposium on Logic in Computer Science**, 2007, pp. 245–254.

[85] NOAM NISAN and AMNON TA-SHMA, *Symmetric logspace is closed under complement*, **Proceedings of the 27th Annual ACM Symposium on Theory of Computing**, 1995, pp. 140–146.

[86] AKIHIRO NOZAKI, TOSHIYASU ARAI, and NORIKO H. ARAI, *Polynomal-size Frege proofs of Bollobás' theorem on the trace of sets*, **Proceedings of the Japan Academy, Series A, Mathematical Sciences**, vol. 84 (2008), no. 8, pp. 159–161.

[87] CHRISTOS H. PAPADIMITRIOU, **Computational Complexity**, Addison Wesley, 1993.

[88] ROHIT PARIKH, *Existence and feasibility in arithmetic*, **The Journal of Symbolic Logic**, vol. 36 (1971), pp. 494–508.

[89] JEFF B. PARIS, W.G. HANDLEY, and ALEX J. WILKIE, *Characterizing some low arithmetic classes*, **Theory of Algorithms** (L. Lovász and E. Smerédi, editor), Colloquia Mathematica Societatis Janos Bolyai, no. 44, North-Holland, 1985, pp. 353–365.

[90] JEFF B. PARIS and ALEX J. WILKIE, *Counting problems in bounded arithmetic*, **Methods in Mathematical Logic**, Lecture Notes in Mathematics, no. 1130, Springer, 1985, pp. 317–340.

[91] STEVEN PERRON, **Power of Non-Uniformity in Proof Complexity**, Ph.D. thesis, University of Toronto, 2008.

[92] FRANCOIS PITT, *A Quantifier-Free String Theory Alogtime Reasoning*, Ph.D. thesis, University of Toronto, 2000.

[93] CHRIS POLLETT, *Structure and Definability in General Bounded Arithmetic Theories*, **Annals of Pure and Applied Logic**, vol. 100 (1999), pp. 189–245.

[94] MICHAEL RABIN, *Digitalized Signatures and Public-Key Functions as Intractable as Factorization*, Technical Report MIT/LCS/TR-212, MIT Laboratory for Computer Science, 1979.

[95] ALEXANDER A. RAZBOROV, *An Equivalence between Second Order Bounded Domain Bounded Arithmetic and First Order Bounded Arithmetic*, **Arithmetic, Proof Theory and Computational Complexity** (Peter Clote and Jan Krajíček, editors), Oxford, 1993, pp. 247–277.

[96] ——, *Bounded Arithmetic and Lower Bounds in Boolean Complexity*, **Feasible Mathematics II** (P. Clote and J. Remmel, editors), Birkhäuser, 1995, pp. 344–386.

[97] OMER REINGOLD, *Undirected ST-Connectivity in Log-Space*, **Proceedings of the 37th Annual ACM Symposium on Theory of Computing**, 2005, pp. 376–385.

[98] SØREN RIIS, *Count(q) does not imply count(p)*, **Annals of Pure and Applied Logic**, vol. 90 (1997), pp. 1–56.

[99] ——, *Count(q) versus the pigeon-hole principle*, **Archive for Mathematical Logic**, vol. 36 (1997), no. 3, pp. 157–188.

[100] WALTER L. RUZZO, *On Uniform Circuit Complexity*, **Journal of Computer and System Sciences**, vol. 22 (1981), pp. 365–383.

[101] STEPHEN SIMPSON, **Subsystems of Second Order Arithmetic**, Springer, 1999.

[102] MICHAEL SIPSER, **Introduction to the Theory of Computation**, second ed., Course Technology, 2005.

[103] RAYMOND SMULLYAN, **Theory of Formal Systems**, Princeton University Press, 1961.

[104] MICHAEL SOLTYS and STEPHEN COOK, *The Proof Complexity of Linear Algebra*, **Annals of Pure and Applied Logic**, vol. 130 (2004), pp. 277–323.

[105] L. J. STOCKMEYER, *The polynomial-time hierarchy*, **Theoretical Computer Science**, vol. 3 (1976), pp. 1–21.

[106] R. SZELEPCSÉNYI, *The method of forced enumeration for nondeterministic automata*, **Acta Informatica**, vol. 26 (1988), no. 3, pp. 279–284.

[107] GAISI TAKEUTI, S_3^i and $V_2^i(BD)$, **Archive for Mathematical Logic**, vol. 29 (1990), pp. 149–169.

[108] ——, *RSUV Isomorphism*, **Arithmetic, Proof Theory and Computational Complexity** (Peter Clote and Jan Krajíček, editors), Oxford, 1993, pp. 364–386.

[109] G. S. TSEITIN, *On the complexity of derivation in propositional calculus*, **Studies in Constructive Mathematics and Mathematical Logic, Part 2** (A. O. Slisenko (Translated from Russian), editor), Consultants Bureau, New York, London, 1970, pp. 115–125.

[110] HERIBERT VOLLMER, **Introduction to Circuit Complexity: A Uniform Approach**, Springer-Verlag, 1999.

[111] CELIA WRATHALL, *Rudimentary predicates and relative computation*, **SIAM J. Computing**, vol. 7 (1978), pp. 194–209.

[112] DOMENICO ZAMBELLA, *Notes on Polynomially Bounded Arithmetic*, **The Journal of Symbolic Logic**, vol. 61 (1996), no. 3, pp. 942–966.

[113] ———, *End Extensions of Models of Linearly Bounded Arithmetic*, **Annals of Pure and Applied Logic**, vol. 88 (1997), pp. 263–277.

INDEX

465

Printed in the United States
By Bookmasters